# Advances in Intelligent Systems and Computing

Volume 326

**Series editor**

Janusz Kacprzyk, Polish Academy of Sciences, Warsaw, Poland
e-mail: kacprzyk@ibspan.waw.pl

## About this Series

The series "Advances in Intelligent Systems and Computing" contains publications on theory, applications, and design methods of Intelligent Systems and Intelligent Computing. Virtually all disciplines such as engineering, natural sciences, computer and information science, ICT, economics, business, e-commerce, environment, healthcare, life science are covered. The list of topics spans all the areas of modern intelligent systems and computing.

The publications within "Advances in Intelligent Systems and Computing" are primarily textbooks and proceedings of important conferences, symposia and congresses. They cover significant recent developments in the field, both of a foundational and applicable character. An important characteristic feature of the series is the short publication time and world-wide distribution. This permits a rapid and broad dissemination of research results.

## Advisory Board

Chairman

Nikhil R. Pal, Indian Statistical Institute, Kolkata, India
e-mail: nikhil@isical.ac.in

Members

Rafael Bello, Universidad Central "Marta Abreu" de Las Villas, Santa Clara, Cuba
e-mail: rbellop@uclv.edu.cu

Emilio S. Corchado, University of Salamanca, Salamanca, Spain
e-mail: escorchado@usal.es

Hani Hagras, University of Essex, Colchester, UK
e-mail: hani@essex.ac.uk

László T. Kóczy, Széchenyi István University, Győr, Hungary
e-mail: koczy@sze.hu

Vladik Kreinovich, University of Texas at El Paso, El Paso, USA
e-mail: vladik@utep.edu

Chin-Teng Lin, National Chiao Tung University, Hsinchu, Taiwan
e-mail: ctlin@mail.nctu.edu.tw

Jie Lu, University of Technology, Sydney, Australia
e-mail: Jie.Lu@uts.edu.au

Patricia Melin, Tijuana Institute of Technology, Tijuana, Mexico
e-mail: epmelin@hafsamx.org

Nadia Nedjah, State University of Rio de Janeiro, Rio de Janeiro, Brazil
e-mail: nadia@eng.uerj.br

Ngoc Thanh Nguyen, Wroclaw University of Technology, Wroclaw, Poland
e-mail: Ngoc-Thanh.Nguyen@pwr.edu.pl

Jun Wang, The Chinese University of Hong Kong, Shatin, Hong Kong
e-mail: jwang@mae.cuhk.edu.hk

More information about this series at http://www.springer.com/series/11156

Viet-Ha Nguyen · Anh-Cuong Le
Van-Nam Huynh
Editors

# Knowledge and Systems Engineering

Proceedings of the Sixth International
Conference KSE 2014

 Springer

*Editors*
Viet-Ha Nguyen
Faculty of Information Technology
VNU University of Engineering and
Technology
Hanoi
Vietnam

Van-Nam Huynh
School of Knowledge Science
Japan Advanced Institute of Science
and Technology
Nomi, Ishikawa
Japan

Anh-Cuong Le
Faculty of Information Technology
VNU University of Engineering and
Technology
Hanoi
Vietnam

ISSN 2194-5357          ISSN 2194-5365    (electronic)
ISBN 978-3-319-11679-2   ISBN 978-3-319-11680-8    (eBook)
DOI 10.1007/978-3-319-11680-8

Library of Congress Control Number: 2014951000

Springer Cham Heidelberg New York Dordrecht London

Printed on acid-free paper

Springer is part of Springer Science+Business Media (www.springer.com)

# Preface

This volume contains papers presented at the Sixth International Conference on Knowledge and Systems Engineering (KSE 2014), which was held in Hanoi, Vietnam, during 9–11 October, 2014. The conference was organized by University of Engineering and Technology, Vietnam National University, Hanoi. The principal aim of KSE Conference is to bring together researchers, academics, practitioners and students in order to not only share research results and practical applications but also to foster collaboration in research and education in Knowledge and Systems Engineering.

This year we received a total of 97 submissions. Each of which was peer reviewed by at least two members of the Program Committee. Finally, there are 51 regular papers chosen beside 3 invited papers for presentation at KSE 2014 and publication in the proceedings. Besides the main track, the conference featured four special sessions focusing on specific topics of interest. The kind cooperation of Marie Christine Ho Ba Tho, Tien Tuan Dao, Sabine Bensamoun, Catherine Marque, Dang Hung Tran, Kenji Satou, Thanh-Phuong Nguyen, Jiuyong Li, Tzung-Pei Hong, Bac Le, Bay Vo, Akira Shimazu, Minh L. Nguyen in organizing these special sessions is highly appreciated.

We would like to express our appreciation to all the members of the Program Committee for their support and cooperation.

We also wish to thank all the authors and participants for their contributions and fruitful discussions that made this conference a success.

Hanoi, Vietnam
October 2014

Viet-Ha Nguyen
Anh-Cuong Le
Van-Nam Huynh

# Organization

## General Chairs

Nguyen Ngoc Binh, Vietnam
Sadaaki Miyamoto, Japan

## Program Chairs

Huynh Van Nam, Japan
Nguyen Viet Ha, Vietnam

## Program Committee

Akira Shimazu, Japan
Anh Cuong Le, Vietnam
Bay Vo, Vietnam
Boontawee Suntisrivaraporn, Thailand
Bui Hung, Vietnam
Catherine Faron Zucker, France
Churn-Jung Liau, Taiwan
Cuong Nguyen, Vietnam
Dang Hung Tran, Vietnam
Duc Dung Nguyen, Vietnam
Gabriele Kern-Isberner, Germany
Hiep Xuan Huynh, Vietnam
Hieu Vo, Vietnam
Hoai Nguyen Xuan, Vietnam
Hoang Truong, Vietnam
Huu Hanh Hoang, Vietnam
Ioannis Parissis, France

Jing Liu, China
Kenji Satou, Japan
Khoat Than, Japan
Lam Bui, Vietnam
Lam Thu Bui, Vietnam
Le Duy Thuc, Australia
Le Hoai Bac, Vietnam
Long-Thanh Ngo, Vietnam
Mai Tran, Vietnam
Marie Christine Ho Ba Tho, France
Marie-Helene Abel, France
Martin Steffen, Norway
Masahiro Inuiguchi, Japan
Mina Ryoke, Japan
Minh Pham, Japan
Ngoc-Thanh Nguyen, Poland
Nguyen Le Minh, Japan

# Contents

**Part I: Keynote Addresses**

Evidential Probabilities for Rough Set in a Case of Competitiveness . . . . .  3
*Hamido Fujita, Yu-Chien Ko*

Interval Analysis for Decision Aiding . . . . . . . . . . . . . . . . . . . . . . . . . . . .  15
*Masahiro Inuiguchi*

Agent Communication and Belief Change . . . . . . . . . . . . . . . . . . . . . . . . .  31
*Satoshi Tojo*

**Part II: KSE 2014 Main Track**

Discriminative Prediction of Enhancers with Word Combinations as
Features . . . . . . . . . . . . . . . . . . . . . . . . . . . . . . . . . . . . . . . . . . . . . . . . . . .  35
*Pham Viet Hung, Tu Minh Phuong*

Improving Acoustic Model for Vietnamese Large Vocabulary
Continuous Speech Recognition System Using Deep Bottleneck
Features . . . . . . . . . . . . . . . . . . . . . . . . . . . . . . . . . . . . . . . . . . . . . . . . . . .  49
*Quoc Bao Nguyen, Tat Thang Vu, Chi Mai Luong*

DFTBC: Data Fusion and Tree-Based Clustering Routing Protocol for
Energy-Efficient in Wireless Sensor Networks . . . . . . . . . . . . . . . . . . . . . .  61
*Nguyen Duy Tan, Nguyen Dinh Viet*

Natural Language Processing for Social Event Classification . . . . . . . . . . .  79
*Duc-Duy Nguyen, Minh-Son Dao, Truc-Vien T. Nguyen*

An Effective NMF-Based Method for Supervised Dimension
Reduction . . . . . . . . . . . . . . . . . . . . . . . . . . . . . . . . . . . . . . . . . . . . . . . . . . .  93
*Ngo Van Linh, Nguyen Kim Anh, Khoat Than*

**Facial Expression Recognition Using Pyramid Local Phase
Quantization Descriptor** .......................................... 105
*Anh Vo, Ngoc Quoc Ly*

**A Community-Based Vietnamese Question Answering System** .......... 117
*Quan Hung Tran, Nien Dinh Nguyen, Kien Duc Do, Thinh Khanh Nguyen,
Dang Hai Tran, Minh Le Nguyen, Son Bao Pham*

**Toward a Rule-Based Synthesis of Vietnamese Emotional Speech** ....... 129
*Thi Duyen Ngo, Masato Akagi, The Duy Bui*

**EasySearch: Finding Relevant Functions Based on API
Documentation** ................................................. 143
*Dinh Huy Tran, Hua Phung Nguyen, Duy Hanh Le*

**A Fast Method for Motif Discovery in Large Time Series Database
under Dynamic Time Warping** ...................................... 155
*Cao Duy Truong, Duong Tuan Anh*

**A Novel Test Data Generation Approach Based Upon Mutation
Testing by Using Artificial Immune System for Simulink Models** ........ 169
*Le Thi My Hanh, Nguyen Thanh Binh, Khuat Thanh Tung*

**A Priority-Based Flow Control Method for Multimedia Data in
Multi-hop Wireless Ad hoc Networks** ............................. 183
*Ngo Hai Anh, Nguyen Tien Lan, Pham Thanh Giang*

**Using Attribute Relationships for Person Re-Identification** ............. 195
*Ngoc-Bao Nguyen, Vu-Hoang Nguyen, Thanh Ngo Duc, Duy-Dinh,
Duc Anh Duong*

**Contingent Information: A Four-Valued Approach** ................... 209
*Seiki Akama, Jair Minoro Abe, Kazumi Nakamatsu*

**On Automating Inference of OCL Constraints from Counterexamples
and Examples** ................................................. 219
*Duc-Hanh Dang, Jordi Cabot*

**Improving Text-Based Image Search with Textual and Visual Features
Combination** .................................................. 233
*Xuan-Son Vu, Thanh Vu, Huong Nguyen, Quang-Thuy Ha*

**Cloud Detection Algorithm for LandSat 8 Image Using Multispectral
Rules and Spatial Variability** ...................................... 247
*Duc Chuc Man, Viet Hung Luu, Van Thang Hoang, Quang Hung Bui,
Thi Nhat Thanh Nguyen*

**Binary Hybrid Particle Swarm Optimization with Wavelet Mutation** .... 261
*Quang-Anh Tran, Quan Dang Dinh, Frank Jiang*

Collaborative Filtering by Co-Training Method .......................... 273
Tran Nhat Quang, Do Thi Lien, Nguyen Duy Phuong

Fast K-Means Clustering for Very Large Datasets Based on
MapReduce Combined with a New Cutting Method ................... 287
Duong Van Hieu, Phayung Meesad

Neural Networks with Hidden Markov Models in Skeleton-Based
Gesture Recognition .................................................. 299
Hai-Son Le, Ngoc-Quan Pham, Duc-Dung Nguyen

Semantic Regions Recognition in UAV Images Sequence ............... 313
Stéphane Lathuilière, Hai Vu, Thi-Lan Le, Thanh-Hai Tran,
Dinh Tan Hung

Fast Optimization of the Pattern Shapes in Board Games with
Simulated Annealing ................................................. 325
Huy Nguyen, Simon Viennot, Kokolo Ikeda

Modelling Timed Concurrent Systems Using Activity Diagram
Patterns ............................................................. 339
Étienne André, Christine Choppy, Thierry Noulamo

SigVer3D: Accelerometer Based Verification of 3-D Signatures on
Mobile Devices ...................................................... 353
Nguyen Ngoc Diep, Cuong Pham, Tu Minh Phuong

New Mechanism of Combination Crossover Operators in Genetic
Algorithm for Solving the Traveling Salesman Problem ............... 367
Pham Dinh Thanh, Huynh Thi Thanh Binh, Bui Thu Lam

A Hybrid Gravitational Search Algorithm and Back-Propagation for
Training Feedforward Neural Networks ............................... 381
Quang Hung Do

Efficient Palmprint Search Based on Database Clustering for Personal
Identification ....................................................... 393
Hoang Thien Van, Thai Hoang Le

Improving Table of Contents Recognition Using Layout-Based
Features ............................................................. 405
Phuc Tri Nguyen, Dang Tuan Nguyen

Predicting the Popularity of Social Curation ......................... 413
Binh Thanh Kieu, Ryutaro Ichise, Son Bao Pham

**Parameter Learning for Statistical Machine Translation Using
CMA-ES** ........................................................ 425
*Viet-Hong Tran, Anh-Tuan Pham, Vinh-Van Nguyen, Hoai-Xuan Nguyen,
Huy-Quang Nguyen*

**Interacting with AutoMed-DM through Layers of Modelling
Abstractions: A Hierarchical, Event-Driven Design** ................... 433
*Duc Minh Le*

**Human Action Recognition Using 2DPCA-DMM Representation and
GA-SVM in Depth Sequences** ..................................... 447
*Vo Hoai Viet, Ly Quoc Ngoc, Tran Thai Son*

**A Study of Feature Combination in Gesture Recognition with Kinect** .... 459
*Ngoc-Quan Pham, Hai-Son Le, Duc-Dung Nguyen,
Truong-Giang Ngo*

**Formalising Concurrent UML State Machines Using Coloured Petri
Nets** .......................................................... 473
*Étienne André, Mohamed Mahdi Benmoussa, Christine Choppy*

**Emotional Facial Expression Analysis in the Time Domain** ............. 487
*Thi Duyen Ngo, Thi Chau Ma, The Duy Bui*

**An Efficient Method for Automated Generating Models of
Component-Based Software** ....................................... 499
*Hoang-Viet Tran, Chi-Luan Le, Quang-Trung Nguyen, Pham Ngoc Hung*

**A Lightweight Formal Approach for Component Reuse** ................ 513
*Khai T. Huynh, Thang H. Bui, Tho T. Quan*

**Part III: KSE 2014 Special Sessions**

**Dynamic Behavior of Uterine Contractions: An Approach Based on
Source Localization and Multiscale Modeling** ........................ 527
*Catherine Marque, Ahmad Diab, Jérémy Laforêt, Mahmoud Hassan,
Brynjar Karlsson*

**Improving the Operability of Personal Health Record System by
Dynamic Dictionary Configuration for OCR** ........................ 541
*Atsuo Yoshitaka, Shinobu Chujyou, Hiroshi Kato*

**Real-Time Rehabilitation System of Systems for Monitoring the
Biomechanical Feedbacks of the Musculoskeletal System** .............. 553
*Tien Tuan Dao, Philippe Pouletaut, Didier Gamet,
Marie Christine Ho Ba Tho*

**Uncertainty Modeling and Propagation in Musculoskeletal Modeling** ....................................... 567
*Tien Tuan Dao, Marie Christine Ho Ba Tho*

**A Comparative Study of Classification-Based Machine Learning Methods for Novel Disease Gene Prediction** ........................... 577
*Duc-Hau Le, Nguyen Xuan Hoai, Yung-Keun Kwon*

**An Efficient Ant Colony Algorithm for DNA Motif Finding** ............ 589
*Hoang X. Huan, Duong T.A. Tuyet, Doan T.T. Ha, Nguyen T. Hung*

**Entity Linking for Vietnamese Tweets** .............................. 603
*Duy K. Van, Huy M. Huynh, Hien T. Nguyen, Vinh T. Vo*

**Using Large N-gram for Vietnamese Spell Checking** ................... 617
*Nguyen Thi Xuan Huong, Tran-Thai Dang, The-Tung Nguyen, Anh-Cuong Le*

**Automatically Learning Patterns in Subjectivity Classification for Vietnamese** ..................................................... 629
*Tran-Thai Dang, Nguyen Thi Xuan Huong, Anh-Cuong Le, Van-Nam Huynh*

**Question Analysis for a Community-Based Vietnamese Question Answering System** ............................................... 641
*Quan Hung Tran, Minh Le Nguyen, Son Bao Pham*

**Using Dependency Analysis to Improve Question Classification** ......... 653
*Phuong Le-Hong, Xuan-Hieu Phan, Tien-Dung Nguyen*

**Unequal Clustering Formation Based on Bat Algorithm for Wireless Sensor Networks** ...................................... 667
*Trong-The Nguyen, Chin-Shiuh Shieh, Mong-Fong Horng, Truong-Giang Ngo, Thi-Kien Dao*

**A New Approach to Multi-variable Fuzzy Forecasting Using Picture Fuzzy Clustering and Picture Fuzzy Rule Interpolation Method** ........ 679
*Pham Huy Thong, Le Hoang Son*

**Author Index** ...................................................... 691

# Part I
# Keynote Addresses

# Evidential Probabilities for Rough Set in a Case of Competitiveness

Hamido Fujita and Yu-Chien Ko

**Abstract.** Probabilistic rough set based on statistics and equivalence relations can provide membership, boundary approximations, subsets dependency, criterion dependency, three-way decision, and so forth. However, the probability is non-determinant due to randomness and users' choices which is subjective in most cases. This research aims to provide determinant probabilities through relevant evidences on three-way decision, called evidential probabilities for rough set (EPRS) to consider the subjective criteria. In this article a research position highlights the boundary regions in terms of subjective criteria showing how subjective criteria may provide a means to narrow the uncertainty by projecting the boundary regions through user preferences in providing better decision precisions. The idea is also projected on prospect theory by showing the adaptable tuning of reference point through decision based subjective preferences. A case study about nations competitiveness is estimated through EPRS and shows the evidential probability can be identified in reality.

**Keywords:** Probabilistic rough set, three-way decision, relevant evidences, competitiveness.

## 1 Introduction

Probabilistic rough set adopts statistics to estimate the probability of a subset of objects with respect to equivalence relations [1, 2, 3, 4, 5, 6, 7, 8]. The technique enables a target subset expressed with rough set approximations and probabilities thus

Hamido Fujita
Software and Information Science, Iwate Prefectural University, Takizawa, Japan
e-mail: issam@iwate-pu.ac.jp

Yu-Chien Ko
Department of Information Management, Chung Hua University, Hsinchu, Taiwan
e-mail: eugene@chu.edu.tw

© Springer International Publishing Switzerland 2015
V.-H. Nguyen et al. (eds.), *Knowledge and Systems Engineering*,
Advances in Intelligent Systems and Computing 326, DOI: 10.1007/978-3-319-11680-8_1

making quantitative explanation possible. However, the probability is non-determinant due to inconsistent choices [9, 10] or uncertain judgments [1, 3, 5]. Three-way decision intends to adjust random errors to be balanced among positive, negative, and boundary regions [11, 12, 13]. However, the randomness is not the human behavior and these balance adjustments does not focus on minimum uncertainty. This research concerns that randomness might not suit estimation of human behaviors. Therefore, we propose evidential probabilities for rough set (EPRS) which intend to give quantitative measures for approximations beyond randomness to approaching minimum uncertainty though subjective projection on objective attribute. In other words a boundary region in three way decisions is projected using subjective aspiration based on experts' preferences. EPRS is designed as a bridge from three way decision method to the probability rough set. Several kinds of generalized sets can be associated with three-valued logics: rough sets [14], conditional events [15], shadowed sets [16], and interval sets [17]. Similar mathematical frameworks using these approaches may have different interpretation. Such three-valued sets lead to three-way decision processes [18] that leave room for a non-committal attitude that differs from interpretation just limited to acceptance and rejection. In this paper we will use the term objective criteria which are related to those criteria in conventional decision making based on statistical analysis and coherently related to the problem. Subjective criteria are those criteria which are sensitive to the situation bounded by the criteria due to mental relation to expert like sentimental relation to the expert experience or user like the mental interest to certain situation. Generally these issues examined by decision making community in terms of uncertainty. Or it is examined based on expert consensus in reaching common weight on criteria expressed in fuzzy terms, like linguistics weight values. Uncertainty also contributed to represent these weight values with non-membership values and hesitant values as well. All the stated representation are expressing these criteria in objective manner therefore such representation are gradually innovated. However, the variation of provided solution is still narrow in providing fitting subjective solution to the needs of the user and the experts' specific needs reflected on problems' characteristics. In practice situation fitting are come from experts subjective knowledge on the situation. Such knowledge in conventional decision making is stated as similarity computation based on different type of distance measurement computed in relation to objective analysis of criteria. Objective characteristics of the problem can be useful for such type of computation in decision making, however such approaches are not accurate in providing appropriate decisions when some uncertainty exists and also complex situation when objective criteria weight are dependent on other criteria in subjective manner. Usually, competitiveness is not random but purposed due to goals, desires, and so forth. We define the competitiveness criteria as subjective. The determinant boundaries of competitiveness approximations thus are hardly identified through randomness. In a case of dominating nations for example, people might want to know the importance degrees of criteria toward the top ranks. We think that competitiveness boundaries of dominance can be approximately determined through subjective projection on a subset defined as users' needs. The subset is called a subjective target, for instance, the top ranks. Usually, experts specify criteria weight in a subjective manner through personal experience and mental characteristic. In facts, most techniques in decision making do not assure what objective weights for criteria are thus providing probability estimations. Some of the probabilistic methods are based on nonlinear optimizations which are complex and not easy to solve. Especially, the preference relations between criteria composed of orders demand transformation of numeric computations to reveal knowledge inside. In PRS research domain, experts define thresholds values for numeric computation. However, threshold determination is a crucial task usually based on objective computation such as randomness, balance, and so forth. Therefore, we define the threshold values of a subjective target for each criterion in terms of EPRS, which gives the relevant evidence to probabilities of approximations. Due to that [19] can achieve thresholds calculations by Bayesian risk decision and three-way decisions, EPRS can project a better relevant approximation for both subjective and objective criteria. The methodology of EPRS quantitatively specifies the appropriate criteria weights in relation to subjective (for example based on user preferences) or objective situations. The technical advantage of EPRS lies on that evidence can give better explanation on a classified target than randomness no mater criteria are subjective or objective. The subjective criteria are like judgments or preferences made by decision makers. The objective criteria are often subjective to other data set: They are usually functional to a finite set of criteria; each criterion has an association to the target. This is the reason that we claim the subjective methodology is more important than the objective in estimation analysis. The proposed methodology, EPRS, in Decision making is resemble to the prospect theory [20], which can reflect the reference point to estimate subjective gains and loss. Rough set and Prospect theory: In any system the interpretation of its' objects is a representation in defining a set of utility functions, and each function have a set of criteria with attributes values. There are some attributes or criteria that have imprecise definition values, (cannot be precisely expressed with crisp attributes). The uncertainty can be expressed as roughness which is studied by Pawlak in what is called rough set theory. A system is rough when the criteria's features are represented in two regions; upper approximation regions and lower approximation region. The boundary regions are for those criteria with features that are not precise, so it is called boundary region, refers to the unknowns, or the possible values. In [21] there are different interpretations to the three valued region set of feature attributes. These interpretations are informal, possible, unknown, undefined, half true, irrelevant and inconsistent. We are providing in this interpretation another one we called it subjective. This is to project the objects' feature on other features that belongs to the two other regions such that to find a preference relation that can have some undefined feature to be relevant to other feature in the two valued regions, using subjective projection of three valued region attributes on other regions feature to find possible feature that can be more relevant through ontological context specified by subjective projection of mental aspiration of the decision maker, so that undefined is used for providing suitable partial function [21]. Such projection is epistemic. The projection used on two valued regions through Subjectivness is related to the state of information concern the propositions evidential relation represented as joint probability of feature extracted

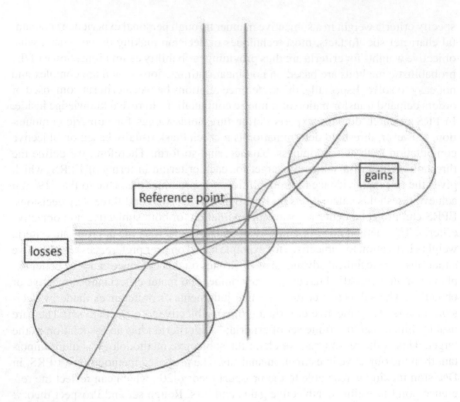

**Fig. 1** Reference point in prospect thery can have different values

as set preferences of rough set values. Subjectivity definition has two parts: First, a set of criteria contributing to the reference point in a convergence scheme. Second, a set of constraints joining the reference points. One point to note: the reference point is often used to provide the subjective relation in the upper appximation. Reference point adaparation is subjective to the outcomes [22]. We think the the adpation is evidentail to experts aspiration which is impirtant to reflect different setting for reference point [23]. The potential advantages of EPRS can povide better prespective siuated in a subjective context in the actual worlds. Initial information needed for such methodology is few, i.e., criteria preferences and the target subset only, and its application scopes are wide [23]. We think that the current stage of social networking analysis (SNA), mobile technologies and cloud computing can provide such subjective criteria from Meta Learning [24] or Knowledge Discovery for the model in Fig. 1. We think that the subjective information gain from SNA would provide better understanding to the criteria and contribute to resolve uncertainty issues in classification. The results projected from EPRS are very different from the objective on actual situations. The evidential probability is the belief decision matrix in which each attribute is described by probability distribution using belief structure in the three regions. The EPRS can capture uncertainty in subjective decision.

The subjective perspective is a classification of a collection of constraints. The constraints are of two types representing subjective perspective: positive relation (participate) and the negative relation (preclude). Recently [25] TODIM (interactive and Multiple Criterion Decision Making) methodology based on Prospect theory is developed, in which criteria of multi value functions are used to measure the dominance degree of alternative among itself. The model is an approximation projection on reference point, which can be considered as epistemological on gains and losses. The reference point is the boundary regions than would affect the upper approximation and lower approximation, and treated as the subjective definition of threshold values in achieving the better qualitative approximation for ranking criteria. It is a sort of subjective granules that can provide fine coarser criterion reduction for achieving better granularity. Incremental learning would provide better transition from one subjective situation to another situation by taking use of the results and this would reduce the complexity. In this research we intend to solve the evident proicturebabilities for these criteria instead of random probabilities. The case study is based on WCY 2012. The remainder of this paper is organized as follows: Section 2 presents the design and implementation of EPRS. Section 3 addresses application results of EPRS on competitiveness and Section 4 presents technical discussions. Finally, concluding remarks are stated to close the paper.

## 2 Evidential Probabilities for Rough Set (EPRS)

This section has two parts. One is about the techniques bridging relevant evidences and probability rough set [1, 2, 11, 26]. In this part, definition, model, and implementation of EPRS are presented. The other is about the extension technique of EPRS on WCY 2012 [27] to identify the evidential probabilities for criteria. A verification of the evidential probabilities is carries out through classifying top ranking nations. The results evidential probabilities are not only determinant but also able to identify dominating nations in criterion perspectives. The information system and definitions of EPRS is defined in 2.2 and the dataset of this research is presented in 2.1.

### 2.1 Dataset

International Institute for Management Development (IMD) annually publishes WCY, a well-known report which ranks and analyzes how a nation's environment can create and develop sustainable enterprises. WCY is a product cooperating with fifty-four partner institutes worldwide. Its ranking considers broad perspectives by gathering the latest and most relevant data on the subject and by analyzing the policy consequence. The dataset include 59 nations, 4 consolidated factors, and 20 criteria in Table 1 [28].

The dataset of this research is collected from WCY 2012, which adopts all criteria and nations, i.e., 20 criteria and 59 nations (objects shown by x). The top ten nations symbolized as a set Y including Canada, Germany, Hong Kong, Norway, Qatar,

**Table 1** Four factors and twenty criteria of WCY-IMD 2012

| Economic Performance | | Business Efficiency | |
|---|---|---|---|
| $q_1$ | Domestic Economy | $q_{11}$ | Productivity and Efficiency |
| $q_2$ | International Trade | $q_{12}$ | Labor Market |
| $q_3$ | International Investment | $q_{13}$ | Finance |
| $q_4$ | Employment | $q_{14}$ | Management Practices |
| $q_5$ | Prices | $q_{15}$ | Attitudes and Values |
| Government Efficiency | | Infrastructure | |
| $q_6$ | Public Finance | $q_{16}$ | Basic Infrastructure |
| $q_7$ | Fiscal Policy | $q_{17}$ | Technological Infrastructure |
| $q_8$ | Institutional Framework | $q_{18}$ | Scientific Infrastructure |
| $q_9$ | Business Legislation | $q_{19}$ | Health and Environment |
| $q_{10}$ | Societal Framework | $q_{20}$ | Education |

Singapore, Sweden, Switzerland, Taiwan, and USA, which will be used to validate the proposed EPRS.

## 2.2 Definitions of EPRS

EPRS is built on information system containing objects, criteria, and users' defined subset Y, i.e., the top ten nations in 2012. The information system makes computation available thus approximation possibly being solved quickly and accurately. The information system and definitions of EPRS is presented in the followings.

**Definition 1:** The information system of EPRS
$IS\_EPRS = (U, A, Q, Y_p)$ Where $U = \{y | y = 1, ...n\}$ n is the number of nations, $A \subseteq U \times U$, $Q = \{q_1, q_2, ...q_m\}$ is a set of criteria, m is the number of criteria, $Y_p$ is a user defined set like the top ten nations $\{1^{th}, 2^{th}, ..n^{th}\}$ in WCY, $U = Y_{dominating} \smile Y_{dominated}$ has two partitions complement to each other and no intersection between them, "*dominating*" means positive region for approximation , "*dominated*" means negative region.

**Definition 2:** Approximations of probabilities
The probability of an approximation is defined as:
$P_*(Y_{dominating}) = P(\underline{apr}(Y_{dominating})) \leq P(Y_{dominating}) \leq P(\overline{apr}(Y_{dominating})) = P^*(Y_{dominating})$
where $P_*(Y_{dominating}), P(\underline{apr}(Y_{dominating})), P(\overline{apr}(Y_{dominating}))$ and $P^*(Y_{dominating})$ are the approximations of the set $P(Y_{dominating})$, and $P(\underline{apr}(Y_{dominating})) = \frac{|P(\underline{apr}(Y_{dominating}))|}{|U|}$.

**Definition 3:** The induction rule of EPRS

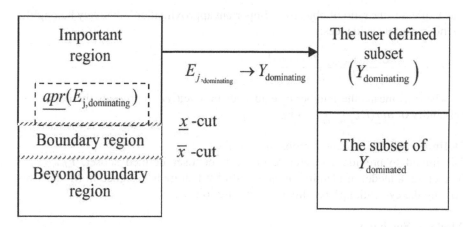

**Fig. 2** A conceptual induction of EPRS

The dependency of an criterion to the user defined subset, $Y_{dominating}$, is used as a key to identify the approximations of probability which is shown in Fig. 2. The technique is based on an induction rule which demands the best combination between sufficient and necessary conditions of an implication. The calculation of EPRS will be presented in Definition 4. Fig. 2 is the conceptual model.

The important region in Fig. 1 contains objects (nations) having higher or equal performance than $\underline{x}$ no matter the nations belong to $Y_{dominating}$ or not. The boundary region contains nations belonging to $Y_{dominating}$ but performing lower than $\underline{x}$ or nations belonging to $Y_{dominating}$ but performing at least as $\underline{x}$. The beyond boundary region has nations perform less than $\underline{x}$. $E_{j'_{dominating}}$ represents a subset having equivalence performance with respect to criterion $q_j$.

According to the induction model of EPRS in Fig. 2, three measures related to EPRS are presented below. Evidence-accuracy rate $(\alpha')$ [28,29]: An accuracy rate presents the ratio of 'Important approximation' to 'upper approximation,' i.e., the degree of the properly classified evidence relative to the possible evidences, and is defined as:

$$(\alpha') = \frac{|\underline{apr}(E_{j'_{dominating}}) \rightarrow Y_{dominating})|}{|\overline{apr}(E_{j'_{dominating}}) \rightarrow Y_{dominating})|}$$

$\alpha'$ for a logical implication represents the degree of necessary condition of 'Important approximation' in the upper approximation. Evidence-coverage rates $(CR')$ [29,30]: A coverage rate expresses the ratio of 'Important approximation' relatively belonging to the user defined subset, and is defined as:

$$(CR') = \frac{|\underline{apr}(E_{j'_{dominating}}) \rightarrow Y_{dominating})|}{|(Y_{dominating})|}$$

$CR'$ represents the degree of sufficient condition that 'Important approximation' influences the user defined subset. Evidence-certainty rate $(Cer')$ [30]: A certainty rate expresses the ratio of objects in 'Important approximation' relatively belonging to the important evidences:

$$(Cer') = \frac{|apr(E_{j_{I_{dominating}}} \to Y_{dominating})|}{|(E_{j_{I_{dominating}}})|}$$

where $|.|$ means the number of evidences in a set $Cer'$ represents the degree of reliability of $apr(E_{j_{I_{dominating}}} \to Y_{dominating})$

**Definition 4:** The evidential probability $(g')$
We intend to provide a clearer feature for the user defined subset, $Y_{dominating}$. Therefore, a model of identifying the significant feature is designed in Model 1 by solving the evidential probability $(g'')$ for criterion $g_j$.

**Model I: Solving** $g''$
Max $g''_j = Cer' \times CR' \times \alpha'$ s.t

$$(Cer') = \frac{|apr(E_{j_{I_{dominating}}} \to Y_{dominating})|}{|(E_{j_{I_{dominating}}})|}, \qquad (CR') = \frac{|apr(E_{j_{I_{dominating}}} \to Y_{dominating})|}{|(Y_{dominating})|},$$

$$(\alpha') = \frac{|apr(E_{j_{I_{dominating}}} \to Y_{dominating})|}{|\overline{apr}(E_{j_{I_{dominating}}} \to Y_{dominating})|}.$$

## 3   Application Results of EPRS on Competitiveness

The results of applying on WCY 2012 to identify the best feature for the user defined top ten nations are presented in Table 2.

According to the results in Table 2, we can obtain the approximations of probabilities by backwardly solving $P_*(Y_{dominating}), P(aprY_{dominating}), P(\overline{apr}(Y_{dominating}))$ and $P^*(Y_{dominating})$, i.e., $P(\underline{apr}(Y_{dominating})) = P(\overline{apr}(Y_{dominating}))$.

So users can get a determinant probability for the featured subset, $Y_{dominating}$.

## 4   Discussion

The bridge for probability rough set to classification rules is technical focus of EPRS which provides the best feature for the defined subset. This can be verified with the

**Table 2** Application results

| Economic Performance | | Government Efficiency | | Business Efficiency | | Infrastructure | |
|---|---|---|---|---|---|---|---|
| $g''_1$ | **0.08** | $g''_6$ | 0.07 | $g''_{11}$ | **0.08** | $g''_{16}$ | 0.03 |
| $g''_2$ | 0.05 | $g''_7$ | 0.03 | $g''_{12}$ | 0.02 | $g''_{17}$ | 0.07 |
| $g''_3$ | 0.05 | $g''_8$ | 0.06 | $g''_{13}$ | **0.08** | $g''_{18}$ | 0.04 |
| $g''_4$ | 0.02 | $g''_9$ | 0.05 | $g''_{14}$ | **0.08** | $g''_{19}$ | 0.05 |
| $g''_5$ | 0.02 | $g''_{10}$ | 0.04 | $g''_{15}$ | 0.05 | $g''_{20}$ | 0.03 |

information entropy due to the less value of two partitions than that of three partitions. The entropy function proposed by Shannon is quantitatively presented as:

$H(P_\pi) = - \sum_{\pi=1}^{3} \frac{|Y'_\pi|}{|U|} \log \frac{|Y'_\pi|}{|U|}$ where $Y'_\pi$ means a partition of nations with respect to an

criterion, $\pi$ means partitions, $\pi = 1$ for positive, $\pi = 2$ for boundary, and $\pi = 3$ for negative. The choice of the finest or the coarsest depends on analysis requirements, specific or featured characteristics. The coarsest will give the most number of sufficient objects and the finest can give the most specific characteristics with the least number of objects. In this research the entropy function with respect to individual criterion is used to verify the best partitions which can clearly explain the dominance competitiveness. The membership of each nation in individual criterion can also be obtained by backwardly from the evidential probabilities. The reliability of the application results can be seen in rules below.

if $EPRS(y) \geq 67.85$ then $x \in$ the top ten nations

where $Cer(utility \rightarrow Y_p) = 1$, $CR(utility \rightarrow Y_p) = 1$, $\alpha(utility \rightarrow Y_p) = 1$ and

$EPRS(y) = - \sum_{j=1}^{m} g''_j r_{yj}$ representing the utility for nation by multiplying observa-

tions $(r_{yj})$ and the evidential probabilities $(g''_j)$ of criteria.

## 5 Concluding Remarks

This research solves the evidential probabilities by induction rules of rough set which named EPRS. The proposed method provides useful probability for approximations and makes utility function available in classification. Also the thresholds of approximations, membership of elements with respect to a user defined feature, the entropy of target subset can be backwardly solved in showing subjective criteria can have important impact in narrowing the uncertainty and vagueness. The projection of using Subjectivness in specifying the reference point would contribute in providing better decision accuracy and precision taken into account experts' aspiration. The case study shows competitiveness can be illustrated by the evidential probabilities which quantitatively present the importance of criteria for nations. The future work can explore other types of the evidential probabilities especially related to subjectivity, objectivity, risks, and so forth.

## References

[1] Pawlak, Z., Wong, S.K.M., Ziarko, W.: Rough sets: probabilistic versus deterministic approach. International Journal of Man-Machine Studies 29(1), 81–95 (1988)
[2] Yao, Y.: Probabilistic approaches to rough sets. Expert Systems 20(5), 287–297 (2003)
[3] Yao, Y.: Probabilistic rough set approximations. Int. J. Approx. Reasoning 49(2), 255–271 (2008)
[4] Ziarko, W.: Probabilistic approach to rough sets. Int. J. Approx. Reasoning 49(2), 272–284 (2008)
[5] Yao, Y.: Three-way decisions with probabilistic rough sets. Inf. Sci. 180(3), 341–353 (2010)

[6] Williamson, J., Wheeler, G.: Evidential Probability and Objective Bayesian Epistemology. In: Bandyopadhyay, P.S., Forster, M. (eds.) Handbook of the Philosophy of Statistics. Elsevier (2011)

[7] Yao, Y.: The superiority of three-way decisions in probabilistic rough set models. Inf. Sci. 181(6), 1080–1096 (2011)

[8] Yang, T.-J., Yao, X.-P., Li, Y.Y.: An axiomatic characterization of probabilistic rough sets. Inf. Sci. 55(1), 130–141 (2014)

[9] Blaszczynski, J., Greco, S., Slowinski, R.: Multi-criteria classification - a new scheme for application of dominance-based decision rules. European Journal of Operational Research 181(3), 1030–1044 (2007)

[10] Greco, S., Matarazzo, B., Slowinski, R.: Parameterized rough set model using rough membership and bayesian confirmation measures. Int. J. Approx. Reasoning 49(2), 285–300 (2008)

[11] Deng, X., Yao, Y.: Decision-theoretic three-way approximations of fuzzy sets. Inf. Sci. 279, 702–715 (2014)

[12] Liang, D., Liu, D.: Systematic studies on three-way decisions with interval-valued decision-theoretic rough sets. Inf. Sci. 276, 186–203 (2014)

[13] Hu, B.Q.: Three-way decisions space and three-way decisions. Inf. Sci. 281, 21–52 (2014)

[14] Ciucci, D., Dubois, D.: Truth-functionality, rough sets and three-valued logics. In: ISMVL, pp. 98–103. IEEE Computer Society (2010)

[15] Algebras, S., Walker, E.A.: Conditional events, and three-valued logic. IEEE Transaction of Systems, 1699–1707 (1994)

[16] Pedrycz, W.: From fuzzy sets to shadowed sets: Interpretation and computing. Int. J. Intell. Syst. 24(1), 48–61 (2009)

[17] Yao, Y.: Interval sets and interval-set algebras. In: IEEE ICCI, pp. 307–314 (2009)

[18] Yao, Y.: An outline of a theory of three-way decisions. In: Yao, J., Yang, Y., Słowiński, R., Greco, S., Li, H., Mitra, S., Polkowski, L. (eds.) RSCTC 2012. LNCS, vol. 7413, pp. 1–17. Springer, Heidelberg (2012)

[19] Yao, Y.Y., Wong, S.K.M., Lingras, P.: A decision-theoretic rough set model. In: Ras, Z.W., Zemankova, M., Emrich, M.L. (eds.) Methodologies for Intelligent Systems, 5: Proc. of the Fifth International Symposium on Methodologies for Intelligent Systems, pp. 17–24. North-Holland, New York (1990)

[20] Tversky, D., Kahneman, D.: A Prospect theory: An analysis of decision under risk. Econometrica 47, 263–292 (1979)

[21] Ciucci, D., Dubois, D.: A map of dependencies among three-valued logics. Inf. Sci. 250, 162–177 (2013)

[22] Arkes, H.R., Hirshleifer, D., Jiang, D., Sonya Lim, S.: Reference point adaptation: Tests in the domain of security trading. Organizational Behavior and Human Decision Processes 105, 67–81 (2008)

[23] Fan, Z.-P., Zhang, X., Chen, F.-D., Liu, Y.: Multiple attribute decision making considering aspiration-levels: A method based on prospect theory. Computers & Industrial Engineering 65(2), 341–350 (2013)

[24] Fujita, H., Ko, Y.-C.: Subjective weights based meta-learning in multi-criteria decision making. In: Fodor, J., Fullér, R. (eds.) Advances in Soft Computing, Intelligent Robotics and Control. Topics in Intelligent Engineering and Informatics, vol. 8. Springer International Publishing (2014)

[25] Lima, M.M.P.P., Gomes, L.F.A.M.: Todim: Basics and application to multicriteria ranking of projects with environmental impacts. Foundations of Computing and Decision Sciencesg 16, 113–127 (1992)

[26] Ko, Y.-C., Fujita, H., Tzeng, G.-H.: A simple utility function with the rules-verified weights for analyzing the top competitiveness of wcy 2012. Knowl.-Based Syst. 58, 58–65 (2013)
[27] IMD. World Competitiveness Yearbook. Institute Management Development, Lausanne, Switzerland (2012),
http://worldcompetitiveness.com/OnLine/App/Index.htm
[28] Pawlak, Z.: Rough set approach to knowledge-based decision support. European Journal of Operational Research 99(1), 48–57 (1997)
[29] Greco, S., Matarazzo, B., Slowinski, R.: Rough sets theory for multicriteria decision analysis. European Journal of Operational Research 129(1), 1–47 (2001)
[30] Pawlak, Z.: Rough sets, decision algorithms and bayes' theorem. European Journal of Operational Research 136(1), 181–189 (2002)

# Interval Analysis for Decision Aiding

Masahiro Inuiguchi

**Abstract.** The interval analysis for decision aiding based on the possibility theory is introduced. The interval analysis provides a new paradigm of data analysis which is based on the idea that the variability of the data is not always caused by the error but by the intrinsic variety of the systems outputs. From this idea, several methods for data analysis have been proposed and obtained results in a different point of view from the conventional analysis. First, the interval regression analysis is briefly introduced dividing into two cases: the case of crisp data and the case of interval data. Then the application to Analytic Hierarchy Process (AHP) called Interval AHP is described. This method can be seen as the AHP counterpart of the interval regression analysis for crisp data. A proper method for the comparison between alternatives is proposed. The obtained dominance relation between alternatives is not always a total order but a preorder. Finally, the extension of Interval AHP to Group Interval AHP is described. This method can be seen as the AHP counterpart of the interval regression analysis for interval data. We describe three models of Group Interval AHP, i.e., perfect incorporation model, partial incorporation model and seeking common ground model. In each model, we obtain the dominance relations between alternatives which are usually preorders but have different meanings. The relationship among the three models is described.

**Keywords:** interval analysis, interval regression, AHP, group decision making.

## 1 Introduction

The interval analysis provides a new paradigm of data analysis which is based on the idea that the variability of the data is not always caused by the error but by the

Masahiro Inuiguchi
Graduate School of Engineering Science, Osaka University
Toyonaka, Osaka, 560-8531, Japan
http://www-inulab.sys.es.osaka-u.ac.jp/

© Springer International Publishing Switzerland 2015                                          15
V.-H. Nguyen et al. (eds.), *Knowledge and Systems Engineering,*
Advances in Intelligent Systems and Computing 326, DOI: 10.1007/978-3-319-11680-8_2

intrinsic variety of the systems outputs. This idea was suggested by Tanaka et al. [1] in the proposal of fuzzy linear regression analysis. This idea was applied by the members of his research group to other types of regression analyses [2, 3, 4, 5], GMDH [6], neural networks [7], the analysis of portfolio data [8], and so on. These proposals were gradually accepted by the researchers in fuzzy sets and systems so that their original paper became frequently cited.

The idea is closely related to the possibility theory [9] because the intrinsic variety of the systems outputs can be seen as the possible range of output values. Moreover, their proposal of fuzzy linear regression analysis for fuzzy data has a similarity to rough sets [10] because dual models corresponding to possibility and necessity of modal logic are calculated in both methods. Because the determination of a proper membership function comes often up for debate, they propose interval linear regression models and applied to Analytic Hierarchy Process [11]. The degeneration of a fuzzy number to an interval does not lose the essence of their original proposals but redounds the explicitness of the models.

In this paper, we introduce the interval analysis for decision aiding. First we briefly review the interval regression analysis [1] for crisp data and then interval regression analysis [2, 3] for interval data. They are the fundamental models of the interval analysis in this paper. For crisp data, a narrowest interval regression function is obtained so as to cover all given data. For interval data, three different approaches were proposed. One is to obtain a narrowest interval regression function including all given interval data. The second is to obtain a widest interval regression function so that the estimated intervals are included in all given interval data. The third one is to obtain a narrowest interval regression function intersecting all given interval data. Some properties are given.

Then we describe Interval AHP [12]. Interval AHP corresponds to interval regression analysis for crisp data. Components of the given pairwise comparison matrix showing the relative importance between criteria or between alternatives correspond to crisp data. The narrowest interval weights are obtained so that all given degrees of relative importance are included in the range of the degrees of relative importance calculated from the interval weights. The interval weights are assumed to be normalized. Thus normalization constraints for interval weights are added. In this approach, the conflict in the given pairwise comparison matrix is assumed to occur by the intrinsic variety of the weights. A proper dominance relation under obtained interval weights is newly defined.

Finally, the Interval AHP [12] is used in group decision making. Group Interval AHP corresponds to interval regression analysis for interval data. Each decision maker provides a pairwise comparison matrix. Then a normalized interval weight vector is obtained for each decision maker by Interval AHP. Those interval weight vectors corresponds to interval data of interval regression analysis. Then three approaches are conceivable. By the nature of group decision making, we give proper interpretations of three models and extend them introducing the concepts of compromise and equality among decision makers. Corresponding to three approaches, three different dominance relations are defined. The relationships among three models are described.

This paper is organized as follows. In Section 2, we describe the interval regression analysis giving a necessary introduction of interval calculations. Interval AHP is described in Section 3. In Section 4, Group Interval AHP is explained. Finally, some concluding remarks are given in Section 5.

## 2  Interval Regression Analysis

### 2.1  Crisp Data Case

Given a set of observed input-output data, $(x_i, y_i)$, $i \in N = \{1, 2, \ldots, n\}$, a suitable $\alpha = (\alpha_0, \alpha_1, \ldots, \alpha_m)^T$ has been estimated under the assumption that $y$ is represented by $\alpha^T x$, where $x_i = (1, x_{i1}, \ldots, x_{im})^T$. $i \in N$ and $x = (1, x_1, \ldots, x_m)^T$. In the conventional regression analysis, each observed value $y_i$ is assumed to include an error $e_i$ as shown in

$$y_i = \alpha^T x_i + e_i, \ i \in N. \tag{1}$$

Parameter vector $\alpha$ is estimated so as to minimize the sum of square errors,

$$\text{minimize} \sum_{i=1}^{n} e_i^2 = \sum_{i=1}^{n} (y_i - \alpha^T x_i)^2. \tag{2}$$

However, we may encounter cases where the variability of the data is not always caused by the error but by the intrinsic variety of the systems outputs. For example, the values obtained by human evaluations and outputs of human-machine systems would be such cases because human cannot detect a small difference as is known by the just noticeable difference and human activity is influenced by the working atmosphere. Moreover, when the systems outputs are influenced by mutable unconsidered indirect factors such as subtle fluctuation of environmental factors, material condition and/or machine condition, we may think that the variability of the data is not always caused by the error but by the intrinsic variety of the systems outputs due to the unconsidered factors. From this point of view, we may estimate the range of possible variation of systems outputs from a given data. Then we estimate an interval linear function $A^T x$ from data set $(x_i, y_i)$, $i \in N$, where $A = (A_0, A_1, \ldots, A_m)^T$ and $A_j = [a_j^L, a_j^R]$ $(a_j^L \leq a_j^R)$, $j = 0, 1, \ldots, m$. Let $a_j^C = (a_j^L + a_j^R)/2$ and $a_j^W = a_j^R - a_j^L \geq 0$. Then interval $A_j = [a_j^L, a_j^R]$ can be characterized also by $a_j^C$ and $a_j^W$. Then we denote $A_j = [a_j^L, a_j^R]$ also by $\langle a_j^C, a_j^W \rangle$. The interval linear function value of $A^T x$ is defined by

$$A^T x = \left[ \sum_{j: x_j \geq 0} a_j^L x_j + \sum_{j: x_j < 0} a_j^R x_j, \ \sum_{j: x_j \geq 0} a_j^R x_j + \sum_{j: x_j < 0} a_j^L x_j \right]$$

$$= \left\langle \sum_{j=0}^{m} a_j^C x_j, \sum_{j=0}^{m} a_j^W |x_j| \right\rangle. \tag{3}$$

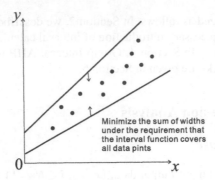

**Fig. 1** Interval regression analysis for crisp data

For a given data set $(x_i, y_i)$, $i \in N$, from the idea that all $y_i$'s are intrinsic varieties of systems output, we require $A^T x$ to satisfy

$$y_i \in A^T x_i, \ i \in N. \tag{4}$$

There are many interval linear functions satisfy requirement (4). Wider interval linear functions are easier to satisfy requirement (4) and too wide interval linear functions would not express the give data well. Then, an interval linear function with minimal widths satisfying requirement (4) is estimated from the given data set $(x_i, y_i)$, $i \in N$. Eventually, the following linear function is minimized under requirement (4):

$$\sum_{i=1}^{n} \sum_{j=0}^{m} a_j^W |x_{ij}|. \tag{5}$$

The function (5) shows the sum of the widths of interval linear function values of $x_i$, $i \in N$. This approach is depicted in Figure 1.

This interval function estimation problem is reduced to the following linear programming problems (see [1]):

$$\begin{aligned}
\text{minimize} \quad & \sum_{i \in N} \sum_{j=0}^{m} a_j^W |x_{ij}|, \\
\text{subject to} \quad & \sum_{j=0}^{m} a_j^C x_{ij} - \frac{1}{2} \sum_{j=0}^{m} a_j^W |x_{ij}| \leq y_i, \ i \in N, \\
& \sum_{j=0}^{m} a_j^C x_{ij} + \frac{1}{2} \sum_{j=0}^{m} a_j^W |x_{ij}| \geq y_i, \ i \in N, \\
& a_j^W \geq 0, \ j = 0, 1, \ldots, m.
\end{aligned} \tag{6}$$

## 2.2 Interval Data Case

The interval regression analysis can be extended to the case of interval data. In this subsection, we describe three approaches to interval regression analysis for interval

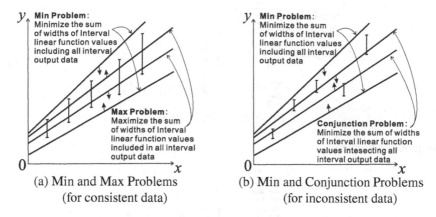

(a) Min and Max Problems
(for consistent data)

(b) Min and Conjunction Problems
(for inconsistent data)

**Fig. 2** Interval regression analysis for interval data

data [2, 3], $(x_i, Y_i)$, $i \in N$, where $Y_i = [y_i^L, y_i^R]$, $i \in N$. In the case of interval data, we consider three approaches. The three approaches, Min problem, Max problem and Conjunction problem, are depicted in Figure 2. As shown in Figure 2, in Min problem, we calculate an interval linear function with minimal widths including all interval output data, i.e., $A^T x_i \supseteq Y_i$, $i \in N$. In Max problem, we calculate an interval linear function with maximal widths whose function values are included in all interval output data, i.e., $A^T x_i \subseteq Y_i$, $i \in N$. In Conjunction problem, we calculate an interval linear function with minimal widths intersecting all interval output data, i.e., $A^T x_i \cap Y_i \neq \emptyset$, $i \in N$.

Those estimation problems are formulated in the following three linear programming problems:

Min problem:

$$\text{minimize} \sum_{i \in N} \sum_{j=0}^{m} a_j^W |x_{ij}|,$$

$$\text{subject to} \sum_{j=0}^{m} a_j^C x_{ij} - \frac{1}{2} \sum_{j=0}^{m} a_j^W |x_{ij}| \leq y_i^L, \ i \in N,$$

$$\sum_{j=0}^{m} a_j^C x_{ij} + \frac{1}{2} \sum_{j=0}^{m} a_j^W |x_{ij}| \geq y_i^R, \ i \in N,$$

$$a_j^W \geq 0, \ j = 0, 1, \dots, m,$$

(7)

Max problem:

$$\text{maximize} \sum_{i \in N} \sum_{j=0}^{m} a_j^{W} |x_{ij}|,$$

$$\text{subject to} \sum_{j=0}^{m} a_j^{C} x_{ij} - \frac{1}{2} \sum_{j=0}^{m} a_j^{W} |x_{ij}| \geq y_i^{L}, \ i \in N,$$

$$\sum_{j=0}^{m} a_j^{C} x_{ij} + \frac{1}{2} \sum_{j=0}^{m} a_j^{W} |x_{ij}| \leq y_i^{R}, \ i \in N, \qquad (8)$$

$$a_j^{W} \geq 0, \ j = 0, 1, \dots, m,$$

Conjunction problem:

$$\text{minimize} \sum_{i \in N} \sum_{j=0}^{m} a_j^{W} |x_{ij}|,$$

$$\text{subject to} \sum_{j=0}^{m} a_j^{C} x_{ij} - \frac{1}{2} \sum_{j=0}^{m} a_j^{W} |x_{ij}| \leq y_i^{R}, \ i \in N,$$

$$\sum_{j=0}^{m} a_j^{C} x_{ij} + \frac{1}{2} \sum_{j=0}^{m} a_j^{W} |x_{ij}| \geq y_i^{L}, \ i \in N, \qquad (9)$$

$$a_j^{W} \geq 0, \ j = 0, 1, \dots, m.$$

When the given interval data set includes a conflict, i.e., there is no hyperplane $y = \alpha^{T} x$ intersecting all interval output data $Y_i$, $i \in N$ (no hyperplane satisfying $\alpha^{T} x_i \in Y_i$, $i \in N$), there is no solution in Max problem. In such cases, we consider Conjunction problem. On the contrary, when there exists $A^{T} x$ such that $\sum_{i \in N} \sum_{j=0}^{m} a_j^{W} |x_{ij}| > 0$, Conjunction problem has a lot of solutions whose objective function values are zero, i.e., multiple optimal solutions exist. Any of such solutions does not well explain the given interval data. This is why we draw two figures (a) and (b) in Figure 2. Therefore, Max problem and Conjunction problem are complementary, while Min problem is meaningful in any cases.

Many variations of the interval regression analysis are proposed in the literature [4, 5]. Moreover, this analysis is applied to several real world data set and obtained results from a different view from the conventional regression analysis (see [3, 13]). The idea is applied to other methods [6, 7, 8] for data analysis.

## 3 Interval AHP for Decision Support

The idea of the interval regression analysis is applied to Analytic Hierarchy Process (AHP) [11] which is a decision support method in the setting of multiple criteria. In AHP, the decision problem is structured hierarchically as criteria and alternatives. At each node except leaf nodes of the hierarchical tree, a weight vector for criteria or for alternatives is obtained from a pairwise comparison matrix given by a decision maker. The $(i, j)$ component of the matrix shows the relative importance of the $i$-th criterion/alternative over the $j$-th criterion/alternative.

In this section, we first describe the estimation of a weight vector $w = (w_1, w_2, \dots, w_n)^{T}$ from a given pairwise comparison matrix,

$$A = \begin{bmatrix} 1 & \cdots & a_{1n} \\ \vdots & a_{ij} & \vdots \\ a_{n1} & \cdots & 1 \end{bmatrix}, \tag{10}$$

where we assume the reciprocity, i.e., $a_{ij} = 1/a_{ij}$, $i, j \in N$. Since the $(i, j)$ component $a_{ij}$ of $A$ shows the relative importance of the $i$-th criterion/alternative over the $j$-th criterion/alternative, theoretically, we have $a_{ij} = w_i/w_j$, $i, j \in N$. This is why we assume the reciprocity, $a_{ij} = 1/a_{ij}$, $i, j \in N$. Moreover, if $a_{ij}$, $i, j \in N$ are obtained exactly, the strong transitivity $a_{ij} = a_{il}a_{lj}$, $i, j, l \in N$ should be satisfied. However, human evaluation is not very accurate so that the strong transitivity is satisfied. In the conventional approach [11], $a_{ij}$, $i, j \in N$ are assumed to be approximations of $w_i/w_j$. $w$ is estimated as the normalized eigen vector corresponding to the maximal eigen value, because the nonnegativity of the eigen vector corresponding to the maximal eigen value is guaranteed by Perron-Frobenius theorem. There are other ways to estimate $w$ such as the least square method [14], the geometric mean method [15] and so on. To evaluate the consistency of the given pairwise comparison matrix, the following consistency index is used:

$$C.I. = \frac{\lambda_{\max} - n}{n - 1}, \tag{11}$$

where $\lambda_{\max}$ is the maximal eigen value of $A$. If $C.I.$ is not greater than 0.1, it is often considered that the obtained vector $w$ is acceptable.

To this estimation problem, the idea of interval analysis can be applied. In this approach, we assume that the decision maker' evaluation is not very accurate to be expressed by a unique weight vector $w$ but intrinsically vague (by hesitation) so that the weight vector has some range. Therefore, the unsatisfaction of the strong transitivity is not regarded as inconsistency but due to the intrinsic variety of possible weight vectors. Accordingly, we consider an interval weight vector $W = (W_1, W_2, \ldots, W_n)^T$ instead of a weight vector $w$, where $W_i = [w_i^L, w_i^R]$, $i \in N$ and $w_i^L \leq w_i^R$, $i \in N$. To fit the given pairwise comparison matrix, we require the interval weight vector $W$ to satisfy

$$\frac{w_i^L}{w_j^R} \leq a_{ij} \leq \frac{w_i^R}{w_j^L}, \ i, j \in N, \ i < j. \tag{12}$$

We note that by the reciprocity, $a_{ij} = 1/a_{ij}$, $i, j \in N$ and $a_{ii} = 1$, $i \in N$, we only consider $i, j \in N$ such that $i < j$. Moreover, corresponding to normalization condition of $w$ in the conventional AHP, we require the interval weight vector $W$ to satisfy

$$\sum_{j \in N \setminus i} w_j^R + w_i^L \geq 1, \ i \in N, \tag{13}$$

$$\sum_{j \in N \setminus i} w_j^L + w_i^R \leq 1, \ i \in N, \tag{14}$$

where $N \setminus i = N \setminus \{i\} = \{1, 2, \ldots, i - 1, i + 1, \ldots, n\}$. (13) and (14) ensure that, for any $w_i^\circ \in W_i$, there exist $w_j \in W_j$, $j \in N \setminus i$ such that $\sum_{j \in N \setminus i} w_j + w_i^\circ = 0$ (see [16]).

Namely, any values in $W_i$, $i \in N$ are meaningful (there is no ineffective subarea in $W$).

Under conditions (12), (13) and (14) as well as $\varepsilon < w_i^L \leq w_i^R$, $i \in N$, we calculate a suitable $W$, where $\varepsilon$ is a very small positive number. The wider each $W_i$ is, the easier $W_i$, $i \in N$ satisfy (12), (13) and (14). And the narrower interval weights give more clear preferences in the comparison of alternatives. Then we minimize the following total widths of interval weights $W_i$, $i \in N$:

$$d = \sum_{i \in N} (w_i^R - w_i^L). \tag{15}$$

Consequently, the interval weight vector $W$ is estimated by solving the following linear programming problem:

$$\begin{aligned}
\text{minimize } d &= \sum_{i \in N} (w_i^R - w_i^L), \\
\text{subject to } \frac{w_i^L}{w_j^R} &\leq a_{ij} \leq \frac{w_i^R}{w_j^L}, \ i, j \in N, \ i < j, \\
\varepsilon &\leq w_i^L \leq w_i^R, \ i \in N, \\
\sum_{j \in N \setminus i} w_j^R + w_i^L &\geq 1, i \in N, \\
\sum_{j \in N \setminus i} w_j^L + w_i^R &\leq 1, i \in N.
\end{aligned} \tag{16}$$

Once we obtain an optimal solution to Problem (16) for each pairwise comparison matrix given at each node except leaf nodes of the hierarchical tree, we can compare the alternatives. Let $X_i = (X_{i1}, X_{i2}, \ldots, X_{im})^T$, $i \in N$ be the interval weights of alternatives obtained for a pairwise comparison matrix in view of the $i$-th criterion, where we define $X_{ip} = [x_{ip}^L, x_{ip}^R]$, $i \in N$, $p \in M = \{1, 2, \ldots, m\}$. Tanaka et al. [12] proposed to obtain the overall interval priority $Y_p$ of the $p$-th alternative by

$$Y_p = [y_p^L, y_p^R] = \left\{ y_p = \sum_{i \in N} w_i x_{ip} \ \middle| \ w_i \in W_i, \ x_{ip} \in X_{ip}, \ i \in N, \ p \in M \right\}. \tag{17}$$

By interval calculations, we obtain

$$y_p^L = \sum_{i \in N} w_i^L x_{ip}^L, \quad y_p^R = \sum_{i \in N} w_i^R x_{ip}^R. \tag{18}$$

Then they proposed to apply the following dominance relation between alternatives $alt_p$ and $alt_q$:

$$alt_p \succsim alt_q \text{ if and only if } y_p^L \geq y_q^L \text{ and } y_p^R \geq y_q^R. \tag{19}$$

This dominance relation is a preorder, i.e., it satisfies the reflexivity, $alt_p \succsim alt_p$, $p \in M$ and the transitivity, $alt_p \succsim alt_q$ and $alt_q \succsim alt_r$ imply $alt_p \succsim alt_r$. The preorder is considered to give the information of the decision maker's preference considering the intrinsic vagueness (hesitation) of human evaluations.

The definitions of $y_p^L$ and $y_p^R$ are not proper because it does not consider the normalization condition of the weight vector $w$. We may modify those by solving following continuous knapsack problems as pointed out by Guo and Tanaka [17]:

$$y_p^L = \min\left\{\sum_{i \in N} w_i x_{ip}^L \ \middle|\ \sum_{i \in N} w_i = 1,\ w_i^L \le w_i \le w_i^R,\ i \in N\right\}, \qquad (20)$$

$$y_p^R = \max\left\{\sum_{i \in N} w_i x_{ip}^R \ \middle|\ \sum_{i \in N} w_i = 1,\ w_i^L \le w_i \le w_i^R,\ i \in N\right\}. \qquad (21)$$

The comparison of alternatives by (19) is not robust for $w \in W$, $x_i \in X_i$, $i \in N$. Namely, even when $alt_p \succsim alt_q$ of (19) holds, we may have some combination $(w, x_1, \ldots, x_n)$ satisfying $w \in W$, $x_i \in X_i$, $i \in N$ and $\sum_{i \in N} w_i x_{iq} > \sum_{i \in N} w_i x_{ip}$ (suggesting that $alt_q$ is possibly better than $alt_p$). Similar to the authors' proposal in [16], the author proposes to define the dominance relations between alternatives $alt_p$ and $alt_q$ by solving the following continuous knapsack problem:

$$d_{pq} = \min\left\{\sum_{i \in N} w_i(x_{ip}^L - x_{iq}^R) \ \middle|\ \sum_{i \in N} w_i = 1,\ w_i^L \le w_i \le w_i^R,\ i \in N\right\}. \qquad (22)$$

Namely, we define $alt_p \succsim^N alt_q$ if and only if $d_{pq} \ge 0$. Relation $\succsim^N$ is reflexive and transitive, i.e., a preorder. This dominance relation is usually less effective than that by (19).

The conventional AHP is useful if the given information is consistent enough, it wrings out a weak order otherwise. On the other hand, Interval AHP is useful independently of the consistency and it gives reasonable assessments although they are rather hesitant to give a crisp dominance reflecting the vagueness of human evaluations. The proposed dominance relation is useful for the robust evaluations of alternatives.

Tanaka et al. [18], Wang [19] and Guo [20] have extended Interval AHP to the case where interval pairwise comparison matrix is given. As Interval AHP is the AHP counterpart of the interval regression analysis for crisp data, the extended Interval AHP is the AHP counterparts of the interval regression analysis for interval data.

## 4 Group Interval AHP

### 4.1 Application to Group Decision Making

Entani and Inuiguchi [21, 22] were preliminarily applied Interval AHP to group decision making. As for the applications of the conventional AHP to group decision making, there are two approaches for aggregating individual opinions into a group opinion: one is first to aggregate individual pairwise comparison matrices and then to calculate a group weight vector, and the other is first to calculate individual weight vectors and then to aggregate them to a group weight vector. Recently Entani and Inuiguchi [16] have proposed three approaches to Group Interval AHP. These approaches are basically from the latter viewpoint of group AHP but individual interval weight vectors are not fixed when the group interval weight vector is calculated. They suppose the usage of the proposed approach at the beginning of the negotiation among group members. Therefore, all individual never change their original pairwise comparison matrices. The group interval weight vector is obtained by requesting members equally to tolerate the diversity/reduction of their interval weights.

Because as in Interval AHP, each individual has an interval weight vector corresponding to her/his given pairwise comparison matrix, Group Interval AHP become similar to the interval regression analysis for interval data. Indeed, Group Interval AHP can be also seen as an AHP counterpart of the interval regression analysis for interval data. As there are three approaches in the interval regression analysis for interval data, there are three approaches to Group Interval AHP.

In what follows, we assume there are $s$ decision makers. Let $S = \{1, 2, \ldots, s\}$. We assume the weight vector $x_i = (x_{i1}, x_{i2}, \ldots, x_{im})$ in criterion $i$ is obtained commonly among all decision makers. Then we apply Interval AHP only to determine the interval weights on criteria. The individual pairwise comparison matrix of the $k$-th decision maker is denoted by $A_k$ and its $(i, j)$-component is denoted by $a_{kij}$. Let $\hat{d}_k$ be the optimal value of Problem (16) for the $k$-th decision maker ($k \in S$). The individual interval weight of the $k$-th decision maker on the $i$-th criterion is denoted by $W_{ki} = [w_{ki}^L, w_{ki}^R]$, $k \in S$, $i \in N$. Then the individual interval weight vector is denoted by $W_k = (W_{k1}, W_{k2}, \ldots, W_{kn})^T$. The group interval weight on the $i$-th criterion is denoted by $W_i = [w_i^L, w_i^R]$, $i \in N$. The group interval weight vector is denoted by $W = (W_1, W_2, \ldots, W_n)^T$. Let $\mathcal{W}^N$ be a set of normalized interval weight vectors. Moreover, let $\mathcal{W}(A_k)$ be a set of interval weight vectors compatible with the preference information given by pairwise comparison matrix $A_k$. An interval weight vector $W$ satisfying (12) is said to be compatible with pairwise comparison matrix $A$.

## 4.2 Perfect Incorporation of All Individual Opinions

The first approach is a perfect incorporation of all individual opinions [16]. In this approach, the group interval weights include all individual interval weights and thus, this approach is corresponding to Min problem of the interval regression analysis for interval data. Namely, $W$ is calculated so as to

$$\text{minimize} \sum_{i \in N} w(W_i), \tag{23}$$

under constraints,

$$\begin{cases} W_i \supseteq W_{ki}, \ i \in N, \ \sum_{i \in N} w(W_{ki}) \leq \hat{d}_k, \ k \in S, \\ W \in \mathcal{W}^N, \ W_k \in \mathcal{W}^N \cap \mathcal{W}(A_k), \ k \in S, \end{cases} \tag{24}$$

where $w(W_i) = w_i^R - w_i^L$ $w(W_{ki}) = w_{ki}^R - w_{ki}^L$ are the widths of $W_i$ and $W_{ki}$, respectively. We note that we have $\sum_{i \in N} w(W_{ki}) = \hat{d}_k$, because $W_{ki}, k \in S, i \in N$ compose an optimal solution to Problem (16) and that $\sum_{i \in N} w(W_{ki}) \leq \hat{d}_k$ implies that $W_{ki}, k \in S$ compose an optimal solution to Problem (16). Because it is possible that many optimal solutions exist in Problem (16), we treat $W_k, k \in S$ as variable interval vectors. After obtaining an optimal $W$, using (22), we obtain the group dominance relation between alternatives. This group dominance relation shows which dominance between two alternatives is supported by all conceivable weight vectors of all decision makers.

However, the group dominance relation is too weak (almost all dominances are negative) when the individual opinions are not very similar. This is because $W_i, i \in N$ are too wide. Then we may request all decision makers to reduce their individual interval weights $W_{ki}$ to $V_{ki}, k \in S, i \in N$. Therefore, in the modified problem, $W$ is calculated so as to minimize $\sum_{i \in N} w(W_i)$ under the constraints,

$$
\begin{cases}
W_i \supseteq V_{ki}, \ V_{ki} \subseteq W_{ki}, \ k \in S, \ i \in N, \\
\sum_{i \in N} w(W_{ki}) \leq \hat{d}_k, \ \sum_{i \in N} w(V_{ki}) \geq \hat{d}_k(1 - r_1), \ k \in S, \\
W, V_k \in \mathcal{W}^N, \ k \in S, \ W_k \in \mathcal{W}^N \cap \mathcal{W}(A_k),
\end{cases}
\tag{25}
$$

where $V_k = (V_{k1}, V_{k2}, \dots, V_{kn})^T$ and $r_1 \in [0, 1]$ is the maximum reduction rate. This problems is reduced to a linear programming problem. If $V_k$ is accepted by the $k$-th decision maker ($k \in S$), the group dominance relation obtained by (22) with $W$ is meaningful. It shows which dominance between two alternatives is supported by all conceivable weight vectors in the reduced interval weight vectors of all decision makers.

## 4.3 Partial Incorporation of All Individual Opinions

The second approach is a partial incorporation of all individual opinions [16]. In this approach, the group interval weights intersect all individual interval weights and thus, this approach is corresponding to Conjunction problem of the interval regression analysis for interval data. Namely, $W$ is calculated so as to minimize $\sum_{i \in N} w(W_i)$ under the constraints,

$$
\begin{cases}
W_i \cap W_{ki} \neq \emptyset, \ i \in N, \ \sum_{i \in N} w(W_{ki}) \leq \hat{d}_k, \ k \in S, \\
W \in \mathcal{W}^N, \ W_k \in \mathcal{W}^N \cap \mathcal{W}(A_k), \ k \in S.
\end{cases}
\tag{26}
$$

After obtaining an optimal $W$, using (22), we obtain the group dominance relation between alternatives. This group dominance relation shows which dominance between two alternatives is supported by at least one conceivable weight vector of each decision maker. Namely all decision maker can somehow agree in the dominance because one of his possible opinions support this dominance.

The difference between (24) and (26) is only their first constraints, i.e., $W_i \supseteq W_{ki}$ versus $W_i \cap W_{ki} \neq \emptyset$. By this weakened constraint, the partial incorporation approach has narrower $W$ than the perfect incorporation approach. Nevertheless, when individual opinions are different to a certain extent, $W_i$, $i \in N$ become sufficiently wide and the group dominance relation is weak.

In such cases we may request all decision makers to diverse their acceptable individual interval weight vectors by relaxing the optimality to Problem (16). Namely, the $k$-th decision maker accept suboptimal interval weight vector $W_k$ as her/his individual opinion. In this modified problem, $W$ is calculated so as to minimize $\sum_{i \in N} w(W_i)$ under the constraints,

$$\begin{cases} v_{ki} \cap W_i \neq \emptyset, \ v_{ki} \in W_{ki}, \ k \in S, \ i \in N, \\ \sum_{i \in N} w(W_{ki}) \leq (1+r_2)\hat{d}_k, \ \sum_{i \in N} v_{ki} = 1 \ k \in S, \\ W \in \mathcal{W}^N, \ W_k \in \mathcal{W}^N \cap \mathcal{W}(A_k), \ k \in S, \end{cases} \tag{27}$$

where $r_2 \geq 0$ is the relaxation rate. The introduction of the normalized weight vectors $v_k = (v_{k1}, v_{k2}, \ldots, v_{kn})^T$, $k \in S$ guarantees that the existence of normalized weight vector in the intersection of $W_k$ and $W$, $k \in S$. If some $A_k$ satisfies the strong consistency, the optimal value of Problem (16) becomes zero. In this case, we substitute a predetermined number $d^{\min} > 0$ for $\hat{d}^k$ considering human evaluation is vague as is known by the just noticeable difference. This problems is reduced to a linear programming problem. If $W_k$ is accepted by the $k$-th decision maker ($k \in S$), the group dominance relation obtained by (22) with $W$ is meaningful. It shows which dominance between two alternatives is supported by at least one weight vector in the compromised interval weight vector of each decision maker.

### 4.4 Seeking Common Ground among All Decision Makers

When all individual opinions are similar, we may have non-empty intersection $\bigcap_{k \in S} W_k$. In this case, a normalized interval vector $W \subseteq \bigcap_{k \in S} W_k$ can be seen as the common ground of all individual opinions. Such $W = (W_1, W_2, \ldots, W_n)^T$ can be obtained easily by (see Entani and Inuiguchi [16])

$$w_i^L = \min_{k \in S} w_{ki}^L, \quad w_i^R = \min_{k \in S} w_{ki}^R, \ i \in N. \tag{28}$$

However, in the real world, unfortunately, we usually have $\bigcap_{k \in S} W_k = \emptyset$. To seek common ground among all decision makers, we should request individuals to accept suboptimal interval weight vectors to Problem (16) as their individual opinions. Suppose that all decision makers accept interval weight vectors $100r_3\%$ worse than the optimal one as their individual opinions, $W$ is calculated so as to maximize $\sum_{i \in N} w(W_i)$ under the constraints,

$$\begin{cases} W_i \subseteq W_{ki}, \ k \in S, \ i \in N, \\ \sum_{i \in N} w(W_{ki}) \leq (1+r_3)\hat{d}_k, \ k \in S, \\ W \in \mathcal{W}^N, \ W_k \in \mathcal{W}^N \cap \mathcal{W}(A_k), \ k \in S, \end{cases} \tag{29}$$

where $r_3 \geq 0$ is the relaxation rate. We note that the objective function is maximized because we would like to have a wide common ground. This problems is also reduced to a linear programming problem. If $W_k$ is accepted by the $k$-th decision maker ($k \in S$), the group dominance relation obtained by (22) with $W$ is meaningful. It shows which dominance between two alternatives is supported by all weight vectors in the common ground. The wider $W_i$, $i \in N$ are, the more stable the agreed dominances between alternatives obtained by (22) are.

## 4.5  Relationships among Three Approaches

Three approaches with the diversity/reduction of individual interval weights are related one another. The perfect incorporation model with $r_1 = 1$ is equivalent to the partial incorporation model with $r_0 = 0$. A group interval weight vector as an optimal solution to the partial incorporation model whose optimal value is zero is a feasible solution to the seeking common ground model with $r_3 = r_2$.

Those relationships can be understood from a viewpoint of group decision making. At the beginning, decision makers want to reflect all individually conceivable (acceptable) weight vectors to the group interval weight vector so that they can accept the resulting group dominance relation without hesitation. Namely, The perfect incorporation model with $r_1 = 0$ is adopted. However, they may realized that their desires are too strong to obtain an useful group dominance relation. Then they gradually increase $r_1$ to reduce their individually conceivable (acceptable) weight vectors to enhance the usefulness of group dominance relation.

If the group dominance relation is not sufficiently useful while $r_1$ is increased to one, they may start to enlarge the set of their individually conceivable (acceptable) weight vectors so that they use the partial incorporation model and increase $r_2$. They agree that they reflect only one of their individually conceivable (acceptable) weight vectors to the group opinion at the worst.

If they are succeeded to share common ground in their opinions, i.e., the optimal value of the partial incorporation model vanishes, by enlarging the set of their individually conceivable (acceptable) weight vectors, decision makers may further seek a larger common ground. This seeking process can be performed by increasing $r_3$ in the seeking common ground model. The decision makers know which dominances between alternatives are stable and what dominances are conceivable as agreeable ones. They have many candidates of agreeable weak orders on alternatives. The final selection from those candidates can be selected by any way such as voting.

In this paper, we do not consider the preference revision of each decision maker. The preference can be revised by learning different viewpoints through discussion with other decision makers in the real world. The preference revision can be introduced by using the idea of distance-based interval regression analysis [23].

## 5   Concluding Remarks

In this paper, the interval analysis based on the idea that the variability of the data is not always caused by the error but by the intrinsic variety of the systems outputs. As the applications of this idea, interval regression analysis, Interval AHP and Group Interval AHP are described. Those approaches are not replacing but complementary to the conventional approaches. The idea can be applied to many of the conventional methods for data analysis. Fortunately, the reduced problems of the applications of this idea are often simple and tractable. Moreover, by such applications, we obtain results in different viewpoints from the conventional one.

**Acknowledgement.** This work was partially supported by JSPS KAKENHI Grant Number 26350423.

## References

[1] Tanaka, H., Uejima, S., Asai, K.: Linear regression analysis with fuzzy model. IEEE Transactions on Systems, Man, and Cybernetics SMC-12(6), 903–907 (1982)

[2] Tanaka, H., Watada, J.: Possibilistic linear systems and their application to the linear regression model. Fuzzy Sets and Systems 27(3), 275–289 (1988)

[3] Tanaka, H., Nagasaka, K., Hayashi, I.: Interval regression analysis by possibilistic measures. The Japanese Journal of Behaviormetrics 16(1), 1–7 (1988) (in Japanese)

[4] Tanaka, H., Lee, H.: Interval regression analysis by quadratic programming approach. IEEE Transactions on Fuzzy Systems 6(4), 473–481 (1998)

[5] Guo, P., Tanaka, H.: Dual models for possibilistic regression analysis. Computational Statistics and Data Analysis 51(1), 253–266 (2006)

[6] Hayashi, I., Tanaka, H.: The fuzzy GMDH algorithm by possibility models and its application. Fuzzy Sets and Systems 36(2), 245–258 (1990)

[7] Ishibuchi, H., Tanaka, H.: Fuzzy regression analysis using neural networks. Fuzzy Sets and Systems 50(3), 257–265 (1992)

[8] Tanaka, H., Guo, P., Burhan Türksen, I.: Portfolio selection based on fuzzy probabilities and possibility distributions. Fuzzy Sets and Systems 111(3), 387–397 (2000)

[9] Zadeh, L.A.: Fuzzy sets as a basis for a theory of possibility. Fuzzy Sets and Systems 1(1), 2–28 (1978)

[10] Pawlak, Z.: Rough Sets. International Journal of Computer and Information Sciences 11(5), 341–356 (1982)

[11] Saaty, T.L.: The Analytic Hierarchy Process. McGraw-Hill, New York (1980)

[12] Sugihara, K., Tanaka, H.: Interval evaluations in the analytic hierarchy process by possibilistic analysis. Computational Intelligence 17(3), 567–579 (2001)

[13] Tanaka, H., Shimomura, T., Watada, J., Asai, K.: Fuzzy linear regression analysis of the number of staff in local government. In: Bezdek, J.C. (ed.) Analysis of Fuzzy Information, vol. III, pp. 191–203 (1987)

[14] Hwang, C.-L., Yoon, K.: Multiple Attribute Decision Making: Methods and Applications. Springer, Berlin (1981)

[15] Rao, R.V.: Decision Making in Manufacturing Environment Using Graph Theory and Fuzzy Multiple Attribute Decision Making Methods, vol. 2. Springer, London (2013)

[16] Entani, T., Inuiguchi, M.: Interval Evaluation Models Obtained from Individual Pairwise Comparisons for Group Decision Aiding in Multiple Criteria Environments (submitted for publication, 2014)

[17] Guo, P., Tanaka, H.: Decision making with interval probabilities. European Journal of Operational Research 203(2), 444–454 (2010)

[18] Sugihara, K., Ishii, H., Tanaka, H.: Interval priorities in AHP by interval regression analysis. European Journal of Operational Research 158(3), 745–754 (2004)

[19] Wang, Y.-M., Elhag, T.M.S.: A goal programming method for obtaining interval weights from an interval comparison matrix. European Journal of Operational Research 177(1), 458–471 (2007)

[20] Guo, P., Wang, Y.: Eliciting dual interval probabilities from interval comparison matrices. Information Sciences 190, 17–26 (2012)

[21] Entani, T., Inuiguchi, M.: Aggregation of group members' opinions based on two principles. In: Proceedings of 2010 IEEE Word Congress on Computational Intelligence, pp. 604–609 (2010)

[22] Entani, T., Inuiguchi, M.: Group decision directly from group members' judgments. In: Proceedings of 2010 World Automation Congress, IFMIP, vol. 237 (2010)

[23] Inuiguchi, M., Tanino, T.: Interval linear regression methods based on Minkowski difference – A bridge between traditional and interval linear regression models. Kybernetika 42(4), 423–440 (2006)

# Agent Communication and Belief Change

Satoshi Tojo

**Abstract.** The notion of multi-agents, including the communication between them, is one of the most fundamental issues to be formalized in artificia intelligence. The communication, however, is not a simple message passing. First, a rational agent should send logically consistent contents in the situation; or, he/she could be an intentional liar. Secondly, there must be a communication channel between agents, e.g., an address of the message recipient. Thirdly, the message can be publicly announced, i.e., there can be simultaneous multiple recipients; otherwise the message passing becomes a personal communication. Finally, the message recipient must adequately maintain the consistency of their belief, that is, as a result of message passing, the recipient must revise his/her belief to be logically consistent. In this talk, I overview the various researches concerning logical representation of communication and belief change, especially in terms of modal logic, where belief change is realized by the restriction of accessibility to some possible worlds. Thereafter, I show some applications of the formalization, such as logical puzzles.

Satoshi Tojo
Japan Advanced Institute of Science and Technology, Nomi, Japan

© Springer International Publishing Switzerland 2015                                   31
V.-H. Nguyen et al. (eds.), *Knowledge and Systems Engineering,*
Advances in Intelligent Systems and Computing 326, DOI: 10.1007/978-3-319-11680-8_3

# Agent Communication and Belief Change

Abstract. The notion of multi-agent ...

# Part II
# KSE 2014 Main Track

# Discriminative Prediction of Enhancers with Word Combinations as Features

Pham Viet Hung and Tu Minh Phuong

**Abstract.** Identification of enhancer regions is important for understanding the regulation mechanism of gene expression. Recent studies show that it is possible to predict enhancers using discriminative classifiers with generic sequence features such as k-mers or words. The accuracy of such discriminative prediction largely depends on the ability of the models to capture not only the presence of predictive k-mers (words), but also spatial constraints on clusters of such k-mers. In this paper, we propose a method that first selects the most important word features and then use combinations of such words, which satisfy certain spatial constraints, as additional features. Experiments with real data sets show that the proposed method compares favorably with a state-of-the-art enhancer prediction method in terms of prediction accuracy.

**Keywords:** Enhancer prediction, SVM, feature extraction, TFBS combination.

## 1 Introduction

The regulation of gene expression plays a fundamental role in cell differentiation and responses of cells to various conditions. There are several levels, at which the expression of genes is regulated, the most important of which is transcriptional regulation. At the transcriptional level, the expression of genes is regulated by transcriptional factors (TFs) that recognize and bind to short DNA sequence motifs, known as transcription factor binding sites (TFBSs). To provide stronger signals for TFs, TFBSs often occur near each other in DNA regions, which are called *cis*-regulatory modules (CRM). CRMs that enhance the expression of genes from distance are called *enhancers*. Identification of enhancers is important for understanding the mechanisms of gene expression regulation.

Pham Viet Hung · Tu Minh Phuong
Department of Computer Science,
Posts and Telecommunications Institute of Technology, Hanoi, Vietnam
e-mail: {hungpv,phuongtm}@ptit.edu.vn

© Springer International Publishing Switzerland 2015                                       35
V.-H. Nguyen et al. (eds.), *Knowledge and Systems Engineering,*
Advances in Intelligent Systems and Computing 326, DOI: 10.1007/978-3-319-11680-8_4

With the current technologies, laboratory methods for enhancer identification are available. Usually, this consists of two steps. First, chromatin immunoprecipitation (ChIP) technique is used to detect signatures of specific TFs associated with activities of enhancers. Then, microarray hybridization or massive parallel sequencing is used to decode the enhancers involved in these activities [14]. This approach allows recognition of enhancers with high accuracy but is resource-intensive, thus limiting its use in practice. A faster and more cost-effective approach is to use computational techniques for predicting enhancers [15], which is the focus of this paper. Methods of this type take advantage of the availability of sequence and other genomic data to recognize enhancers.

A large group of enhancer prediction methods rely on analysis of sequence data. Methods of this group mainly use two strategies. The first strategy is to use predetermined TFBSs, for example from TFBS databases or by running a motif finding algorithm, and search for clusters of these TFBSs [1]. These methods depend on the availability of known TFs and their motifs. The second strategy is discriminative, i.e. using classification algorithms with confirmed enhancers as training data to differentiate between enhancers of a specific type and non-enhancers [12, 7].

Sequence-based enhancer prediction approaches rely on the assumption that similar sequence content is associated with similar binding events, which are in turn associated with similar gene expression. It is well known that TF binding is sequence-specific, i.e. each TF recognizes and binds to a specific short DNA motif (TFBS) of length up to 10 base-pairs (bp). It is also observed that the presence of a motif is not a guarantee for binding and a large fraction of motifs are false positives, i.e. are not associated with binding events. For binding events to occur, TFBSs usually cluster together in enhancers to provide stronger signals for TFs. Previous studies show that there are certain constraints on motif types, motif numbers, and relative motif location within those clusters [5, 14]. For example, pair or triple of TFs of certain types have been observed to bind to closely located TFBSs to co-regulate the expression of some genes [18]. Thus, to successfully predict enhancers, classifier-based methods should be able to model such constraints.

In this paper, we follow the classification-based approach and use SVM classifiers to discriminate between enhancer and non-enhancer sequences. Our method is similar to one presented in [12, 7, 16], which also use SVM for this problem. The difference is that we introduce a new type of features that explicitly count for the constraints on types, distance and order of combinations of words ($k$-mers). To make counting of word combinations tractable, we use feature selection to reduce the set of words to consider. We develop an algorithm to count the number of word combinations that satisfy certain spatial constraints on the locations of words. In this paper, only word pairs are considered but experiments show that this leads to improvement in prediction accuracy in many cases. In experiments using several enhancer datasets from three species (human, mouse, and *Caenorhabditis elegans*), our method compares favorably with a state-of-the art enhancer prediction method. An analysis of $k$-mer pairs extracted from the models learned by the method also provides interesting patterns of TFBS clusters from experimented datasets. Importantly, this is achieved without significant increase in computational complexity.

## 1.1 Related Work

A significant number of computational methods for CRM prediction, in general, and enhancer prediction, in particular, have been developed. Some methods rely on the assumption that CRMs, as a functional regions, are more conserved between related species than background, non-functional regions. These methods search for conserved regions as putative CRMs by combining phylogenetic footprinting with sequence information [11]. For certain types of CRMs, this conservation-based approach can make predictions with high accuracy. However, it is known that many CRMs that are not highly conserved, for which this approach produces poor predictions [15].

Another group of methods rely on signals contained in genomic sequences to make predictions. Early methods of this group search for clusters of known TFBSs within a sequence window [1]. Other methods construct probabilistic models of CRMs, for example in forms of HMMs, and search for regions that fit the learned models with high probabilities. Window clustering require databases of confirmed TFBSs while probabilistic methods require only positive examples or no examples at all but tend to produce many spurious predictions.

Recently, a new, discriminative approach has been developed. Methods of this approach take as input both positive (CRM sequences) and negative examples (sequences that are believed to be not CRMs) and build a classification model to discriminate positive sequences from negative ones. Several studies have shown the ability of discriminative methods in predicting enhancers of complex organisms such as mammals, which is challenging for other approaches [7, 9, 4, 12]. Besides the availability of training data, the success of classification-based approach depends, to a large degree, on the selection of appropriate features to represent sequences, or, more generally, on the selection of appropriate similarity measures [4] and kernels [12] between sequences. While some methods of this group use confirmed motifs as features [12], thus depends on the availability of TFBS databases, other methods extract features from input sequences, making them easier to use in practice [7, 4, 16]. Our proposed method is similar to [7, 4, 16] in that it uses generic sequence features. However, we introduce additional features by explicitly counting the numbers of word combinations. In this way, our approach is able to incorporate different constraints on the presence of motifs within enhancers.

## 2 Preliminaries

In this work, we follow the discriminative approach to predict enhancers. Specifically, we train Support Vector Machines (SVMs) classifiers to differentiate between enhancer (positive) and non-enhancer (negative) sequences. SVMs have been the technique of choice in many enhancer prediction methods due to their superior accuracy and flexibility in dealing with different types of biological data such as sequences and interaction networks. The success of SVMs classifiers depends, to a large extent, on choosing appropriate features or kernels. In this section, we review popular features for sequence data, which serve as basis for our new types of features.

**The Spectrum and Mismatch Kernels.** In general, when classifying biological sequences with SVMs, one needs to measure the similarity between each pair of sequences and use it as kernel. This process normally involves alignment mechanism of some kind, which makes the similarity computation expensive. Leslie *et al* [8] proposed a simple type of features and respective kernel for sequences that is alignment-free: the *spectrum kernel*. A sequence $s$ of length $l$ with alphabet $\alpha$ ($\alpha = \{A,C,T,G\}$ for DNA sequences) is scanned and the numbers of occurrences of words of length $k$, or $k$-mers is used to build a feature vector for $s$ (in this paper, "word" and "$k$-mer" are interchangeable). For DNA sequences, this method creates a feature vector with $4^k$ elements corresponding to $4^k$ distinct $k$-mers. The inner product of two such feature vectors is calculated as the spectrum kernel function of corresponding sequences. This kernel is alignment-free since the similarity of two sequences could be computed without using any alignment, hence making such kernel computational efficient.

The spectrum kernel can be extended to incorporate partial matches of $k$-mers, which is important in comparison of sequences with less conserved motifs. Such variation of the spectrum kernel is known as *mismatch kernel*. A $(k,m)$ mismatch kernel considers two $k$-mers the same if they have no more than $m$ mismatches.

The spectrum and mismatch kernels are simple to calculate and have been shown to deliver satisfactory results in certain cases [7]. Both types of kernels measure the similarity based on the co-occurrences of $k$-mers in a pair of sequences independent of $k$-mers' positions. However, it is well known that, in many cases, for binding events to occur certain constrains on the locations, and orders of motifs (TFBS) should be met. For example, some enhancers consist of pairs or triples of instances of the same or different motifs that are located near each other (within tens of bp), and the spectrum and mismatch kernel may not work well in these situations. To model such constrains, we present novel kernels that explicitly take into account the relative locations of $k$-mers.

## 3  Word Combination Features

In this section, we introduce a new type of features and kernel that explicitly incorporate location and order constrains on occurrences of $k$-mers or words, which we call *word combination feature*(WCF). Basically, the values of such features are the numbers of times each pair of $k$-mers co-occurs within a sequence and also satisfies certain location constrains. For example, a possible feature is the number of times $k$-mer A co-occurs with $k$-mer B and the distance between the two instances are less than a predefined threshold. There are two main obstacles in using such features. First, the number of $k$-mer pairs is very large, resulting in very high-dimensional feature space. With $k = 6$, for example, there are thousands of possible $k$-mers and millions of their pairs. Second, calculating such features may be time-consuming and thus requires the development of efficient algorithms.

In what follows, we describe solutions for these two problems. Specifically, to reduce the number of $k$-mer pair features, we select a small set of important $k$-mers and compute pair features only for $k$-mer from this set. Then, we introduce an algorithm that calculate such features with complexity linear to the sequence length.

## 3.1 Selection of Important K-mers

Although there are $4^k$ possible $k$-mers, only a small number of them are important in the sense that they are predictive of an enhancer. By focusing only on these important features, we can reduce the number of $k$-mer pairs to a manageable size. In this work, we use two methods to select important $k$-mers based on feature selection ability of linear SVMs and AdaBoost [3].

**Selecting Important $k$-mers with SVMs.** A linear SVM uses a linear score function of the form $f(x) = \sum_{i=0}^{N} w_i x_i$ to calculate a score, which is then thresholded to decide the class label. In this function, $x_i$ is $i$-th feature and $w_i$ is its weight learned from training data. The larger the absolute value of $w_i$, the higher contribution of $i$-th feature to the score function, and thus the more important it is. Our selection method works as follows. First, we train a SVM using the spectrum kernel, i.e. with all $k$-mers. Once the SVM has been trained, we sort $k$-mers based on their weights, and then use three strategies to select most important ones: 1) select $k$-mers with top positive weights (SVM+), 2) $k$-mers with top negative weights (SVM−), and 3) combined list of $k$-mers with top positive and negative weights (SVM+−). Here, "top negative weights" mean negative weights with highest absolute values.

**Selecting Important $k$-mers with AdaBoost.** AdaBoost [3] is a special case of boosting algorithms. In general, boosting works by combining many weak classifiers (each has a slightly better prediction accuracy than choosing at random) to produce a strong classifier. Each of these weak classifiers could be as simple as a decision stump model. The method of learning is an iterative process of growing an ensemble of weak classifiers, each time adding one more. AdaBoost is adaptive since subsequent classifier added are selected to focus on examples mis-classified by previous classifiers.

In this work, we use AdaBoost with decision stumps as weak classifiers. A decision stump $ds(i, \theta)$ is an one-level decision tree, which has the form "*class = positive if $i \geq \theta$; and class = negative otherwise*". In each iteration, AdaBoost selects $i$ and $\theta$ so that the most number of training samples are classified correctly. Therefore, the algorithm tends to select most predictive features in early iterations, and multiple times. We use this property to select predictive $k$-mers as follows. We train AdaBoost on the training data using all $k$-mers as features. Each time a $k$-mer is selected in a decision stump, it is added into the set of important $k$-mers. This process ends when a desired number of distinct $k$-mers has been selected.

## 3.2 Calculating Combination Features

Now, we describe an efficient algorithm to extract features that are combinations of important $k$-mers selected in the previous step. Recall that we use features that are numbers of times pairs of important $k$-mers or two instances of a same $k$-mer occur within a predefined distance $d$. When counting these features, we do not consider order of words, and we do not differentiate between a word and its reverse complement. In other words, pairs of $k$-mers A and B with A appear before B or B before A or pair of A and reverse complement of B will all be counted as one feature.

To count the number of such pairs, first we map each important $k$-mer and its reverse complement to an unique index. For example, 100 $k$-mers will be assigned indexes from 1 to 100 while their reverse complements will be assigned indexes from -1 to -100. This step produces a $k$-mers dictionary called $k\_mers\_dict$ that allows our algorithm to compress the DNA sequence to an array of indexes so that comparison will be faster.

The second preprocessing step produces another dictionary allowing the feature extraction step to accurately detect each feature. In this step, we determine the set of index pairs by calculating all possible pair combinations of $k$-mers indexes. Then, combinations corresponding to the same feature are then bagged to form the features mapping.

This second step produces a feature index dictionary called $feat\_index\_dict$ that mapping pairs of $k$-mer indexes to feature indexes. Each genomic sequence is then processed to extract features by using the algorithm presented in Algorithm 1.

In the worse case scenario we will have to analyze at most $(L-k)*(D+k)$ $k$-mers indexes with L being the length of DNA sequence and D is the maximum distance of $k$-mers pair.

**Combining Features.** Once combination features are calculated, we remove features that does not appear in any training sample. The remaining features are then normalized to sum to one and the resulting feature vector is concatenated to spectrum kernel's feature vector to form a new feature vector. Note that, the two set of features are normalized separately. We then normalize again after combining these features so they contribute equally to the result. For any two sequences, the inner product of their feature vectors forms the kernel value.

# 4 Experiments and Results

## 4.1 Data and Settings

### 4.1.1 Datasets

We evaluated the effectiveness of the proposed method on datasets containing enhancers for multiple TFs and their co-factors as well as histone marks from human,

**Data**: $D$, $k$, $k\_mers\_dict$, $feat\_index\_dict$ and $seq$
**Result**: $feat\_vector$
scan $seq$ to produce $k$-mers list $k\_mers\_list$;
**for** $i = 0$ **to** $length(k\_mers\_list)$ **do**
  $kmer = k\_mers\_list[i]$;
  find index $ki$ for $kmer$ in $k\_mers\_dict$;
  $index\_array[i] = ki$;
**end**
**for** $i = 0$ **to** $length(index\_array)$ **do**
  $idx1 = index\_array[i]$;
  **if** $idx1 \neq 0$ **then**
    **for** $j = 0$ **to** $i + D + k$ **do**
      $idx2 = index\_array[j]$;
      **if** $idx2 \neq 0$ **then**
        find index $fi$ for $idx1 : idx2$ in $feat\_index\_dict$;
        $feat\_vector[fi] + +$;
      **end**
    **end**
  **end**
**end**

**Algorithm 1.** Algorithm for counting word combination features

mouse, and C. elegans. Specifically, as positive datasets, we used datasets provided by Yanez-Cuba *et al* [18] and Fletez-Brant *et al* [2].

Yanez-Cuba *et al* [18] have compiled several sets of enhancers based on data from previous work. The datasets include enhancers for the following TFs and histone marks: TAL1 (from [10]), HNF4A,GATA6, CDX2, and H3K4me2 (from [17]), for different cell lines. The original datasets contain only short fragments corresponding to ChIP peaks. Therefore, for each fragment, we extended from 300 to 500 bp in both directions to get an enhancer sequence of approximately 1000 bp, and used these sequences to form positive sets. To generate negative sets, we use the method by Lee *et al* [7]. This method generates null sequences (without enhancers) by randomly selecting DNA sequences from the same genome and have the same length and repeat fraction distributions. The Kmer-Svm Server [2] implements this method and we use this server to generate negative datasets for all our experiments. In the following section, we will use this collection of datasets (referred to as the *first collection*) for exploring different settings of our method as well as for comparison with existing methods.

The second collection of datasets is the same as used by Fletez-Brant *et al* [2]. These datasets contain enhancers obtained through ChIP-seq or DNase-seq experiments for several TF binding sites: ESRRB (in mouse ES cells), GR (in mouse 3134 and AtT20 cells), EWS-FLI (in human EWS502 and HUVEC cells). The datasets were provided with both positive and negative sequences, extended from peak fragments appropriately and we used them as is. In our experiments, we used the datasets from this second collection only for comparison of our and other methods.

### 4.1.2 Experiment Settings

We implemented the proposed method in Matlab using the built-in SVM algorithm with linear kernel and parameter $C = 1$. All experiments were performed with hexamers ($k = 6$). This value of $k$ has been proved to deliver the best performance [7]. For each set of data, we ran several experiments, varying the following parameters: method of selecting $k$-mers, number of selected $k$-mers, maximum distance between two $k$-mers in a valid feature.

The performance of the classifier was judged by two metrics: the area under the ROC curve (AUROC) and the PR curve (AUPRC). The AUROC is the area under the ROC which is a curve plotting true positive rate (sensitivity) against false positive rate (1-specificity) at different SVM score thresholds. It measures the probability that a randomly selected positive sample will score higher than a randomly selected negative sample. The PR curve plots Precision against Recall and AUPRC could be interpreted as what is the probability that a sequence really contains enhancer if the classification said so. The ROC could yield a better performance for a classifier in the case of imbalance training examples where as the PR curve directly assesses the accuracy of positive predictions.

All classifiers were evaluated with 5-fold cross validation protocol, in which a classifier is trained with four fifth of the data set and tested on the rest. The AUC and PR scores are averaged over the five folds.

## 4.2 Results

### 4.2.1 The Effect of Feature Selection Methods

The first experiment was designed to verify the effect of feature selection methods on prediction accuracy. Recall that for this experiment, we used only datasets from the first collection. The SVM−based and AdaBoost-based feature selection method were run to select 100 most important hexamers. For the SVM−based method, all three strategies were used. Selected hexamers were then used to produce combination features with distance between two hexamers not exceeding 100 bp. Table 1 summarizes AUROC and AUPRC scores for five datasets using the four selection methods. In all tables, (j), (p) and (d) stand for jurkat, proliferating and differentiated cell lines respectively. As shown, the selection method with the best AUC values is SVM+. SVM+ achieved the highest AUROC scores in four and the best AUPRC scores in three out of five datasets. The second best SVM+− achieved the highest scores in just one case. The results also show that using $k$-mers with negative weights as features is harmful for prediction accuracy. The SVM− option, i.e. using only negative weight $k$-mers, achieved the worst accuracy. In general, features selected by SVM yield higher AUC scores than by AdaBoost, making it more convenient to use SVM for both feature selection and subsequent classification.

**Table 1** AUC scores for different feature selection methods

| Method | SVM+ | | SVM+− | | SVM− | | AdaBoost | |
|---|---|---|---|---|---|---|---|---|
| | ROC | PR | ROC | PR | ROC | PR | ROC | PR |
| **TAL1(j)** | **0.9133** | **0.5846** | 0.9088 | 0.5829 | 0.8302 | 0.3847 | 0.9046 | 0.5602 |
| **HNF4A(d)** | 0.8395 | 0.3903 | **0.8416** | **0.4002** | 0.7748 | 0.2899 | 0.8368 | 0.3811 |
| **GATA6(p)** | **0.9754** | **0.8218** | 0.9734 | 0.8161 | 0.8992 | 0.5774 | 0.9726 | 0.8113 |
| **CDX2(d)** | **0.8594** | **0.4254** | 0.8547 | 0.4222 | 0.7936 | 0.3338 | 0.8492 | 0.4190 |
| **H3K4me2(d)** | **0.7976** | 0.2952 | 0.7959 | 0.2996 | 0.7721 | 0.2633 | 0.7966 | **0.3068** |

## 4.2.2 The Effect of Feature Number and Distance

In the second experiment, we used SVM+, the method that has delivered the most accurate results, and experimented with different feature numbers and distance values. The number of selected $k$-mers ($N$) was set to 10, 30, 50, 100, and the maximum distance ($D$) between two $k$-mers was set to 10, 30, 50, 100, and 200 bp. Due to space limit, Tab. 2 show only highest AUC values and corresponding $N$ and $D$.

As shown, $N = 100$ yielded superior AUPRC scores for all five datasets, although it achieved the best AUROC scores for only two out of five datasets. The best AUROC scores for the other three datasets were achieved with $N = 30$, although the difference in AUROC scores for $N$=30, 50, and 100 is not statistically significant (according to paired T-tests with threshold 0.05). The best AUROC and AUPRC scores were achieved with $D = 100$ on three datasets and $D = 30$ on two other datasets. The difference in the best values of $D = 100$ may be attributed to the variability in spatial constraints for different enhancer types, as reported previously [18]. Overal, the combination of $N = 100$ and $D = 100$ provides the best results and will be used in the remaining experiments.

**Table 2** The best AUROC and AUPRC scores and corresponding $N$ and $D$ for SVM+ feature selection method

| Dataset | Best AUROC | N | D | Best AUPRC | N | D |
|---|---|---|---|---|---|---|
| **TAL1(j)** | 0.9133 | 100 | 100 | 0.5846 | 100 | 100 |
| **HNF4A(d)** | 0.8518 | 50 | 200 | 0.4141 | 50 | 200 |
| **GATA6(p)** | 0.9769 | 30 | 100 | 0.8265 | 50 | 200 |
| **CDX2(d)** | 0.8594 | 100 | 100 | 0.4301 | 50 | 200 |
| **H3K4me2(d)** | 0.8050 | 30 | 30 | 0.3109 | 10 | 200 |

### 4.2.3    Comparison with Existing Methods

We compared our method with the method by Lee *et al* [7] (referred to as SK). In that method, linear SVMs with spectrum kernels are used to predict enhancers from genomic sequences. Experiments have shown that the method achieves state-of-the-art prediction accuracy in predicting mouse enhancers [7, 2] and therefore we used only the method by Lee *et al* in our comparison. The authors of that method provides an implementation of the method at http://kmersvm.beerlab.org, which we used with default parameters in our experiments. The size of negative sets was set to 10000 sequences. The comparison was performed in the datasets from both collections. Our method (called WCF) was run with SVM+, $N = 100$, and $D = 100$.

Table 3 summarizes the average AUROC and AUPRC scores of the methods. For all the datasets from the first collection, the proposed method outperforms SK in terms of both AUROC and AUPRC. The improvement of AUROC is from nearly 2% (for H3K4me2(d)) to 5%(TAL1(j)). The improvement of AUPRC is more substantial, which is more than 10% in two cases (TAL1(j) and GATA6(p)). For the datasets from the second collection, two methods achieve comparable results. More precisely, the proposed method perform worse than SK in three out of five datasets, however the differences are negligible ($< 0.5\%$) and not statistically significant (paired t-test). A possible explanation for the difference in performance of our method with two data collections is the differences in organization of TF binding sites in two cases. Since our method explicitly model the constraints on relative locations of combinative binding sites, it would be more suitable for enhancers with such constraints, which seem to be the first case, while it does not influence the results when such constraints are not tight, and the second collection may be such a case. Overall, our method outperforms SK substantially in half the cases and performs comparably in the other cases.

### 4.2.4    Analysis of Important Features

After training, a linear SVM outputs a weight vector, each element of which corresponds to an input feature. Features with larger absolute weights are more important because they contribute more to the final score. Following Lee *et al* [7], we ask if features with large positive weights are also biologically meaningful. For each dataset, we list the features, both single and combination, with the highest positive weights and find corresponding (if any) TFs in databases of known TFBSs. Due to space limit, we show only top 10 features for CDX2(d) (Tab. 4). As shown, nine out of 10 top features are combination ones, suggesting that for this dataset, combination features are more predictive than single ones. More importantly, most highly ranked words are known TFBSs for CDX2, GATA, HNF4A, FOXA, AP-1, suggesting that combinations of these motifs are important for CDX2 binding events to occur. To verify this, we compare the results with previous findings. Using experimental methods, Verzi *et al* [17] found that CDX2 TF partners with distinct motifs during different cell states. Specifically, GATA motifs are found close to the binding site of CDX2 during proliferating state while CDX2 binding site regions specific to

**Table 3** Comparison of enhancer prediction with Spectrum Kernel (SK) [7] and Word Combination Feature Kernel (WCF)

| Kernel | WCF | | SK | | Differences | |
|---|---|---|---|---|---|---|
| | AUROC | AUPRC | AUROC | AUPRC | AUROC | AUPRC |
| **TAL1(j)** | 0.9133 | 0.5846 | 0.8678 | 0.4785 | +0.0455 | +0.1062 |
| **HNF4A(d)** | 0.8395 | 0.3903 | 0.8164 | 0.3526 | +0.0231 | +0.0377 |
| **GATA6(p)** | 0.9754 | 0.8218 | 0.9499 | 0.7180 | +0.0255 | +0.1038 |
| **CDX2(d)** | 0.8594 | 0.4254 | 0.8276 | 0.3895 | +0.0318 | +0.0359 |
| **H3K4me2(d)** | 0.7976 | 0.2952 | 0.7849 | 0.2902 | +0.0128 | +0.0050 |
| **EWS502** | 0.9612 | 0.9527 | 0.9640 | 0.9570 | -0.0028 | -0.0043 |
| **HUVEC** | 0.9621 | 0.9610 | 0.9600 | 0.9590 | +0.0021 | +0.0020 |
| **3134** | 0.8934 | 0.8701 | 0.8970 | 0.8740 | -0.0036 | -0.0039 |
| **Att20** | 0.9051 | 0.7769 | 0.9050 | 0.7840 | +0.0001 | -0.0071 |
| **ESRRB** | 0.9148 | 0.9282 | 0.9160 | 0.9310 | -0.0012 | -0.0028 |

**Table 4** Top 10 features with highest positive weights as returned by SVM for CDX2 (d) binding sites

| Features | | Reverse complements | | SVM weights | Known TF(s) | |
|---|---|---|---|---|---|---|
| CATAAA | CTTATC | TTTATG | GATAAG | 15.804 | CDX2 | GATA |
| AGGGCA | CATAAA | TGCCCT | TTTATG | 14.891 | HNF4A | CDX2 |
| CAAACA | CAAAGG | TGTTTG | CCTTTG | 14.613 | FOXA | HNF4A |
| AATAAA | GACTCA | TTTATT | TGAGTC | 14.590 | CDX2 | AP-1 |
| ATAAAA | CTTATC | TTTTAT | GATAAG | 14.394 | CDX2 | GATA |
| GCCCCA | GGCCCC | TGGGGC | GGGGCC | 13.826 | | |
| AGAGAG | | CTCTCT | | 13.471 | GATA | |
| AGTCAT | CATAAA | ATGACT | TTTATG | 13.053 | AP-1 | CDX2 |
| CAAAGG | TAAACA | CCTTTG | TGTTTA | 13.007 | HNF4A | CDX2 |
| CATAAA | CCACCC | TTTATG | GGGTGG | 12.905 | CDX2 | |

differentiated cell show a significant enrichment of HNF4A, AP-1 and FOXA motifs. These are almost the same motifs we found. The agreement between motif sets found by our method and reported by Verzi *et al* provides evidence that the motifs of highly ranked combination feature are biologically meaningful, and the proposed method can be used to get insight of enhancer organization.

## 5 Conclusion

We have presented a novel method for enhancer prediction using only sequence data. Based on generic features in forms of words extracted from genomic sequences, we introduce a new type of features that are pairs of words satisfying certain constraints on their locations. We have developed a fast feature extraction method that combines feature selection with a fast word pair counting algorithm. In a comparison with a leading method, using such word pairs as additional features for SVM classifiers has resulted in improvements of prediction accuracy as measured by AUC values of ROC and PR in half the cases while does not affect the accuracy in the others. The most important word combination features found by SVMs are biologically meaningful, thus providing additional information about enhancer content and structure. In this work, we consider only pairs of words and one type of spatial constraints (distance). However, the method can be extended to consider other types of constraints as well as combinations with more than two words to cover cases with more complex enhancer organization.

## References

[1] Bailey, T.L., Noble, W.S.: Searching for statistically significant regulatory modules. Bioinformatics 19(suppl. 2), ii16–ii25 (2003)
[2] Fletez-Brant, C., Lee, D., McCallion, A.S., Beer, M.A.: kmer-SVM: a web server for identifying predictive regulatory sequence features in genomic data sets. Nucleic Acids Res. 41(Web Server issue), W544–W556 (2013)
[3] Freund, Y., Schapire, R.E.: A Decision-Theoretic Generalization of On-Line Learning and an Application to Boosting. Journal of Computer and System Sciences 55(1) (1997)
[4] Göke, J., Schulz, M.H., Lasserre, J., Vingron, M.: Estimation of Pairwise Sequence Similarity of Mammalian Enhancers with Word Neighbourhood Counts. Bioinformatics 28(5), 656–663 (2012)
[5] Kim, T., Hemberg, M., Gray, J.M., Costa, A.M., Bear, D.M., Wu, J., Harmin, D.A., Laptewicz, M., Barbara-Haley, K., Kuersten, S., et al.: Widespread transcription at neuronal activity-regulated enhancers. Nature 465, 182–187 (2010)
[6] Langmead, B., Trapnell, C., Pop, M., Salzberg, S.L.: Ultrafast and memoryefficient alignment of short DNA sequences to the human genome. Gen. Biol. 10, R25 (2009)
[7] Lee, D., Karchin, R., Beer, M.A.: Discriminative prediction of mammalian enhancers from DNA sequence. Gen. Res. 21(12), 2167–2180 (2011)
[8] Leslie, C., Eskin, E., Noble, W.S.: The spectrum kernel: A string kernel for SVM protein classification. In: Proc. of Pac. Symp. Biocomput. 2002 (2002)
[9] Leung, G., Eisen, M.B.: Identifying cis-regulatory sequences by word profile similarity. PLoS One 4, e6901 (2009), doi:10.1371/journal.pone.0006901
[10] Palii, C.G., Perez-Iratxeta, C., Yao, Z., Cao, Y., Dai, F., Davison, J., Atkins, H., Allan, D., Dilworth, F.J., Gentleman, R., et al.: Differential genomic targeting of the transcription factor TAL1 in alternate haematopoietic lineages. EMBO J. 30, 494–509 (2011)
[11] Pierstorff, N., Bergman, C.M., Wiehe, T.: Identifying cis-regulatory modules by combining comparative and compositional analysis of DNA. Bioinformatics 22, 2858–2864 (2006)

[12] Schultheiss, S.J., Busch, W., Lohmann, J.U., Kohlbacher, O., Ratsch, G.: KIRMES: Kernel-based identification of regulatory modules in euchromatic sequences. Bioinformatics 25(16), 2126–2133 (2009)

[13] Sinha, S., He, X.: MORPH: probabilistic alignment combined with hidden Markov models of cis-regulatory modules. PLoS Comput. Biol. 3, e216 (2007)

[14] Spitz, F., Furlong, E.E.M.: Transcription factors: from enhancer binding to developmental control. Nature Reviews Genetics 13, 613–626 (2012)

[15] Su, J., Teichmann, S.A., Down, T.A.: Assessing Computational Methods of Cis-Regulatory Module Prediction. PLoS Comput. Biol. 6(12), e1001020 (2010)

[16] Thanh, H.V., Phuong, T.M.: Enhancer Prediction Using Distance Aware Kernels. In: Proc. of RIVF 2013 (2013)

[17] Verzi, M.P., Shin, H., He, H.H., Sulahian, R., Meyer, C.A., Montgomery, R.K., Fleet, J.C., Brown, M., Liu, X.S., Shivdasani, R.A.: Differentiation-Specific Histone Modifications Reveal Dynamic Chromatin Interactions and Partners for the Intestinal Transcription Factor CDX2. Developmental Cell 19, 713–726 (2010)

[18] Yanez-Cuna, J.O., Dinh, H.Q., Kvon, E.Z.: Uncovering cis-regulatory sequence requirements for context specific transcription factor binding. Genome Research 22, 2018–2030 (2012)

[19] Zhong, M., Niu, W., Lu, Z.J., Sarov, M., Murray, J.I., Janette, J., Raha, D., Sheaffer, K.L., Lam, H.Y.K., Preston, E., et al.: Genome-wide identification of binding sites defines distinct functions for Caenorhabditis elegans PHA-4/FOXA in development and environmental response. PLoS Genet. 6, e1000848 (2010)

# Improving Acoustic Model for Vietnamese Large Vocabulary Continuous Speech Recognition System Using Deep Bottleneck Features

Quoc Bao Nguyen, Tat Thang Vu, and Chi Mai Luong

**Abstract.** In this paper, a method based on deep learning for extracting bottleneck features for Vietnamese large vocabulary speech recognition is presented. Deep bottleneck features (DBNFs) is able to achieve significant improvements over a number of base bottleneck features which was reported previously. The experiments are carried out on the dataset containing speeches on Voice of Vietnam channel (VOV). The results show that adding tonal feature as input feature of the network reached around 20% relative recognition performance. The DBNF extraction for Vietnamese recognition decrease the error rate by 51%, compared to the MFCC baseline.

**Keywords:** Deep bottleneck features, Vietnamese automatic speech recognition, Neural network.

## 1 Introduction

Vietnamese is a syllabic tonal language with six lexical tones, which are very important to decide on the word meanings. A change in tone can lead to the change in word meaning, which can be vastly different from the original word. For this reason, tone recognition is an essential part of the tonal speech recognition system, especially for large vocabulary continuous system. Previous studies [1][2][3] showed efforts toward Vietnamese large vocabulary continuous speech recognition with tone modeling approaches to manifest the tonal structure of Vietnamese or applying tonal features that present tone information. However, their systems did not employ the full range of state-of-the-art techniques for acoustic model.

Quoc Bao Nguyen · Chi Mai Luong
University of Information and Communication Technology, Thai Nguyen University
e-mail: nqbao@ictu.edu.vn, lcmai@ioit.ac.vn

Tat Thang Vu
Institute of Information Technology, Vietnam Academy of Science and Technology
e-mail: {alfred.hofmann,ursula.barth,ingrid.haas,
    frank.holzwarth}@springer.com,
    vtthang@ioit.ac.vn

© Springer International Publishing Switzerland 2015                                     49
V.-H. Nguyen et al. (eds.), *Knowledge and Systems Engineering*,
Advances in Intelligent Systems and Computing 326, DOI: 10.1007/978-3-319-11680-8_5

Acoustic model is an important module in the automatic speech recognition system (ASR. Acoustic model is used to model the acoustic space of input feature. The state-of-the-art acoustic models for speech recognition utilize a statistical pattern recognition framework called HMM/GMM (Hidden Markov Model/Gaussian Mixture Model)[4] with short time spectral input features. Although the HMM/GMM approach has been effective in capturing speech patterns, it has several inherent limitations. For example, speech feature vectors at different frames are assumed to be statistically independent given the state sequence. This assumption ignores the obvious correlation between neighboring frames to achieve mathematical simplification. Hence, discriminative models may be a more appropriate choice for acoustic modeling.

Many researchers have been trying to incorporate the power of artificial neural networks (ANNs) in acoustic modeling to improve performance over the traditional HMM/GMM approach. There are two main approaches for incorporating artificial neural networks (ANNs) in acoustic modeling today, namely hybrid systems and tandem systems. In the former, a neural network is trained to estimate the emission probabilities for HMM [5]. In contrast, the later use neural networks to generate discriminative features as input values for the common combination of GMM and HMMs. This is done by training a network to predict phonetic targets, and then either using the estimated target probabilities [6] or the activations of a narrow hidden layer ("bottleneck features", BNF [7]). Those features are usually used as standard input features for modeling with GMMs.

Recently, deep learning algorithms that deal with training deep neural networks (DNNs), consisting of many hidden layers, have been successfully applied in acoustic modeling [8]. A popular approach is to pre-train individual layers as restricted Boltz-mann machines (RBMs), which are unsupervised generative models [9]. Ideally, the pre-training procedure initializes the network parameters in a space that is beneficial for subsequent supervised training towards the actual classification task [10]. Gehring et al. demonstrated that denoising auto-encoders are applicable for modeling speech data and initializing deep networks as well [11].

The purpose of this study is to improve acoustic model for Vietnamese speech recognition using deep bottleneck features as in [11] with the combination of acoustic feature and tonal feature as neural network input. We also show the way to extract the pitch feature using modified algorithm which can achieve large improvement. Furthermore, we also optimize proposed deep bottleneck feature architecture from [11] to know how much additional gains are obtained. The rest of this paper is organized as follows:

- In the next section, a brief description Vietnamese phonetics is given.
- In Section 3, we showed the way to extract the pitch features.
- In Section 4, we briefly describe the deep neural network architecture for bottleneck feature extraction proposed in [11].
- In Sections 5 and 6, the experiments setup and results are presented.
- Finally, conclusions and future research are given in the last section.

## 2   Structure of Vietnamese Language

Vietnamese is a monosyllabic tonal language. Each Vietnamese syllable can be considered as a combination of initial, final and tone components. These components include an initial sound, a medial sound, a nucleus sound, a final sound and a tone [1] . In total Vietnamese has 21 initial and 155 final components. The total number of pronounceable syllables in Vietnamese is about 19.000 but only about 7.000 syllables (with and without tone) are used in daily language [2]. Vietnamese vowels are implosive. It has six tones, five of which have marks, called diacritical marks or tonal marks. The tones differ in pitch, length, and melody.

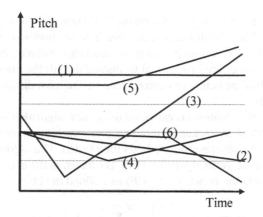

**Fig. 1** Variations of F0 in the Vietnamese Tones

Figure 2 shows the variations in Vietnamese tones as can be observed, the level tone has a high onset and consistent steady state. Similarly, falling tone has a low onset and gradual falls toward the end. Rising tone is characterized by a medium onset and gradually increases towards the end. The curve tone is differentiated by its low onset and the minimum at two-thirds within the syllable duration. Broken tone is recognized from the high onset and the minimum at the middle of the syllable. Drop tone has a low onset and a rapid fall to the end of the syllable.

In this study, we decided to choose the tone modeling approach for Vietnamese speech recognition as described in [1][12]. In so-called tonal phoneme, each tonal phone (vowels, diphthongs etc.) contains tone information by combining correlative tonal marks as shown in the Table 1.

**Table 1** Examples of creating tonal phoneme

| English | Telex | Tone | Phoneme |
|---------|---------|------|------------|
| Zero | khoong | 1 | kh oob ngzb |
| Boat | thuyeenf | 2 | th w ief nzf |
| Act | dieenx | 3 | d iex nzx |
| Seven | baayr | 4 | b aar izr |
| Four | boons | 5 | b oos nzs |
| Spot | munj | 6 | m uj nzj |

## 3 Pitch Feature Extractions

In so-called tonal languages, e.g. Vietnamese, Cantonese, Chinese (phonetic) pitch carries phonological (tone) information and needs to be modeled explicitly. To detect tones, one needs to detect rising, falling, or otherwise marked pitch contours. By themselves, pitch features are insufficient to distinguish all the phonemes of a language, but pitch (absolute height or contour) can be the most distinguishing feature between two sounds.

In this work, the pitch features is extracted using new algorithm described in [13] which is called Kaldi pitch tracker. It improves pitch tracking as measured by Gross Pitch Error, versus the off-the-shelf methods they tested. The algorithm is a highly modified version of the getf0 algorithm [14]. Basically, getf0 is based on the Normalized Cross Correlation Function (NCCF) as defined in (1):

$$NCCF(\tau) = 1 / \frac{1}{\sqrt{e_n e_\tau}} \sum_{n=0}^{N-\tau-1} x(n)x(n+\tau),\qquad(1)$$

where $e_i = \sum_{n=i}^{i+N-1} x^2(n)$, x(n) is the input speech sample, N is the length of the speech analysis window, $\tau$ is the lag number in range between 0 and N-1.

A Kaldi pitch tracker algorithm is based on finding lag values that maximize the NCCF. The most important change from getf0 is that rather than making hard decisions about voicing on each frame, all frames are treated as voiced to be continuous and allow the Viterbi search to naturally interpolate across unvoiced regions.

The output of the algorithm is 2-dimensional features consisting of pitch in Hz and NCCF on each frame, and then the output is post-processed for use as features for ASR, produces 3-dimensional features consisting of pov-feature, pitch-feature and delta-pitch-feature:

1. pov-feature is warped NCCF. This method was designed to give the feature a reasonably Gaussian distribution. Let the NCCF on a given frame be written c. If $-1 \leq c \leq 1$ is the raw NCCF, the output feature be $f = 2((1.0001 - c)^{0.15} - 1)$.
2. pitch-feature is feature that on each time t, subtraction of a weighted average pitch value, computed over a window of width 151 frames centered at $t$ and weighted by the probability of voicing value $p$. Where p is obtained by

plotting the log of *count*(*voiced*)/*count*(*unvoiced*) on the Keele database [15] as a function of the NCCF

3. and delta-pitch-feature is delta feature computed on raw log pitch.

## 4   Deep Bottleneck Features

In this section, we briefly describe the deep neural network architecture for bottleneck feature extraction proposed in [11] and depicted in Figure 2. The network consists of a variable number of moderately large, fully connected hidden layers and a small bottleneck layer which is followed by an additional hidden layer and the final classification layer. The architecture differs from setups previously described, where the bottleneck layer has been placed in the middle of a deep network [16], [17] or added as a second model trained on the output values of the original network [18].

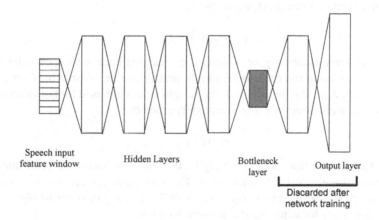

**Fig. 2** Deep Network Architecture

The hidden layers in front of the bottleneck are initialized using unsupervised, layer-wise pre-training. Thanks to their success in the deep learning community, restricted Boltzmann machines have become the default choice for pre-training the individual layers of deep neural networks used in speech recognition. Gehring et al. demonstrated that denoising auto-encoders [19] which are straight-forward models that have been successfully used for pre-training neural architectures for computer vision and sentiment classification [20] are applicable to speech data as well.

We follow their training scheme and initialize the hidden layers as denoising auto-encoders, too. Like regular auto-encoders, these models consist of one hidden layer and two identically-sized layers representing the input and output values. The network is usually trained to reconstruct its input at the output layer with the goal to generate a useful intermediate representation in the hidden layer. In denoising auto-encoders, the network is trained to reconstruct a randomly corrupted version of its input, which can be interpreted as a regularizing mechanism that facilitates the learning of large and over complete hidden representations [19].

For denoising auto-encoders working on binary data (i.e. grayscale images or sigmoid activations of a previous hidden layer), Vincent et al. proposed the use of masking noise for corrupting each input vector [19] in order to extract useful features, the network is forced to reconstruct the original input from a corrupted version, gen-erated by adding random noise to the data. This corruption of the input data can be formalized as applying a stochastic process $q_D$ to an input vector x:

$$\tilde{x} \sim q_D(\tilde{x}|x) \tag{2}$$

This approach also used in our work is to apply masking noise to the data by setting every element of the input vector to zero with a fixed probability. Then the corrupted input first maps (with an encoder) to the hidden representation $y$ using the weight matrix $W$ of the hidden layer, the bias vector b of the hidden units and a non-linear activation function $\sigma_y$ as follows:

$$y = \sigma_y(W\tilde{x} + b) \tag{3}$$

The latent representation y or code is then mapped back with a decoder into reconstruction z using the transposed weight matrix and the visible bias vector c. Because In a model using tied weights, the weight values are used for both encoding and decoding, again through a similar transformation $\sigma_z$:

$$z = \sigma_z(W^T y + c) \tag{4}$$

The parameters of this model (namely $W, W^T, b, c$) are optimized such that the average reconstruction error is minimized. The reconstruction error can be meas-ured by the cross-entropy error objective as defined in (5) in order to obtain the gradients necessary for adjusting the network weights.

$$L_H(x,z) = \sum_i x_i \log z_i + (1 - x_i) \log z_i \tag{5}$$

When training a network on speech features like MFCCs, the first layer models real valued rather than binary data, so the mean squared error $L_2(x,z) = \sum_i (x_i - z_i)^2$ is selected as the training criterion. In this work, we also apply masking noise to the first layer, although other types of noise could be used as well [19].

After a stack of auto-encoders has been pre-trained in this fashion, a deep neural network can be constructed. The bottleneck layer, an additional hidden layer and the classification layer are initialized with random weights and connected to the hidden representation of the top-most auto-encoder. While all out hidden units use sigmoid nonlinearities, the classification layer output is obtained with the softmax activation function. The resulting network is then trained with supervision to esti-mate either context-independent or context-dependent HMM tri-phone states. For this last training step, errors are obtained with the cross-entropy function. Finally, the last two layers of the network can be discarded as the units in the bottleneck layer provide the final features used for training standard Gaussian mixture (GMM) acoustic models.

# 5 Experiment Setups

## 5.1 Corpora Description

The data used in our experiments is the Voice of Vietnam (VOV) data which is a collection of story reading, mailbag, new reports, and colloquy from the radio pro-gram the Voice of Vietnam. There are 23424 utterances in this corpus including about 30 male and female broadcasters and visitors. The number of distinct syllables with tone is 4923 and the number of distinct syllables without tone is 2101. The total time of this corpus is about 19hours. We separated into training set of 17 hours and decoding set of 2 hours. The data is in the wav format with 16 kHz sampling rate and analog/digital conversion precision of 16bits. The language model in this experiment is a tri-gram model which is trained by using all of transcriptions in the training data.

## 5.2 Baseline Systems

Baseline HMM/GMM systems were performed with the Kaldi developed at Johns Hopkins University [21]. We extracted two sets of acoustic features to build baseline acoustic models. Those are Mel-frequency cepstral coefficient (MFCC), which is popular in speech recognition applications and perceptual linear predictive cepstrum (PLP). In both feature extraction, 16-KHz speech input is coded with 13-dimensional MFCCs with a 25ms window and a 10ms frame-shift. Each frame of the speech data is represented by a 39-dimensional feature vector that consists of 13 MFCCs with their deltas and double-deltas. Nine consecutive feature frames are spliced to 40dimensions using linear discriminant analysis (LDA) and maximum like-lihood linear transformation (MLLT) that is a feature orthogonalizing transform, is applied to make the features more accurately modeled by diagonal-covariance Gaussians.

We extracted pitch features using Kaldi pitch extractor, starting from audio 16-KHz speech input. We low-pass the signal to 1 kHz; Our pitch is 3-dimensional features consisting of pov-feature, pitch-feature and delta-pitch-feature as described in section 3. All models used 5500 context-dependent state and 96000 Gaussian mixture components. The baseline systems were built, follow a typical maximum likelihood acoustic training recipe, beginning with a flat-start initialization of context-independent phonetic HMMs, followed by tri-phone system with 13-dimensionalMFCCs or PLP plus their deltas and double-deltas and ending with tri-phone system with LDA+MLLT.

## 5.3   Network Training

In our experiments, we extracted deep bottleneck features from MFCCs, PLP and their combination with pitch features (MFCC+pitch, PLP+pitch). The network input for these features was pre-processed using the approach in [22] that is called splicing speaker-adapted features.

During supervised fine-tuning, the neural network was trained to predict context-dependent HMM states (there were about 4600 states in our experiments).For pre-training the stack of auto-encoders in the architecture at section 4, mini-batch gradient descent with a batch size of 128 and a learning rate of 0.01 was used. Input vectors were corrupted by applying masking noise to set a random 20% of their elements to zero. Each auto-encoder contained 1024 hidden units and received 1 million updates before its weights were fixed and the next one was trained on top of it.

The remaining layers were then added to the network, with the bottleneck consisting of 39 units. Again, gradients were computed by averaging across a mini-batch of training examples; for fine-tuning, we used a larger batch size of 256.The learning rate was decided by the "newbob" algorithm: for the first epoch, we used 0.008 as the learning rate, and this was kept fixed as long as the increment in cross-validation frame accuracy in a single epoch was higher than 0.5%. For the subsequent epochs, the learning rate was halved; this was repeated until the increase in cross-validation accuracy per epoch is less than a stopping threshold, of 0.1%. After each epoch, the current model was evaluated on a separate validation set, and the model performing best on this set was used in the speech recognition system afterwards.

For these experiments we used GPUs for training of auto-encoder layers and neural networks using the Theano toolkit. Training on 17 hours of VOV data took about a 18 hours for the architecture of network we used (around 9 million parameters).

## 5.4   Bottleneck Features

The 39 output values of the bottleneck layer of 11 adjacent frames were stacked and reduced to a 40-dimensional feature vector using LDA+MLLT. Using this feature vector, a tri-phone context-dependent system was trained starting from a monophone context-independent MFCC or PLP baseline system as described in section 5.2

## 6   Experiment Results

### 6.1   Baseline

Table 2 lists the performance of baseline systems on the Vietnamese language in terms of the word error rate (WER) with MFCC and PLP features. The MFCC num-ber is 21.25% WER that is slightly better the one with PLP feature.

**Table 2**  Baseline Recognition performance for the Vietnamese system with MFCC and PLP features

| Features | WER (%) |
|----------|---------|
| MFCC     | 21.25   |
| PLP      | 22.08   |

## 6.2  DBNF vs BNF

In [23], we applied BNF for Vietnamese speech recognition using MLP net-work using 5 layers without using any pre-training technique. The WER of the BNF systems are around 14% that presents at first column of Table 3.

**Table 3**  DBNFs vs BNFs Recognition performance for the Vietnamese system with MFCC and PLP input features

| Features | BNF WER(%) | DBNF WER(%) |
|----------|------------|-------------|
| MFCC     | 14.09      | 13.12       |
| PLP      | 14.20      | 13.63       |

While in this experiments, we applied DBNF as described in section 4 for Vietnamese speech recognition. As we can see the results at second column of Table 3 the DBNF numbers are about 1% absolute better than BNF number. The DBNF was trained using well-performing configuration for Tagalog in [24] with 9 layers that contained 5 auto-encoder layers were pre-trained.

## 6.3  Optimizations on DBNF

In this section, we describe and evaluate the some optimizations proposed in order to improve the performance of the DBNF for Vietnamese recognition. First, we experimented with neural network input features using the combination of acoustic features and pitch features as in [12]. It can be seen in Table 4 that adding pitch fea-tures helped the recognition performance to increases by 20% relative (a drop in WER from 13.12% to 10.58% or from 13.63% to 10.76%).

The size of the context window that the DBNF extraction network is able to observe directly influences the frame-level accuracy during training. In [24], we did experiments that contains 80 hours of training data and found that increasing the context window to 13 frames (130 ms) makes the best recognition performance. In these experiments, we only have 17 hours of training data. We thus varied the window size in order to investigate whether the improved accuracy would result

**Table 4** DBNFs Recognition performance for the Vietnamese system with the combination of MFCC and PLP with pitch as input features

| Features   | DBNF WER |
|------------|----------|
| MFCC+pitch | 10.58%   |
| PLP+pitch  | 10.76%   |

in more useful bottleneck features. The numbers in Table 5 were obtained using DBNFs trained on MFCC + Pitch features with 1000 units per hidden layer and 5 auto-encoder layers. Reducing context window to 5 frames (50 ms) increased the recognition performance by 1.4% relative (a drop in WER from 10.58% to 10.43%). Further contraction resulted in worse recognition performance.

**Table 5** Influence of varying the size of the input context window

| Frames | 3        | 5          | 7        | 9        | 11       |
|--------|----------|------------|----------|----------|----------|
| WER    | 10.60 %  | **10.43 %** | 10.44 %  | 10.58 %  | 10.97 %  |

Third, we performed further architectural optimizations and varied the number of auto-encoder layers placed in front of the bottleneck. As shown in Table 6, the best result could be achieved by using 4 layers, with no additional gains obtained by adding further layers.

**Table 6** Error rates for DBNFs trained with different numbers of auto-encoder(AE) layers

| AE layer | 2        | 3        | 4          | 5        | 6        | 7        |
|----------|----------|----------|------------|----------|----------|----------|
| WER      | 11.21 %  | 10.85 %  | **10.40 %** | 10.43 %  | 10.63 %  | 10.48 %  |

# 7 Conclusion

In this work, we have improved acoustic model for Vietnamese recognition using deep bottleneck features and have shown its ability to achieve significant improvements over a number of base bottleneck features which was reported previously [23]. It was shown that adding pitch features as input feature that obtained from new algorithm increased the performance by 20%. We have also evaluated several optimizations to the DBNFs architectures. The gains were achieved by reducing the number of hidden layers of the network and the context window of accessible input features. The DBNF extraction was tuned on a small-sized VOV speech corpus, which increased the relative improvement in word error rate over the MFCC baseline to 51%.

In the future, we intend to apply a hybrid HMM/DNN on top of DBNFs as well as multilingual network training approaches to improve acoustic model for Vietnamese recognition system.

# References

[1] Vu, T.T., Nguyen, D.T., Luong, M.C., Hosom, J.-P.: Vietnamese large vocabulary continuous speech recognition. In: INTERSPEECH (2005)

[2] Quang, N.H., Nocera, P., Castelli, E., Van Loan, T.: A novel approach in continuous speech recognition for vietnamese. In: SLTU (2008)

[3] Vu, N.T., Schultz, T.: Vietnamese large vocabulary continuous speech recognition. In: Proc. Automatic Speech Recognition and Understanding (ASRU), Merano, Italy. IEEE (December 2009)

[4] Rabiner, L.R.: A tutorial on hidden markov models and selected applications in speech recognition. Proceedings of the IEEE, 257–286 (1989)

[5] Bourlard, H.A., Morgan, N.: Connectionist Speech Recognition: A Hybrid Approach. Kluwer Academic Publishers, Norwell (1993)

[6] Hermansky, H., Ellis, D.P.W., Sharma, S.: Tandem connectionist feature extraction for conventional hmm systems. In: Proc. ICASSP, pp. 1635–1638 (2000)

[7] Grezl, F., Karafiat, M., Kontair, S., Cernocky, J.: Probabilistic and bottle-neck features for lvcsr of meetings. In: 2007 IEEE International Conference on Acoustics, Speech and Signal Processing (ICASSP), pp. V–757–IV–760. IEEE (2007)

[8] Seide, F., Li, G., Yu, D.: Conversational speech transcription using context-dependent deep neural networks. In: Proc. Interspeech 2011, pp. 437–440 (2011)

[9] Hinton, G.E., Osindero, S., Teh, Y.-W.: A fast learning algorithm for deep belief nets 18, 1527–1554 (2006)

[10] Erhan, D., Bengio, Y., Courville, A., Manzagol, P.-A., Vincent, P., Bengio, S.: Why does unsupervised pre-training help deep learning? J. Mach. Learn. Res. 11, 625–660 (2010)

[11] Gehring, J., Miao, Y., Metze, F., Waibel, A.: Extracting deep bottleneck features using stacked auto-encoders. In: ICASSP 2013, Vancouver, CA, pp. 3377–3381 (2013)

[12] Metze, F., Sheikh, Z.A.W., Waibel, A., Gehring, J., Kilgour, K., Nguyen, Q.B., Nguyen, V.H.: Models of tone for tonal and non-tonal languages. In: ASRU, pp. 261–266. IEEE (2013)

[13] Ghahremani, P., BabaAli, B., Povey, D., Riedhammer, K., Trmal, J., Khudanpur, S.: A pitch extraction algorithm tuned for automatic speech recognition. In: 2014 IEEE International Conference on Acoustics, Speech and Signal Processing (ICASSP). IEEE Signal Processing Society (to appear, May 2014)

[14] Talkin, D.: A robust algorithm for pitch tracking (RAPT). In: Klein, W.B., Palival, K.K. (eds.) Speech Coding and Synthesis. Elsevier (1995)

[15] Plante, F., Meyer, G.F., Ainsworth, W.A.: A pitch extraction reference database. In: EUROSPEECH. ISCA (1995)

[16] Yu, D., Seltzer, M.L.: Improved bottleneck features using pretrained deep neural networks. In: INTERSPEECH, pp. 237–240 (2011)

[17] Tüske, Z., Schlüter, R., Ney, H.: Deep hierarchical bottleneck mrasta features for IVCSR. In: ICASSP, pp. 6970–6974 (2013)
[18] Sainath, T.N., Kingsbury, B., Ramabhadran, B.: Auto-encoder bottleneck features using deep belief networks. In: 2012 IEEE International Conference on Acoustics, Speech and Signal Processing (ICASSP), pp. 4153–4156 (2012)
[19] Vincent, P., Larochelle, H., Bengio, Y., Manzagol, P.-A.: Extracting and composing robust features with denoising autoencoders. In: ICML 2008, pp. 1096–1103 (2008)
[20] Glorot, X., Bordes, A., Bengio, Y.: Domain adaptation for large-scale sentiment classification: A deep learning approach. In: Proceedings of the 28th International Conference on Machine Learning (ICML 2011), pp. 513–520 (2011)
[21] Povey, D., Ghoshal, A., Boulianne, G., Burget, L., Glembek, O., Goel, N., Hannemann, M., Motlicek, P., Qian, Y., Schwarz, P., Silovsky, J., Stemmer, G., Vesely, K.: The kaldi speech recognition toolkit. In: IEEE 2011 Workshop on Automatic Speech Recognition and Understanding. IEEE Signal Processing Society, IEEE Catalog No.: CFP11SRW-USB (December 2011)
[22] Rath, S.P., Povey, D., Vesely, K., Cernocky, J.: Improved feature processing for deep neural networks. In: INTERSPEECH, pp. 109–113. ISCA (2013)
[23] Nguyen, V.H., Luong, C.M., Vu, T.T.: Applying bottle neck feature for vietnamese speech recognition, pp. 379–388 (2013)
[24] Nguyen, Q.B., Gehring, J., Kilgour, K.: A Waibel, "Optimizing deep bottleneck feature extraction." in. In: 2013 IEEE RIVF International Conference on Computing and Communication Technologies, Research, Innovation, and Vision for the Future (RIVF), pp. 152–156 (November 2013)

# DFTBC: Data Fusion and Tree-Based Clustering Routing Protocol for Energy-Efficient in Wireless Sensor Networks

Nguyen Duy Tan and Nguyen Dinh Viet

**Abstract.** Energy efficiency is a unique challenge for designing routing protocols in wireless sensor network (WSN) to extend the lifetime of the entire network because sensor nodes are power constrained devices. To solve this problem, in this paper, we propose a data fusion and tree-based clustering routing algorithm (DFTBC). DFTBC designing consists of two main works. In the first work, tree-based clustering routing technique is used to connect the nodes in a cluster to form a minimum spanning tree, where each node communicates only with its nearest neighbor using distance and remaining energy of nodes to decide which node will be the cluster head (CH). In the second work, we fuse one or more packets to generate a packet with a smaller size based on the Dempster-Shafer and Slepian-Wolf theory. Our simulation results show that the network lifetime with using of our proposed protocol can be improved about 350% and 35% compared to low-energy adaptive clustering hierarchy (LEACH) and power-efficient gathering in sensor information system (PEGASIS), respectively.

**Keywords:** Wireless Sensor Networks, energy-efficient, routing protocol, tree-based clustering, data fusion.

## 1 Introduction

Wireless sensor network (WSN) with hundreds or thousands of micro-sensor nodes can be randomly deployed in a large geographical region for various applications such as military, environment monitoring, intelligent home and so on [9]. Sensor

Nguyen Duy Tan · Nguyen Dinh Viet
Dept. of Networks and Computer Communications
University of Engineering and Technology,
VNU-HN Hanoi, Vietnam
e-mail: tanndhyvn@gmail.com,
        vietnd@hn.vnn.vn

© Springer International Publishing Switzerland 2015                                    61
V.-H. Nguyen et al. (eds.), *Knowledge and Systems Engineering,*
Advances in Intelligent Systems and Computing 326, DOI: 10.1007/978-3-319-11680-8_6

nodes have a small size, low cost, limited bandwidth, processor abilities and random access memory (RAMs). Especially, the constrained battery power at each node affects the lifetime of entire network and becomes chief challenge for designing routing protocols in WSN. Tree-based clustering routing and data fusion protocols [5, 6, 16] are the most significant techniques to reduce energy consumption and extend the lifetime of sensor networks. If Tree-based routing protocol greatly reduces the communication distance between nodes by constructing tree-based clusters, then data fusion will eliminate the redundant data collected from different sensors to get more accurate information. In addition, data compression also is a good solution for saving battery power by decreasing the amount of bits to be transmitted, however, the common compression methods as Huffman; Lempel-Ziv [10] are not suitable for sensor nodes because they requires the large storage and strong processor capability. Distributed Source Coding (DSC) technique [15], which is founded by Slepian-Wolf theorem, states the lossless data compression of two correlated sources using side information, which may be data in the past or sensed data of its neighbors, to encode source information. DSC technique is one of the most appropriate methods for WSN to save energy because limited processing capability and storage of sensor nodes. Moreover, in a sensor network, sensor nodes often are densely deployed in a sensor field, so this correlation condition can be satisfied easily.

Heinzelman. W. B. et al [4, 9] proposed centralized-LEACH (LEACH-C), in which sensor nodes are organized into several clusters. Each cluster elects a leader called CH to transmit data to base station (BS); other nodes (cluster-members) will send collected data to its CH. The advantage of the clustering is energy consumption balance among sensor nodes hence lengthened network lifetime.

An improvement of LEACH and LEACH-C algorithm were proposed by Lindsey et al called PEGASIS. PEGASIS is a basic chain-based routing protocol [1, 8, 11, 12], in which, sensor nodes only connect and communicate with a nearest neighbor to form a chain. In order to transmit data to the BS, PEGASIS chooses a node to become CH, which has randomly location in the chain. The performance of PEGASIS is better than LEACH [8]; however, there are still some limitations in this protocol. Firstly, the CH is selected at random location in chain, (no considering the remaining energy and location of the BS). Secondly, some "long links" inevitability still exists between neighboring nodes in PEGASIS, which is the cause of unbalance of energy consumption distribution among nodes. Moreover, the transmission phase of PEGASIS may become high delay and a bottleneck at the CH node since the CH is a single node in "long chain".

To solve that problem, in this paper, we propose a new routing protocol, namely DFTBC (data fusion and tree-based clustering) based on LEACH-C, which achieves higher energy and bandwidth utilization efficiency by reducing communication distance between nodes and the amount of data transmitted. In DFTBC, each cluster will select a CH, which considers the energy residual and the distances between candidate nodes and the BS, and then, the nodes in each cluster build the minimum spanning tree with the root as the CH using the Prim algorithm. In addition, the theory of Dempster-Shafer and Slepian-Wolf also used to aggregate and compress correlative data along the tree to reduce amount of transferred data. Our simulation results show that the network lifetime of DFTBC can be also extended to about 350% and 35% in comparison with LEACH and PEGASIS, respectively. The remainder of this paper is organized as follows. Section 2 presents the framework and Section 3 describes DFTBC in detail. In Section 4, we evaluate and analyze the simulation results. Finally, we conclude the paper in Section 5.

## 2 The Framework

In this section, we discuss some techniques in which data compression model and energy consuming model is clearly analyzed that are important for designing of our tree-based clustering routing protocol.

### 2.1 The Network Model

In our system model, we assume that a sensor network consist of N uniformly micro-sensor nodes deployed densely within a target field to periodically monitor the environment and a BS (i.e., sink), whose location is far away from the square sensing area and equipped with an unlimited energy resource. We simplify a few reasonable assumptions in the network model as follows:

- All micro-sensor nodes have the same capabilities of sensing area, processing of data and each node can change the transmission power to communicate with BS directly.
- All links are symmetric; all nodes contain the same initial battery power and cannot recharge batteries. Every micro-sensors and the BS are stationary after deployment.

### 2.2 Energy Consuming Model

As can be seen in Fig. 1, a typical architecture of a micro-sensor node consists of four main components: a data processor unit; a micro-sensor unit; a radio communication subsystem that contains receiver/transmitter electronics, antenna and amplifier; and power battery unit [9, 11]. In this paper, we mainly consider energy dissipation from transmitting and receiving data because our concern is developing an energy efficient routing protocol to improve network lifetime. In addition, energy dissipated during data fusion in the parent nodes is also taken into account. We use a simplified power model discussed in [7, 12] for energy consuming of radio communication. In order to transmit $q - bit$ data between two nodes with distance $d(a,b)$, the energy consumption is computed as follows:

$$E_{TX}(q,d) = \begin{cases} q.E_{elec} + q.E_{friss}d^2 & , if\, d < d_{co} \\ q.E_{elec} + q.E_{tworay}d^4 & , if\, d > d_{co} \end{cases} \quad (1)$$

Where $E_{elec}$ is a fixed dissipating energy to run the transmitter or receiver electronics, $E_{friss}$ and $E_{tworay}$ are the unit energy required for the transmit amplifier to ensure

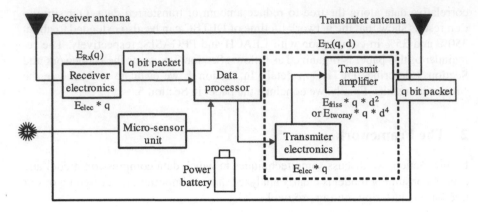

**Fig. 1** Energy dissipation model of a typical micro-sensor node architecture

that an acceptable signal-to-noise ratio at the receiver can be reliable, in both the free space and the two rays ground propagation model that depends on the distance, respectively, and $d_{co}$ is the crossover distance used in NS2 [3, 13] in our simulation scenario as follows:

$$d_{co} = \sqrt{\frac{(4.\pi)^2.l.h_t^2.h_r^2}{\lambda^2}} = \sqrt{\frac{E_{friss}}{E_{tworay}}} \qquad (2)$$

Where $\lambda$ is the wavelength; $l$ is the system loss value, $h_t$ and $h_r$ are the heights of the transmitter and receiver antennas, respectively. The assumed values for energy parameters used in simulation are presented in Table 3.1 and the other parameters are set as follows: $l = 1$, $h_t = h_r = 1.5$ (m) and $\lambda = 0.328227$(m), $d_{co} = 86.1424$ (m), [3, 13]. For receiving a data packet with $q - bit$, the radio dissipates energy:

$$E_{RX}(q) = q.E_{elec} \qquad (3)$$

Therefore, the energy consuming of a parent node in tree for a round is described as follows:

$$E_{parent}(i) = nc.E_{RX}(q) + E_{DF}(q) + E_{TX}(q, d(i, j)) + E_G(q) + E_S(q) \qquad (4)$$

In this equation, nc is the number of children of parent node $i$, which transmits sensing data packet to node $i$; $E_{DF}(q)$ is energy consuming for data fusion with $q - bit$ data. $E_G(q)$ and $E_S(q)$ are the energy consumption in sensing environment and generating a packet data, and the standby mode, respectively, and that is usually a constant for fixed. If node $i$ is CH node, the distance will be $d(i, BS)$, and the energy consuming for a child node as follows:

$$E_{child}(j) = E_{TX}(q, d(i, j)) + E_G(q) + E_S(q) \qquad (5)$$

## 2.3 Distributed Source Coding Analysis

One foundation of the DSC techniques bases on Slepian-Wolf theorem, which is one of the most appropriate methods of saving energy transmission in WSNs [15]. To achieve lossless data compressing of correlated data from many sensor nodes with a total rate at least equal to (or greater than) the joint entropy by reducing amount of data transmitted. Given two distributed sources X and Y generate $u$ bits binary data as shown in Fig. 2. Suppose that the data of $X$ and $Y$ are equiprobable or highly correlated in the following case: the Hamming distance between $X$ and $Y$ is $dH(X,Y) \leq 1$ or $Pr(X_i = 0) = Pr(X_i = 1) = Pr(Y_i = 0) = Pr(Y_i = 1) = 0.5$, $i = 0, \ldots, u$. The above supposition can be stated informally as follows: in two bit strings generated by X and Y, there is at most one bit position where two bits of two bit strings are different. If $Y$ is available to both the encoder and the decoder as side information, which consists of $u$ bits, we can represent and transmit $X \oplus Y$ into $H(X|Y)$ bits per sample instead of transmitting $X$ directly, where $H(X|Y)$ and $Pr(X)$ indicate the conditional entropy function and the probability distribution density function of random source $X$, respectively. At the decoder, $X$ can be recovered as $(X \oplus Y) \oplus Y$.

Without loss of generality, we assume that Slepian-Wolf coding of $X$ and $Y$ sources are equiprobable and uniformly distributed $2^u$ samples (u = 7 bits binary per sample). Let $m = u - k$, where k is an integer, then $H(X) = H(Y) = 7$ bits per sample, $X \oplus Y \in \{[0000000], [0000001], \ldots, [1000000]\}$, we only take $8 = 2^3$ value difference. Therefore, $H(X|Y) = m = 3$ bits, and $H(X,Y) = 7 + 3 = 10$ bits per pair of sample for joint decoding, where $H(.)$ denotes the entropy function. We can allocate $2^3$ different values corresponding to $8 = 2^3$ subsets, which is indexed by $Z_{XXX}$, respectively, as follows: $\{Z_{000} = [0000000], Z_{001} = [0010000], Z_{010} = [0100000],$

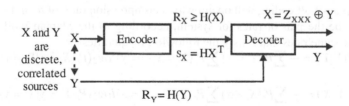

**Fig. 2** Independent encoding and joint decoding of two correlated data sources X and Y

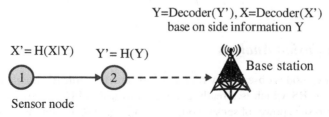

**Fig. 3** Implementation distributed source coding in WSN

$Z_{011} = [0000100]$, $Z_{100} = [1000000]$, $Z_{101} = [0000001]$, $Z_{110} = [0001000]$, $Z_{111} = [0000010]$} because the Hamming distance between $X$ and $Y$ is less than or equal to one ($dH(X,Y) = 1$). We consider the transmitted data of source $Y$ as $\{1001110\}$, $X$ as $\{1001100\}$ and $X \oplus Y = [0000010]$, corresponding to $Z_{111}$. In fact, the side information $Y$ is perfectly available at the decoder (e.g. BS or sink) but the collaboration of encoders (that is the sensor, which generates $X$) is difficult to implement in sensors node, as shown in Fig. 2 and 3. The $Y$ is only available to the decoder but not the encoder, thus, how can we compress two sources $X$ and $Y$ into totally $(u+m)$ bits as the above example with lossless decoding? To solve this problem, we can use the H, a $m$ by $u$ parity check matrix, which is given by $[I_m : P^T]_{m \times u}$ in $GF(2)$ (Galois Field), where $P^T$ is the transpose of the $(u-m)$ by $m$ matrix $P$, as above. The encoder simply computes the syndrome $s_X = H.X^T \in Z_{XXX}$ of the source $X$. For

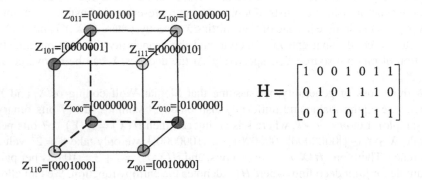

$Z_{011}=[0000100]$     $Z_{100}=[1000000]$

$Z_{101}=[0000001]$     $Z_{111}=[0000010]$

$Z_{000}=[0000000]$     $Z_{010}=[0100000]$

$$H = \begin{bmatrix} 1 & 0 & 0 & 1 & 0 & 1 & 1 \\ 0 & 1 & 0 & 1 & 1 & 1 & 0 \\ 0 & 0 & 1 & 0 & 1 & 1 & 1 \end{bmatrix}$$

$Z_{110}=[0001000]$     $Z_{001}=[0010000]$

a received syndrome $s_X$, the decoder will compute $s_Y = H.Y^T$ together with all subsets $Z_{XXX}$, which has syndrome $[s_X \oplus s_Y]^T = Z_{XXX}$. As above instance, $s_X = [001]^T$, $s_Y = [110]^T$, $[s_X \oplus s_Y]^T = [111]$. And thus, $X = Z_{111} \oplus Y = [1001100]$ as desired.

The Slepian-Wolf coding will obtain with a compression ratio of $u : m$ for $X$. Let $R_X$ and $R_Y$ be the pair of rates for system, according to the Slepian-Wolf coding achieves the same rates that satisfy the following inequalities:

$$R_X \geq H(X|Y) = -\sum_y P_r(Y = y) \sum_x P_r(X = x|Y = y) log_2(P_r(X = x|Y = y)) \quad (6)$$

$$R_Y \geq H(Y|X) = -\sum_x P_r(X = x) \sum_y P_r(Y = y|X = x) log_2(P_r(Y = y|X = x)) \quad (7)$$

and

$$R_X + R_Y \geq H(X,Y) = -\sum_x \sum_y P_r(X = x, Y = y) log_2(P_r(Y = y, X = x)) \quad (8)$$

## 2.4  Data Fusion Analysis

In WSN, sensor nodes observe environment and send periodically highly correlated data packets to BS which is directly connected to user's PC or through the Internet. In order to save power of sensor nodes, we can use DSC techniques as described

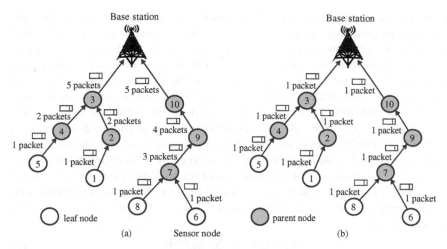

**Fig. 4** The model of data transmission in tree, (a) without data fusion and (b) with data fusion

above. However, it is much wasteful energy for transmitting the redundant collected data packets by sensor nodes. Data fusion is a well known technique to obtain more accurate information and energy efficiency using Dempster-Shafer theory [2, 6]. In this section, we consider that the data packet is fussed through tree-based topology as illustrated in Fig. 4. The CHs and parent nodes will fuse one or more data packets from different sensor nodes to generate a single packet, hence, the amount of data transmitted in tree and the BS will be greatly reduced. We assume there are ten nodes those must transmit data packet to BS along tree as can be seen in Fig. 4; the total dissipated energy can be calculated for a round a long tree as bellows:

$$E_a = 15E_{RX}(q) + 25E_{TX}(q, d(i, j)) + 10E_G(q) + 10E_S(q) \tag{9}$$

$$E_b = 8E_{RX}(q) + 10E_{TX}(q, d(i, j)) + 6E_{DF}(q) + 10E_G(q) + 10E_S(q) \tag{10}$$

Where $E_a$ and $E_b$ are total energy consumption without data fusion and with data fusion, respectively. We also assume that the packet size of each node is $q = 1$ bit and the distance between every two nodes is equal, $d = 10m$.

$$E_a = 40E_{elec} + 25E_{fiss}d^2 + 10E_G + 10E_S \tag{11}$$

$$E_b = 18E_{elec} + 10E_{fiss}d^2 + 6E_{DF} + 10E_G + 10E_S \tag{12}$$

Therefore, it is clearly that the energy dissipation without data fusion is greater than about two times compared to data fusion. Hence, data fusion will be helpful to efficiently use energy and prolong the network lifetime. However, how to fuse one or more correlated data packets to generate a single packet as shown in Fig. 4(b)? To deal with that problem, we can utilize Dempster-Shafer theory. According to Dempster-Shafer reasoning, all possible mutually exclusive events of the same kind are enumerated in "the frame of power set $\Theta$". In our system, we assume that each sensor node will allocate its sensed value about environment into power set $\Theta$ as its belief, for instance, $\Theta = \{A, B, \{A, B\}, \oslash\}$ that can be "$value - A$", "$value - B$", "either $value - A$ or $value - B$", or "neither $value - A$ nor $value - B$ and $empty$". Let $m_i$ be "probability mass function" of the $i-th$ sensor $(id = i)$, the probability, which the detected value is "$value - A$", is denoted by a "confidence interval" as $[Belief(A), Plausibility(A)]$. The lowest area of the confidence interval is the belief confidence "$Belief(A)$", which is the sum of the probability mass of elements (evidence) $m(PA)$ that supports the given element "$value - A$" (which are subsets of A, including itself).

$$Belief_i(A) = \sum_{P_A \subseteq A} m_i P(A) \tag{13}$$

The upper area of the confidence interval is the plausibility confidence, which is the sum of all the probability mass of the sets that intersect with the set A.

$$Plausibility_i(A) = 1 - \sum_{P_A \cap A = \oslash} m_i P(A) \tag{14}$$

For each possible element (e.g. A or B) of the power set, Dempster has given a rule of combining for fusing two nodes, sensor $i - th's$ observation $m_i$ and sensor $j-th's$ observation $m_j$ as follow:

$$[m_i \oplus m_j](A) = \frac{\sum_{P_A \cap P_A' = A} m_i P(A) m_j P(A')}{1 - \sum_{P_A \cap P_A' = \oslash} m_i P(A) m_j P(A')} \tag{15}$$

To fuse n sensor node, this combining rule can be used for sensor $j - th$ and sensor $k - th$ by iteration equation (15). Although, Dempster-Shafer theory is a suitable method to solve data fusion problem, however, the time complexity for combining rule of Dempster-Shafer algorithm is $O(v^n)$, where v is the number of possible value at one time. Therefore, Zeng et al have been proposed an event-based data aggregation algorithm (LEECF) based on matrix analysis [16]. The authors have also proved that the same result as Dempster-Shafer rule combination can be achieved in a much less time. Let $W = \{w_{n,v}\} n \times v$, n by v matrix represents the confident allocation of $n$ sensor nodes, which is observed target environment with v probability mass function, $w_{i,j}$ is the confident of $j-th$ target probability mass function given by the $i-th$ sensor, and the sum confident of $w$ of one sensor should be 1, equivalently, we have $w_{i1} + w_{i2} + \ldots + w_{iv} = 1$ $(i = 1, \ldots, n)$. We permute one row in $W$ and multiply by another row: $w_i^T \times w_j = [w_{i1} + w_{i2} + \ldots + w_{iv}]^T \times [w_{j1}, w_{j2}, \ldots, w_{jv}]$, we will obtain a new $n$ by $v$ matrix $U$ :

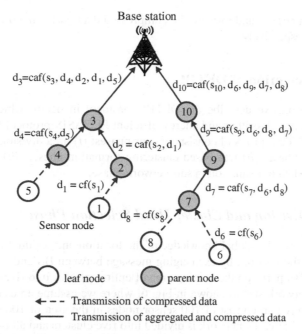

**Fig. 5** An example of the tree-based clustering data fusion

$$
U = \begin{bmatrix}
w_{i1} \times w_{j1} & w_{i1} \times w_{j2} & \dots \dots & w_{i1} \times w_{jv} \\
w_{i2} \times w_{j1} & w_{i2} \times w_{j2} & \dots \dots & w_{i2} \times w_{jv} \\
\dots & \dots & \dots \dots & \dots \\
\dots & \dots & \dots \dots & \dots \\
w_{iv} \times w_{j1} & w_{iv} \times w_{j2} & \dots \dots & w_{iv} \times w_{jv}
\end{bmatrix}
$$

Where the elements of main diagonal of U is the result of confident from $n$ sensor and sum of other element that differ main diagonal is the uncertainty factor of evident, $F$.

$$
F = \sum_{p \neq q} w_{ip} \times w_{jq} \quad (p, q = 0, \dots, v) \tag{16}
$$

and

$$
M_j = \frac{u_{jj}}{(1 - F)} \quad (j = 0, \dots, v) \tag{17}
$$

Fig. 5 gives an example of the tree-based clustering data fusion schemes, where each leaf sensor node $i - th$ will generate the original sensing data packet $s_i$ which is compressed by the function $cf(s_i)$. The parent and CH nodes $j - th$ will decompress and aggregate the data sent from the child nodes with its own sensing data $s_j$ by the fusion function $caf(.)$, and forward the compressed data packet $d_j$ to upper parent node and BS, respectively.

## 3 The Description of DFTBC

In this subsection, we describe the DFTBC protocol in detail, which bases on LEACH-C routing algorithm and energy-efficient PEGASIS protocol [9, 14]. The detail of the DFTBC protocol consists of four phases; ($i$) area division and cluster head selection phase, ($ii$) tree-based clustering formation phase, ($iii$) data fusion phase and ($iv$) data transmission in sub-network phase.

### 3.1 Area Division and Cluster Head Selection Phase

Firstly, BS will get the global knowledge of the location and residual energy of all nodes alive in the network by exchanging message between BS and sensor nodes. And then, the BS partition the sensing area of entire network into five clusters that is equal sub-network size as shown in Fig. 6, where we assume an example of the network topology that consist of 100 sensor nodes in the area of $100 \times 100$ square meters. In this figure, the network is divided into five clusters and all nodes in each cluster are organized into a minimum spanning tree.

In each round of DFTBC, the BS will select the CH for each cluster, which has remaining energy greater than $E_{average}$ and maximum cost function as follows:

$$E_{average} = \frac{\sum_{i=1}^{nn} E_{residual}(i)}{nn} \tag{18}$$

Where nn and $E_{residual(i)}$ are the total number of nodes alive in the cluster and the residual energy of candidate node $i - th$ at the current time in considering cluster, respectively.

$$cost(i) = \left( \frac{w_1}{w_2} \times \frac{E_{residual}(i)}{d(i, BS)} \right) \tag{19}$$

Where $d(i, BS)$ is the geographic distance from the candidate node $i - th$ to the BS which is computed as follows:

$$|d(a,b)| = \sqrt{(x_a - x_b)^2 + (y_a - y_b)^2} \tag{20}$$

Furthermore, the coefficient of cost factors $w_1$ and $w_2$ must be set to appropriate values based on the characteristics of the WSN's scale. When $w_1$ is greater than or equal $w_2$, it means that the residual energy of candidate node is more important than distance from the node to BS when selecting the CH.

### 3.2 Tree-Based Clustering Formation Phase

In this phase, the minimum spanning tree will be constructed for five clusters base on Greedy algorithm to solve the undirected weight graph problem. In order to get

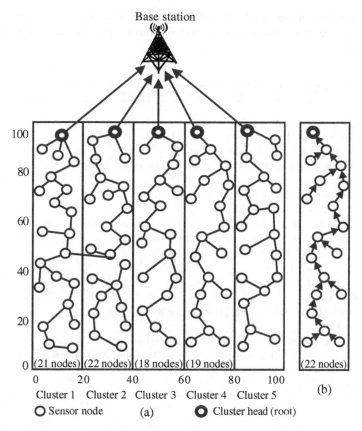

**Fig. 6** (a) 100-node in 100m×100m network, (b) an example of the minimum spanning tree

---

**Algorithm 1** Area Division and Cluster Head Selection Phase

---
1: **for** each node in {List of the nodes alive} **do**
2:    Transmit HELLO message, which contain its ID, remaining energy and location, to BS
3: **end for**
4: **if** (round = 1) **then**
5:    Divide entire network into five clusters with equal size sub-network as shown in Fig. 6(a).
6: **end if**
7: **for** each cluster (sub-network) in network **do**
8:    Calculate average energy as equation (18)
9:    Choose cluster head node, which has maximum cost function as equation (19)
10: **end for**
11: go to Algorithm 2.

---

the minimum sum of the weight, which is the distance between the nodes together
in cluster, we use Prim algorithm to support us constructing the minimum spanning
tree-cluster.

---

**Algorithm 2** Tree-based Clustering Formation Phase

---
    label[CH] ← 0
2:  label[others node] ← MAX
    {TREE[BS]} ← BS
4: **while** {List of the nodes alive} ≠ ∅ **do**
      Search node $i$ in {List of the nodes alive}, whose label[node $i$] is minimum
6:    *endNode* ← node $i$
      **for** each node $j$ in {List of the nodes alive} **do**
8:      **if** (node $j$ ≠ *endNode*) **then**
         **if** (label[node $j$] > d(node $j$, *endNode*)) **then**
10:         label[node $j$] ← d(node $j$, *endNode*)
          {TREE[node $j$]} ← *endNode*
12:      **end if**
      **end if**
14:    **end for**
      Discard the *endNode* in {List of the nodes alive}
16: **end while**
    Create TDMA schedule for all nodes in each cluster
18: Broadcast the TDMA schedule and the information about minimum tree-
based clustering in network
**return** {TREE}

---

## 3.3  Data Fusion Phase

After selecting CH and constructing minimum spanning tree, each CH could create a
TDMA schedule and distribute them to each member in its cluster to transmit data.
The sensor nodes start collecting, fusing and compressing data as in Algorithm 3
below:

## 3.4  Data Transmission Phase

After DFTBC has completed the CH selection and the tree-base clustering formation
phase above, the data packet transmissions are started by nodes within each cluster
in their TDMA schedule. Firstly, the leaf nodes from farthest in each tree will start
transmitting to their parent node along the tree. The parent nodes receive the data,
decompress and aggregate this data with its own sensed data, compress and forward
it to the upper level node in the tree as shown in Fig. 6(b). Whenever the CH node
receives all the data, it will transmit the data to the BS after compressing it with
the same way. After an interval time, the next round will be restarted by reselecting

---

**Algorithm 3** Data Fusion Phase

---

    **if** (node $i$ in {TREE} is parent node) **then**
        **if** (the data packet need decompressing) **then**
3:          **for** each 10 bits in compressed data packet **do**
              $Y \leftarrow$ high 7 bits as side information
              $s_X \leftarrow$ lower 3 bits
6:             Calculate $s_Y \leftarrow H * Y^T$
              Calculate $X \leftarrow [s_X \oplus s_Y]^T$
              Search $X$ in $Z_X$ which has $Z_X \oplus Y = X$
9:             Append $Y$ and $X$ into data packet decompression
          **end for**
        **end if**
12:      Calculate and allocate probability mass function base on received data into W matrix
        **for** each sensor p, which connected by node $i$, in {TREE} **do**
          Calculate F function value as equation (16)
15:      **end for**
        **for** each row j of W matrix **do**
          Calculate $M_j \leftarrow u_{jj}/(1 - F)$ as equation (17)
18:      **end for**
    **end if**
    **if** (the data packet need compressing) **then**
21:      **for** each 14 bits in the data packet **do**
          $Y \leftarrow$ high 7 bits as side information
          $X \leftarrow$ lower 7 bits
24:         Calculate $s_X \leftarrow H * X^T$
          Append $Y$ and $s_X{}^T$ into new packet
      **end for**
27: **end if**

---

cluster heads as well as reconstructing minimized spanning trees in each cluster for a new round.

# 4 Evaluation and Simulation Results

## 4.1 Simulation Parameters

To evaluate the performance of DFTBC, we simulated DFTBC, LEACH and PEGASIS by network simulator ns-2 (v.2.34) [3, 13] using the same scenes with the parameters that are described in Table 3.1, [3, 11, 14].

## 4.2 Simulation Results

Fig. 7(a) shows the simulation result of the total number of nodes alive in network with the number of rounds. Here, the Y axis indicates the number of nodes alive

**Table 1** Detailed parameters for Simulations

| No.Item | Parameters Description | Value |
|---------|------------------------|-------|
| 1 | Simulation area | 100m x 100m |
| 2 | Network size | 100 nodes |
| 3 | $E_{elec}$ (*Radio electronics energy*) | 50 $nJ/bit$ |
| 4 | $E_{amp}$ (*Radio amplifier energy*) | 100 $pJ/bit/m^2$ |
| 5 | $E_{fs}$ (*Radio free space*) | 0.013 $pJ/bit/m^4$ |
| 6 | $E_{init}$ (*Inital energy of node*) | 2J |
| 7 | Energy model | Battery |
| 8 | Packet size | 500 bytes |
| 9 | Simulation time | 3600s |
| 10 | Base station at | 50,175 |
| 11 | Channel type | Channel/wireless channel |
| 12 | Antennae model | Antenna/omniantenna |
| 13 | Number of cluster | 5 |

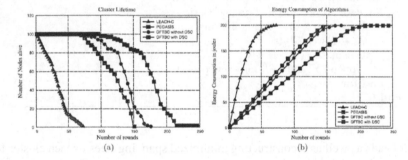

**Fig. 7** a) Number of nodes alive with the change of rounds. b) Energy consumption by three protocols.

and the X axis represents the lifetime of the nodes in the WSN according to the number of rounds. In Fig. 7(a), we can obviously observe that the DFTBC have more rounds or longer network lifetime than about 350% and 35% compared to LEACH and PEGASIS, respectively. Furthermore, Fig. 7(a) also describes the comparison between the DFTBC with compression and without compression.

In Fig. 7(b), we illustrate energy consumption for three protocols with the change of rounds. Based on results shown in Fig. 7(b), it is clearly observable that the DFTBC achieved better energy efficiency and increased the WSN lifetime compared to LEACH and PEGASIS.

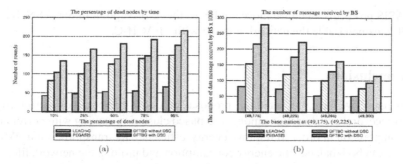

(a)                                                    (b)

**Fig. 8** a) The percentage of dead nodes when BS at (49,175). b) The number of message received at the BS when position changes

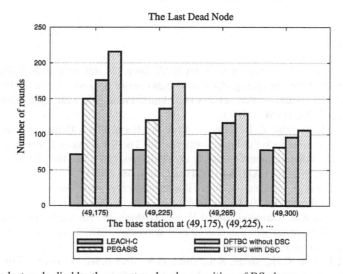

**Fig. 9** The last node died by three protocols when position of BS changes

As illustrated in Fig. 8(a), the percentage of dead nodes of three protocols is analyzed with the change of rounds. We can see that DFTBC with the minimum energy consumption achieved better energy efficiency and prolonged the lifetime of network compared to LEACH and PEGASIS.

Fig. 8(b) shows the total number of received messages at different positions of the BS. It seems there is a relative decrease in the number of received messages by the BS with all protocols when we move the BS to the farthest point in the WSN but our proposed protocol still outperforms the existing LEACH and PEGASIS algorithm protocols.

In Fig. 9, we show the last dead node when the location of BS changes. It is clear that there is also a relative decrease the number of rounds with DFTBC and

PEGASIS protocols when we move the BS to the farthest point in the WSN, but, it is still remained stable the network lifetime with LEACH-C.

## 5 Conclusions

In this paper, we have presented our proposal of a new tree-based clustering routing protocol and collaboration among tree-based clustering routing with correlative data compression in order to improve energy-efficient data transportation in WSN. DFTBC can both balance the energy consumption and prolong the network lifetime by constructing minimized tree in each cluster with Greedy algorithm and fusing, compressing sensed data before sending packet to the BS. Simulation results show that the energy efficiency of DFTBC is better than that of PEGASIS at about 35% in case of large network (100 nodes and 100m x 100m).

## References

[1] Ahn, K.S., Kim, D.G., Sim, B.S., Youn, H.Y., Song, O.: Balanced chain-based routing protocol (BCBRP) for energy efficient wireless sensor networks. In: Ninth IEEE International Symposium on Parallel and Distributed Processing with Applications Workshops, pp. 227–231 (May 2011)

[2] Fontani, M., Bianchi, T., De Rosa, A., Piva, A., Barni, M.: A Dempster-Shafer framework for decision fusion in image forensics. In: The IEEE International Workshop on Information Forensics and Security, Iguacu Falls, pp. 1–6 (December 2011)

[3] Heinzelman, W.: MIT uAMPS LEACH ns Extensions (2004),
http://www.ece.rochester.edu/research/wcng/code/index.htm
(accessed February 29, 2014)

[4] Heinzelman, W.B., Chandrakasan, A.P., Balakrishnan, H.: An application-specific protocol architecture for wireless microsensor networks. IEEE Transactions on Wireless Communications 1(4), 660–670 (2002)

[5] Kim, K.T., Lyu, C.H., Moon, S.S., Youn, H.Y.: Tree-based clustering (TBC) for energy efficient wireless sensor networks. In: The IEEE 24th International Conference on Advanced Information Networking and Applications Workshops, Perth, Australia, pp. 680–685 (April 2010)

[6] Li, J., Luo, S., Jin, J.S.: Sensor data fusion for accurate cloud presence prediction using Dempster-Shafer evidence theory. Sensors Open Access Journal 10(10), 9384–9396 (2010)

[7] Li, Y., Yu, N., Zhang, W., Zhao, W., You, X., Daneshmand, M.: Enhancing the performance of LEACH protocol in wireless sensor networks. In: IEEE Conference on Computer Communications Workshops, Shanghai, China, pp. 223–228 (April 2011)

[8] Lindsey, S., Raghavendra, C.S.: PEGASIS: power-efficient gathering in sensor information system. In: IEEE Aerospace Conference Proceedings, pp. 1125–1130 (March 2002)

[9] Muruganathan, S.D., Ma, D.F., Bhasin, R.I., Fapojuwo, A.O.: A centralized energy-efficient routing protocol for wireless sensor networks. IEEE Radio Communications 43(3), 8–13 (2011)

[10] Sayood, K.: Lossless Compression Handbook. Academic Press, California (2003)

[11] Sen, F., Bing, Q., Liangrui, T.: An improved energy-efficient PEGASIS-Based protocol in wireless sensor networks. In: Eighth International Conference on Fuzzy Systems and Knowledge Discovery, Shanghai, China, pp. 2230–2233 (July 2011)

[12] Tian, Y., Wang, Y., Zhang, S.F.: A novel chain-cluster based routing protocol for wireless sensor networks. In: Proceedings of the Second International Workshop on Education Technology and Computer Science, Shanghai, China, pp. 2456–2459 (September 2007)

[13] VINT Project. The network simulator - NS2 (1997),
http://www.isi.edu/nsnam/ns (accessed February 2, 2014)

[14] Xinhua, W., Sheng, W.: Performance comparison of LEACH and LEACH-C protocols by ns2. In: Ninth International Symposium on Distributed Computing and Applications to Business, Engineering and Science, pp. 254–258 (August 2010)

[15] Xiong, Z., Liveris, A.D., Cheng, S.: Distributed Source Coding for sensor networks. IEEE Signal Processing Magazine 21(5), 80–94 (2004)

[16] Zeng, B., Wei, J., Hu, T.: An energy-efficient data fusion protocol for wireless sensor network. In: The 10th International Conference on Information Fusion, Quebec, Canada, pp. 1–7 (July 2007)

# Natural Language Processing for Social Event Classification

Duc-Duy Nguyen, Minh-Son Dao, and Truc-Vien T. Nguyen

**Abstract.** In this paper, a simple but effective method for social event detection based mainly on natural language processing is introduced. Meanwhile existing approaches use many typical text-classification methods and disregard the importance of language characteristics, the proposed method exploits such language characteristics from text items in social metadata (e.g. title, description and tag) to leverage social event detection. First and foremost, we analyze the specific characteristics of natural language in social media to choose the most suitable features. Second, we employ common natural language processing techniques along with machine learning methods to extract features and perform classification. As a result, we experienced the F1 score higher than the results of related works that used state-of-the-art methods. The proposed method proves the significance of understanding language characteristics in building social event classification programs. It also offers good clues to improve existing works on social event detection.

**Keywords:** Social Event Classification, Social Event Detection, social network, text classification, language processing, natural language characteristics.

## 1 Introduction

Just within decades, we have witnessed the dramatical development of social networks (SNs) where users can share their data and interact with their's friends

Duc-Duy Nguyen · Minh-Son Dao
University of Information Technology
Vietnam National University HCMC, Vietnam
e-mail: nguyenduy.uit@gmail.com, sondm@uit.edu.vn

Truc-Vien T. Nguyen
University of Lugano, 6900 Lugano, Switzerland
e-mail: thi.truc.vien.nguyen@usi.ch

© Springer International Publishing Switzerland 2015
V.-H. Nguyen et al. (eds.), *Knowledge and Systems Engineering,*
Advances in Intelligent Systems and Computing 326, DOI: 10.1007/978-3-319-11680-8_7

data easily and quickly. For example, Facebook[1], Twitter[2], LinkedIn[3], Instagram[4] comfort their users very well with various and attractive features such as personal information sharing (create, update, public profiles), media creation and sharing (message, photo, video, file. etc.), personal network management, applications. These advantages bring SN a unique ability to connect billions of users worldwide and establish online communities where user connected each other by friendship [1].

The interesting research presented in [2] reveals the fact that media content from SNs can effectively reflect the problems or events, since majority of topics are headline news or message from famous people/organization. Beside, recent researches concern with a possibly next generation of SN known as Social Life Network (SLN)[8], inspired by the idea of connecting people and their interested resources. Specifically, SLN helps people not only connect to each other and sharing experiments, but also provide them to efficient means to keeping up with real-time and real-life resource (many of whom come under the form of events). Therefore, there is a tight relation between social data and events they represent for.

This dramatically increases an emergent demand of controlling social media resources to facilitate the evolution of SN and user-centric web technology [2]. Obviously, social event classification play an undeniable importance in detecting, organization and mining events, not only to the SN today, but also to the future of SN. Therefore social event detection has become a hot topic within a decades. In [4], social event is defined as follow: *"the events that are planned by people, attended by people and for which the social multimedia are also captured by people.".* That leads to the trend of analyzing social media to understand social events.

Since many social media contents today come with accompanying metadata (time-stamps, tags, geographic tags, uploader's ID, etc.), these information itself partially reflects the insight of the contents. Therefore, we point our target to exploit tag and title to show how effective they are on solving social event classification problem.

To tackle the challenges, we consider SEC as a supervised text classification problem. First of all, we specify the characteristics of the Social Media natural languages and explain how it affects the performance of many typical text classification methods. Second, we proof our arguments by proposing a method based on Social Media characteristics.

We select features based on characteristics of Social Media natural language. Then we extract selected features using common Natural Language Processing tools. After that, we use a bag-of-words like model for simplifying representation metadata of a media content before employing Support Vector Machine as supervised learning method. For experiments, we apply our methods to MediaEval 2013 dataset [4]

---

[1] www.facebook.com

[2] www.twitter.com

[3] www.linked.com

[4] www.instagram.com

for Social Event Detection since it is an up-to-date dataset from a populated SN. For evaluation, we use F1 score from precision and recall as evaluation metric.

**Contribution:** One of the most significant characteristics of SNs is free-style text contents. SNs is known as the place where users can be unique by using their own styles. Therefore, languages used in SNs differs from normal language used in real world. Unfortunately, many existing works apply complex text classification technique on solving SEC did neglect or pay less attention on characteristics of social media natural language. Therefore, the proposed method tries to explore and exploit these characteristic to increase the accuracy of social event detection as well as the speed of processing stage in order to cope with real-time processing requirements.

## 2  Related Works

In [3], the authors compute semantic relatedness between the synset representing the tags and each of the categories. They used Lin – a WordNet-based relatedness similarity measure proposed by Dekang Lin in [5]. Lin measure bases on information content of the least common subsumer (LCS). It sums the information content of the individual concepts then scale this sum; the result is the information content of the LCS. For those media contents that encounter the same Lin measure result, they use another constrain (Date, Time) to determine the best result. This method gain the event F1 score 7.48%. The method represented in [6] give the F1 score of 22.01% of even/non-event classification. This method first used 1000 most frequent tags then applied 200 latent topics as features. The concept of which the detector produced the highest prediction score is assumed to be the choice.

Another team employs machine learning models for supervised learning and classification. In [7], the authors uses title, description and keywords of metadata to build classifier for each event type. They employed linear *support vector machine* classifier to classify test samples. Instead of classifying media content individually, they first apply constraint-based Spatiotemporal clustering to cluster media contents, then apply classification on groups of items. This method gained F1 score 37%. The best methods for social event classification from metadata is reported in [10], where the authors extracted the following features from text of title, description and tag: original word, word in lower-case, prefixes, suffixes, part-of-speech, orthographic characteristics and ontological feature. They also employed *support vector machine* as supervised machine learning method and *disambiguation* to Wikipedia[5] to optimize the performance. As a result, they experimented the F1 score 47.61%.

All the above works used typical text classification technique. They treat text from SN normally, no clear-concern over characteristics of natural language in Social Media. In addition, many of these techniques [6] [7] [8] are complex and require much time and resources, therefore they are not applicable for real-time applications.

---

[5] www.wikipedia.com

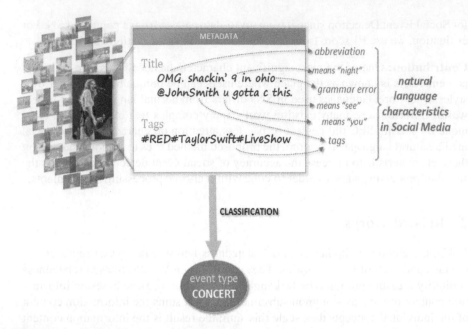

**Fig. 1** Social Even Classification on natural language of Social Media

# 3   Methodology

The proposed method focuses on understanding the characteristics of Social Media natural language (verbal) and apply these characteristics on proposing a simple and effective method to solve SEC problem. In section 3.1, we discuss characteristics of Social Media natural language from causes to effects. In section 3.2, we show how we apply these characteristics on text feature selection. In section 3.3, we represent our document representing model and the learning model that we used to build the classifier.

## 3.1   Natural Language Characteristics in Social Media

Language is an interactive and social phenomenon that centralize to the inter-personal communication between individuals in a society [9]. Thus, language is always dynamically affected and evaluated with the changes of its society. In SN, its cross-culture and up-to-date environment push the evolution of language obviously. In this section, we turn our attention to find out the characteristics of English language which is widely used by SN users. There are many factors that make up the characteristics of social media natural language today. Five main factors are concerned as follow:

- **Fast and short message composing habit**: modern mobile internet technologies and smart devices such as smartphone, tablet PC, etc., are providing fast and convenient methods that help people are easily connected to SNs anywhere. For instance, a man can visit Facebook when he is in the elevator, waiting for the bus. Their messages, often short and impact, contain many words under shortened form, slang or novel words. In addition, some SNs limit the maximum number of character of the message, for instance, Twitter set tweet maximum number of character to 140, hence, users have to constrict their message and discard grammar rules and unimportant words.

- **Young users' ages**: a research by Pew Research Center (PRC) shows that by September 1st 2013, 90% of online American adult ages 18 -29 are SN users. For adult ages 30-49 and 50-64, 87% and 65% of them do, respectively. An earlier PRC research reveals that 81% of American online teens use some kind of SN by September 2012. These statistics could by different by country and time, however it show that a great number of young people are using SN everyday. Notably, these young user often use slang and informal words in their communication; thus the text in Social Media today contain lots of noises.
- **Open and multicultural environment**: SNs attracted billions of people who came from different societies, have different educational backgrounds, speak different languages and interest in very different things. This make SN become a multicultural environment, especially the language usage.
- **From anonymity, equality to freedom**: Anonymity is one of the most important characteristic that SN inherited from its ascendant - the internet. Specifically, a person can partly hide his/her real identify; or reveal things that they want other people think he/she is. In addition, SN provide an environment where people can equally express their ideas, no matter who they are. These two characteristics give the users the feeling of freedom, hence many people quickly express their ideas without caring about language mistakes.

The above factors make up characteristics of language in social media, included:

1. **Lots of Noise**: many words in Social Media are dialect or meaningless.
2. **Abnormal vocabulary**:

    - **English-based Creole Languages**: words that derive from English languages. For example: in Singlish (i.e. colloquial Singaporean English) the word **then** often be written or pronounced **den**; or in Hawaiian Pidgin English the word **baby** often be replaced by **behbeh**.
    - **High Diversity**: The diversity of English in social media come from it multi-cultural environment. As a result, (1) English slang and "half-English" slang are widely used. For example, the English slang **"Lose The Plot"** means to go crazy or suffer mental illness. Also, there are some special English slang which combine an alphabet character and a non-English character/word, (2) Multilingual environment allows user to use their own language and vocabulary, and (3) Many abbreviation are used. Such as in Figure 1, the word **OMG** is abbreviation of "Oh my God!".

3. **Abnormal Grammar**

- **Weak Grammar Constrains**: As we mentioned early this section, fast and short message composing habits make user dismiss or misuse grammatical rules, such as capitalize proper nouns, tense, punctuation marks usage, etc.
- **Specific Structures**: There are some considerable sentences that we can see everywhere on SN, such as "Like us on Facebook.", "Forgot your password?" or "Ask me anything!". These specific sentences focus on easy to understand and remember, therefore the structure was simplified, often violate common grammar rules.

4. **Other Characteristics**:

- **Hashtag**: indicates the topic/category of the post.
- **At-tag**: indicates another user.
- **Emotion**: such as ":)", ":b", "(:", etc.

## 3.2  Feature Selection Based on Characteristics of Natural Language in Social Media

Based on characteristics that explained in section 3.1, we come with three assumptions:

1. The dictionary or list of words on which feature extraction carries out must be very rich. Ideally, it should be an independent dictionary consist all English and Non-English words, slang, aberration, dialect, etc.. A too small dictionary (such as the work [6]) could lead to low effective feature representation. However, at the moment, there are no such a perfect dictionary; so the alternative choices could be conducting the dictionary from the training data itself to get as many words as possible. Therefore, a bag-of-word like model could be a good choice.
2. Many grammar-related features, such as parse tree, do not work effectively on social media language, due to its abnormal grammar.
3. For statistics approach, words should be transform into its base form to fill the gap between differences. The diversity as explained in 3.1 comes from abnormal vocabulary or error on grammar need to be considered. For example, words **car, cars, car's, cars', Car, CAr** should be understand as **car**. Without this step, all the words will be treat individually, therefore no similarity between them can be seen. This is also an important step before applying Similarity compute using a dictionary/knowledge-based like WordNet[6]. Discarding this step may lead to the low performance problem en-countered by [3], because all term that is not under the base form can not be found in WordNet. Likewise, the above problem also limited the effectiveness of ontology extraction [10].

The above assumptions help us propose a simple but suitable set of text features, included:

---

[6] `wordnet.princeton.edu`

1. $w_i$ the original word in title, description or tag of the metadata.
2. $lc_i$ is $w_i$ in lowercased form.
3. $le_i$ is base form of $w_i$, after applied lemmatization.
4. $ne_i$ is the named-entity recognition(NER) result of $w_i$. It determines whether $w_i$ indicates time, location, organization, person, money, percent, date or other.
5. $p_i$ is part-of-speech-tag of the word $w_i$, using Penn TreeBank tag set [11].
6. $or_i$ is orthographic characteristic of the word $w_i$, this determine whether the word is all upper-cased, initial letter upper-cased, all lower-cased or others.

Beside, we use a stopword list from [12] to eliminate valueless words. For instance, the word data could be extracted according to our features. In this case, $w_i$ is data, $lc_i$ is data, $le_i$ is datum, $ne_i$ is $O$ (means other), $p_i$ is NNS, and $or_i$ is *ALL_LOWER_CASED*.

Within the proposed features, the use of feature 1 indicates that we consider all the word in text, feature 2, 3, 4 support our assumption of transforming the word into general or base form, and while feature 5, 6 help us extract basic grammatical features in weak grammar constrains.

## 3.3  Bag-of-Words Like Model

In section 3.1, we explained the abnormal grammar in social media natural language, included weak grammar constrains. Obviously, a sentence with weak grammar constrain accompany with various tags is very similar to an orderless set of words. Therefore, we reuse with some modification the idea of bag-of-words, a model represented by Salton & McGill in 1983 as a model for orderless document representation.

Bag-of-words Model:

Bag-of-words (BoW) is a model for representing document (a sentence, a document) using frequencies of words from a dictionary. BoW is very effective to represent documents, and it completely disregards the order or structure of the words. For example, we have two document. Document 1 is "He bought a new car." and Document 2 is "He loves his car. She loves it too!". The dictionary is now contain 10 words that structured as:
{ "He": 1, "bought": 2, "a": 3, "new": 4, "car": 5, "loves": 6, "his": 7, "she": 8, "it": 9, "too": 10 }
Then document 1 and document 2 represent by the vector [1,1,1,1,1,0,0,0,0,0] and [1,0,0,0,1,2,1,1,1,1], respectively.

The Proposed BoW-like Model:

In order to use up the advantages of traditional BoW model, we reused its idea, but for multi-dictionaries. That is, we concatenate the vectors representing the document by each dictionary to make the only feature vector for the whole document. We built six dictionaries from training data, they are:

- **Original word dictionary**: contains all words of documents in training data.
- **Lowercase word dictionary**: contain all words of documents in training data under the form of lowercased.
- **Base word dictionary**: contains base form of all words of documents in training data.
- **NER dictionary**: contains all NER results: "TIME", "LOCATION", "ORGA-NIZATION", "PERSON", "MONEY", "PERCENT", "DATE" and "OTHER".
- **Part-of-speech tag dictionary**: contains 40 Penn Treebank Part-of-speech tag appear in training data.
- **Orthographic dictionary**: contains orthographic feature: "ALL_LOWER_-CASED", "ALL_UPPER_CASED", "INITIAL_LETTER_UPPER_CASED" and "OTHER".

The document is represented by using each of the above five dictionaries, then concatenating the result to have the final feature vector of the document.

Support Vector Machine:

Support Vector Machine (SVMs) is a highly successful supervised learning model represented by V.N. Vapnik in [13] [14]. It based on Structural Risk Minimization principle that represented Vapnik in 1995. In [15], Joachims also proved that SVMs are very suitable for the solving text categorizing, from both theoretical and empirical approach. Therefore, we employ SVM as supervised learning models for solving our problem. We use linear SVM and optimize parameter with the validation set. Finally, we get the best performance at **C** (the trade-off between training error and margin) values $2^{15}$.

## 4   Experimental Results

The purpose of experiments is to verify the effectiveness of the approach that a good concern over natural language of Social Media is important to solve SEC problem. We use the dataset, evaluation method and compare the result to the works at MediaEval 2013 [4]. MediaEval is a benchmarking initiative, whose reputation is widely accepted by the research community. Its datasets for tasks are updated every year and it uses many standard measurement for evaluation results.

Dataset:

We use the dataset from MediaEval 2013 because it is up-to-date and was collected from Instagram, one of the most popular SN today. The training set included 27,754 pictures, which was collected from $27^{th}$ and $29^{th}$ of April 2013. The test set was collected from $7^{th}$ and $13^{th}$ of May 2013 consist of 29,411 pictures. The picture came with some of metadata, but not all.

Evaluation Tools:

We evaluate our result with the ground truth annotated manually by MediaEval 2013. We use standard F1-Score from Precision and Recall as evaluation measure. Precision and Recall are common measurements for evaluating binary classification in pattern recognition and information retrieval base on the comparison between the expected result and the result generated by classifier. These values could be computed by the following equations:

Precision:

$$Precision = \frac{tp}{tp+fp}$$

Recall:

$$Recall = \frac{tp}{tp+fn}$$

where

- **tp** (true positive) is total number of items that were correctly labeled as positive.
- **fp** (false positive) is total number of items that were incorrectly labeled as positive
- **fn** (false negative) is total number of items that were incorrectly labeled as negative.

To explore the harmonic mean between these two values, people often use F score, whose general form is:

F-beta (where Beta are non-negative real values)

$$F_\beta = (1+\beta)^2 \cdot \frac{Precision.Recall}{\beta^2.Precision+Recall}$$

F1 is a special case of the above equation, with can be calculated by: F-measure (F1):

$$F1 = 2 \cdot \frac{Precision.Recall}{Precision+Recall}$$

Experimental Results and Discussions

Table 1 shows the result of our proposed method for event/non-event classification. We experiment F1 score up to 47.94%. Although we gain a good Precision (74.83%), the Recall is not really good (35.26%), but acceptable comparing to others.

Table 2 denotes the comparison between the proposed method and corresponding works that use the same dataset provided by task organizers of MediaEval 2013 for Social Event Detection task. The table also shows that works used Machine Learning methods in learning and classification gain a good result, compared with the remaining works.

Table 3 shows the result of our approach for multi-class classification. We classify every document in the testing set to decide to which event type it is belong. There are 8 predefined event types: conference, fashion, concert, sports, protest, other, exhibition, "theater_dance" and one special type named "non_event".

**Table 1** Proposed Method Results

|           | Result |
|-----------|--------|
| **Precision** | 0.7483 |
| **Recall** | 0.3526 |
| **F1 (event)** | 0.4794 |

**Table 2** Comparison Results

|                     | F1    |
|---------------------|-------|
| **Proposed method** | 47.94 |
| **VIT[3]**          | 7.48  |
| **QMUL[7]**         | 37    |
| **CERTH-1[6]**      | 22.01 |
| **UNITN[10]**       | 47.61 |

We experimented very different results by event type due to the imbalance of the training set.

Computational Complexity:

We setup a working environment as denoted in Table 4.

Table 5 illustrates the time for training as well as classifying of the proposed method. The results show that the proposed method definitely has an ability to run on on-line mode.

It should be noted that for training set, we extract feature and update dictionary continuously; and for test set, we load dictionary directly from a storage, so that it takes time to look up the whole dictionary.

## 5 Conclusion

The simple but effective method that takes into the benefit of characteristics of natural language in social media to leverage the accuracy and speed of social event classification is introduced. Distinct from existing methods that carry on typical text-classifications, the proposed method can cope with short, ambiguous, and non-standard English language often written in social data. Moreover, the proposed method successfully proves that a better understanding in characteristics could help on making smarter and more effective program for solving Social Event

**Table 3** Multi-class classification Results

| Event | Precision | Recall | F1 |
|---|---|---|---|
| *conference* | 0.4196 | 0.662 | 0.5137 |
| *fashion* | 0.1071 | 0.5 | 0.1764 |
| *concert* | 0.76.5 | 0.4013 | 0.5264 |
| *non_event* | 0.8829 | 0.9668 | 0.9229 |
| *sports* | 0.1675 | 0.1349 | 0.1495 |
| *protest* | 0.5942 | 0.7069 | 0.6457 |
| *other* | 0.1632 | 0.0851 | 0.1119 |
| *exhibition* | 0.4818 | 0.1384 | 0.215 |
| *theater_dance* | 0.75 | 0.2727 | 0.4 |

**Table 4** Working Environment

| | |
|---|---|
| **Number of CPU(s)** | 4 |
| **Number of tenths of physical CPUs** | 12 |
| **Amount of Memory** | 32768 MB |
| **Disk Space Size** | 200 GB |
| **CPU usage** | 25% |

Classification problem. The major advantage of the proposed method is using all simple and easy-to-understand natural language processing technique but gain a good performance. The experimental results shows that the proposed method can beat up-to-date related methods on standard benchmark MediaEval 2013.

In the future, we intend to improve the method by deeper understanding about social media natural language and apply these knowledge intelligently. Since SN are available for people all around the world, a new and very hard question is how to solve the SEC problem in multilingual context. Besides, it is important to examine the method on real-time systems and applications to process the stream of social media content in SN.

**Table 5** Computational Complexity Results

|  | Features Extraction | Features Extractions | Training | Classification |
|---|---|---|---|---|
| **Time** | 8 mins 37 secs *(all training set)* | 9 mins 14 secs *(all test set)* | 24 mins 20 ses | 1.70004420115 $\mu$secs *(each social media)* |
| **Time** | 0.018627946 sec *(average time for one social media)* | 0.01883649 sec *(average time for (for each social media)* |  |  |

# References

[1] Ellison, N.B., et al.: Social network sites: Definition, history, and scholarship. Journal of Computer-Mediated Communication 13(1), 210–230 (2007)

[2] Kwak, H., Lee, C., Park, H., Moon, S.: What is Twitter, a social network or a news media? In: Procs of the 19th Int. Conf. on World Wide Web, pp. 591–600 (2010)

[3] Gupta, I., Gautam, K., Krishna, C.: VIT@ MediaEval 2013 Social Event Detection Task: Semantic Structuring of Complementary Information for Clustering Events. In: MediaEval (2013)

[4] Reuter, T., et al.: Social Event Detection at MediaEval 2013: Challenges, datasets, and evaluation. In: Procs. of the MediaEval 2013 Multimedia Benchmark Workshop, Barcelona, Spain, October 18-19 (2013)

[5] Dekang, L.: An information-theoretic definition of similarity. Journal of ICML 98, 296–304 (1998)

[6] Schinas, E., et al.: CERTH@ MediaEval 2013 Social Event Detection Task. In: Procs. of the MediaEval 2013 Multimedia Benchmark Workshop, Barcelona, Spain, October 18-19 (2013)

[7] Brenner, M., Izquierdo, E.: Social event detection and retrieval in collaborative photo collections. In: Procs of the 2nd ACM International Conference on Multimedia Retrieval (2012)

[8] Sakaki, T., et al.: Earthquake shakes Twitter users: real-time event detection by social sensors. In: Procs. of the 19th international conference on World Wide Web, pp. 851–860 (2010)

[9] Duranti, A., Goodwin, C.: Rethinking context: Language as an interactive phenomenon, vol. (11). Cambridge University Press (1992)

[10] Nguyen, T.T.V., Dao, M.S., Mattivi, R., De Natale, F.: Event Detection from Social Media: User-centric Parallel Split-n-merge and Composite Kernel. In: Procs. of ICMR 2014 Workshop on Social Events in Web Multimedia (SEWM) (2014)

[11] Marcus, M., et al.: Building a large annotated corpus of English: The Penn Treebank. Journal of Computational Linguistics 19(2), 313–330 (1993)

[12] Lewis, D., et al.: Rcv1: A new benchmark collection for text categorization research. The Journal of Machine Learning Research 5, 361–397 (2004)

[13] Vapnik, V.: The nature of statistical learning theory. Springer (2000)

[14] Vapnik, V.: Statistical learning theory (adaptive and learning systems for signal processing, communications and control series). John Wiley & Sons, A Wiley-Interscience Publication, New York (1998)

[15] Joachims, T.: A statistical learning learning model of text classification for support vector machines. In: Procs. of the 24th Annual International ACM SIGIR Conference on Research and Development in Information Retrieval, pp. 128–136 (2001)

# An Effective NMF-Based Method for Supervised Dimension Reduction

Ngo Van Linh, Nguyen Kim Anh, and Khoat Than

**Abstract.** Sparse topic modeling is a potential approach to learning meaningful hidden topics from large datasets with high dimension and complex distribution. We propose a sparse NMF-based method for supervised dimension reduction which aims to detect the particular topics of each class. Beside exploiting constraint convex combination of the hidden topics for each instance, our method separably learns among classes to extract interpretable and meaningful class topics. Our experimental results showed the effectiveness of our approach via significant criteria such as separability, interpretability, sparsity and performance in classification task of large datasets with high dimension and complex distribution. Our obtained results are highly competitive with state-of-the-art NMF-based methods.

**Keywords:** Non-negative matrix factorization, supervised dimension reduction, sparse algorithm, classification.

## 1 Introduction

Recently, progresses of data collection techniques and storage capabilities have led to an information overload in most fields and humans working in those fields have to face large datasets with high dimension and complex distribution. Such datasets present new challenges in data analysis because many algorithms in data mining cannot deal with these datasets. Therefore, dimension reduction techniques play a crucial role in machine learning and data mining. For datasets with high dimension and complex distribution, transforming the data into a lower dimension space is a practical solution [1]. Some topic models such as LSA, LDA and NMF for automatically learning hidden topics from large datasets can be potential approaches to dimension reduction.

Ngo Van Linh · Nguyen Kim Anh · Khoat Than
Hanoi University of Science and Technology, Hanoi, Vietnam
e-mail: {linhnv, anhnk, khoattq}@soict.hust.edu.vn

© Springer International Publishing Switzerland 2015                               93
V.-H. Nguyen et al. (eds.), *Knowledge and Systems Engineering,*
Advances in Intelligent Systems and Computing 326, DOI: 10.1007/978-3-319-11680-8_8

NMF [2] is one of the most popular effective dimension reduction methods. By assuming that each object is a non-negative combination of basic parts, NMF tries to approximate a data matrix $X$ by the product $F * G$ of two matrices. Up to now, by using different divergence functions [3, 4], many NMF-based methods have been proposed for representation of object in dimension reduction and classification. Moreover, some authors [5, 6, 7] use constraints and regularizations to adapt to requirements in data analysis problems with different data types. Fisher-based supervised NMF methods (FNMF) was proposed by incorporating Fisher discriminant constraints into NMF to extract more discriminative features [8]. Moreover, Lee et.al.[9] proposed a semi-supervised version of NMF (ssNMF) which performs better than FNMF in supervised feature extraction. ssNMF extracts discriminative features by incorporating the full data matrix and the class label matrix into NMF. Depending on weights, this method can behave as the standard (unsupervised) NMF or a fully-supervised NMF. Extensive numerical experiments confirmed that using label information in NMF performed better in the task of feature extraction. However, FNMF and ssNMF do not help to understand and interpret each label. And current learning methods for those supervised models are often expensive, which is problematic with large data of high dimensions.

Recently, learning sparse representation has been well studied in topic modeling to find meaningful topics with few nonzero elements [10, 11]. Specially, by adding a convexity constraint into NMF in which each data instance is a convex combination of the latent topics, the simplicial NMF [10] have more advantages than other sparse NMF such as easy interpretability, low complexity. However, due to being unsupervised method, it does not exploit label information. Therefore, it would be less preferable for supervised dimension reduction.

In fact, data instances in a class usually present some special topics that are different from other classes. Hence, class topics are discriminative and they can help us separate classes. In this paper, we propose a sparse NMF-based method for supervised dimension reduction which aims to detect the particular topics of each class. Beside exploiting constraint convex combination of the hidden topics for each instance, our method separably learns among classes. Therefore, our method not only inherits excellent advantages of simplicial NMF but also discovers the hidden topics within each class separately. Thence, our method finds out a low-dimensional space which is discriminate among classes. In addition, it can understand and interpret about each label when exploiting the hidden topics. Moreover, by using sparse algorithm with the convex combination assumption, it has lower complexity when projecting data to the low-dimensional space. Our obtained results are highly competitive with state-of-the-art NMF-based methods.

This paper is organized as follows: in Section 2, we introduce an NMF-based method for supervised dimension reduction and explain why our method is good. Extensive experiments to evaluate the effectiveness of our method are presented in Section 3. Finally, we provide some concluding remarks in Section 4.

## 2 An NMF-Based Method for Supervised Dimension Reduction

In this section, we introduce math formulation for supervised dimension reduction with non-negative matrix factorization. We assume that dataset $X = \{X_1, X_2, ..., X_M\}$ consists of $M$ $N$-dimensional vectors, where each vector represents a data instance. Let $Y = \{Y_1, Y_2, ..., Y_M\}$ be the set of $L$ labels for which $Y_i$ is the label of $X_i$. Non-negative matrix factorization (NMF) [3] algorithm decomposes matrix $X$ to two non-negative matrices $F \in \mathbb{R}_+^{M \times K}$ and $G \in \mathbb{R}_+^{K \times N}$, respectively, $X \approx F * G$, where $K$ is the number of hidden topics. The factorization turns out a optimization problem with the objective function is $||X - F * G||_F^2$, where matrices $G$ and $F$ are latent components matrix and coefficient matrix. $F$ is considered to be low dimensional representations for dataset $X$. However, unsupervised models do not consider discrimination when learning hidden topics. This new low-dimensional space based on the best reconstruction is unnecessarily good, since NMF lacks to exploit local structure of labeled data in each class in order to find out the discriminative space.

### 2.1 Non-Negative Matrix Factorization with Labels

Motivated by some advantages in simplicial NMF [10], we assume that each instance $X_m$ is the convex combination of hidden topics in supervised dimension reduction. Namely, $X_m \approx F_m * G$, subject to constraint $\sum_{k=1}^{K} F_{m,k} = 1$. Hence, we construct a method to detect the low-dimensional space which has discriminative property among classes. By using training data, our method finds out the latent components matrix $G$ that contains information about relationship between original space and new discriminative topical space. After that, $G$ is used to project all instances in original space to that discriminative topical space as unsupervised model. Projection is implemented by optimizing $F$ when fixed $G$. $F$, data in low-dimensional space, will be the input of classification algorithm. Consequently, two problems, finding out discriminative projection from training set and projecting each instance to new space, will be solved in our method.

For the first problem, it is natural to assume that data instances in a class present some special topics that help to distinguish classes. In order to reinforce the separability of classes, we represent each class to $V$ topics, where $V$ is a parameter in method. Finding good representation helps to understand and interpret the meaning of labels. We construct representation $F$ into $L$ parts corresponding to $L$ labels, $F = [F^1, F^2, ...., F^L]$, where $F^l \in \mathbb{R}_+^{M \times V}$ is characteristic matrix for label $l \in \{1, 2, ..., L\}$. Each instance $X_m$ is projected to

$$F_m = \{f_{m,1}, ..., f_{m,V}, f_{m,V+1}, ..., f_{m,2*V}, ..., f_{m,(L-1)*V+1}, ..., f_{m,L*V}\} \quad (1)$$

where $\{f_{m,l*(V-1)+1}, ..., f_{m,l*V}\}$ explains the hidden topics of label $l$.

Hence, we construct a new method namely NMF with labels (LNMFs). LNMFs is a specific form of NMF where each instance $X_m$ is the convex combination of hidden topics which are related to labels $X_m \approx \sum_{k=1}^{L*V} f_{m,k} * G_k$, subject to $\sum_{k=1}^{L*V} f_{m,k} = 1$ for every instance. Specially, to prominently display class topics, our method separately learns each class. When $X_m$ has label $l$, the particular topics of other labels are not considered. Namely, $f_{m,k} = 0$, where $k \neq \{l * (V - 1) + 1, ..., l * V\}$ if $Y_m = l$. That helps to achieve the special topics for each label.

In order to find an approximation factorization $X \approx F * G$, we consider the following optimization problem: Minimize objective function

$$D(X || F * G) = \sum_{m=1}^{M} ||X_m - F_m * G||^2 \qquad (2)$$

with respect to $F$ and $G$, subject to the constraints $F, G \geq 0$, $\sum_{k=1}^{L*V} f_{m,k} = 1$ and $f_{m,k} = 0$, where $k \neq \{l * (V - 1) + 1, ..., l * V\}$ if $Y_m = l$.

For solving LNMFs, we employ iterative multiplicative updates consisting of two steps: an inference step and a learning step. In the inference step, we optimize $F$ given $G$, to find out discriminative representation of data $X$. In the learning step, we optimize latent components matrix $G$ given $F$. We illustrate to find out discriminative projection from training data in algorithm 1.

**Require:** Dataset $X = \{X_1, X_2, ..., X_M\}$, labeled set $Y = \{Y_1, Y_2, ..., Y_M\}$, and
    parameter $V$
**Ensure:** Latent components matrix $G$
    Initialize randomly $G$
    **repeat**
        **Inference step:**
        Each instance $X_m$ in training set, optimize $F_m$ for given $G$ in $D(X_m || F_m * G)$,
        subject to the constraints $F_m \geq 0$, $\sum_{k=1}^{L*V} f_{m,k} = 1$, and $f_{m,k} = 0$, where
        $k \neq \{l * (V - 1) + 1, ..., l * V\}$ if $Y_m = l$.
        **Learning step:**
        Optimize $G$ for given $F$
    **until** Convergence

**Algorithm 2.** LNMFs

The second problem, after detecting the low-dimensional space and the projection, all instances are projected on new space. The result $F$ of the projection is low dimensional representations for dataset $X$. By fixed G, objective function is

$$D(X || F * G) = \sum_{m=1}^{M} ||X_m - F_m * G||^2 \qquad (3)$$

subject to the constraints vector $f_m \in \mathbb{R}_+^{L*V}$, $\sum_{k=1}^{L*V} f_{m,k} = 1$, where $X_m \in X = \{X_{train}, X_{test}\}$.

Problem becomes inference for each instance. Moreover, projection needs to ensure low complexity when projecting each instance to new space. It is obviously solved in next section. We design general algorithm with LNMFs and classification (Algorithm 2).

## 2.2 Inference Step

In this section, we focus to the optimization problem in the inference step for each instance. The learning step can be easily optimized by solving non-negative least squares constraints problem [12]. We concentrate on solving inference step in which we gain some interesting contributions. The inference step for each instance in algorithm 1 and 2 is convex optimization problem. Concisely, $F_m$ is replaced by $f_m$. For each labeled instance, objective function can be rewritten as

$$D(f_m) = \sum_{n=1}^{N} (X_{m,n} - \sum_{k=1}^{L*V} f_{m,k} * G_{k,n})^2 \tag{4}$$

subject to the constraints vector $f_m \in \mathbb{R}_+^{L*V}$, $\sum_{k=1}^{L*V} f_{m,k} = 1$, and $f_{m,k} = 0$, where $X_m$ belongs training set and $k \neq \{l * (V - 1) + 1, ..., l * V\}$ if $Y_m = l$ in Algorithm 1 and

$$D(f_m) = \sum_{n=1}^{N} (X_{m,n} - \sum_{k=1}^{L*V} f_{m,k} * G_{k,n})^2 \tag{5}$$

subject to the constraints vector $f_m \in \mathbb{R}_+^{L*V}$, $\sum_{k=1}^{L*V} f_{m,k} = 1$, where $X_m \in X = \{X_{train}, X_{test}\}$ in Algorithm 2.

In the inference step for both Algorithm 1 and 2, we project each instance to topical space. In fact, the number of topics that is talked into a instance is not large. It is suitable when the representation of instance in topical space is sparse. Therefore, we approach optimization $F$ in inference step following sparse method.

**Require:** Dataset $X = \{X_{train}, X_{test}\}$, labeled set $Y = \{Y_{train}\}$, and parameter $V$
**Ensure:** $Y_{test}$
　　**Learn discriminative space:**
　　Get $G$ from LNMFs($X_{train}, Y_{train}, V$)
　　**Project dataset to new space:**
　　Each instance $X_m$ in dataset, optimize $F_m$ for given $G$ in $D(X_m || F_m * G)$,
　　subject to the constraints $F_m \geq 0$, $\sum_{k=1}^{L*V} f_{m,k} = 1$, where $F = \{F_{train}, F_{test}\}$.
　　**Classify:**
　　$Y_{test} = ClassificationAlgorithm(F_{train}, Y_{train}, F_{test})$
　　　　**Algorithm 3.** LNMFs for Classification

Recently, sparse approximation algorithms has been various excellent results [13, 14, 15, 10]. The Frank-Wolfe algorithm [13] has many attractive qualities. That algorithm solves the optimization problem, convex minimization over simplex,

$$f^* = \arg\min_{f \in \Delta} D(f) \tag{6}$$

where $D(f) : \mathbb{R}_+^K \rightarrow \mathbb{R}$ is convex objective function in the unit simplex $\Delta$, namely, $f_k \geq 0$ and $\sum_{k=1}^{K} f_k = 1$. It achieves a linear rate of convergence and effective solution. It can control the sparsity by trading off sparsity against quality.

Specially, it is worth noting that the objective function of LNMFs with its constraints is suitable to adapt for the Frank-Wolfe algorithm. That is significant advantage when comparing with other supervised non-negative matrix factorization models. The inference step, using the Frank-Wolfe algorithm, is detailed in the Algorithm 3 which is designed to implement in Algorithm 1. Similarly, it is easy to design the inference step in Algorithm 2.

**Require:** Objective function $D(f_m)$
**Ensure:** $f_m$ that minimizes $D(f_m)$ over $\Delta$: $f \geq 0$, $\sum_{k=1}^{L*V} f_{m,k} = 1$, and $f_{m,k} = 0$,
    where $k \neq \{l * (V-1) + 1, ..., l * V\}$ if $Y_m = l$
    Initialize $f_m^0 = argmin_i D(e_i)$, where $i \in \{l * (V-1) + 1, ..., l * V\}$ and $e_i$ is the $i$th
    elementary vector.
    **for** $t = 1, 2...,$ **do**
        $i' = argmin_i \frac{\delta D(f_m^t)}{\delta f_{m,i}^t}$, where $i \in \{l * (V-1) + 1, ..., l * V\}$.
        $\alpha' = argmin_{\alpha \in [0,1]} D(f_m^t + \alpha * (e_{i'} - f_m^t))$
        $f_m^{t+1} = f_m^t + \alpha' * (e_{i'} - f_m^t))$
    **end for**

**Algorithm 4.** Inference with the Frank-Wolfe algorithm

## 2.3   Why Is the Method Good?

In this section, we explain why our method is good. We emphasize that the inference algorithm in LNMFs have some enjoyable qualities that lack of other NMF method. Those are advantage to expect good result.

### The Discriminative Property of Data in the Topical Space

By using label information in training set, LNMFs detects the difference topics which are particular for each label which is separably presented in $V$ specific topics. Those topics conserve the characteristic of each class and avoid noisy information from other classes. It helps to reduce overlap among classes. Therefore, topical space learned by our method is discriminative among classes and projecting data into the discriminative space achieves good representations. Furthermore, we consider whether similarity among labels influences on our algorithm. That is not difficult problem in LNMFs. it can assess similarity among topics in diverse classes based on latent components $G$. Similarity between topic $i$ and $j$ is computed by similarity the $i$th and the $j$th row in $G$.

### Understanding and Interpretability

Those are prominent property of LNMFs comparing with other NMF models. By adding constraint, we can discover topics which are discriminative for each label. In each instance, the sum of topical coefficients always equals to 1 thanks to which it is a probabilistic distribution over the special topic via labels. According to that, the contribution of the special topics on instances is understood by values of coefficients. Moreover, the latent component $[G]_{L*V \times N}$ shows connection between original features and special topics via labels. Therefore, it is easy to find out some original features associated with each label in order to understand meaning of that label. For example, we easily find which words express well the meaning of a label in text dataset.

## Low Complexity in the Inference Step

Low complexity is important criterion to evaluate dimension reduction methods. By exploiting the Frank-Wolfe algorithm, our method has lower complexity when comparing other NMF-based methods. We evaluate the complexity of the inference algorithm in algorithm 1 and 2. For each instance,

$$\frac{\delta D(f_m)}{\delta f_m} = 2(f_m * G - X_m) * G^T \tag{7}$$

According to [10], complexity of inference algorithm 1 and 2 are $O(T.M.(V.S(N) + N))$ and $O(T.M.(L.V.S(N) + N))$, where $S(N)$ is the number of non-zero elements in the latent component $G$ after $T$ iterations whereas the inference complexities in NMF requires $O(M.N^2)$. When $S(N) \ll N$, it is advantage to compare other methods. Moreover, our algorithm is easy to parallel and distribute when the inference algorithm implements on each instance.

## Fast convergence and Effective Solution of the Inference Algorithm

By adding the constraint of convex combination and flexibly exploiting the Frank-Wolfe algorithm, the inference algorithm converges at a linear rate. Moreover, at each iteration, the algorithm finds a good approximate solution. It is detailed by theorem [13].

**Theorem 1:** *Let $f$ be a twice differentiable convex function over $\Delta$ and denote $C_f = \frac{1}{2} \sup_{y,z \subset \Delta; \bar{y} \in [y,z]} (y \quad z) \nabla^2 f(\bar{y}).(y-z)^T$. After $t$ iterations, the Frank-Wolfe algorithm will find an approximate solution $x_t$ with at most $(t+1)$ non-zeros coefficients which satisfy*

$$f(x_t) - \min_{x \in \Delta} f(x) \leq \frac{4C_f}{t+1} \tag{8}$$

## Sparsity

It is worth noting that the content of each data instance only presents several topics. It means that representation in topical space always has sparse property. Dimension reduction method is good if it finds out sparse representation, in which each non-zero component play a important role to explain data. On the other hand, theorem 1 shows that it is easy to directly control the number of non-zeros in topical representations by trading off sparsity against quality in the Frank-Wolfe algorithm. After $t$ iterations at most $t+1$ out of $L * V$ topics in $f_m$ are non-zero. In each iteration, the Frank-Wolfe algorithm either increases a coefficient of a old topic or finds out a new topic which is appropriate for that instance. Therefore, the Frank-Wolfe algorithm is suitable for finding to optimization solution that ensures both quality and sparsity.

# 3  Experiments

In this section, we describe some experiments to evaluate the effectiveness of our method in supervised dimension reduction. We will show some clear evidences about the discriminative property and sparse property of data in the topical space. Our method is compared with unsupervised and supervised dimension reduction methods which are based on non-negative matrix factorization.

- **Simplicial NMF** [10] is unsupervised NMF which finds out a sparse solution by using the Frank-Wolfe algorithm.
- **Supervised NMF (sNMF)** separably learns among classes the same our method. However, it lacks of convex combination constraint of LNMFs and uses gradient descent in order to solve optimization problem as NMF.
- **Semi-supervised NMF (ssNMF)** [9] exploits label information by different way. By jointly incorporating the data matrix and the label matrix into NMF, they extracts more discriminative topical space. The tradeoff parameter between the data matrix and the label matrix of ssNMF is selected 1 the same their experiments.

In simplicial NMF and ssNMF, parameter $K$ is the number of hidden topics as in NMF. In sNMF and LNMFs, parameter $V$ is the number of hidden topics which explain to each label.

We used four high dimensional text datasets from the UCI repository data. A summary of all the datasets is shown in table 1. The balance of a dataset is defined as the ratio of the number of documents in the smallest class to the number of documents in the largest class. So a value close to 1(0) indicates a very (un)balanced dataset. In our experiments, we randomly select 80 percent of dataset to train in methods.

**Table 1** Summary of text datasets (for each dataset, $nd$ is the total number of documents, $nw$ is the total number of terms, $k$ is the number of classes)

| Data | $nd$ | $nw$ | $k$ | Balance | training set |
|------|------|------|-----|---------|--------------|
| hitech | 2301 | 10080 | 6 | 0.1924 | 1840 |
| la1 | 3204 | 31472 | 6 | 0.290 | 2563 |
| la2 | 3075 | 31472 | 6 | 0.32670 | 2460 |
| ohscal | 11161 | 11465 | 10 | 0.430 | 8929 |

## The Discrimination Property of Data in the Topical Space

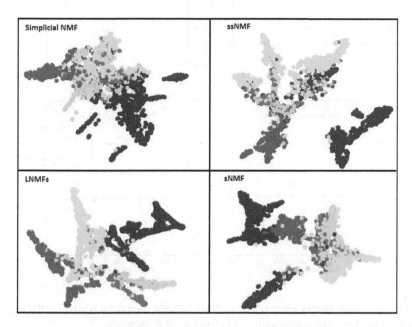

**Fig. 1** Discrimination of the low-dimensional space learned by simplicial NMF, sNMF, ss-NMF, LNMFs. Hitech was the dataset for visualization. Data points in the same class have the same color. These embeddings were done with t-SNE [16]

Firstly, in order to illustrate how class separation in topical space, we will visualize the dataset which is projected by those NMF methods. In this experiment, dataset hitech and the parameters $K = 90$, $V = 5$ are used to learn the new space. Result is shown in Fig.1. Observationally, the projection of simplicial NMF confuses the dataset among classes. Due to lacking of label information, it has been projected a number of data onto incorrect classes. Conversely, by exploiting label information, other projections separate more cleanly among classes. Specially, the projection of LNMFs is most discriminative. Because of the sparse property of representation in topical space, LNMFs only exploits sparse algorithm to find out special topics via each label. Therefore, LNMFs finds out the good representation which reduces noise. Albeit sNMF and ssNMF use label information, the representations of dataset in topical space are dense. For this reason, the representations contain noisy that influences to visualization.

Secondly, we use quality of classification to quantify the goodness of those dimensional reduction methods. The representation of dataset in low-dimensional topical space is input for classification method as Algorithm 2. Multiclass SVM [17] with sigmoid kernel fuction is selected to classification. We run those algorithms with difference parameters $K$ and $V$. Due to differently meaning of those, we depict

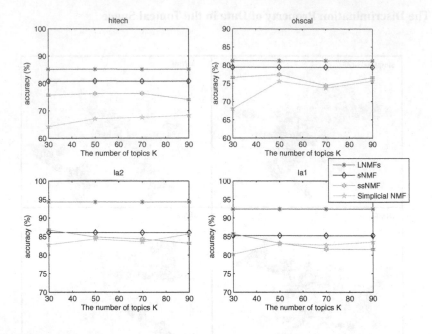

**Fig. 2** Classification quality when LNMFs, sNMF, ssNMF, simplicial NMF play the role of dimension reduction. The parameter $V = 5$ is fixed when $K$ increases.

two figures to illustrate classification quality (Fig.2 and Fig.3). Observing Fig.2, our method consistently achieved the best performance. Previously, discussion exposes that dataset which are projected by LNMFs owns properties: class separation and reduction noise. Due to exploiting labeled data and unlabeled data in learning topical space, ssNMF is not as good as sNMF, in classification. Moreover, Fig.3 shows that LNMFs is more stable than sNMF when parameter $V$ changes. It is explained that the number of topics in each label is not large and LNMFs easily detects those topics by greedy approximation in the Frank-Wolfe algorithm. In each iteration, it either only selects at most a topic which is the most potential or reinforces the value of old topics. The dimensional coefficients corresponding to noisy topics have little value or zero.

## Sparsity

In order to show the sparsity of solutions, we compute the percentage of zero coefficients of representation in topical space. Fig.4 shows that the sparsity of representation increases following the number of topics in each label. Albeit the number of topics increases, our method finds out suitable topics whereas noisy topics are eliminated. It is a cause why classification result in our algorithm is stable.

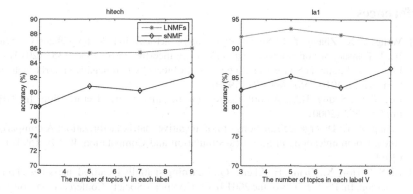

**Fig. 3** Classification quality when LNMFs, sNMF play the role of dimension reduction. The parameter $V$ increases.

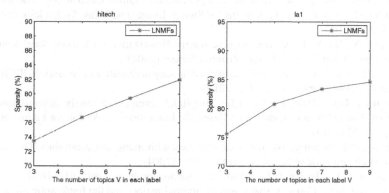

**Fig. 4** The sparsity of LNMFs as the parameter $V$ increases

## 4 Conclusion

In this paper. we have proposed a new formulation for supervised NMF, NMF via labels (LNMFs). This formulation considers NMF as topic model, in which each data instance is modeled as a convex combination of latent topics. By detecting the special topics of labels, our method can understand and interpret them. Moreover, we theoretically proved that our algorithms have low complexity in inference and sparsity in representation. Our experimental results showed the effectiveness of our approach via significant criteria such as separability, interpretability, sparsity and performance in classification task of large datasets with high dimension and complex distribution. Our obtained results are highly competitive with state-of-the-art NMF-based methods.

# References

[1] Wang, Y.-X., Zhang, Y.-J.: Nonnegative matrix factorization: A comprehensive review. IEEE Transactions on Knowledge and Data Engineering 25(6), 1336–1353 (2013)

[2] Lee, D.D., Seung, H.S.: Learning the parts of objects by non-negative matrix factorization. Nature 401(6755), 788–791 (1999)

[3] Lee, D.D., Seung, H.S.: Algorithms for non-negative matrix factorization. In: NIPS, pp. 556–562 (2000)

[4] Zhang, Z.-Y.: Divergence functions of non negative matrix factorization: A comparison study. Communications in Statistics-Simulation and Computation 40(10), 1594–1612 (2011)

[5] Li, S.Z., Hou, X., Zhang, H., Cheng, Q.: Learning spatially localized, parts-based representation. In: Proceedings of the 2001 IEEE Computer Society Conference on Computer Vision and Pattern Recognition, CVPR 2001, vol. 1, pp. I-207. IEEE (2001)

[6] Choi, S.: Algorithms for orthogonal nonnegative matrix factorization. In: IEEE International Joint Conference on Neural Networks, IJCNN 2008 (IEEE World Congress on Computational Intelligence), pp. 1828–1832. IEEE (2008)

[7] Li, H., Adal, T., Wang, W., Emge, D., Cichocki, A.: Non-negative matrix factorization with orthogonality constraints and its application to raman spectroscopy. The Journal of VLSI Signal Processing Systems for Signal, Image, and Video Technology 48(1-2), 83–97 (2007)

[8] Wang, Y., Jia, Y.: Fisher non-negative matrix factorization for learning local features. In: Proc. Asian Conf. on Comp. Vision. Citeseer (2004)

[9] Lee, H., Yoo, J., Choi, S.: Semi-supervised nonnegative matrix factorization. IEEE Signal Processing Letters 17(1), 4–7 (2010)

[10] Nguyen, D.K., Than, K., Ho, T.B.: Simplicial nonnegative matrix factorization. In: IEEE International Conference on Research, Innovation and Vision for Future, RIVF, pp. 47–52 (2013)

[11] Hoyer, P.O.: Non-negative matrix factorization with sparseness constraints. The Journal of Machine Learning Research 5, 1457–1469 (2004)

[12] Lawson, C.L., Hanson, R.J.: Solving least squares problems, vol. 161. SIAM (1974)

[13] Clarkson, K.L.: Coresets, sparse greedy approximation, and the frank-wolfe algorithm. ACM Transactions on Algorithms (TALG) 6(4), 63 (2010)

[14] Than, K., Ho, T.B.: Fully sparse topic models. In: Flach, P.A., De Bie, T., Cristianini, N. (eds.) ECML PKDD 2012, Part I. LNCS, vol. 7523, pp. 490–505. Springer, Heidelberg (2012)

[15] Than, K., Ho, T.B., Nguyen, D.K., Khanh, P.N.: Supervised dimension reduction with topic models. Journal of Machine Learning Research - Proceedings Track 25, 395–410 (2012)

[16] Van der Maaten, L., Hinton, G.: Visualizing data using t-sne. Journal of Machine Learning Research 9(11) (2008)

[17] Keerthi, S.S., Sundararajan, S., Chang, K.-W., Hsieh, C.-J., Lin, C.-J.: A sequential dual method for large scale multi-class linear svms. In: Proceedings of the 14th ACM SIGKDD International Conference on Knowledge Discovery and Data Mining, pp. 408–416. ACM (2008)

# Facial Expression Recognition Using Pyramid Local Phase Quantization Descriptor

Anh Vo* and Ngoc Quoc Ly

**Abstract.** Facial expression recognition is a challenging and interesting problem. It has many potential and important applications in data-driven animation, human computer interaction (HCI), social robots, deceit detection and behavior monitoring. In this paper, we present the novel descriptors Pyramid local phase quantization (PLPQ). The effective of our proposed descriptor is evaluated by facial expressions recognition very efficiently and with high accuracy. On the other hand, the proposed framework extracts texture features in a pyramidal fashion only from the perceptual salient region of the face thereby our proposed framework achieved reduction in computation time of feature extraction and improved accuracy. There with the proposed framework achieved accuracy of 96.7% on extended Cohn-Kanade (CK+) posed facial expression database for six basic emotions and exceed the state-of–the-art methods for expression recognition using texture features.

**Keywords:** Facial Expression Recognition, Local Phase Quantization, Support Vector Machine, Salient Facial Region.

## 1  Introduction

The early 1970s, facial expression analysis has attracted some attention in the community of psychology. In the communities of computer vision and pattern recognition, most of automatic facial expression analysis work only focused on identifying

Anh Vo
Faculty of Information Technology, Ton Duc Thang University, Ho Chi Minh City, Viet Nam
e-mail: vhanh@it.tdt.edu.vn

Ngoc Quoc Ly
Faculty of Information Technology, University of Science, Ho Chi Minh City,
Viet Nam
e-mail: lqngoc@fit.hcmus.edu.vn

* Corresponding author.

© Springer International Publishing Switzerland 2015                               105
V.-H. Nguyen et al. (eds.), *Knowledge and Systems Engineering,*
Advances in Intelligent Systems and Computing 326, DOI: 10.1007/978-3-319-11680-8_9

an input facial image or sequence as one of six basic emotions or recognition of expression thought facial action coding system (FACS). Recognize facial expression in real-time with high reliability is a difficult task such as variability in pose, illumination, blur and the way people show expression across cultures are some of the parameters that make this task difficult. The general approach to facial expression recognition system consists of three steps (Fig 1): face detection, feature extraction and facial expression recognition.

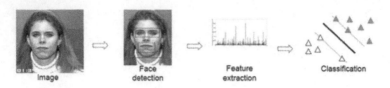

**Fig. 1** Basic structure of facial expression recognition system pipeline

The essential step for successful facial expression recognition is how to construct an effective facial representation from face images. According to we noticed the more information of facial expression is appeared in the eyes, eyebrows and mouth region than other region so that region is called "salient region" where "salient" means most noticeable or most important. Conclusions drawn from the psycho-visual experimental study suggests that for some expression (happiness, sadness, surprise) only one facial region is salient, for example facial of surprise expression has two salient regions but mouth is the most important region according to the human visual system (HVS) while for other expressions (anger, fear, disgust) two facial region are salient (e.g. facial of anger expression become difficult to determine if only from mouth region) [9]. The facial expression recognition framework in this paper reduces the feature vector dimensionality whereby suitable for real-time application because it minimized computational complexity.

Steps of the proposed framework are as follows:

Step 1: The first localizes salient facial regions using Viola-Jones object detection (Viola and Jones) [8]. This algorithm is selected because it is considered the fasted and most accurate pattern recognition method for facial region detection [5].

Step 2: Then, the feature from the mouth region extracted by PLPQ descriptors, PLPQ feature vector of 2560 dimensions . The first classifier performed on the basic of extract features to make two groups of emotion. First group emotion include happiness, sadness and surprise with one perceptual salient while the second group emotion include anger, fear and disgust with two perceptual salient.

Step 3: If input vector is classified in the first group emotion, then it is classifier either as happiness, sadness or surprise by feature vector extracted PLPQ from mouth region.

Step 4: If input vector is classified in the second group emotion, the framework ex-tract PLPQ feature from eyes region and concatenates them with the already extracted PLPQ features from the mouth region, feature vector of 5120 dimensions.

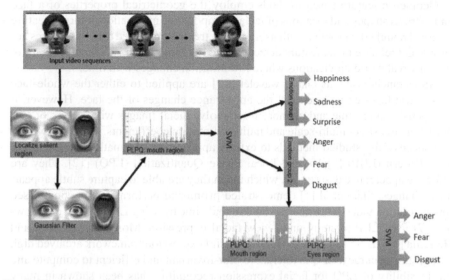

**Fig. 2** The outline of the framework [9] use the proposed descriptor

Then, concatenates feature is fed to the second classifier for the final classifier.

In specific, the main contributions of this paper are summarized as follows: (a) we propose pyramid local phase quantization (PLPQ). The proposed descriptor is a simple and computationally efficient which it show improved performance for facial expression recognition in real–time. On the other hand, it reduce the intrinsic effect of noise even so it robust to image blurring and shift. (b) We proposed a framework which applied our proposed descriptor only from the visually salient facial region and it represents stimuli by its local texture (LPQ) and the spatial layout of the texture. Finally, the experiment results show that PLPQ feature provide just as good or better performance so are very promising for real-world applications otherwise it prove the system based on LPQ generally achieve higher accuracy rate than those based on LBPs not only for the problem of FACS Action Unit analysis [2] but also for the problem of basic emotion recognition. This paper is structured as follows: Section 2 reviews related works. The basic principle of static appearance descriptors Local Phase Quantization, and our proposed extensions PLPQ were presented in Section 3. Section 4 presents the classification technique used in this work. Experiment and results are given in Section 5. Section 6 presents the conclusion.

## 2 Related Works

There are two approaches to extract facial features: geometric feature-based methods or appearance-based methods.

Geometric feature based methods employ the geometrical properties of a face and extracts shapes and locations of facial components information to form a feature vector. Thought the problem with using geometric feature-based usually requires accurate and reliable facial feature detection and tracking which is difficult to achieve in many real world applications where illumination changes with time.

Appearance-based, as Gabor wavelet [14] are applied to either the whole-face or specific face-region to extract the appearance changes of the face. However, it is expensive both time and memory to convolve facial images with a bank of Gabor filters to extract multi-scale and multi orientation coefficients. The other widely and successfully studied methods to extract appearance information are Local Binary Pattern (LBP) [7][10] and Local Phase Quantization (LPQ) [12]. They are effective appearance descriptors, which mean they are able to capture subtle appearance changes. Shan et al [1] demonstrated promising performance in compressed low-resolution video sequence captured in real-time by using LBP. Moore and Bowden [7] used LBP to study multi-view facial expression. Moreover, Ojansivu and Heikkila [12] using LPQ feature for their facial expression framework archived high recognition accuracy. LPQ is tolerant to blur-invariant and efficient to compute and the feasibility of LPQ for facial expression recognition has been shown in many existing works and Jiang et al. [1] proved in theirs experiment that AU detection systems based on LPQ generally achieve higher accuracy rate than those based on LBPs. In the other hand, there are a number of works extending the conventional descriptors to the pyramid transform domain. Yang et al. [3] proposed a new pyramid Gabor features during Khan et al [9] applied a pyramid Local Binary Pattern (PLBP) for emotion recognition. Recently, Jiang et al. [2] using Block-based Pyramid Appearance descriptors for the problem of Facial Action Detection, they are represented the face image in an image pyramid by different spatial resolutions. For each pyramid level, the image is divided into regions. The dense appearance descriptor features extracted from each region, and each level of the pyramid, are concatenated into a single, spatially enhanced feature histogram. However the face image is represented in an image pyramid by different spatial resolution make expensive computation time so that in this paper, we proposed PLPQ descriptor only from the salient region of the face image to reduce computation time.

## 3 Feature Extraction

### 3.1 Gaussian Filter

Gaussian filter is the basic filter are used in image processing because they have a property that their support in the frequency domain. This comes about from the Gaussian being its own Fourier Transform, Gaussian filter that is used to blur image and remove noise. The idea of Gaussian filter use this 2-D distribution as a point-spread function, and this is achieved by convolution, the Gaussian function in 2D defined as

$$G(x,y) = \frac{1}{2\pi\sigma^2} e^{-\frac{x^2+y^2}{2\sigma^2}} \tag{1}$$

Where $\sigma$ is the standard deviation of the distribution, G(x,y) is the Gaussian distribution in 2D.

## 3.2 Local Phase Quantization

The Local Phase Quantization (LPQ) operator was proposed by Ojansivu and Heikkila [12] as a texture descriptor that is robust to image blurring. The descriptor uses local phase information extracted using the 2-D DFT, a short-term Fourier transform (STFT) computed over a rectangular M-by-M neighborhood $N_x$ at each pixel position x of the image f(x) defined by

$$F(u,x) = \sum_{y \in N_x} f(x-y) e^{-j2\pi yu^T} = w_u^T f_x \tag{2}$$

Where $w$ is the basic vector of the 2-D DFT at frequency u and $f_x$ is the vector containing all $M^2$ samples from $N_x$. The STFT is efficiently evaluated for all image position $x \in \{x_1, \dot{,} x_N\}$ using simply 1-D convolutions for the rows and columns successively. The local Fourier coefficients are computed at four frequency points: $u_1 = [a,0]^T, u_2 = [0,a]^T, u_3 = [a,a]^T, u_4 = [a,-a]^T$, where a is a sufficiently small scalar. For each pixel position this results in a vector $F_x = [F(u_1,x), F(u_2,x), F(u_3,x), F(u_4,x)]$. The phase information in the Fourier coefficients is recorded by examining the signs of the real and imaginary parts of each component in $F_x$. This is done by using a simple scalar quantizer.

$$q_x = \begin{cases} 1 & \text{if } g_x \leq 0 \text{ is true} \\ 0 & \text{otherwise} \end{cases} \tag{3}$$

Where $g_j(x)$ is the $j^{th}$ component of the vector $F_x = [Re\{F_x\}, Im\{F_x\}]$. The resulting eight bit binary coefficients $g_j$ are represented as integers using binary coding:

$$f(x)_{LPQ} = \sum_{j=1}^{8} q_j 2^{j-1} \tag{4}$$

As a result, a histogram of these values from all positions is composed and used as a 256-dimensional feature vector in classification. It can be shown that in quantization the information is maximally preserved if the samples to be quantized are statistically independent.

## 3.3 Pyramid Local Phase Quantization

We propose a novel descriptors PLPQ (Pyramid of local phase quantization) to extended LPQ operator. PLPQ descriptors can be represented local texture and the spatial layout of texture information for face representation.

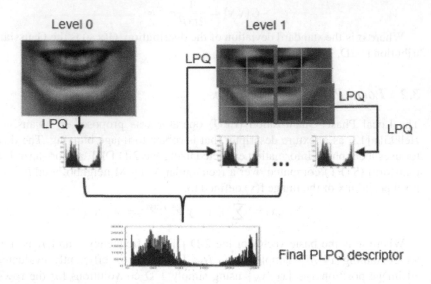

**Fig. 3** The pyramid local phase quantization representation

In the Fig 3, the visually salient facial region is represented in a pyramid image by dividing at level 1 has $2^l$ and overlapped finer spatial subregion $(R_0, ;R_{m-1})$. It can be observed Fig 2. Histograms extracted from each region at the same levels are concatenated into a final histogram. It can be defined as

$$H_{ij} = \sum_l \sum_{xy} I\{f_l(x,y) = i\}I((x,y) \in R_l) \qquad (5)$$

Where $l = 0,..,m-1, i = 0,...,n-1$; $n$ is the number of different labels produced by the LPQ operator and $I(A) = \begin{cases} 1 \ if\ A\ is\ true \\ 0\ otherwise \end{cases}$

To improve efficiency texture information of the face we associated Gaussian filter and LPQ/ PLPQ descriptor. Steps of the processing as following: First, the enter image is applied Gaussian filter. Second, the image result is extracted LPQ/PLPQ features. In our experiment, we choose σ=2 and window size is 5x5, with PLPQ descriptor, we selected level l=1 and overlap is 0.5 at the same level as we created pyramid up to level 0 and extracted 256 LPQ feature dimension, at the level 1, we create 9 sub-region image and extracted 256 LPQ feature dimension. Finally, we have a feature with length of 2560 (10x256) only from one of the visually salient facial region. On the other hand, if we use LPQ, a final feature vector have length of 256 from one of the salient facial region.

## 4 Classification

A previous successful technique to facial expression classification is Support Vector Machine, so in this work, we adopted SVM as alternative classifiers for expression recognition with six basic emotions. As a powerful machine learning technique for data classification, SVM performs an implicit mapping of data into a higher dimensional feature space, and then finds a linear separating hyper-plane with the maximal margin to separate data in this higher dimensional space. Given a training set of labeled examples $(x_i, y_i), i = 1, ..., l$ where $x_i \in R^n$ and $y_i \in \{1, -1\}$, a new test example $x$ is classified by the following function:

$$f(x) = sgn(\sum_{i=1}^{l} \alpha_i y_i K(x_i, x) + b) \tag{6}$$

Where $\alpha_i$ are Lagrange multipliers of a dual optimization problem that describe the separating hyper-plane, $K(.,.)$ is a kernel function, and $b$ is threshold parameter of hyper-plane. The training sample $x_i$ with $\alpha_i > 0$ is called support vectors, and SVM finds the hyper-lane that maximizes the distance between the support vectors and the hyper-plane. Given a non-linear mapping $\phi$ that embeds the input data into the high dimensional space, kernels have the form of $K(x_i, x_j) = \langle \phi(x_i), \phi(x_j) \rangle$. SVM allows domain-specific selection of the kernel function. The new kernel being proposed, the most frequency used kernel functions are the linear, polynomial, and Radial basic function (RBF) kernels. SVM makes binary decisions, so the multiclass classification here is accomplished by using the one-against-rest technique, which trains binary classifiers to discriminate one expression from all others, and outputs the class with the largest output of binary classification. The kernels we compared are following:

Linear kernel: $K(x_i, x_j) = x_i . x_j$

Polynomial kernel: $K(x_i, x_j) = (x_i . x_j)^d$

RBF kernel: $K(x_i, x_j) = exp(-\frac{\|x_i - x_j\|^d}{2\sigma^2})$

## 5 Experiment and Results

In our experiments, we evaluated 309 image sequences from the extended Cohn-Kanade (CK+) database [4] which have one of the six universal expression. Observers include both male and female aging from 20 to 45 years with normal or corrected to normal vision. All the observer were naïve to the purpose of an experiment. Most of the methods in literature report their performance on this database, so this experiment could be considered as the benchmark experiment for facial expression recognition framework.

The proposed descriptor have compared memory and time cost of feature extraction processing in Table 1. All the algorithm are run in Matlab environment on a

Window 64 bit with CPU Intel Core (TM) i7 2.0 Hz and Window 8.1 having 4GB
of RAM. The average execution time for the feature extraction is 0.02s. Generally
the drawback of using approach is that it increases the size of the feature dimension
and increases the computational cost in Jiang et al. [2] create pyramid before applying LPQ operator by down sampling original image i.e. scale-space representation
and then dividing region in each pyramid level. However, that makes the block size
in each layer which capture too detailed information lead to increases the size of
the feature dimension and increases the computational cost, whereas we propose
to create the spatial pyramid by dividing the stimuli into finer spatial sub-region by
itera-tively doubling the number of divisions in each dimension beside in upper level
0, we use overlap to capture local appearance information while maintaining its shift
and scale robustness, therefore our proposed descriptor is computationally more efficient and reduces memory consumption because it do not require to store same
image in different resolution. We achieve the task of multi-scale analysis and recognition rate higher other earlier proposed methods in Table 3 although the proposed
descriptor is feature dimension and time for the feature extraction higher Khan [9].

**Table 1** Comparison of time and memory consumption

|                                | Jiang et al. (2012) | Shan et al (2009)b[1]. | Bartlett et al. (2003)[16] | Khan (2013) [9] | Ours (PLPQ) |
|--------------------------------|---------------------|------------------------|----------------------------|-----------------|-------------|
| Memory (feature dimension)     | 12,800              | 42,650                 | 92,160                     | 590             | 5120        |
| Time (feature extraction time) | 0.45s               | 30s                    | –                          | 0.01s           | 0.02s       |

**Table 2** Comparison between LBP, PLBP and PLPQ feature based on different SVM kernels

| Kernel      | Shan et al.(2009)[1] (LBP) | Khan (2013) [9] (PLBP) | Ours (PLPQ) |
|-------------|----------------------------|------------------------|-------------|
| Linear      | 91.5%                      | 93.5%                  | 95%         |
| Polynomial  | 91.5%                      | 94.7%                  | 94.7%       |
| Rbf         | 92.6%                      | 94.9%                  | 96.7%       |

The best result were reached using 10-fold cross validation technique and grid
search to find optimized parameter for C and $\gamma$ which reported results of the experiment is obtained with C = 300 and $\gamma$ = 1. We performed experiments with the

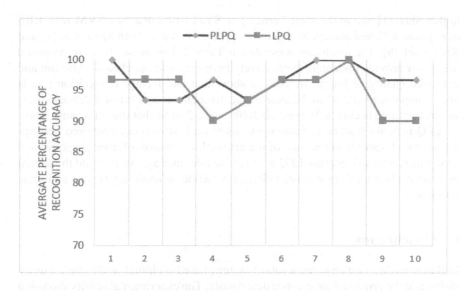

**Fig. 4** Recognition performance of based PLPQ, LPQ

**Table 3** Comparison with the state of the art methods for posed expression

|  | Sequence num | Class num | Performance measure | Recog. rate (%) |
|---|---|---|---|---|
| Zhao and Pietikainen [15] | 374 | 6 | 10 Fold | 96.26 |
| Kotsia et al (2008) [6] | 374 | 6 | 5 Fold | 94.5 |
| Tian (2004) [11] | 375 | 6 | 5 Fold | 93.8 |
| Yang et al (a) [13] | 352 | 6 | 66 split | 92.3 |
| Ours method (LPQ) | 309 | 6 | 10 Fold | 94.7 |
| Ours method (PLPQ) | 309 | 6 | 10 Fold | 96.7 |

Gaussian filter use $\sigma$ parameter is 2 and window size is 5x5. All image sequences are divided into ten subsets. At each iteration, nine of the ten subsets are used to create training set and one is testing. Therefore no data from a subject appears in both the training and testing set. In Fig 4 we compared LPQ, PLPQ descriptor and noticed PLPQ descriptor have best recognition rate. Beside to compare with Shan et al. [1] and Khan [9] work, we conducted experiments on 6 class basic expression recognition using SVM with different kernel. Our system obtained average recognition rate is 94.7% for SVM polynomial linear kernel is equal to Khan [9] and higher Shan [1] and in the same kernel. For SVM with linear and SVM with RBF our system achieved average recognition accuracy is higher both Shan et al [1] and Khan work [9]. The results are presented in Table 2. This proves that our proposed descriptor provide discriminate effectively between different class of emotion and enhanced quality of image. In Table 3 show that our average recognition rate is 96.7% improved better than the state of the art methods in term of average expression recognition accuracy. In general, Table 2 and 3 show that the framework based on LPQ is better than other framework based on LBP and our proposed descriptor improved better than the state of the art methods in term of average expression recognition accuracy because LPQ features are tolerance against blur and the spatial pyramid capture local appearance information while maintaining its shift and scale robustness.

## 6 Conclusions

In this work, we implemented a robust system based on initial study about human vision and the proposed appearance descriptors. The experimental results show that the proposed system using the appearance descriptor based on LPQ appearance descriptors generally achieves higher accuracy rate than the systems based on LBP appearance descriptors for the problem of the basic emotion recognition. Extracted features using the proposed Pyramidal Local Phase Quantization operator and Gaussian filter have strong discriminative as recognition result for six emotions. All in the experimental result show that our descriptor PLPQ outperformed for the problem of basic emotion recognition. On the other hand, our system reduced feature extraction computational time by using localizes salient facial regions and making two groups of expression so that can be used for real-time applications. In the future, we plan to improve the performance of the proposed frame-work by concatenated dynamic information and the appearance descriptors beside we investigate our system is efficiently with other database.

## References

[1] Shan, C., Gong, S., McOwan, P.: Facial expression recognition based on local binary patterns: A comprehensive study. Image and Vision Computing 27(6), 803–816 (2009)
[2] Jiang, B., Valstar, M., Pantic, M.: Facial Action Detection using Block-based Pyramid Appearance Descriptors. In: Proc. ASE/IEEE Int'l Conf. on Social Computing, Amsterdam, pp. 429–434 (2012)
[3] Yang, D., Jin, L., Yin, J., Zhen, L., Huang, J.: Facial expression recognition with pyramid gabor features and complete kernel fisher linear discriminant analysis. International Journal of Information Technology 11(9), 91–100 (2005)
[4] Kanade, T., Cohn, J., Tian, Y.: Comprehensive database for facial expression analysis. In: Proceedings of the Fourth IEEE International Conference on Automatic Face and Gesture Recognition (FG 2000), Grenoble, France, pp. 46–53 (2000)

 [5] Kolsch, M., Turk, M.: Analysis of rotational robustness of hand detection with viola–jones detector. In: 17th International Conference on Pattern Recognition, vol. 3, pp. 107–110 (2004)
 [6] Kotsia, I., Zafeiriou, S., Pitas, I.: Texture and shape information fusion for facial expression and facial action unit recognition. Pattern Recognition 41, 833–851 (2008)
 [7] Moore, S., Bowden, R.: Local Binary Patterns for Multi-view Facial Expression Recognition. Computer Vision and Image Understanding 115(4), 541–558 (2011)
 [8] Viola, P., Jones, M.: Robust real-time object detection. International Journal of Computer Vision (2010)
 [9] Khan, R., Meyer, A., Konik, H., Bouakaz, S.: Framework for reliable, real-time facial expression recognition for low resolution images. Pattern Recognition Letters 34(10), 1159–1168 (2013)
[10] Ojala, T., Pietikainen, M., Harwood, D.: A comparative study of texture measures with classification based on featured distribution. Pattern Recognition 29(1), 51–59 (1996)
[11] Tian, Y.: Evaluation of face resolution for expression analysis. In: Computer Vision and Pattern Recognition Workshop
[12] Ojansivu, V., Heikkilä, J.: Blur insensitive texture classification using local phase quantization. In: Elmoataz, A., Lezoray, O., Nouboud, F., Mammass, D. (eds.) ICISP 2008 2008. LNCS, vol. 5099, pp. 236–243. Springer, Heidelberg (2008)
[13] Yang, P., Liu, Q.: Metaxas.: Exploring facial expression with compositional features. In: IEEE Conference on Computer Vision and Pattern Recognition (2010)
[14] Zang, Z., Lyons, M.J., Schuster, M., Akamatsu, S.: Comparison between geometry-based and Gabor-wavelets-based facial expression recognition using muti-layer perceptron. In: IEEE International Conference on Automatic Face & Gesture Recognition (FC) (1998)
[15] Zhao, G., Pietikainen, M.: Dynamic texture recognition using local binary patterns with an application to facial expression. IEEE Transaction on Pattern Analysis and Machine Intelligence 29, 915–928 (2007)
[16] Bartlett, M., Littlewort, G., Fasel, J., Movellan, R.: Real time face detection and facial expression recognition: development and applications to human computer interaction. In: Conference on Computer Vision and Pattern Recognition Workshop (2003)

# A Community-Based Vietnamese Question Answering System

Quan Hung Tran, Nien Dinh Nguyen, Kien Duc Do, Thinh Khanh Nguyen, Dang Hai Tran, Minh Le Nguyen, and Son Bao Pham

**Abstract.** Most recent Vietnamese QA systems have not considered so far in using the data crawled from the community web services as a useful resource. In this paper, we take into accounts the community-based resource to build a Vietnamese question answering system named VnCQAs. Our system comprises of three modules for building the database of question-answer pairs, analyzing questions and choosing the best answer respectively. Experimental results show that our system achieves promising performances.

## 1 Introduction

Nowadays, the community web services play a crucial role in significantly supporting human users to seek desired responses, especially in technology domain. Users often pose their queries on Yahoo! Answer, technology web forums or Facebook for finding helps as well as personal experience-based advice from others. However, queries are often complex and contain multiple sub-questions whilst others' feedbacks and comments miss valuable information or only deal with a part of these queries. For example, a question *"Có nên mua Samsung Galaxy S4 không?" (should I buy Samsung Galaxy S4?)* expects the answer about individual opinions instead of the specifications of the phone itself.

Quan Hung Tran · Nien Dinh Nguyen · Kien Duc Do · Thinh Khanh Nguyen ·
Dang Hai Tran · Son Bao Pham
Faculty of Information Technology
University of Engineering and Technology
Vietnam National University, Hanoi
e-mail: {quanth_55,niennd_55,kiendd_55,thinhnk_55,
    dangth,sonpb}@vnu.edu.vn

Minh Le Nguyen
School of Information Science
Japan Advanced Institute of Science and Technology
e-mail: nguyenml@jaist.ac.jp

© Springer International Publishing Switzerland 2015    117
V.-H. Nguyen et al. (eds.), *Knowledge and Systems Engineering,*
Advances in Intelligent Systems and Computing 326, DOI: 10.1007/978-3-319-11680-8_10

Assuming that we have a collection of users' queries from community web services and a corresponding collection of feedbacks and comments, building a community-based question answering (cQA) system to return a best answer for each user's whole query raises a challenge issue. It is because that the task is under the key research problems of how to construct the database of question-answer pairs, how to analysis questions from users' queries, and how to produce a best answer. Regarding to these problems, some researches concern about question identification [5, 16, 6, 15], question similarity [2], question generation [17], question analysis [10, 9], answer summarization [3] and answer re-ranking [13].

At this time, most recent Vietnamese QA systems have not considered so far in using community web services as a useful resource for such researches. Existing Vietnamese QA systems [8, 14, 7, 12] are usually rule/grammar-based ones and utilizes structured databases or crawled web-pages. Additionally, there is a Vietnamese QA system that used community data as described in the Dang et al. [1] 's research. The Dang et al. 's system responds to a new question by finding the similar questions from Yahoo Answer. However, this system did not return the answers of those similar questions, and the reported accuracy was not high.

In this paper, we present a community-based Vietnamese question answering system, namely VnCQAs. Our system solves the issue of domain adaptability and inability to be able to answer complex questions. Furthermore, our VnCQAs system uses machine learning techniques to obtain high accuracy. Our system contains three main modules: Database Construction, Question Analysis and Answer Selection, which are responsible for building the database, analyzing questions and choosing the best answer, respectively. Figure 1 illustrates an example[1] with the input question: *"nên mua Ipad hay laptop" (should I buy Ipad or latop?)*. The output includes the best available answer and related questions. Users can find the answers of the related questions by clicking the corresponding links.

The paper is presented as follows. In the section 2, we introduce the overview of the whole system and describe the modules. We describe the experimental results in section 3. The conclusion and future work are presented in section 4.

## 2 System Architecture

In this section, we introduce the VnCQAs system architecture (as displayed in Figure 1) and briefly describe all modules in the system. When a new question is presented to the system, our system finds the most similar questions from the database. The answers of these similar questions are called candidate answers. These candidate answers are then processed to output the final answer.

We build a system with three modules: Database Construction, Question Analysis, Answer selection.

*Database Construction* module is responsible for building the database of question-answer pairs.

---

[1] The online demonstration is available at: http://150.65.242.39:8080/VNQA/

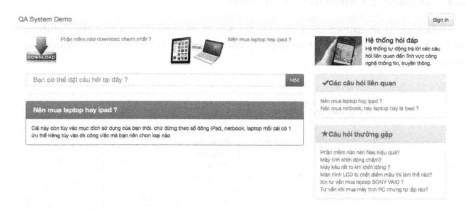

**Fig. 1** The system user interface

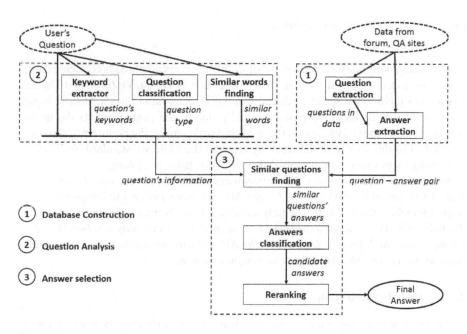

**Fig. 2** The system architecture

*Question analysis* module analyzes the questions and gives useful information about the question such as keywords, questions types and words' synonyms.

*Answer selection* module processes the candidate answers and return the final answers for the given question.

## 2.1 Database Construction Module

This module is to extract the question-answer pairs for constructing the database from the community data by two steps: question detection and answer detection (as shown in Figure 3).

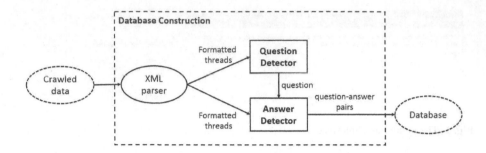

**Fig. 3** The database construction module

Our main community sources of question-answer pairs are threads collected from some famous technology forums in Vietnamese such as Vatgia, VnZoom and Tinhte. These sites include the series of threads, in which each thread has a specific topic and it is further divided into posts. Typically one of the posts presents the question, and some other posts contain the answer. Furthermore, the community data we crawled from the sites have different layouts and therefore we standardize the data by parsing them to our predefined XML format for later processing.

Figure 4 give the question: *"Mấy anh cho em hỏi tại sao khi em vào My computer hay bị treo vài giây"* (*Why when I enter My computer folder, the computer stops responding for a few second*). The only suitable answer is the last post: *"cũng có thể do phần mềm AV và cũng có thể là do 1 phần mềm nào trong máy nên bạn kiểm tra lại máy xem nhé."* (*maybe because of the AV software or maybe because of some other software, you should check your computer again*).

### 2.1.1 Question Detection

In the question detection, we use a machine learning model to classify whether a post is the question post or not. Our features used for the machine learning are sequence patterns which are based on the generalized form of text. E.g., a sentence: *"Subnet Mask dùng để làm gì?"* (*What is Subnet Mask used for?*) can be represented in sequence form as follows: *"Np V E làm gì"*, in which Np, V, E are part of speech (POS) tags of the respective words.

We define the question words that also appear many times over questions e.g., *giúp(help), phân biệt (distinguish), đánh giá (evaluate), làm sao (how to do), tại sao (why).* These question words are kept in their original form and arranged into 18 groups namely:

```
<thread>
        <title>Mấy anh cho em hỏi tại sao khi em vào My computer hay bị treo vài giây</title>
        <post>
            <user>xtduong</user>
            <text>Mong các anh chị giúp em cách khắc phục</text>
        </post>
        <post>
            <user>vnz870152</user>
            <text>treo máy luôn hay 1 thời gian thôi bạn chụp hình đầy đủ task manager mục
processes xem</text>
        </post>
        <post>
            <user>quanghuy259</user>
            <text>cũng có thể do phần mềm AV và cũng có thể là do 1 phần mềm nào trong máy nên
bạn kiểm tra lại máy xem nhé.</text>
        </post>
</thread>
```

**Fig. 4** Thread XML example

- q0000: gì (what)
- q0001: nào, nào là (which)
- q0002: ai (who)
- q0003: đâu (where)
- q0004: hay (or)
- q0005: sao, vì sao, tại sao (why)
- q0006: làm sao (how to do)
- q0007: làm gì (what to do)
- q0008: ra sao, thế nào, như thế nào (how)
- q0009: bao nhiêu (how many), bao lâu (how long), bao xa (how far)
- q0010: không (not), chưa (not yet)
- q0100: giúp(help), tư vấn(advise), dạy(teach), hướng dẫn(instruct), chỉ dẫn(instruct)
- q0101: hỏi (ask), thắc mắc (worry)
- q0102: khắc phục (overcome)
- q0103: vấn đề (problem)
- q0104: cách (solution)
- q0105: so sánh (compare), đánh giá (evaluate)
- q0106: phân biệt (distinguish)

Other words in question are replaced by their POS tags to make the sequence more general. The sequence patterns are extracted by using Prefix Span algorithm [15]. After that, we select the patterns that contain the question words. We then apply the method called *"Multiple minimum supports"* [4] to guarantee the quality of patterns.

### 2.1.2 Answer Detection

After finding the questions from the previous step, we detect the corresponding answers for each question by classifying the remaining posts through using a SVM model with a set of features:

- Is the post belonged to the author of the thread?
- Does the post contain quote of the questioner?
- Does the post contain quote of other users?
- The relative position of the post in the thread.
- The relative length of the post compared to others in the thread.
- Similarity between the post and the detected question.
- The proportion of noun, verb and pronoun that the post contains.

If the post is from the question's owner, it is unlikely to be the answer. Otherwise, the remaining post which contains the quote of the questioner often has a high possibility to be the answer.

## 2.2 Question Analysis Module

The question analysis module aims to extract important information from the questions for finding the similar questions in the later module. In this module, we investigate three steps: question classification, question keyword identification, and similar word identification as presented in Figure 5. Figure 2 shows an example of data extracted by the question analysis module from the question: *"Làm thế nào để tạo vùng nhớ ảo thay thế RAM"* (How to create virtual memory to replace RAM).

### 2.2.1 Question Classification

We classify questions because a question that is classified as a different type from the original question is unlikely to be a similar question. Moreover, the question type also provides the constraints for verifying the answers.

We categorize the questions into 3 classes: *Fact, Solution and Explanation* by using the machine learning method with a set of features: *Unigrams, Bigrams, and Similarity*. The unigram and bigram features are calculated as the boolean value, while the value of the similarity feature which represents a measure of similarity between two questions is estimated by using the phrasal overlap.

### 2.2.2 Keyword Identification

The questions which are likely to be similar usually have the same set of keywords. Besides, many questions in online forums and QA sites contain the unnecessary words and phrases, removing these helps to improve the ability of finding similar questions.

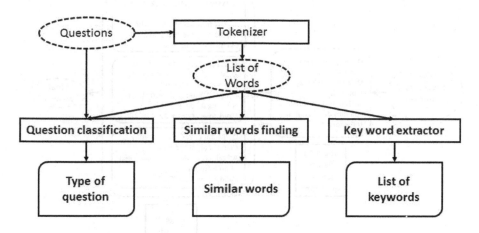

**Fig. 5** The question analysis module

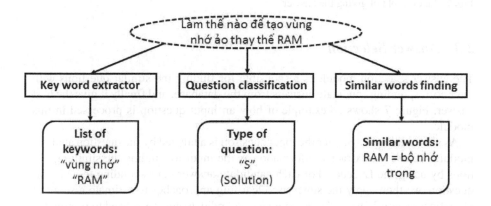

**Fig. 6** An example of analyzing question

The keyword identification aims to find the most important words in a question. We compute a score for each word appearing in the question corpus by using *term frequency - inverse document frequency (tf*idf)* weighting scheme. Then we use a threshold to determine whether the word is a keyword or not.

### 2.2.3 Similar Word Identification

Regarding to the performance of our system for finding the similar questions, we also use a synonym dictionary to return the words that has the same meaning with the original words in the input question.

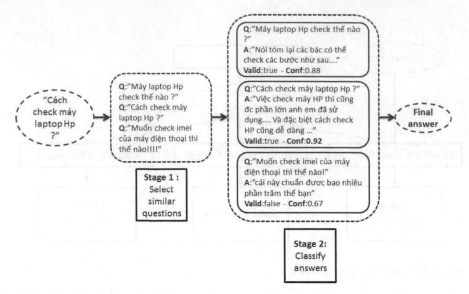

**Fig. 7** An example of giving the answer

## 2.3 Answer Selection

The answer selection module is responsible for finding the similar questions with their corresponding candidate answers from the database, and finally it give the best answer. Figure 7 shows an example of how an input question is processed in this module.

As shown in Figure 8, after the input question is analyzed by the question analysis module, we use the extracted information as the input for finding the similar questions by using the Lucene[2]. For each candidate answer corresponding to a similar question, we then apply the supervised learning approaches to estimate a score of classification confidence. The score for each candidate answer is used to re-rank the list of candidate answers, and finally the candidate answer with the highest score is selected as the final answer.

We consider the following triplet: (Qnew, Qpast, A), where Qnew is the original question, Qpast is the similar question and A is the candidate answer for Qpast. Each triplet is classified as *satisfied* if the answer A can be used to respond to the question Qnew. Otherwise, the triplet is classified as *unsatisfied*. We employ the supervised learning approaches with a set of features:

- Text length
- Number of question marks
- Number of stopwords
- IDF statistics
- Query clarity

---

[2] http://lucene.apache.org/

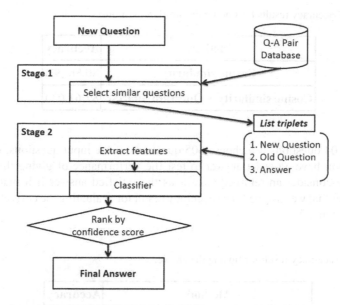

**Fig. 8** The answer selection module

- Cosine similarity
- Topic model

## 3 Evaluation

### 3.1 *Experimental Result*

We evaluate our system by the results of finding the similar questions and giving the correct answer for each test question.

We collect the community data from three famous technology forums in Vietnamese: Vatgia, VnZoom and Tinhte. For each question, we assign a score arranged from 0 to 4 to each candidate answer corresponding to the question, in which the exact answers are given the score of 4, the irrelevant answers are given the score of 0.

The evaluation data for finding similar questions consists of 1704 questions obtained from the database construction module. These questions are checked by hand to ensure that each question is not similar to other questions. Then we paraphrase each question into 3 different versions with the same meaning. The paraphrased questions with the corresponding candidate answers are indexed into Lucene as mentioned in section 2.3. We use 1704 original questions as the input questions for testing our system performance, if the returned question is one of 3 paraphrased questions, we evaluate this as a good result, and otherwise we count it as a bad result. The accuracy result for finding the similar questions is presented in table 1.

**Table 1** The accuracy results for finding the similar questions

| Method | Accuracy |
|---|---|
| **Cosine similarity** | 80.86 (%) |
| **Cosine similarity + Question analysis** | **87.61** (%) |

From 1704 questions, we choose 605 questions as the input questions, in which each question have an exact answer to test the performance of giving the correct answer. We consider an returned answer as the satisfied answer if it matches the exact answer that we assigned. The accuracy result for evaluating the correct answers is shown in table 2.

**Table 2** The accuracy result for finding the answer

| Method | Accuracy |
|---|---|
| **Baseline using the default of Lucene** | 59.66 (%) |
| **Our approach** | **71.19** (%) |

## 4   Conclusion

In this paper, we proposed the community-based question answering system for Vietnamese. Our system consists of three modules: database construction, question analysis and answer selection. The database construction module is used for creating the database of question-answer pairs, in which each question corresponds to the candidate answers. The question analysis module is responsible for extracting useful information such as keywords, question types and synonyms. The answer selection module takes the extracted information from the input question for finding the similar questions in the database, and then re-rank the list of corresponding candidate answers to give the best answer. Experimental results are promising, where the question analysis module helps to improve the accuracy from 80.86% to 87.61% in finding the similar questions, and the answer selection module get the accuracy of 71.19% that is 11.53% higher than the baseline using the default of Lucene.

In the future, we will extend the question analysis module by using other additional features based on the dependency tree [11]. We will also expand the database to be able to deal with a wide range of questions and improve the answer selection module.

**Acknowledgment.** This work is partially supported by the Research Grant from Vietnam National University, Hanoi No. QG.14.04.

# References

[1] Son, D.T., Dung, D.T.: Apply a mapping question approach in building the question answering system for vietnamese language. In: Proceedings of the Conference on Green Technology and Sustainable Development (2012)

[2] Bernhard, D., Gurevych, I.: Answering learners' questions by retrieving question paraphrases from social q&a sites. In: Proceedings of the Third Workshop on Innovative Use of NLP for Building Educational Applications, pp. 44–52. Association for Computational Linguistics (June 2008)

[3] Chan, W., Zhou, X., Wang, W., Chua, T.-S.: Community answer summarization for multi-sentence question with group l1 regularization. In: Proceedings of the 50th Annual Meeting of the Association for Computational Linguistics. Long Papers, vol. 1, pp. 582–591 (2012)

[4] Jindal, N., Liu, B.: Identifying comparative sentences in text documents. In: Proceedings of the 29th Annual International ACM SIGIR Conference on Research and Development in Information Retrieval, SIGIR 2006, pp. 244–251 (2006)

[5] Li, B., Jin, T., Lyu, M.R., King, I., Mak, B.: Analyzing and predicting question quality in community question answering services. In: Proceedings of the 21st International Conference Companion on World Wide Web, WWW 2012 Companion, pp. 775–782 (2012)

[6] Li, B., Si, X., Lyu, M.R., King, I., Chang, E.Y.: Question identification on twitter. In: Proceedings of the 20th ACM International Conference on Information and Knowledge Management, CIKM 2011, pp. 2477–2480 (2011)

[7] Nguyen, A.K., Le, H.T.: Natural language interface construction using semantic grammars. In: Ho, T.-B., Zhou, Z.-H. (eds.) PRICAI 2008. LNCS (LNAI), vol. 5351, pp. 728–739. Springer, Heidelberg (2008)

[8] Nguyen, D.Q., Nguyen, D.Q., Pham, S.B.: A Vietnamese Question Answering System. In: Proceedings of the 2009 International Conference on Knowledge and Systems Engineering, KSE 2009, pp. 26–32 (2009)

[9] Nguyen, D.Q., Nguyen, D.Q., Pham, S.B.: A Semantic Approach for Question Analysis. In: Jiang, H., Ding, W., Ali, M., Wu, X. (eds.) IEA/AIE 2012. LNCS (LNAI), vol. 7345, pp. 156–165. Springer, Heidelberg (2012)

[10] Nguyen, D.Q., Nguyen, D.Q., Pham, S.B.: Systematic Knowledge Acquisition for Question Analysis. In: Proceedings of the International Conference Recent Advances in Natural Language Processing 2011, pp. 406–412 (2011)

[11] Nguyen, D.Q., Nguyen, D.Q., Pham, S.B., Nguyen, P.-T., Le Nguyen, M.: From Treebank Conversion to Automatic Dependency Parsing for Vietnamese. In: Métais, E., Roche, M., Teisseire, M. (eds.) NLDB 2014. LNCS, vol. 8455, pp. 196–207. Springer, Heidelberg (2014)

[12] Nguyen, D.T., Hoang, T.D., Pham, S.B.: A vietnamese natural language interface to database. In: Proc. of the 2012 IEEE Sixth International Conference on Semantic Computing, pp. 130–133 (2012)

[13] Surdeanu, M., Ciaramita, M., Zaragoza, H.: Learning to rank answers on large online QA collections. In: Proceedings of ACL 2008. HLT (June 2008)

[14] Tran, V.M., Nguyen, V.D., Tran, O.T., Pham, U.T.T., Ha, T.Q.: An experimental study of vietnamese question answering system. In: International Conference on Asian Language Processing, IALP 2009, pp. 152–155 (December 2009)

[15] Wang, K., Chua, T.-S.: Exploiting salient patterns for question detection and question retrieval in community-based question answering. In: Proceedings of the 23rd International Conference on Computational Linguistics, COLING 2010, pp. 1155–1163 (2010)

[16] Yang, L., Bao, S., Lin, Q., Wu, X., Han, D., Su, Z., Yu, Y.: Analyzing and predicting not-answered questions in community-based question answering services. In: AAAI (2011)

[17] Zhao, S., Wang, H., Li, C., Liu, T., Guan, Y.: Automatically generating questions from queries for community-based question answering. In: Proceedings of 5th International Joint Conference on Natural Language Processing, pp. 929–937 (2011)

# Toward a Rule-Based Synthesis of Vietnamese Emotional Speech

Thi Duyen Ngo, Masato Akagi, and The Duy Bui

**Abstract.** This paper presents a framework used to simulate four basic emotional styles of Vietnamese speech, by means of acoustic feature transplantation techniques applied to neutral utterances. First, it describes some analyses of acoustic features of Vietnamese emotional speech, accomplished to find the relations between prosodic, voice quality variations and emotional states in Vietnamese speech. Then the target pitch profiles together with duration, energy and spectrum constraints were obtained by applying rules which were inferred from the analysis results and based on the idea that when some emotional speech is synthesized from neutral speech, acoustic features are modified more in some syllables, instead of uniformly modified in all syllables. From there, neutral speech were morphed to produced synthesized speech with emotions. Results of perceptual tests show that emotional styles were well recognized.

**Keywords:** rule-based synthesis, acoustic feature, speech morphing, Vietnamese, emotional speech.

## 1 Introduction

Emotional speech synthesis is a technique that performs some manipulations on acoustic feature parameters of the voice data to produce human perception of different emotions [23]. There are two main approaches used to synthesize normal speech: formant synthesis and concatenative synthesis. These two techniques are also used to synthesize emotional speech. Formant synthesis is a technique which entirely uses rules on the acoustic correlation of various speech sounds to create emotional speech data (i.e. [2, 3]). The speech quality of these methods is relatively worse than that of

Thi Duyen Ngo · The Duy Bui
University of Engineering and Technology, Vietnam National University, Hanoi
e-mail: {duyennt,duybt}@vnu.edu.vn

Masato Akagi
School of Information Science, Japan Advanced Institute of Science and Technology,
Nomi, Japan
e-mail: akagi@jaist.ac.jp

© Springer International Publishing Switzerland 2015                                     129
V.-H. Nguyen et al. (eds.), *Knowledge and Systems Engineering,*
Advances in Intelligent Systems and Computing 326, DOI: 10.1007/978-3-319-11680-8_11

state-of-the-art concatenative methods but using this technique gives us the ability to flexibly control a large number of parameters related not only to voice source but also to vocal tract. Concatenative synthesis concatenates pre-recorded human speech segments to create synthetic voice (i.e [24, 4]). The characteristic of this synthesis approach is the fact that it can not explicitly model the acoustic properties of emotions but the quality of synthesized emotional speech is much better than that resulted from formant technique. To synthesize emotional speech, there is another technique called speech morphing that is used quite often recently (i.e. [25, 21, 8]). This technique also uses pre-recorded speech of complete sentences and can be used to manipulate pre-recorded voice. Normally, a neutral voice is used as the source and some speech conversion methods or other techniques are used to convert the source voice to the target voice. The target voice is emotional speech. Because speech morphing technique use neutral voices as inputs, the naturalness of synthetic speech is partly promised. Besides, using some speech modification techniques may give us a high degree of flexibility and control over acoustic parameters. Most of the previous works synthesized emotional speech by modifying acoustic parameters at the phrase-level. This maybe leads to the decrease of the naturalness of synthesized emotional speech because of the fact that acoustic features vary more in some syllables of phrases instead of uniformly changes in all syllables when the emotional state in the speech changes.

Being able to perform attempts to add emotions to synthesized speech, on the acoustic aspect, we need to have detailed knowledge on how acoustic characteristics in voice are related to emotions. The review of literature has shown that there are two types of acoustic cues which have great influence on emotional state in speech. One is related to the prosody and the other is related to the voice quality. As a monosyllabic and tonal language, Vietnamese has particular features in comparison with the one of European language (polysyllabic languages). Up to now, there have been some proposed works on prosody and voice quality of Vietnamese speech. Some researches on synthesizing Vietnamese speech have been also brought out. Le [15] brought out and proved five hypotheses for Vietnamese speech's duration basing on analysing 36 file of 20.815 words read by the broadcasters from several distinctive regions in Vietnam. According to [26], factors which impact on the duration of a Vietnamese phonetic unit are the position, the pitch, and the structure of that unit. In [9], Vietnamese compounds and phrasal constructions were investigated for phonetic correlates of lexical stress; acoustic and perceptual characteristics of Vietnamese compound words and their phrasal counterparts were reported. In [13, 14], Le described some results of researches on acoustic features of Vietnamese speech to help for synthesising Vietnamese speech from text. Mac [17] presented a study on Audio-Visual prosodic attitudes in Vietnamese; it showed the relative contribution of audio, visual, and audio-visual information in attitude perception and how native and non-native listeners recognize and confuse the attitudes. An analysis on speech prosody was also carried out in order to further validate the results of the perception experiments, and to bring out some prosodic characteristics of Vietnamese social affects. However, almost all proposed researches have focused on Vietnamese neutral speech; there have been very few ones focusing on Vietnamese emotional speech.

In this paper, we presents a framework used to simulate four basic emotional styles of Vietnamese speech, by means of acoustic feature transplantation techniques applied to neutral utterances. First, we describes some analyses of acoustic features related to the prosody and the voice quality of Vietnamese emotional speech, accomplished to find the relations between prosodic and voice quality variations and emotional states in Vietnamese speech. Specifically, a multi-style Vietnamese emotional speech database was recorded and analysed to verify the correlations and to quantify, for the emotional styles, the prosodic, voice quality feature variations with respect to the neutral situation. The database consisted of Vietnamese sentences uttered in five different styles: neutral, happiness, sadness, cold anger, and hot anger. According to the analysis results, for each speaker of the database, a set of prosodic and voice quality variation coefficients was produced for each emotional style. After that, the target pitch profiles together with duration, energy and spectrum constraints were obtained by applying simple rules which were inferred from the sets of variation coefficients and based on the idea that when some emotional speech is synthesized from neutral speech, acoustic features are modified more in some syllables, instead of uniformly modified in all syllables; this idea makes the synthesized emotional speech more realistic because of the fact that when the emotional state in speech changes, acoustic features vary more in some syllables instead of uniform changes in all syllables. From there, neutral speech were morphed to produced synthesized speech with emotions.

The rest of the paper is organized as follows. Section 2 presents the composition and acquisition of the emotional speech database. Then, in Section 3, we describe the acoustic feature extraction phase and the analysis results. After that, Section 4 shows our processes of synthesizing Vietnamese emotional speech and the results of the evaluation test. Finally, Section 5 shows conclusion.

## 2  Emotional Speech Database

The emotional speech database which was used for investigating Vietnamese prosodic and voice quality features consisted of Vietnamese utterances produced by two professional Vietnamese actors, one male and one female. The two actors were asked to produce utterances using five different styles. They had to utter 19 sentences in four emotional styles that were: happiness, cold anger, sadness, and hot anger. Besides, they also recorded the same 19 sentences in a neutral way. Consequently, each sentence has one utterance in each of the five styles, for both male and female voices. Therefore, there is a total of 190 utterances in the database - a half for the male voice and the other for the female voice. Sentences were about 8 words long and well representative of the Vietnamese phonetic alphabet. Most of them were non-sense and had no semantic emotional content and therefore could not influence the actors provoking any particular emotional attitude. During the recording sessions, the actors had to simulate each of emotional styles in turn, and a director was always present to control their pronunciation and their prosody to avoid emphatic performances. Signals were recorded in a sound-proof room, high quality

**Table 1** Emotional speech database recognition rates

|        |           | neutral | happy  | cold angry | sad    | hot angry |
|--------|-----------|---------|--------|------------|--------|-----------|
| Male   | neutral   | 95.37%  | 15.3%  | 6.11%      | 4.29%  | 0%        |
|        | happy     | 3.98%   | 82%    | 0.44%      | 0%     | 0%        |
|        | cold angry| 0%      | 2.7%   | 86.23%     | 1.57%  | 0%        |
|        | sad       | 0.65%   | 0%     | 0%         | 94.14% | 0%        |
|        | hot angry | 0%      | 0%     | 7.22%      | 0%     | 100%      |
| Female | neutral   | 94.44%  | 10.24% | 2.44%      | 6.86%  | 0%        |
|        | happy     | 2.3%    | 84.56% | 2.14%      | 0%     | 0%        |
|        | cold angry| 0%      | 5.2%   | 87.41%     | 2.3%   | 0%        |
|        | sad       | 3.26%   | 0%     | 0%         | 90.84% | 0%        |
|        | hot angry | 0%      | 0%     | 8.1%       | 0%     | 100%      |

microphone and digital acquisition equipments were used, and waveforms were digitally acquired with sampling frequency of 22050 Hz and quantization of 16 bit.

A perceptual test has been proposed to 12 volunteers in order to evaluate the corpus, and verify if it could be successfully used to extract style dependent rules. 12 students, native Vietnamese speakers, average age 21 years old, with normal hearing participated in the experiment. Subjects were asked to rate the 190 utterances according to the perceived degree of each of the 5 emotional speech styles. The results of the evaluation test are reported in the confusion matrix of Table 1. Recognition rates were quite high even if some confusion values occurred.

## 3  Acoustic Feature Extraction

### 3.1  Acoustic Cues Related to Emotional Speech

The determination of considering which acoustic cues should be measured is conducted by reviewing previous studies and basing on particular prosodic features of Vietnamese. As above-mentioned, according to much previous work, there is a general agreement that there are two types of acoustic cues which are always considered as the most important factors correlated to the emotional states in speech; one is involved with prosody and the other is related to voice quality. As a monosyllabic and tonal language, Vietnamese has particular features in comparison with the one of European language (polysyllabic languages). The Vietnamese prosody is related to rhythm between words in a word groups or in a compound words, while into-nation has the global effect on the whole sentence. There is no need to change the intonation of a word in order to highlight it because each word has its own meaning thanks to one of the six tones. Moreover, not like polysyllable languages, there is no question to emphasize a syllable in a Vietnamese word because each word has only one syllable. Each Vietnamese syllable could be considered as a combination of Initial, Final and Tone. The tone is associated with the syllable as a whole; it plays an important role on entire syllable. However, tonal features are not as explicit as other features in speech signal. In the first consonant, we can hear a little of the tone. Tone becomes clearer in rhyme and finished completely at the end of the sylla-ble. The pervading phenomenon determines the non-linear nature of tone. So, with monosyllabic language like Vietnamese, a syllable can not easily be separated into small acoustic parts like European languages.

**Prosody**

Prosody is essentially a collection of factors that control the pitch, loudness, and rate of speaking. The variations of intonation, rhythm, stress pattern, belong to what we call the prosody of a sentence. Depending on the emotional states of the speaker, a sentence can be uttered with different prosodic characteristics. There-fore, the prosodic variations in an utterance have great influences on the emotions expressed in speech [12]. This is the reason why prosody is an important factor needed to be investigated in finding acoustic feature variations related to emotional states in speech. In the acoustic aspect, the acoustic cues which are considered sig-nificant for prosody are largely extracted from fundamental frequency (F0), power, and duration. These features are described below.

*Fundamental frequency:* Most previous works which studied to find which types of acoustic cues were related to emotional states in speech found that the F0 contour had a deeply effect on emotional states in speech (see [5] for summary). In our work, fundamental frequency was investigated at utterance level as well as syllable level. At the utterance level, three acoustic parameters which were quantitative studied were highest pitch (HP), average pitch (AP) and pitch range (PR). In addition, aver-age pitch of syllables were also examined. These acoustic parameters were chosen to be investigated because the variations in their values fairly well represent the change in intonation and pitch contour of the utterance.

*Duration:* There have been proposed works which showed an effect of duration on emotional states in speech, in different languages [7, 18, 20, 22]. Acoustic parameter related to duration, which were investigated in our work were total length of utterance (TL), mean of pause length (MPAU), consonant length (CL), and ratio between consonant length and vowel length (RCV). These four parameter were chosen because their variations could express almost all change in rhythm of the utterance.

*Power:* The relationship between the power envelope and emotional states in speech has been reported in a number of proposed researches (e.g. [7, 16, 22]). Similar to fundamental frequency, at the utterance level, three acoustic parameters which were quantitative studied were maximum power (HPW), average power (APW) and power range (PWR); average power of syllables were also examined. The variations in these parameters' values could well represent the change in of the power contour of the speech signal.

**Voice Quality**

In addition to prosody, voice quality is another acoustic cue that researchers have much focused on. Voice quality refers to the auditory impression that a listener gains while hearing a speech; it is represented by the spectrum content of the speech signal. Although the spectrum is a not-as-well investigated feature compared to F0 movement, it has been found to have a relationship with expressive speech perception (e.g. [16, 18, 22]). In the following, parameters measured in spectral that are considered related to voice quality are described.

*Formants:* Theoretically, there are an infinite number of formants, but only the lowest three or four are of interest for practical purposes. In the study by Ishii and Campbell [10], they examined the correlation between acoustic cues and some categories of voice quality and speaking manners. Their results suggested that formants were found to be one of the most influential parameters of voice quality. Therefore, formants are one of the meaningful acoustic cues that should be investigated. In our work, three first formants F1, F2, F3 were examined.

*Spectral shape:* Spectral shape can provide cues related to aspects of voice quality, for example, H1−A3. This is the ratio of amplitude of first harmonic (H1) relative to that of the third-formant spectral peak (A3), and has been used by Hanson [6] to characterize spectral tilt (ST). As indicated by Manezes and Maekawa [19] H1-A3 is glottal cycle characteristics, i.e., speed of closing of vocal folds. Generally, spectral tilt (H1-A3) is related to the brightness of voice quality. A spectrum with a flatter spectral tilt, stronger energy in high frequency region, represents a brightness to the voice while a spectrum with a sharper spectral tilt, weaker energy in the high frequency region, represents a darkness to the voice. Therefor, in our research, spectral tilt (ST) was chosen to be investigated.

## 3.2   Acoustic Feature Extraction Phase

In this subsection, we describe the acoustic feature extraction phase. The F0 contour, the power envelope, and the spectrum were calculated by using STRAIGHT [11] with a FFT length of 1024 points and a frame rate of 1ms. The sampling frequency was 22050 Hz. Time duration was manually specified with the partly support of WaveSurfer [1].

| phone | m | ê | | t | i | n | | c | ô | h | ù |
|---|---|---|---|---|---|---|---|---|---|---|---|
| phone NO. | -1 | 1 | 2 | 0 | 3 | 4 | 5 | 0 | 6 | 7 | 8 | 9 |
| timePoint | 182 | 223 | 331 | 392 | 403 | 476 | 537 | 594 | 616 | 716 | 778 | 895 |
| vowel | -1 | 2 | 1 | 0 | 2 | 1 | 2 | 0 | 2 | 1 | 2 | 1 |

**Fig. 1** An example of time segmentation

At the utterance level, a total of 14 acoustic parameters were calculated and analysed in order to find the relations between prosodic, voice quality variations and emotional state in Vietnamese speech. Average pitch and average power at the syllables level were also examined. Specifically, the extraction phase was performed as follow: For each utterance, firstly, the F0 information was extracted using STRAIGHT [11]. Then from this information, some acoustic parameters related to the F0 contour were measured. These parameters were highest pitch (HP), average pitch (AP), and pitch range (PR); average pitch of syllables were also measured. After that, the power envelope was measured in a way similar to that for the F0 contour. Power information was firstly extracted using STRAIGHT [11] and then acoustic parameters related to the power envelope were calculated. The acoustic parameters considered were: maximum power (HPW), average power (APW), and power range (PWR), average power of syllables. Next, with the duration, for each utterance, the information of time segmentation was manually measured first. The measurement included phoneme number, time (ms), and vowel. The duration of all phonemes (time), both consonants and vowels, as well as pauses, were manually specified with the partly support of WaveSurfer [1]. Figure 1 illustrate an example of time segmentation. In the table, the first row indicates the phonemes, in there the empty cells correspond with pause phases in the utterance. The second row represents the order of the phonemes, noted by -1 before the first phoneme; in this row pause phases in the utterance are marked with cells having the value of zero. The third row indicates the start time of the next phoneme and also represents the end time of the current phoneme. The fourth row shows whether the phonemes are consonant or vowel: 1 – vowel, 2 – consonant, the cells having zero value correspond with pause phases. Basing on this table of time segmentation, the following parameters related to duration were measured: mean of pause lengths (MPAU), total length (TL), consonant length (CL), ratio between consonant length and vowel length (RCV). Finally, for spectrum, formants (F1, F2, F3) and spectral tilt (ST) were measured. Formant measures were taken approximately at the vowel midpoint of the vowels. The sampling frequency of the speech signal was set at 10 kHz. The spectrum was obtained by using STRAIGHT and three formants (F1, F2, and F3) were calculated with LPC-order 12. The reason why the sampling frequency was reduced to 10kHz is that when this frequency was 10kHz, maximally 5 spectral peaks which the LPC procedure tried to distribute along the frequency axis according to the presence of energy in a particular frequency band were located between 0 and 5kHz, which is an important range for speech sounds, especially vowels. The locations of these peaks were an estimation of the formants. Spectral tilt was

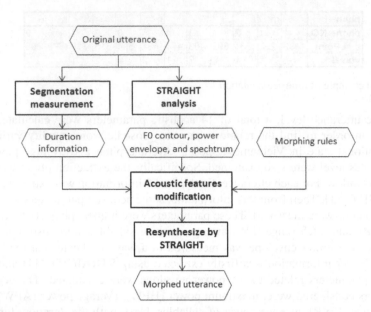

**Fig. 2** Speech morphing process using STRAIGHT

calculated from H1-A3, where H1 is the amplitude of the first harmonic and A3 is the amplitude of the strongest harmonic in the third formant.

After performing the above extraction phase, with each of 190 utterance in the emotional speech database, we had a set of 14 values corresponding to the 14 acoustic parameters at the utterance level. From these 190 sets, for these parameters of each emotional style, the values of variation coefficients with respect to the baseline (neutral style) were calculated. As a result, we had 152 sets of 14 values of variation coefficients. In which there were 19 sets for each of four emotional styles (happiness, sadness, cold anger, hot anger), for each of the two speakers. After that, with each pack of these 19 sets, clustering was carried out. Then, the cluster which had the largest number of sets was chosen. Finally, from the chosen cluster, the mean values of variation coefficients corresponding to 14 parameters of each emotional style were calculated. These values are reported in Table 2 and 3. Actually, different analysis results were found among speakers' voices. These differences were due to the fact that the speakers expressed emotions in different ways and with different intensities.

At the syllable level, when examining variations of F0, we found that in all four styles, mean F0 of syllables belonging the word/compound word at the end of the utterances change more than that of other syllables, for both two speakers. More specifically, in the happy, cold angry, and hot angry styles, mean F0 of syllables belonging the word/compound word at the end of the utterances increased with respect to the neutral case more than that of other syllables. Meanwhile, in the sad style, mean F0 of these syllables more decreased with respect to the neutral case.

**Table 2** Mean variations of prosodic parameters for four emotional styles with respect to the neutral one

|        |       | happy   | sad     | cold angry | hot angry |
|--------|-------|---------|---------|------------|-----------|
| Male   | HP    | 9.28%   | -2.25%  | 8.60%      | 15.12%    |
|        | AP    | 8.09%   | -4.60%  | 6.17%      | 15.22%    |
|        | PR    | 31.46%  | 18.66%  | 15.05%     | 32.00%    |
|        | APW   | 11.04%  | -3.81%  | 16.04%     | 19.74%    |
|        | HPW   | 20.81%  | -5.84%  | 13.90%     | 10.01%    |
|        | PWR   | 11.53%  | -3.26%  | 22.19%     | 23.77%    |
|        | MPAU  | -6.46%  | 66.86%  | 50.86%     | 59.80%    |
|        | CL    | -4.96%  | 9.47%   | -10.36%    | -1.15%    |
|        | RCV   | -7.50%  | -2.13%  | -11.72%    | 2.84%     |
|        | TL    | -2.50%  | 15.23%  | 0.64%      | -12.35%   |
| Female | HP    | 12.23%  | -0.66%  | 9.09%      | 14.37%    |
|        | AP    | 7.75%   | -2.10%  | 6.99%      | 13.92%    |
|        | PR    | 51.57%  | 28.53%  | -11.51%    | 48.34%    |
|        | APW   | 17.21%  | -4.98%  | 21.45%     | 27.72%    |
|        | HPW   | 7.96%   | -6.61%  | 28.97%     | 28.86%    |
|        | PWR   | 12.61%  | -8.15%  | 15.79%     | 20.36%    |
|        | MPAU  | -3.00%  | 43.95%  | -17.03%    | 37.86%    |
|        | CL    | -3.15%  | 22.00%  | -2.12%     | -0.07%    |
|        | RCV   | -10.24% | -9.87%  | -8.23%     | 1.57%     |
|        | TL    | 3.55%   | 16.92%  | 2.20%      | -5.98%    |

**Table 3** Mean variations of voice quality parameters for four emotional styles with respect to the neutral one

|        |     | happy   | sad      | cold angry | hot angry |
|--------|-----|---------|----------|------------|-----------|
| Male   | F1  | 2.80%   | -3.21%   | 6.29%      | 10.26%    |
|        | F2  | 1.38%   | 1.88%    | -4.05%     | -1.99%    |
|        | F3  | 1.42%   | -1.17%   | -1.84%     | 5.29%     |
|        | ST  | -15%    | 6.50%    | 7.55%      | -57%      |
| Female | F1  | 9.99%   | -13.54%  | 10.82%     | 20.23%    |
|        | F2  | 15.43%  | -1.87%   | -4.21%     | -8.87%    |
|        | F3  | 2.17%   | -2%      | -4.23%     | 1.87%     |
|        | ST  | -14%    | 5.33%    | 6.23%      | -43%      |

Similar to F0, in all four styles, we also realized that average power of syllables belonging the word/compound word at the end of the utterances also change more than that of other syllables. They more increased in the happy, cold angry, hot angry styles and more decreased in the sad style.

## 4 Speech Synthesis and Evaluation Test

It is a fact that when the emotional state changes, acoustic features vary more in some syllables of phrases instead of uniformly changes in whole phrases. In polysyllabic languages (i.e. Japanese, English) these syllables are usually accent syllables. In almost previous works, acoustic features were analyzed and modified at the phrase level, this maybe leads to the decrease of the naturalness of synthesized emotional speech. Vietnamese is a monosyllabic language and it does not have accent syllables. However, when the emotional state in Vietnamese speech changes, acoustic features do not uniformly change in all syllables, as indicated in the analysis results in Section 3.2. In our work, we used speech morphing technique to produce Vietnamese emotional speech. Our speech morphing process is presented in Figure 2. Fistly, STRAIGHT [11] is used to extract F0 contour, power envelope, and spectrum of the neutral speech signal while segmentation information was measured manually. Then, acoustic features in terms of F0 contour, power envelope, spectrum and duration were modified basing on morphing rules inferred from the sets of variation coefficients in Table 2 and 3. These modifications were carried out with taking account of variations of acoustic features at the syllable as indicated in Section 3.2.

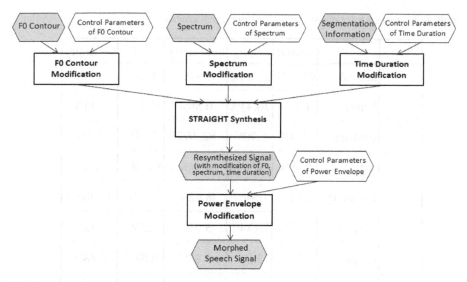

**Fig. 3** Acoustic Feature Modification Process

Finally, emotional speech is synthesized from the modified F0 contour, power envelope, spectrum and duration using **STRAIGHT**. The modifications are carried out according to the flow presented in Figure 3.

**Evaluation Test**

A perceptual experiment was also conducted on morphed utterances. This experiment was carried out in a way similar to the one in Section 2. The results are showed in Table 4; recognition rate of the happy style is slightly lower than that of others, but overall recognition rates of synthesized emotional speech is high. This results suggest that the framework is effective, and that the used rules are appropriate.

## 5   Conclusion

This paper has presented the framework used to simulate four basic emotional styles of Vietnamese speech. It produced results of some analyses of the prosody and voice quality of Vietnamese emotional speech. The analyses were perform on an emotional speech database which consists of Vietnamese sentences uttered in five different styles. The relations between prosodic, voice quality variations and emotional states in Vietnamese speech were obtained by investigating the variations of prosodic and voice quality features of Vietnamese emotional speech in comparison with those of neutral speech. According to the analysis results, a set of prosodic and voice quality variation coefficients was produced for each emotional style. Then, the target pitch profiles together with duration, energy and spectrum constraints were obtained by applying simple rules which were inferred from the sets of variation coefficients and based on the idea that when some emotional speech is synthesized

**Table 4**  Evaluation rates of the synthesized emotional speech

|  |  | neutral | happy | cold angry | sad | hot angry |
|---|---|---|---|---|---|---|
| Male | neutral | 93.33% | 22.54% | 7.81% | 4.21% | 0% |
| | happy | 4.82% | 70.44% | 0.88% | 0.44% | 0.44% |
| | cold angry | 0.44% | 5.70% | 82.46% | 7.37% | 3.2% |
| | sad | 0.79% | 0.88% | 0.44% | 87.28% | 0% |
| | hot angry | 0.61% | 0.44% | 8.42% | 0.7% | 95.96% |
| Female | neutral | 92.81% | 20.44% | 5.96% | 5.26% | 0% |
| | happy | 5% | 73.86% | 0.88% | 0.88% | 0.88% |
| | cold angry | 0.44% | 5.26% | 84.04% | 7.02% | 3.86% |
| | sad | 1.36% | 0.44% | 0% | 86.84% | 0% |
| | hot angry | 0% | 0% | 9.12% | 0% | 95.26% |

from neutral speech, acoustic features are modified more in some syllables, instead of uniformly modified in all syllables; this idea makes the synthesized emotional speech more realistic because of the fact that when the emotional state in speech changes, acoustic features vary more in some syllables instead of uniform changes in all syllables. From there, neutral speech were morphed to produced synthesized speech with emotions. Results of perceptual tests show that emotional styles were well recognized.

# References

[1] Wavesurfer, http://www.speech.kth.se/wavesurfer/index.html
[2] Burkhardt, F.: Emofilt: the simulation of emotional speech by prosody-transformation. In: Proc. of Interspeech (2005)
[3] Cahn, J.E.: The generation of affect in synthesized speech. Journal of the American Voice I/O Society, 1–19 (1990)
[4] Edgington, M.: Investigating the limitations of concatenative synthesis. Eurospeech (1997)
[5] Erickson, D.: Expressive speech: Production, perception and application to speech synthesis. Acoust. Sci. & Tech. 26, 317–325 (2005)
[6] Hanson, H.: Glottal characteristics of female speakers: acoustic correlates. J. Acoust. Soc. Am. 101, 466–481 (1997)

[7] Huttar, G.L.: Relations between prosodic variables and emotions in normal american english utterances. Journal of Speech and Hearing Research 11, 481–487

[8] Inanoglu, Z., Young, S.: A system for transforming the emotion in speech: Combining data-driven conversion techniques for prosody and voice quality. In: Proc. of Interspeech (2007)

[9] Ingram, J., Nguyen, T.: Stress, tone and word prosody in vietnamese compounds. In: Proceedings of the 11th Australian International Conference on Speech Science & Technology, pp. 193–198 (2006)

[10] Ishii, C.T., Campbell, N.: Analysis of acoustic-prosodic features of spontaneous expressive speech. In: Proceedings of 1st International Congress of Phonetics and Phonology, p. 19 (2002)

[11] Kawahara, H., Masuda-Katsuse, I., de Cheveigne, A.: Restructuring speech representations using a pitch adaptive time-frequency smoothing and an instantaneous-frequency-based f0 extraction: possible role of a repetitive structure in sounds. Speech Communication 27, 187–207 (1999)

[12] Kent, R.D., Read, C.: Acoustic Analysis of Speech. Singular Publishing Group, San Diego (1992)

[13] Le, H.M., Le, K.H.: Analysis and synthesis for duration feature of vietnamese. In: The 6th National Conference in Information Technology, Thainguyen, Vietnam (2003)

[14] Le, H.M., Quach, T.N.: Some results in phonetic analysis to vietnamese text-to-speech synthesis based on rules. Journal on Information and Communication Technology (2006)

[15] Lê, T.-H., Nguyen, A.-V., Truong, H.V., Van Bui, H., Lê, D.: A study on vietnamese prosody. In: Nguyen, N.T., Trawiński, B., Jung, J.J. (eds.) New Challenges for Intelligent Information and Database Systems. SCI, vol. 351, pp. 63–73. Springer, Heidelberg (2011)

[16] Leinonen, L.: Expression of emotional-motivational connotations with a one-word utterance. J. Acoust. Soc. Am. 102, 1853–1863 (1997)

[17] Mac, D.K., Castelli, E., Aubergé, V., Rilliard, A.: How vietnamese attitudes can be recognized and confused: Cross-cultural perception and speech prosody analysis. In: International Conference on Asian Language Processing, pp. 220–223 (2011)

[18] Maekawa, K.: Phonetic and phonological characteristics of paralinguistic information in spoken japanese. In: Proc. Int. Conf. Spoken Language Processing, pp. 635–638 (1998)

[19] Menezes, C., Maekawa, K., Kawahara, H.: Perception of voice quality in pralinguistic information types: A preliminary study. In: Proceedings of the 20th General Meeting of the PSJ, pp. 153–158 (2006)

[20] Pell, M.D.: Influence of emotion and focus location on prosody in matched statements and questions. J. Acoust. Soc. Am. 109, 1668–1680 (2001)

[21] Goto, M., Unoku, M., Saitou, T., Akagi, M.: Speech-to-singing synthesis: converting speaking voices to singing voices by controlling acoustic features unique to singing voices. In: Proc. WASPAA 2007 (2007)

[22] Wallbott, R., Scherer, H.G., Banse, K.R., Goldbeck, T.: Vocal cues in emotion encoding and decoding. Motivation and Emotion 15, 123–148 (1991)

[23] Wallbott, R., Scherer, H.G., Banse, K.R., Goldbeck, T.: Vocal communication of emotion: a review of research paradigms. Speech Communication 40, 227–256 (2003)

[24] Stallo, J.: Simulating emotional speech for a talking head. Honours Thesis, School of Computing, Curtin University of Technology, Australia (2000)

[25] Tao, J., Kang, Y., Li, A.: Prosody conversion from neutral speech to emotional speech. IEEE Trans. on Audio, Speech and Language Processing 14(2006), 1–19 (2007)
[26] Tran, D.D., Castelli, E., Serignat, J.-F., Le, V.B.: Analysis and modeling of syllable duration for vietnamese speech synthesis. O-COCOSDA (2007)

# EasySearch: Finding Relevant Functions Based on API Documentation

Dinh Huy Tran, Hua Phung Nguyen, and Duy Hanh Le

**Abstract.** This subject proposes a hybrid approach integrating keyword-based and semantic-based search to find relevant functions. Structure and contents of API documentation (API-Doc) are mined in our approach. The approach uses natural language processing technologies to expand queries, exploits the multi-level categorized structure of API-Doc to find candidates, and applies API-Doc's short function descriptions to refine solutions. Our case study with eleven Java programmers demonstrates that our approach is better than the other approaches in precision, recall and F-measure criteria.

**Keywords:** code search, query expansion, keyword weighting, semantic ranking, API documentation.

## 1 Introduction

To develop a new software, programmers usually reuse existing functions provided by programming libraries. According to Singer et al. [22], programmers spend over 40% of their time in searching and reading source code. While programming libraries have grown more numerous and diverse. For instance, there are thousands of classes and more than 20,000 functions in the Java Standard Library [14], and more than 140,000 classes, functions in .NET framework [23].

Search engines, like google, can support programmers to find expected functions. However, the results of the search engines are tutorials, not function definitions. That makes **less experienced programmers** spend more time to search function definitions in local tutorial pages. Furthermore, some unpopular functions, which have very few tutorials, are hard to be found.

Dinh Huy Tran · Hua Phung Nguyen · Duy Hanh Le
Faculty of Computer Science & Engineering,
Ho Chi Minh City University of Technology, Vietnam
268 Ly Thuong Kiet, Ho Chi Minh, Vietnam

© Springer International Publishing Switzerland 2015          143
V.-H. Nguyen et al. (eds.), *Knowledge and Systems Engineering,*
Advances in Intelligent Systems and Computing 326, DOI: 10.1007/978-3-319-11680-8_12

Many specific search engines, called code search engines, have been proposed to find function definitions directly. Most of them match keywords in queries to words in code, comments, names or types of functions. There are some challenges in this traditional approach: function names do not always reflect user's desired requests, code is not well-documented to search, and comments are very little in function definitions. It makes a gap between programmer's requirements normally described in natural language and function definitions represented in programming languages.

To close this gap, a human-friendly structure called API-Doc is applied in recent researches. API-Doc provides function specifications which are written in natural language to describe what the purposes of functions are. API-Doc can be considered as a multi-level structure. For instance, Java API-Doc is a three-level structure, where the first level is the description for category *package*, the second for category *class* in *package* and the third for document *function* inside *class*.

Unfortunately, API-Doc based approaches are normally ineffective because of keyword mismatches between function descriptions and queries. The reason is that function descriptions in API-Doc are very short documents (less than 20 words). In information retrieval, information of short documents is important but very limited [19].

To deal with this problem, this paper proposes a novel function search approach combining keyword-based retrieval and semantic-based ranking. There are three main technologies in our approach: query expansion, keyword weighting, semantic ranking. This approach uses the query expansion technology to add related keywords to a query, uses the keyword weighting technology to retrieve candidates and uses the semantic ranking technology to refine solution.

**Query expansion** is a technique that attempts to reduce the number of keyword mismatches between queries and relevant short documents. Almost query expansion methods [3] find related words from exist dictionaries like Wordnet, Wikipedia, thesaurus, and so forth. Other methods [19] use word co-occurrence statistical information of document set or queries to find related words. Our approach uses the thesaurus-based query expansion method because it is simple and requires no additional resources.

**Keyword weighting** is a technique that measures how important a keyword is. Keyword frequency statistics on documents are used in our approach. In this paper, because function descriptions are short documents with very limited information, we consider keyword frequency statistics not only on documents but also on categories containing the documents. For instance, to determine how important a keyword in *function pow* is, we measure the frequency of the keyword in the description of the function and categories *class math* and *package java.lang* to which the function belongs.

**Semantic ranking** is a technique that calculates semantic similarities between documents and queries. In our approach, we take all word-to-word similarities of every pair document-query into a bipartite graph. After that, based on the semantic similarity graph, a highest similarity score is determined. The time complexity of measuring semantic similarity is $O(document\_length^2 * query\_length^2)$ which is so

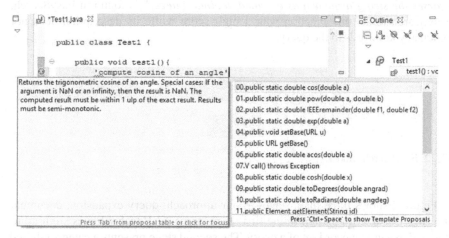

**Fig. 1** A screenshot of EasySearch being used within Eclipse IDE

high and depends on document length, so this technology is only suitable for short documents like Java API-Doc.

Finally, a code search engine, named EasySearch, is implemented and applied on the Java standard library. Figure 1 shows a screenshot of EasySearch running within the Eclipse [17] development environment. This tool is integrated into Eclipse as a plug-in. The right-bottom corner of the screenshot shows a list of relevant functions for the query *"compute cosine of an angle"*.

The contributions of the paper are as follows:

☐ Proposing a novel search method combining keyword-based quick search and semantic-based precise search. This method allows newbie programmers, who rarely know function names or related keywords, to describe the purposes of functions in natural language queries.

☐ Introducing an approach to mine structure and contents of API-Doc. The *n*-level categorized structure of API-Doc is used in keyword weighting. Short function descriptions of API-Doc are considered in semantic comparison.

The outline for the rest of the paper is as follows. Section 2 presents our approach in detail. Section 3 presents the results of this evaluation which is also compared against the performance of previous approaches. Section 4 introduces related approaches and their features. Conclusions and suggestions for future work are presented in Section 5.

## 2 Our Approach

Our goal is to develop a search engine that quickly searches relevant functions based on semantics instead of keywords. For instance, using the following query *"Convert to number"*, an unlikely candidate function *parseInt(String s)* with a description *"Parses the string argument as a signed decimal integer"* is found in EasySearch. EasySearch ranks this function with a high score while its description does not match any keywords to the query.

**Fig. 2** EasySearch Model

Figure 2 shows three main stages in our approach: query expansion, document filtering, document ranking. In the first stage, some NLP technologies are used to find and expand keyword set of a query. The second stage presents a category-based keyword weighting method to measure how a document is suitable for a query. Finally, the filtered documents are sorted based on their semantics in the third stage.

## 2.1 Query Expansion

This stage maximizes probability of finding relevant documents. This stage has two main steps: extracts a keyword set from the query based on stopword filtering and part-of-speech, and expands the keyword set by stemming and synonym.

**Stopword Removal:** Stopwords are the most common words such as *the, is, at, which,* and so forth. According to our word frequency statistic on API-Doc, over 3500 nouns and 500 verbs retrieved from API-Doc of JDK library, there are about 2700 nouns and 400 verbs appeared less than 10 times. Only 6 nouns and 3 verbs appear more than 900 times, there are *string*-1111, *method*-1344, *component*-1594, *value*-2579, *object*-3410, *get*-901, set-1559, *return*-5898. Based on this statistic, we built a stopword list for our application.

**Part-of-Speech Based Filtering:** Prepositions, pronouns, conjunctions, and interjections are often defined as stopword which need to be removed from keyword set. In our approach, Stanford-Parser [15] is used to mark up a word in a natural language sentence (NL Sentence) as corresponding to a particular part-of-speech based on both its definition and context.

**Stemming** is the technology to reduce inflected (or sometimes derived) words to their root form, for instance, the verb *return* is the root form of words *"returns, returning, returned"* after stemming to remove suffixes such as *"-s, -ing, -ed"* respectively. This technology also supports the next stage to change keywords in difference forms into a same form before calculating the similarity between them.

**Synonym Generation:** Finally, keyword set is expanded with synonyms using WordNet [16]. In WordNet, words are grouped into sets of synonyms called synsets.

It mean that the words belonging to the same synset have the similar meaning. We also defined a set of synonyms for API-Doc based on our knowledge. For instance, *append-add-insert* are synonymous in API-Doc scope.

## 2.2 Document Filtering

The main objective of this stage is to filter out irrelevant documents based on a keyword weighting method. Because the time complexity of measuring semantic similarity in the next stage is high, a filtering threshold needs to be considered carefully. If the threshold is really low, runtime may be large because a lot of documents need to be ranked in the next stage. On the contrary, many correct documents may be filtered out.

There are many keyword weighting methods introduced in [25] such as Term Frequency - Inverse Document Frequency (TFIDF), Odds Ratio (OR), Mutual Information (MI), Information Gain (IG), Chi-Squared ($X^2$), Bi-Normal Separation (BNS). In this paper, TFIDF is applied to mine contents of categories and documents. In the remaining methods, BNS is chosen to mine relationship between documents in a same category. According to many experiments [7], which compared the performance of BNS against several other approaches, BNS produced the best results.

For $n$ nested categories $C_1 \subset C_2 \subset ... \subset C_n$, $D$ is a document in the category $C_1$, k is a keyword in a query $Q$. To keep the simplicity of our idea, the document $D$ is considered as a lowest-level category $C_0$. The similarity between the document $D$ and the query $Q$ is calculated as follows:

$$S(Q,D) = \sum_{k \in Q} \sum_{i=0}^{n-1} W(k,C_i,C_{i+1}) \tag{1}$$

$$W(k,C_i,C_{i+1}) = W_{TFIDF}(k,C_i,C_{i+1}) + W_{BNS}(k,C_i,C_{i+1}) \tag{2}$$

$W_{TFIDF}(k,C_i,C_{i+1})$ is the TFIDF method used to find the most important keywords for the document within a category by assessing how often the keyword occurs within the document (TF) and how often in other documents (IDF):

$$W_{TFIDF}(k,C_i,C_{i+1}) = -\log \frac{df(k,C_{i+1})}{N} \times \frac{tf(k,C_i)}{|C_i|} \tag{3}$$

Where $tf(k,C_i)$ is the term frequency of word $k$ within description of subcategory $C_i$, $|C_i|$ is the length of the subcategory, $df(k,C_{i+1})$ is the number of subcategories in category $C_{i+1}$ which the word $k$ appears in, and $N$ is the number of subcategories in the category $C_{i+1}$. Note that, if the subcategory $C_i$ does not have description, $W_{TFIDF}$ equals zero.

$W_{BNS}(k,C_i,C_{i+1})$ is the BNS method used to find the most important keywords based on the difference between two distributions "*how often the keyword occurs in the positive samples*" and "*how often the keyword occurs in negative samples*". Suppose that category $C_{i+1}$ is divided into four part: Part $X$ is the number of documents in subcategory $C_i$ containing k, part $Y$ is the number of documents out of $C_i$ containing $k$, part $Z$ is the number of documents in $C_i$ without $k$, part $T$ is the number of documents that is not in $C_i$ and do not include $k$.

$$W_{BNS}(k,C_i,C_{i+1}) = F^{-1}(\frac{X}{X+Z}) - F^{-1}(\frac{Y}{Y+T}) \tag{4}$$

Where $F^{-1}$ is the inverse Normal cumulative distribution function. As the inverse Normal would be infinite at 0 and 1, Forman limited both distributions to the range [0.0005,0.9995].

## 2.3 Document Ranking

This stage ranks the candidate documents based on semantic comparison. In this stage, each document-query pair is represented in a semantic similarity graph, where all word-to-word similarities of the document-query pair are measured by using a function of the number of matching sequences of keywords. In the semantic similarity graph, nodes represent words, while edges show semantic similarities between these words.

For a bipartite graph $M$, where the query $Q$ and the document $D$ are two sets of disjoint nodes. An element $e_{ij}$ in the $i$-th row and $j$-th column represents the semantic similarity of assigning the $i$-th keyword $q_i$ in $Q$ to the $j$-th keyword $d_j$ in $D$.

$$e_{ij} = Pos(q_i,d_j) * Sim(q_i,d_j) \tag{5}$$

$Pos(x,y)$ is part-of-speech based weight. Different part-of-speeches get a different weight in the final score. In the part of speeches, verb get the highest weight, because recognizing the verb is the most important step in understanding the meaning of a sentence. Verb is the heart of a sentence, and every sentence must have at least one verb. In the rest of types, noun is more important than others because it is the original and central building blocks of a sentence. Note that, if the part of speeches of $x$ and $y$ are different, this weight equals 0.

$Sim(x,y)$ uses WuPalmer [28] method to measure semantic similarity between the words $x$ and $y$. The WuPalmer measure calculates relatedness by considering the depths of the two words $x$ and $y$ in the WordNet taxonomies, along with the depth of the least common subsumer.

$$Sim(q_i,d_j) = \frac{2*N}{N_1+N_2} \tag{6}$$

Where $N_1$ and $N_2$ are distances from the root to nodes $q_i$ and $d_j$ and $N$ is a distance from the root to the closest common ancestor of $q_i$ and $d_j$.

Finally, semantic similarity measure between $Q$ and $D$ is:

$$R(Q,D) = \frac{maximum(M)}{|Q|+|D|} \tag{7}$$

Where $maximum(M)$ is the function to maximize sum of matching weights of the bipartite graph $M$, $|Q|$ and $|D|$ are the lengths of the query $Q$ and the document $D$ respectively.

# 3 Experiment

Three evaluation criteria *precision, recall, F-measure* are used to measure the effectiveness of search engines. *Precision* is the fraction of retrieved documents that are relevant to the find. *Recall* is the fraction of the documents that are relevant to the query that are successfully retrieved. *F-measure* is the harmonic mean of *Precision* and *Recall*.

In our experiment, *Recall* is difficult to measure because it requires the knowledge of the total number of relevant items in the database. To overcome this problem, a human relevance judgment approach [27] is applied to find the relevant document set.

## 3.1 User Study

Our goal is to evaluate how well these participants can find functions that match given tasks using three different search engines: EasySearch, Krugle [13], Ohloh [6]. This evaluation focuses on demonstrating that our approach is better than the other approaches in long queries.

**Participants:** We recruited eleven participants who are the third and fourth year students from the Faculty of Computer Science & Engineering, HCMC University of Technology. The reason for this recruitment is that they are newbie programmers who are less experience with code search. In search tasks, they rarely know function names or related keywords, so they often use many words to express their queries.

**Search Engines:** Krugle and Ohloh, which are popular search engines with the large open source Java project repository, were used to compare to our code search engine EasySearch. The features of three search engines are shown in Table 1.

In our study, the search engines are configured to find Java function definitions as follows:

☐ Krugle: In "*Advanced Search*", choose "*Function Definition Search*". After performing a query, select "*Document Types*" as "*Java*" to filter out non-Java functions.

☐ Ohloh Code: Establish the query form as "$fndef : \langle content \rangle$". The keyword "*fndef*" indicates that search objects are function definitions. After performing a query, select "*Languages*" as "*Java*" to filter out non-Java functions.

**Tasks:** Twenty tasks were designed to find Java function definitions and assigned to the participants. In a task, every participant suggested at least three query sentences which were applied to all search engines. With every solution of a query

**Table 1** Search Engine Features

|  | Ohloh Code | Krugle | EasySearch |
|---|---|---|---|
| Indexed Files | 1 million | 10 millions | 170.000 |
| Java Files | 600.000 | 3.5 millions | 40.000 |
| Search Method | Keyword-based | Keyword-based | Semantic-based |
| Filtering | By file types, class, functions, and interface. | By comments, source code, functions, function calls, and classes. | None |
| Regular Expressions | Yes | No | No |
| Result | Function definition, function call, class definition, ... | Function definition, function call, class definition, ... | Function definition |
| Matching | Syntax recognition of source code | Syntax recognition of source code | Description in API-Doc |

on a search engine, the participants considered top 20 results and selected related functions based on their descriptions. A logger is used to save the participants's selections. Task descriptions should be common and flexible enough to allow the newbie participants to suggest different queries for searching. The difficulty levels of tasks depends on how many words they require to express the queries. The followings are some of our task descriptions: *"How to detect if Socket is closed"*, *"How to capture an event when a mouse button is clicked"*, and *"How to calculate a maximum between two numbers"*.

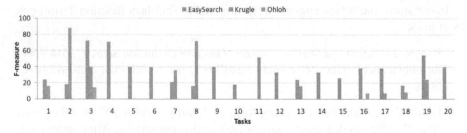

**Fig. 3** F-measure of search engines according to tasks

## 3.2   Result

Figure 3 plotted F-measure values of three search engines for twenty tasks. Twenty clustered columns displayed experiment results of twenty tasks. Every column of a clustered column displayed an F-measure value of a search engine. In Figure 3, EasySearch had the highest F-measure value in most tasks, while the least F-measure value was for Ohloh. The reason is that Krugle and Ohloh are ineffective in long queries while less experienced participants needs many words to describe their questions. According to our statistic on 700 queries of the participants, average length of all queries is 4.52, the longest query length is 11 words, and about 75% of queries are more than 3 words.

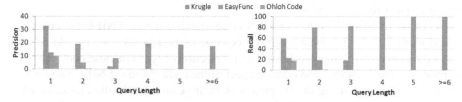

**Fig. 4**  Precision and Recall of search engines based on the query length

In Figure 4, Krugle and Ohloh gave good results with short queries less than three words. With longer queries, most relevant functions were skipped in their top 20 results. When query length was greater than three words, both Krugle and Ohloh did not find any relevant functions for all queries. On the contrary, EasySearch got better solutions in the longer queries.

The superior results of EasySearch over Krugle and Oholh on long queries might come from two reasons. Krugle and Oholh do not recognize which are important keywords in long queries, while EasySearch can retrieve them by using keyword weighting methods. They ineffectively match keywords in a query to source code which is not well-documented, while EasySearch matches keywords based on API-Doc which is a human-friendly document.

## 4   Related Work

In code search, a common approach finds relevant components based on component names, comments and source code. **Codebroker** [29] uses source code and comments to query code repositories to find relevant artifacts. Codebroker is dependent upon the descriptions of documents and meaningful names of program variables and types, and this dependency often leads to lower precision of returned projects. **CodeFinder** [10] index components based on terms extracted from their names and comments. CodeFinder allows a programmer to define new terms or remove exist terms. CodeFinder heavily depends on ad-hoc descriptions of components whose is of lower quality than API-Doc. **SSI** [2] indexes each component based on the identifier names and comments. SSI seeds the index with keywords from API call documentation. SSI ranks source code based on the keyword occurrences in the source code.

Different code mining techniques and tools have been proposed to refine process of code search. **Sando** [8] supports a programmer by recommending queries before starting searching or after searching fails (returns no results). Sando uses synonymy to expand queries, while EasySearch uses both synonymy and hypernymy to compare documents and queries. **S6** [20] finds functions or code fragments based on a programmer's detailed requirements such as description, specification, test cases, and so forth. To find a function definition exactly, S6 requires a programmer to know low-level details of this function. **Hipikat** [5] retrieves functions from a software system's history that may be of relevance to a programmer working on software evolution tasks for that system.

Recently, API-Doc is mined in many code search engines. **SNIFF** [4] collects descriptions in API-Doc to annotate source code in its data set code by adding its own comments. SNIFF then performs the intersection of types in these code chunks to retain the most relevant and common part of the code chunks. SNIFF also ranks these pruned chunks using the frequency of their occurrence in the indexed code base. **Exemplar** [9] considers API-Doc as an integral instrument to expand the range of a query. Exemplar uses relations between API calls in the retrieved projects to compute the level of ranking of the project. **Mica** [23] mines API-Doc to refine the returned results. **Exoa** [12] uses API-Doc to find high-quality code examples.

In function search, a branch of code search, many approaches have been introduced to find relevant functions. **Sourcerer** [1] indexes a function by its name, source code, comments. Sourcerer clusters functions with similar names into a category and entities by their fully qualified names to refine results of search. Sourcerer returns a poor result if a query does not contain any keywords from names of appropriate functions, while EasySearch ranks these functions with high scores even though their descriptions may not match any keywords to the query. **Portfolio** [18] find relevant functions and their usages based on behavior of programmers and associations of terms in functions in the search graph. Portfolio uses *"Pagerank"* and *"Spreading Activation Networks"* to help programmers navigate and understand usages of retrieved functions, while EasySearch uses structure and contents of API-Doc to refine solutions. **FSE** [26] uses API-Doc to find function definitions and generate function calls automatically. FSE considers contents of API-Doc as constraints to filter out irrelevant functions, while EasySearch uses the multi-level categorized structure of API-Doc to retrieve candidates.

Furthermore, many approaches even support programmers to understand how to use components by example code. **Strathcona** [11] uses the structural context of the code under development to query example code. Strathcona uses heuristics to matches the code structure to the example code. **XSnippet** [21], **Prospector** [14] and **PARSEWeb** [24] are all systems designed to provide examples of object instantiation. Their queries take the form "Source object type $\rightarrow$ Destination object type" as input, and relevant function-invocation sequences are suggested as solutions.

# 5   Conclusions and Future Work

This paper proposed an efficient function search approach based on API-Doc. This paper implemented EasySearch, a code search engine that helps programmers to quickly find different functions suitable for their tasks. Some evaluations were performed to demonstrate that our tool was better than the other code search engines in precision, recall and F-measure.

In future work, the context of the code a programmer is currently working on can be used to improve the ranking method. For example, functions belonging to classes declared in the code a programmer is writing may get higher rank in the search result. Moreover, we also plan to mine other contents of API-Doc such as function specifications, parameters, class hierarchy and so forth.

**Acknowledgment.** This paper is funded by the research project T-KHMT-2014-29 from Ho Chi Minh University of Technology. We are immensely grateful to particulars who took the time to participate in our study and help us to prepare this paper.

# References

[1] Bajracharya, S., Ossher, J., Lopes, C.: Sourcerer: An infrastructure for large-scale collection and analysis of open-source code. Science of Computer Programming 79, 241–259 (2014)

[2] Bajracharya, S.K., Ossher, J., Lopes, C.V.: Leveraging usage similarity for effective retrieval of examples in code repositories. In: Proceedings of the Eighteenth ACM SIGSOFT International Symposium on Foundations of Software Engineering, pp. 157–166. ACM (2010)

[3] Carpineto, C., Romano, G.: A survey of automatic query expansion in information retrieval. ACM Comput. Surv. 44(1), 1:1–1:50 (2012)

[4] Chatterjee, S., Juvekar, S., Sen, K.: Sniff: A search engine for java using free-form queries. In: Chechik, M., Wirsing, M. (eds.) FASE 2009. LNCS, vol. 5503, pp. 385–400. Springer, Heidelberg (2009)

[5] Cubranic, D., Murphy, G.C., Singer, J., Booth, K.S.: Hipikat: A project memory for software development. IEEE Trans. Softw. Eng. 31(6), 446–465 (2005)

[6] Duck, B.: Ohloh Code Search (January 2014), http://code.ohloh.net

[7] Forman, G.: BNS feature scaling: an improved representation over tf-idf for svm text classification. In: Proceedings of the 17th ACM Conference on Information and Knowledge Management, pp. 263–270. ACM (2008)

[8] Ge, X., Shepherd, D., Damevski, K., Murphy-Hill, E.R.: How the sando search tool recommends queries. CoRR, abs/1401.6931 (2014)

[9] Grechanik, M., Fu, C., Xie, Q., McMillan, C., Poshyvanyk, D., Cumby, C.: A search engine for finding highly relevant applications. In: Proceedings of the 32nd ACM/IEEE International Conference on Software Engineering, ICSE 2010, vol. 1, pp. 475–484. ACM, New York (2010)

[10] Henninger, S.: Supporting the construction and evolution of component repositories. In: Proceedings of the 18th International Conference on Software Engineering, pp. 279–288. IEEE Computer Society (1996)

[11] Holmes, R., Murphy, G.C.: Using structural context to recommend source code examples. In: Proceedings of the 27th International Conference on Software Engineering, ICSE 2005, pp. 117–125. ACM, New York (2005)
[12] Kim, J., Lee, S., Hwang, S.-W., Kim, S.: Towards an intelligent code search engine. In: AAAI (2010)
[13] Krugle. Krugle (January 2014), http://www.krugle.com
[14] Mandelin, D., Xu, L., Bodík, R., Kimelman, D.: Jungloid mining: Helping to navigate the api jungle. SIGPLAN Not. 40(6), 48–61 (2005)
[15] Manning, C., Klein, D.: The Stanford parser (January 2014), http://nlp.stanford.edu/software/lex-parser.shtml
[16] Princeton University. Wordnet (January 2014), http://wordnet.princeton.edu/
[17] The Eclipse Foundation. Eclipse (January 2014), http://www.eclipse.org/
[18] McMillan, C., Grechanik, M., Poshyvanyk, D., Xie, Q., Fu, C.: Portfolio: finding relevant functions and their usage. In: 2011 33rd International Conference on Software Engineering (ICSE), pp. 111–120. IEEE (2011)
[19] Petri, M., Culpepper, S.J., Moffat, A.: Exploring the magic of wand. In: Proceedings of the 18th Australasian Document Computing Symposium, pp. 58–65. ACM (2013)
[20] Reiss, S.P.: Semantics-based code search. In: IEEE 31st International Conference on Software Engineering, ICSE 2009, pp. 243–253. IEEE (2009)
[21] Sahavechaphan, N., Claypool, K.: Xsnippet: mining for sample code. ACM Sigplan Notices 41(10), 413–430 (2006)
[22] Singer, J., Lethbridge, T., Vinson, N., Anquetil, N.: An examination of software engineering work practices. In: CASCON First Decade High Impact Papers, CASCON 2010, Riverton, NJ, USA, pp. 174–188. IBM Corp. (2010)
[23] Stylos, J., Myers, B.A.: Mica: A web-search tool for finding api components and examples. In: IEEE Symposium on Visual Languages and Human-Centric Computing, VL/HCC 2006, pp. 195–202. IEEE (2006)
[24] Thummalapenta, S., Xie, T.: Parseweb: a programmer assistant for reusing open source code on the web. In: Proceedings of the Twenty-Second IEEE/ACM International Conference on Automated Software Engineering, pp. 204–213. ACM (2007)
[25] Timonen, M., et al.: Term weighting in short documents for document categorization, keyword extraction and query expansion (2013)
[26] Tran, D., Nguyen, H.: API specification-based function search engine using natural language query. In: 2013 International Conference on Computing, Management and Telecommunications (ComManTel), pp. 140–145. IEEE (2013)
[27] Vaughan, L.: New measurements for search engine evaluation proposed and tested. Information Processing & Management 40(4), 677–691 (2004)
[28] Wu, Z., Palmer, M.: Verbs semantics and lexical selection. In: Proceedings of the 32nd Annual Meeting on Association for Computational Linguistics, pp. 133–138. Association for Computational Linguistics (1994)
[29] Ye, Y., Fischer, G.: Supporting reuse by delivering task-relevant and personalized information. In: Proceedings of the 24th International Conference on Software Engineering, pp. 513–523. ACM (2002)

# A Fast Method for Motif Discovery in Large Time Series Database under Dynamic Time Warping

Cao Duy Truong and Duong Tuan Anh

**Abstract.** Finding similar time series in a time series database is one of the important problems in time series data mining. The time series which has the highest count of its similar time series within a range $r$ in a time series database is called the *1-motif*. The problem of time series motif discovery has attracted a lot of attention and is useful in many real world applications. However, most of the proposed methods so far use Euclidean distance to deal with this problem. In this work, we propose a fast method for time series motif discovery which uses Dynamic Time Warping distance, a better measure than Euclidean distance. Experimental results showed that our proposed motif discovery method performs very efficiently on large time serried datasets while brings out high accuracy.

**Keywords:** time series, motif discovery, DTW distance.

## 1 Introduction

Time series data arises in so many applications of various areas ranging from business, finance, medicine, meteorology, environment to government. The problem of finding similar time series in a time series database has attracted a lot of attention in research community. The time series which has the highest count of its similar time series within a range $r$ is called the *1-motif*. There have been several algorithms for time series motif discovery in literature ([7], [8], [12]). However, most of the proposed methods so far use Euclidean distance to deal with this problem. For example, in 2007, Xi et al. proposed a method for motif discovery in a time series database of shapes [13], but they applied SAX discretization and Euclidean distance in this work. However, Euclidean distance is not a suitable measure for working on time

Cao Duy Truong · Duong Tuan Anh
Faculty of Computer Science and Engineering,
Ho Chi Minh City University of Technology,
Ho Chi Minh, Vietnam
e-mail: caoduytruong@hcmunre.edu.vn,
        dtanh@cse.hcmut.edu.vn

© Springer International Publishing Switzerland 2015                                      155
V.-H. Nguyen et al. (eds.), *Knowledge and Systems Engineering,*
Advances in Intelligent Systems and Computing 326, DOI: 10.1007/978-3-319-11680-8_13

series in various fields, for example, multimedia data. In [1], Ding et al. pointed that Dynamic Time Warping (DTW) is the best measure for various kinds of time series data. Nevertheless, it has been suggested many times in the literature that DTW incurs high computational complexity and that is the main obstacle of using DTW in time series data mining tasks.

There have been several research works in speeding up the computation of DTW distance. Most recently, in 2012, Rakthanmanon et al. [9] introduced a suite of techniques for speeding up DTW similarity search on very large time series datasets (consisting of trillions of time series subsequences). However the complexity of similarity search on $N$ time series subsequences is just about $O(N)$ while the complexity of motif discovery on N time series subsequences will cost about $O(N^2)$. Due to this difficulty, so far there is no research work on time series motif discovery under DTW distance.

In this work, we propose a fast method for time series motif discovery using DTW distance. To the best of our knowledge, this is the first method for motif discovery with DTW distance. In this method, besides applying a suite of speedup techniques for DTW computation given in [9], we suggest two improvements: the use of Euclidean distance as upper bound for DTW distance and the use of an index structure for this upper bound in order to accelerate the motif discovery process. We conducted experiments on our proposed method over several different time series datasets. Experimental results showed that our motif discovery method under DTW distance can perform very fast on very large time series datasets while brings out high accuracy.

## 2   Background

### 2.1   Definitions

**Definition 1.** *Time Series*: A time series $T = t_1, \ldots, t_n$ is an ordered set of $n$ real-values measured at equal intervals.

**Definition 2.** *Time Series Database (D)* is an unordered set of $m$ time series possibly of different lengths.

**Definition 3.** *Similar time series*: $Dist(T_i, T_j)$ is a positive value used to measure the difference between two time series $T_i$, and $T_j$, based on some measure method. If $Dist(T_i, T_j) \leq r$, where $r$ is a real number (called range), then $T_i$ is considered as similar to $T_j$.

**Definition 4.** *The Euclidean distance (ED)* between two time series $Q$ and $C$, where $|Q| = |C|$, is defined as

$$ED(Q,C) = \sqrt{\sum_{i=1}^{n} dist(q_i, c_i)}$$
$$\text{where } dist(q_i, c_i) = (q_i - c_i)^2$$

**Definition 5.** *Warping path*: Given two time series $Q$ and $C$, both of length $n$, to compare these two time series using DTW, an n-by-n matrix is constructed, where the $(i_{th}, j_{th})$ element of the matrix is the distance $dist(q_i, c_j)$. A warping path $P$ is a contiguous set of matrix elements that defines a mapping between $Q$ and $C$. The $t^{th}$ element of $P$ is defined as $p_t = (i, j)_t$, so we have:

$$P = p_1, p_2, ..., p_t, ..., p_T \qquad n \leq T < 2n - 1$$

The Euclidean distance between two time series can be seen as a special case of DTW distance where the $k^{th}$ element of P is constrained such that $p_k = (i, j)_k, i = j = k$.

To speed up the DTW distance calculation, all practitioners using DTW constrain the warping path in a global manner by limiting how far it may stray from the diagonal. The subset of matrix that the warping path is allowed to visit is called *warping window* or a *band*. Two of the most frequently used global constraints in the literature are the Sakoe-Chiba band proposed by Sakoe and Chiba, 1978 [11] and Itakura Parallelogram proposed by Itakura, 1975 [2]. Sakoe-Chiba band is the area defined by two straight lines in parallel with the diagonal and Itakura Parallelogram is the area defined by the parallelogram which is symmetric over the diagonal.

In this work, we use the Sakoe-Chiba Band with width $R$. We use DTW distance in matching two time series $Q$ and $C$. Two time series $Q$ and $C$ are similar to each other within range $r$ if $DTW(Q, C) \leq r$. As mentioned in [10] that similarity search on the time series with different lengths using DTW is very hard and meaningless, in our work, if the lengths of time series are different, we apply some interpolation technique to convert them to the same length before doing motif discovery. Here, in this paper we assume that all the time series are of the same length.

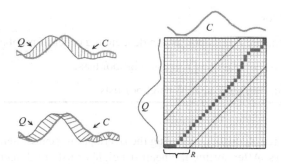

**Fig. 1** (Left) Two time series Q and C which are similar but out of phase: using Euclidean distance (top left) and DTW distance (bottom left). Note that Sakoe-Chiba Band with width R is used to constrain the warping path (right).

**Definition 7.** *Time series motif*: Given a time series database $(D)$, a time series $T_i$ in $D$ is the most-significant motif in D (called the *1-motif* ), if $T_i$ has the highest count of the time series $T_j$ which are similar to $T_i$ in D.

The *K-th* most significant motif in *D* (called *K-motif*) is the time series $T_k$ that has the *K-th* highest count of the time series which are similar to $T_k$ in *D*.

All time series that are similar to the *K-motif* are called instances of that *K-motif*. Note that Definition 7 forces the sets of instances of the motifs to be mutually exclusive.

## 2.2 Brute-Force Algorithm

The outline of the brute-force algorithm for discovering the *K-motif* in a time series database *D* is given in Table 1. This algorithm has the complexity $O(N^2)$.

**Table 1** The brute-force algorithm for discovering motif in a time series database

| **Algorithm. Motif-Brute-Force-DTW** |
| --- |
| 1.       for $i = 1$ to $N$ |
| 2.          for $j = i + 1$ to $N$ |
| 3.             if $DTW(T_i, T_j) \leq r$ then |
| 4.                append $i$ to nearest neighbor list of $j$; |
| 5.                append $j$ to nearest neighbor list of $i$; |
| 6.             end if; |
| 7.          end for; |
| 8.       end for; |
| 9.       sort nearest neighbor lists in descending order of number of instances; |
| 10.      remove overlapping nearest neighbor lists; |
| 11.      return the *top-k* nearest neighbor lists; |

In the brute-force algorithm, removing the overlapping nearest neighbor lists can be done as follows. After sorting the nearest neighbor lists in descending order of the number of instances, we compare repeatedly one of the nearest neighbor lists at the beginning of the order to one of the nearest neighbor lists at the end of the order, if they are overlapping, we remove the nearest neighbor list at the end of the order.

## 2.3 Speeding Up by Using Lower Bound

In this section, before introducing our proposed algorithms for discovering motif in time series under DTW, we describe some techniques to speed up the computation of

DTW distance as well as to reduce the number of times invoking the DTW distance computation. These techniques was introduced in [9].

## Using the Squared Distance

Both DTW and ED have a square root calculation. However, as suggested in [9], if we omit this step, it does not change the relative rankings of nearest neighbors. In this work, we will still use DTW and ED; however, the reader may assume that we mean the squared versions of them.

## Lower Bounding

To speed up the computation of DTW distance between two time series $Q$ and $C$, besides using warping window, researchers proposed some cheap-to-compute lower bound techniques ([3], [5], [6], [14]). In a DTW lower bounding technique, the lower bound of the DTW distance between two time series $Q$ and $C$, denoted by $LB(Q,C)$, must satisfy the following inequality:

$$LB(Q,C) \leq DTW(Q,C)$$

The computation cost of $LB(Q,C)$ is much less than that of $DTW(Q,C)$. If $LB(Q,C) > r$, then $DTW(Q,C) > r$ and therefore we do not have to compute $DTW(Q,C)$ and conclude that the two time series $Q$ and $C$ can not be similar to each other.

In 2001 [5], Kim et al. proposed a lower bounding technique for DTW called *LB_Kim*. LB_Kim extracts a four-tuple feature vector from each time series. The features are the first and last data points of the time series, together with the maximum and minimum values. The maximum squared differences of corresponding features are used as the lower bound. The complexity of *LB_Kim* is $O(n)$. In 2012, Rakthanmanon et al. improved *LB_Kim* to *LB_KimFL*, with the complexity $O(1)$. *LB_KimFL* uses only the distances between the first (last) pair of points from both time series. In this work, we also use the *LB_KimFL* as proposed in [9].

One of the most popular lower bounding techniques for DTW is *LB_Keogh* [3]. *LB_Keogh* has the complexity $O(n)$. In [9], Rakthanmanon et al. use *LB_Keogh* in similarity search on very large time series. In 2009, Lemire proposed the lower bounding technique called *LB_Improved* [6] which is an improved variant of *LB_-Keogh*. Lemire has proved that *LB_Improved* is a tighter lower bound than *LB_-Keogh*. Instead of using *LB_Keogh* to speed up DTW distance computation, in this work we use *LB_Improved*.

We use the two lower bounding techniques (*LB_KimFL* and *LB_Improved*) in a cascade. We first use the *LB_KimFL*. If a candidate is not pruned at this stage, we turn to the *LB_Improved*.

**LB Reversing**

Since $LB(Q,C) \neq LB(C,Q)$, after checking whether $LB(Q,C) < r$ or not, if we can not prune off a candidate, we can continue to compute $LB(C,Q)$ in the same way given in [9].

**Look-Ahead of DTW**

For $1 < k < n$, the following inequality will be satisfied: $DTW(Q_{1:k},C_{1:k}) + LB(Q_{k+1:n},C_{k+1:n}) \leq DTW(Q_{1:n},C_{1:n})$ [9]. Therefore, during the computation of $DTW(Q,C)$ at any time point $k$, if $DTW(Q_{1:k},C_{1:k}) + LB(Q_{k+1:n},C_{k+1:n}) > r$ then we can conclude that the two time series $Q$ and $C$ can not be similar to each other and stop the computation of $DTW(Q,C)$ at that time point.

**Early Abandoning**

During the computation of the lower bound $LB$ or $DTW$, if we note that the current sum of the squared differences between each pair of corresponding data points exceeds the range $r$, we can stop the calculation at that data point and conclude that $DTW(Q,C) > r$ and $Q$ and $C$ are not similar to each other.

**Algorithm MBF-DTW-LB**

We can improve the brute-force algorithm in Table 1 by adding into the algorithm some speedup techniques we have just mentioned above. This new algorithm is called *MBF-DTW-LB* (Motif discovery using Brute Force with DTW and Lower Bounding) in this work.

## 3 The Proposed Algorithms

### 3.1 Speeding Up by Using Upper Bound

**Upper Bound by Euclidean Distance**

Note that $ED(Q,C)$ is a special case of $DTW(Q,C)$, when the warping band $R$ is set to 1. We have the following inequality: $DTW(Q,C) \leq ED(Q,C)$. If $ED(Q,C)$ is within the range $r$, then we have $DTW(Q,C) \leq r$. $ED(Q,C)$ can be used as an upper bound of $DTW(Q,C)$. So the computation of $DTW(Q,C)$ includes the computation of $ED(Q,C)$ at the point $p_k = (i,j)_k, i = j = k$. To speed up the computation of DTW, we first compute $ED(Q,C)$ and during the process of computing $ED(Q,C)$ we store the values $p_k = (i,j)_k, i = j = k$ which may be used in the later computation of $DTW(Q,C)$ if necessary. When the computation of $ED(Q,C)$ finishes, if $ED(Q,C) \leq r$ then we can stop the computation and conclude that $Q, C$ are similar.

**Look-Ahead of Euclidean Distance**

Similar to the look-ahead of DTW, we have the following inequality:

$DTW(Q_{1:n}, C_{1:n}) \leq DTW(Q_{1:k}, C_{1:k}) + ED(Q_{k+1:n}, C_{k+1:n}), 1 < k < n$ Therefore, during the computation of $DTW(Q,C)$ at any time point $k$, if

$DTW(Q_{1:k}, C_{1:k}) + ED(Q_{k+1:n}, C_{k+1:n}) \leq r$, then we can conclude that the two time series $Q$ and $C$ can be similar to each other and stop the computation of $DTW(Q,C)$ at that time point.

**Algorithm MBF-DTW-LB-UB**

We can improve the brute-force algorithm in Table 1 by adding into the algorithm the two speedup techniques we have just mentioned above. This new algorithm is called *MBF-DTW-LB-UB* (Motif discovery using Brute-Force with DTW and Lower Bound - Upper Bound.)

## 3.2   Using Wedge-Tree

Given a set of time series $Cset = C_1, .., C_k$, we can form two new time series $U$ and $L$ using the following formula: $U_i = max(C_{1i}, .., C_{ki})$ and $L_i = min(C_{1i}, .., C_{ki})$. $U$ and $L$ stand for upper and lower bound of $Cset$, respectively. They form the smallest possible bounding envelope that encloses all members of the set $Cset$ from above and below. For notational convenience, we will call the combination $U, L$ a wedge, and denote a wedge as $W : W = \{U, L\}$. We have the following inequality: $DTW(C_i, C_j) \leq ED(C_i, C_j) \leq ED(U, L), \forall C_i, C_j \subset Cset$ For convenience, we denote $W_{UL} = ED(U, L)$.

The expansion of $W = \{U, L\}$ of $Cset$ when inserting a new time series $C$ to $Cset$ can be done simply as follows: $U_i = max(U_i, C_i)$ and $L_i = min(L_i, C_i)$. Similarly, when we have two sets of time series: $Cset_1$, with $W_1 = \{U_1, L_1\}$ and $Cset_2$, with $W_2 = \{U_2, L_2\}$, then the wedge $W = \{U, L\}$ of $Cset_1 \cup Cset_2$ can be computed as follows: $U_i = max(U_{1i}, U_{2i})$ and $L_i = min(L_{1i}, L_{2i})$.

Now we construct a tree structure based on the wedges $W = \{U, L\}$, which is called *wedge-tree*. The structure of wedge-tree is almost similar to that of $R^*$-*Tree*. At each entry in a node, wedge-tree stores a wedge $W$ which is the bounding envelope of all child nodes of this entry. There are two parameters in a wedge-tree: $B$, the maximum number of entries in a non-leaf node and $r$, the distance threshold $W_{UL}$ of all entries in a leaf-node, i.e. all entries in a leaf node must satisfies $W_{UL} \leq r$.

Insertion of a time series $C$ to wedge-tree is as follows. At each level of the wedge-tree, we insert $C$ to the entry whose wedge $W$, $ED(U, L)$ needs the least enlargement to include that time series. Insertion of a time series $C$ to an entry at a leaf node can be done only when the enlargement satisfies the condition $W_{UL} \leq r$; otherwise we have to create a new entry at that leaf node. The node splitting due to a violation of the threshold $B$ can be done by selecting the two entries $E_1$ and $E_2$ in a node such that the wedge $W$ of $E_1 \cup E_2$ is the largest and then based on $E_1$ and $E_2$, we split the node in question into two different nodes.

## 3.3 The MDTW-WedgeTree Algorithm

Based on wedge-tree, we propose an algorithm for motif discovery in a time series database $D = T1, T_2, ..., T_m$. This algorithm is called *MDTW-WedgeTree* (Motif discovery under DTW using Wedge Tree). The algorithm consists of the four following phases:

- Phase 1: [*Constructing wedge-tree*] The algorithm builds the wedge-tree with two parameters $B$ and $r$ by repeatedly inserting time series $T_i$ to the wedge-tree. After constructing the wedge-tree, we obtain the subclusters corresponding to the entries at the leaf nodes whose wedges satisfy the condition $W_{UL} \leq r$. That means the time series in the same subcluster are similar to each other.
- Phase 2: [*Merging the subclusters*] In this phase, we consider every pair of subclusters $S_i$ and $S_j$ and compute $W$ of the set $S_i \cup S_j$. If $W$ satisfies the threshold $r$, $W_{UL} \leq r$, then we merge the two subclusters to a bigger sub-cluster, $S_i \cup S_j$.
- Phase 3: [*Enriching the subclusters*] During the process of building they wedge-tree, since the subcluters must be mutually exclusive and due to the order of insertion of $T_i$ to the tree, two similar time series might be inserted to different subclusters. In this phase, we consider each pair of subclusters obtained after phase 2, $S_i$ and $S_j$, to check if each instance in $S_i$ can be inserted to $S_j$, and vice versa. Details of this phase will be described later.
- Phase 4: [*Discovering motifs*] After phase 3, we sort the subclusters in the descending order of the number of instances, and remove the overlapping subclusters using the technique described in the algorithm in Table 1. The subcluster with the highest number of instances will be the *1-motif*, and the subcluster with the K-th highest number of instances will be the *K-motif*.

### Enriching the Subclusters

For each subcluster $S$, based on the wedge $W = \{U, L\}$ of $S$ and the warping range $R$, we form two new time series $DTW\_U$ and $DTW\_L$: $DTW\_U_i = max(U_{i-R} : U_{i+R})$, $DTW\_L_i = min(L_{i-R} : L_{i+R})$ Keogh et al. in [4] gave the lower bounding measure between an arbitrary time series $Q$ to the subcluster $S$ as follows:

$$LB\_Keogh_{DTW}(Q,W) = \sqrt{\sum_{i=1}^{n} \begin{cases} (q_i - DTW\_U_i)^2 & if\, q_i > DTW\_U_i \\ (q_i - DTW\_L_i)^2 & if\, q_i < DTW\_L_i \\ 0 & otherwise \end{cases}}$$

and proved that: $LB\_Keogh_{DTW}(Q,W) \leq DTW(Q,S_k) \quad \forall S_k \subset S$

Therefore, if we have $LB\_Keogh_{DTW}(Q,W) > r$, then we can conclude that there is no time series in the subcluster $S$ that is similar to $Q$.

For a pair of subclusters $S_i$ and $S_j$, checking if some instances of $S_i$ can be included in $S_j$, and vice versa can be done as follows:

1. In each subcluster $S_i$, we select randomly a time series from this subcluster to be its representative. This time series is denoted as $C\_S_i$.
2. When trying to insert $C\_S_i$ to $S_j$ we check if the wedge of $S_j$ still satisfies the range $r$. If that is the case, all the instances of $S_j$ are similar to $C\_S_i$, and we insert all the instances of $S_j$ to $S_i$. We apply the same check between $C\_S_j$ and $S_i$.
3. This step starts when the check in Step 2 fails. Based on $LB\_Keogh$ lower bound, we check if any instance in the subcluster $S_j$ is similar to $C\_S_i$. If that is the case, we check if each instance in $S_j$ is similar to $C\_S_i$ using the techniques described in Section 3.2. If there is any instance in $S_j$ similar to $C\_S_i$ we insert it to $S_i$. We apply the same check between $C\_S_j$ and $S_i$.

In the algorithm MDTW-WedgeTree, the early abandon technique is applied in all the phases.

# 4 Experimental Evaluation

We evaluate the performance of our proposed algorithms for discovering motif in 13 publicly available datasets downloaded from the web pages [4] [13]. All these datasets are normalized by *z-normalization*. The performance comparison of our proposed algorithms against the brute-force algorithm is in terms of accuracy and efficiency.

We implemented the four algorithms with Microsoft Visual C# and conducted the experiments on an Intel Xeon E5620 2.4GHz, RAM 8GB IBM Server.

We conduct the experiments on the datasets with different range thresholds $r$ for discovering motifs. We use some different values of $r$, ranging from $0.01m$ to $0.1m$, where $m$ is the length of each time series. In all experiments, we select the warping range $R = 0.05m$, and $B = 10$ in building the wedge-tree. As for the MDTW-WedgeTree algorithm, we compute the average of experimental results after running the algorithm 10 times.

## 4.1 Accuracy

Since our proposed method is the first one in time series motif discovery with DTW distance, the brute-force algorithm by Lin et al. [7], given in Table 1, has been considered as the baseline algorithm in evaluating the accuracy of our proposed motif discovery algorithms. The two algorithms MBF-DTW-LB and MBF-DTW-LB-UB which are modified from the brute-force algorithm by adding some heuristics can be considered as bringing out motifs with the accuracy about 100%. As for the MDTW-WedgeTree algorithm with some stochastic factors in selecting the representatives of subclusters, the accuracy of this algorithm in motif discovery might be less than 100%. We evaluate the accuracy of the MDTW-WedgeTree algorithm by

**Table 2** The accuracy ratios (%) of the MDTW-WedgeTree algorithm in discovering 1-motifs

| Dataset/r | Length(points) | 0.02m | 0.04m | 0.06m | 0.08m | 0.1m |
|---|---|---|---|---|---|---|
| Trace | 55000 | 94.7 | 95.8 | 99.1 | 100 | 99.6 |
| DiatomSizeReduction | 111090 | 100 | 100 | 100 | 100 | 100 |
| OSULeaf | 188734 | 100 | 100 | 100 | 100 | 91.3 |
| WordsSynonyms | 244350 | 100 | 97.6 | 95.8 | 90.1 | 96.2 |
| FaceAll | 294750 | 100 | 98.7 | 100 | 96.5 | 100 |
| Symbols | 405960 | 99.8 | 100 | 100 | 100 | 100 |
| Two_Patterns | 640000 | 100 | 100 | 100 | 100 | 100 |
| Wafer | 1088928 | 99.9 | 70.9 | 98.3 | 95.1 | 96.1 |
| Yoga | 1405800 | 99.0 | 98.0 | 86.1 | 79.4 | 88.7 |
| Arrowhead | 3765000 | 97.7 | 95.4 | 98.8 | 97.8 | 99.2 |
| heterogeneous | 5984256 | 100 | 100 | 100 | 97.1 | 97.0 |
| StarLightCurves | 9457664 | 98.8 | 99.1 | 100 | 98.6 | 100 |

considering the accuracy ratio which will be defined as follows. Let $M$ be the set of instances of the *K-motif* discovered by MDTW-WedgeTree and $B$ be the set of instances of the *K-motif* discovered by the brute-force algorithm. The accuracy ratio of MDTW-WedgeTree to the brute-force algorithm is defined by the formula:

$$Acc = \frac{|M \cap B|}{|M|} \times 100\%$$

Table 2 reports the accuracy ratios of the MDTW-WedgeTree algorithm in discovering 1-motifs on several datasets. The experimental results in Table 2 show that the lowest accuracy ratio in all datasets is greater than 70%. That means more than 70% instances of the 1-motif discovered by MDTW-WedgeTree algorithm are exactly the same as those dis-covered by the brute force algorithm. Therefore, we can say that the accuracy of the motifs discovered by the MDTW-WedgeTree algorithm is almost 100% in comparison to those discovered by the brute-force algorithm.

## 4.2  Efficiency

Since our proposed method is the first one in time series motif discovery with DTW distance, we use the brute-force algorithm, given in Table 1, as the baseline to evaluate the efficiency of the proposed algorithms. We evaluate the efficiency of the motif discovery algorithms by considering the ratio of how many times of the distance function must be invoked by the proposed algorithm over the number of times it must be called by the brute-force algorithm. However, in this work we compute the efficiency ratio by using the number of times $dist(q_i, c_j)$ is invoked rather than the number of times $Dist(T_i, T_j)$ is called as in some previous works [4][9]. This new way of computation is especially suitable in the case that we have to compute the lower bound LB as well as the upper bound ED and these computational costs are significant. The efficiency ratio of a motif discovery algorithm A, $Eff(A)$, is defined as follows:

$$Eff(A) = \frac{number\, calls\, of\, dist(q_i,c_j)\, in\, A}{number\, calls\, of\, dist(q_i,c_j)\, in\, brute-force} \times 100\%$$

The algorithm with lower efficiency ratio is better. Table 3 shows the efficiency ratios of the three algorithms against the brute-force algorithm.

From the experimental results in Table 3, we can see that: 1. In average, the three

algorithms (MBF-DTW-LB, MBF-DTW-LB-UB and MDTW-WedgeTree) bring out a one to two order of magnitude speed up over the brute-force algorithm.

2. The efficiency of MBF-DTW-LB-UB is better than MBF-DTW-LB. The reason is that MBF-DTW-LB-UB uses the upper bound $ED(Q,C)$, but this upper bound is also computed in MBF-DTW-LB.

3. In most of the datasets, especially with the large datasets, MDTW-WedgeTree is the best performer. On a few datasets, MDTW-WedgeTree is less efficient than the two other algorithms (e.g. on OSULeaf, WordsSynonyms, FaceAll, and Two_-Patterns), however, the differences among them are insignificant. This case happens when efficiency ratio of MBF-DTW-LB-UB is almost the same as that of MBF-DTW-LB. That means the early abandoning based on $ED(Q,C)$ is not effective and $ED(Q,C)$ always exceeds the range r for all the pairs of time series in the dataset. This case might happens when the range threshold r is too small and this situation brings out the discovered motifs consisting of just a very few instances. On the other hand, if the dataset consists of several similar time series which can be matched by using $ED(Q,C)$ (for example, DiatomSizeReduction dataset) then the use of only lower bound becomes less efficient. For example, on DiatomSizeReduction dataset, Eff(MBF-DTW-LB) = 0.98 while Eff(MBF-DTW-LB-UB) = 0.046 and Eff(MDTW-WedgeTree) = 0.012.

From this analysis, we can conclude that the use of upper bound $ED(Q,C)$ combined with lower bound $LB(Q,C)$ is very efficient on all the datasets. Especially for very large datasets or the datasets in which there are many similar time series within the range r, the MDTW-WedgeTree algorithm is the most efficient.

To know more about efficiency of the three algorithms, we test the runtimes of the algorithms on a very large dataset: StarLightCurves. The runtimes (in minutes) of the three algorithms on StarLightCurves dataset are shown in Table 4. We can see that MDTW-WedgeTree can run faster than MBF-DTW-LB and MBF-DTW-LB-UB.

**Table 3** Efficiency ratios(%) of the three algorithms MBF-DTW-LB, MBF-DTW-LB-UB and MDTW-WedgeTree

| Dataset/r | 0.02m | 0.04m | 0.06m | 0.08m |
|---|---|---|---|---|
| Trace | 17.8 I 7.4 I 6.1 | 20.6 I 6.6 I 4.6 | 22.4 I 6.4 I 3.6 | 23.7 I 6.4 I 3.6 |
| DiatomSizeReduction | 95.4 I 33.9 I 10.0 | 97.8 I 12.6 I 4.0 | 97.9 I 6.6 I 2.0 | 98.0 I 4.9 I 1.3 |
| OSULeaf | 0.7 I 0.7 I 1.0 | 1.8 I 1.7 I 2.1 | 3 .0I 2.8 I 3.2 | 4.3 I 3.9 I 4.4 |
| WordsSynonyms | 1.6 I 1.5 I 2.2 | 2.5 I 2.2 I 3.1 | 3.3 I 2.9 I 3.8 | 4.1 I 3.5 I 4.3 |
| FaceAll | 1.2 I 1.3 I 1.7 | 3.2 I 3.3 I 3.9 | 6.2 I 6.4 I 7.1 | 10.2 I 10.2 I 10.8 |
| Symbols | 13.7 I 7.6 I 5.8 | 17.0 I 6.4 I 4.7 | 18.3 I 5.3 I 4.0 | 19.2 I 4.8 I 3.7 |
| Two_Patterns | 1.7 I 1.7 I 2.6 | 2.3 I 2.3 I 3.6 | 2.8 I 2.8 I 4.4 | 3.2 I 3.2 I 5.0 |
| wafer | 9.1 I 7.8 I 4.8 | 10.7 I 8.7 I 4.1 | 12.3 I 9.5 I 3.8 | 17.0 I 12.9 I 4.7 |
| yoga | 7.6 I 5.1 I 3.9 | 10.8 I 6.7 I 4.4 | 14.0 I 8.7 I 5.1 | 17.2 I 10.7 I 5.7 |
| arrowHead | 8.1 I 6.8 I 3.9 | 13.4 I 10.5 I 5.1 | 17.7 I 13.1 I 5.9 | 21.2 I 14.9 I 6.2 |
| heterogenous | 10.5 I 7.0 I 4.8 | 14.9 I 9.0 I 5.6 | 17.9 I 10.6 I 6.0 | 20.4 I 11.3 I 6.4 |
| StarLightCurves | 15.5 I 10.0 I 8.2 | 21.7 I 11.3 I 7.9 | 25.6 I 11.5 I 7.3 | 28.3 I 11.3 I 6.7 |

**Table 4** Runtime (in minutes) of the three proposed algorithms on the StarLightCurves dataset

| Algorithm/r | 0.02m | 0.04m | 0.06m | 0.08m | 0.1m |
|---|---|---|---|---|---|
| MBF-DTW-LB | 799.5 | 1136.9 | 1688.4 | 1711.4 | 1920.4 |
| MBF-DTW-LB-UB | 539.3 | 610.7 | 616.6 | 592.7 | 581.6 |
| MDTW-WedgeTree | 430.1 | 417.0 | 385.7 | 353.4 | 325.6 |

## 5 Conclusions

In this paper, we propose a method for time series motif discovery under DTW measure. This method combines lower bounding techniques $LB(Q,C)$, upper bounding techniques $ED(Q,C)$ and wedge-tree index structure. Experimental results reveal that our proposed method can run very fast and brings out motifs with high accuracy on very large time series datasets.

As for future works, we plan to adapt the proposed algorithms for discovering motifs as subsequences in a longer time series. Besides, we intend to develop disk-aware version of the proposed algorithms to allow the exploration of truly massive time series datasets stored on hard disk.

# References

[1] Ding, H., Trajcevski, G., Scheuermann, P., Wang, X., Keogh, E.: Querying and mining of time series data: experimental comparison of representations and distance measures. In: Proc. of 34th International Conference on Very Large Data Bases, VLDB (2008)

[2] Itakura, F.: Minimum prediction residual principle applied to speech recognition. IEEE Transactions on Acoustics, Speech, and Signal Processing 23, 67–72 (1975)

[3] Keogh, E.: Exact indexing of Dynamic Time Warping. In: Proc. of 28th International Conference on Very Large Databases, VLDB (2002)

[4] Keogh, E., Wei, L., Xi, X., Lee, S.H., Vlachos, M.: LB_Keogh supports exact indexing of shapes under rotation invariance with arbitrary representations and distance measures. In: Proc. of 32th International Conference on Very Large Data Bases, VLDB (2006)

[5] Kim, S., Park, S., Chu, W.: An index-based approach for similarity search supporting time warping in large sequence databases. In: Proc. of 17th Int. Conf. on Data Engineering, ICDE (2001)

[6] Lemire, D.: Faster retrieval with a two-pass dynamic-time-warping lower bound. Pattern Recognition 42(9), 2169–2180 (2009)

[7] Lin, J., Keogh, E., Patel, P., Lonardi, S.: Finding motifs in time series. In: Proc. of the 2nd Workshop on Temporal Data Mining, at the 8th ACM SIGKDD Int. Conf. on Knowledge Discovery and Data Mining, KDD (2002)

[8] Mueen, A., Keogh, E., Zhu, Q., Cash, S., Westover, B.: Exact discovery of time series motif. In: Proc. of 2009 SIAM Int. Conf. on Data Mining. SIAM (2009)

[9] Rakthanmanon, T., et al.: Searching and mining trillions of time series subsequences under dynamic time warping. In: Proc. of the 18th ACM SIGKDD International Conference on Knowledge Discovery and Data Mining, KDD (2012)

[10] Ratanamahatana, C.A., Keogh, E.: Three myths about Dynamic Time Warping data mining. In: Proc. of 5th SIAM Int. Conf. on Data Mining, SDM (2005)

[11] Sakoe, H., Chiba, S.: Dynamic programming algorithm optimization for spoken word recognition. IEEE Trans. Acoustics, Speech, and Signal Proc. ASSP-26 (1978)

[12] Tanaka, Y., Iwamoto, K., Uehara, K.: Discovery of time series motif from multi-dimensional data based on MDL principle. Machine Learning 58(2-3), 269–300 (2005)

[13] Xi, X., Keogh, E., Li, W., Mafraneto, A.: Finding Motifs in a Database of Shapes. In: Proc. of 7th SIAM International Conference on Data Mining, SDM (2007)

[14] Yi, B., Jagadish, H., Faloutsos, C.: Efficient retrieval of similar time sequences under time warping. In: Proc. of the 14th Int. Conf. on Data Engineering, ICDE (1998)

# A Novel Test Data Generation Approach Based Upon Mutation Testing by Using Artificial Immune System for Simulink Models

Le Thi My Hanh, Nguyen Thanh Binh, and Khuat Thanh Tung

**Abstract.** Software testing is costly, labor intensive, and time consuming activity. Test data generation is one of the most important steps in testing process in terms of revealing faults in software. A set of test data is considered as good quality if it is highly capable of discovering possible faults. Mutation analysis is an effective way to assess the quality of a test set. Nowadays, high level models such as Simulink are widely used to reduce the time of software development in many industrial fields. This also allows faults to be detected at the earlier stages. Verification and validation of Simulink models are becoming vital to users. In this paper, we propose the automated test data generation approach based on mutation testing for Simulink models by using Artificial Immune System (AIS) in order to evolve test data. The approach was integrated into the MuSimulink tool [15]. It has been applied to some different case studies and the obtained results are very promising.

**Keywords:** Software testing, mutation testing, test data generation, artificial immune system, clonal selection algorithm, Simulink.

## 1 Introduction

Software testing is an expensive, tedious and time-consuming activity as it can consumes more than 50% of the total cost of the software development [1] but it is an important means used to assure the quality of software.

Mutation testing proposed by DeMillo *et al.* [2] is a powerful and effective testing technique in order to assess the quality of test suites. Whilst there is much evidence that automated test data generation techniques plays an important role in automating the testing process effectively, however, applying them in the context of mutation

Le Thi My Hanh · Nguyen Thanh Binh · Khuat Thanh Tung
DATIC Laboratory, University of Science and Technology,
The University of Danang, Vietnam
e-mail: {ltmhanh,ntbinh}@dut.udn.vn, thanhtung09t2@gmail.com

© Springer International Publishing Switzerland 2015       169
V.-H. Nguyen et al. (eds.), *Knowledge and Systems Engineering,*
Advances in Intelligent Systems and Computing 326, DOI: 10.1007/978-3-319-11680-8_14

testing has received little attention in the literature. Besides, generating test data for mutation testing has also not attracted a lot of attention from the Evolutionary Testing community. Herein, our objective is to demonstrate the feasibility of applying AIS to mutation testing by using Clonal Selection Algorithm (CSA) for generating and evolving test data.

Model-based design is a development methodology for modern software engineering. It promotes the use of powerful and specialized modeling languages, allowing the engineers to focus on the domain-specific aspects of the system under development [3]. Matlab/Simulink has become a standard formalism that is widely used for modeling, simulating high-level designs of control application in embedded systems engineering. The simulation facilities allow such models to be executed and observed. This property of Simulink turns out to be an advantage for effective dynamic testing at design level [4] .

Generation of test data for Simulink models is the focus of several researches. Zhan [4] built a framework including search-based test data generation for structural testing and mutation testing in Simulink using Simulated Annealing. However, the nature of mutation used in this work is a very simple one; only values on signal lines are perturbed. Ghani *et al.* [5] extended the work of Zhan, they used Simulated Annealing and Genetic Algorithms to generate test data for Simulink models based on branch coverage. Nannan He *et al.* [6] have translated Simulink models into a formal specification and generated test cases by basing on bounded model checking technique and mutation testing. Formal methods are hard to be implemented and require more mathematical knowledge. Translating Simulink models into formal language is possible for small models. However, this becomes difficult and very costly, if the size of the models is increasing. Meng Li and Kumar [7] have translated Simulink models into an Input/Output Extended Finite Automaton, then generated test data. Sangharatna Godboley *et al.* [8] have generated code C/C++ for Simulink models and generated test data using branch coverage for that code. Jungsup Oh *et al.* [9] introduced a messy-GA for transition coverage of Simulink models. Satpathy *et al.* [10] used a randomized directed approach in order to generate test data for Simulink models achieving high structural coverage. This method includes both randomly generating inputs, creating and evaluating condition trees for model transitions. We found that most studies of test data generation for Simulink models used structural coverage criteria. However, in [22], mutation testing was shown to subsume most structural criteria.

AIS is adaptive system, inspired by theoretical immunology and observed immune functions, principles and models, which is applied to problem solving [11]. However, there has been little work on applying it in the context of software testing. In [12], May *et al.* adopted the AIS to evolve test data for Fortran programs based on mutation testing. Pachauri *et al.* [13] used the clonal selection algorithm as software test data generation technique for branch coverage of a program. In our work, we use clonal selection algorithm - a subfield of AIS – to evolve test cases which are randomly initialized to kill as many mutants as possible.

The paper is organized as follows. Section 2 briefly introduces mutation testing, the Simulink environment, and mutation testing for Simulink models. In section 3, we represent the background of Immune System and Clonal Selection Algorithm. In section 4, automatic test data generation approach using the CSA is proposed. The case studies and obtained results are discussed in section 5 and section 6 draws the conclusion, discusses ongoing and future work.

## 2  Mutation Testing for Simulink Models

### 2.1  Mutation Testing

Mutation testing [2] is an effective software testing method, which can simulate software defects systematically and measure the quality of a test set. It works in the following way: a large number of simple faults introduced into the program under test one at a time by using mutation operators. These operators can be seen as representing common faults usually found in software. When applying a mutation operator into model, i.e. inserting a single fault into model, we obtain a faulty model, which is called a mutant. Then, test data should be generated to expose the introduced fault. A test suite is considered good if it contains tests that are able to distinguish a large number of these mutants from the original design. If a mutant can be distinguished from the original model by at least one of the test cases in the test set, the mutant is considered to be killed. Otherwise the mutant is alive. Generally there are two possible reasons why a mutant has been left alive. The first is that the used test cases were not capable of revealing fault, whilst the second is which there is no such data. In the second case the mutant model is functionally equivalent to the original one and called equivalent mutants. After executing the mutant and original models against the test sets, the proportion of mutants killed out of all non-equivalent mutants is called Mutation Score (MS) and is formally defined as follows:

$$MutationScore = \frac{K}{T - E} \tag{1}$$

where $K$ is the number of mutants that has been killed, $T$ is the total number of mutants, and $E$ is the number of equivalent mutants.

### 2.2  Simulink

Simulink [14] is a part of Matlab tool suite used for modeling, simulating and analyzing dynamic systems. Simulink models are widely used in design and implementation of embedded systems, including aerospace, automobile and electronics systems. Simulink models contain functional blocks with input and output ports to interconnect them. These interconnections define the data flows between the blocks and set up equations relating the involved interface variables. Blocks are the functional units built-in, used to generate, manipulate and output signals. A block can be a parent container containing other blocks, each modeling a subsystem or subfunctionality. Each block has a number of parameters, for example, initial value and ranges, which control the operation of the block. Simulink plays an increasingly important role in industry, and the verification and validation of Simulink models is becoming vital to users [5]. Thereby, automatic test data generation for designs in Simulink has a significantly meaning. In our study, we perform test data generation for Simulink models that contain basic blocks in pre-defined libraries such as commonly user blocks, continuous, discrete, logic and relational operations, math operations, sinks and sources.

## 2.3  Mutation Testing for Simulink Models

Since mutation testing was proposed, it has been applied for many programming languages such as Fortran, Ada, C, Java, C#, etc. In this study, we adopt mutation testing for designs in Simulink. Applying mutation testing for Simulink models consists of generating mutants, executing mutants, analyzing results and generation of test suites. If this process is manually done, it will require too much time. Hence, we designed and implemented MuSimulink tool to automate this process. The design details of this tool were presented in [15]. Concerning mutant generation, we use the mutation operators which were proposed in our previous work [16] to introduce systematic faults into original Simulink model. The mutation operators considered in this paper are given in Table 1.

We can state that the most computation-expensive parts of the mutation process are mutant execution, original model execution, output comparison and test data generation. Mutant execution is done once for each mutant and each test data. This includes the interpretation of a mutant on a test case, the comparison of the mutant output with original ouput, and, if they differ, the killing of the mutant. Since the mutants is independently executed of each other, thus, we proposed the parallel computing approach [17], using Parallel Computing Toolbox of Matlab, to improve mutant and original model execution. Mutant interpretation and output comparison are delegated to the node processors, making each node processor responsible for a portion of the total number of mutants. In particular, the host processor distributes mutants to each node processor and the node processors interpret mutants for each test case. The empirical results in [17] shown that the mutants execution time was significantly reduced. In this paper, we focus on only improving test data generation.

## 3  Background of Immune System

### 3.1  Artificial Immune System

Artificial Immune System is computational method inspired by the process, mechanisms, characteristics of learning and memory of the biological immune system. They also express interaction between antibody and antigen which is called as affinity. De Castro and Timmis [11] proposed a general framework for the AIS which is described in Fig. 1. In order to build the AIS, the first step is to determine the application domain or target function. This knowledge will help specify the way to represent the components of the system. The affinity measure also depends on the representation of AIS. There are many possible affinity measures such as Hamming

**Table 1** Mutation Operators

| Operator | Depcristion |
|----------|-------------|
| VNO | Variable Negation Operator |
| VCO | Variable Change eperator |
| TRO | Type Replacement Operator |
| CCO | Constant Change Operator |
| CRO | Constant Replacement Operator |
| SCO | Statement Change Operator |
| SSO | Statement Swap Operator |
| DCO | Delay Change Operator |
| ROR | Relational Operator Replacement Operator |
| AOR | Arithmetic Operator Replacement Operator |
| ASR | Arithmetic Sign Replacement Operator |
| LOR | Logical Operator Replacement Operator |
| BRO | Block Removal Operator |
| SRO | Subsystem Replacement Operator |

and Euclidean distance metrics. Each of these distance measures have their own bias and affinity function must therefore be selected with great care as it can affect the overall performance and the final solution obtained of the system [18]. The final layer is the selection of appropriate immune system-based algorithms. In this framework, immune algorithms include negative and positive selection, clonal selection, and immune network algorithms.

## 3.2 Clonal Selection Algorithm

The Clonal Selection Theory

The clonal selection principle is the whole process of antigen recognition, cell proliferation and differentiation into memory cell [19]. It establishes the idea that only cells recognizing the antigens are selected to proliferate and differentiate into effector cells. The selected cells are subject to an affinity maturation process, which improves their affinity to the selective antigens. When an animal is exposed to an antigen, some subpopulation of its bone marrow derived B-cells can recognize a

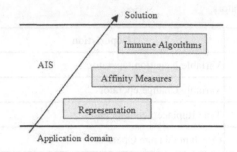

**Fig. 1** AIS layered framework [11]

non-self antigen with a certain affinity. These cells are selected to proliferate and produce antibodies in high volumes. The antibodies are soluble forms of the B-cell receptors which are released from the B-cell surface to against the invading non-self antigen [20]. The result of this process is to create a set of clones identical to the parent cell. Then, these clones undergo a hyper-mutation process that result in they mature into non-dividing antibody secreting cells, named plasma cells. Besides proliferating and differentiating into plasma cells, the activated B-cells that produce higher affinity antibodies are preferentially selected to become memory cells [11] with long life spans. Memory cells circulate through the blood, lymph and tissues and when the antigen is encountered at a later date, clonal expansion of specific memory B-cells can take place and eliminate the antigen.

Clonal Selection Algorithm

Clonal selection algorithm has taken inspiration from the antigen driven affinity maturation process of B-cells and the associated hyper-mutation mechanism. There are two important features of affinity maturation in B-cells which are applied to develop the AIS called CLONALG [21]. The first feature is that the proliferation of B-cells is proportional to the affinity of the antigen that binds it, thus the higher the affinity, the more clones that are produced. Secondly, the mutations suffered by the antibody of B-cell are inversely proportional to the affinity of the antigen it binds. The CLONALG generates a population of $N$ antibodies, each specifying a random solution for the optimization process. In the each iteration, some of the best existing antibodies are selected, cloned and mutated in order to construct a new candidate population. New antibodies are then evaluated and certain percentage of the best antibodies is added to the original population. Finally, a number of worst antibodies of previous generations are replaced with new randomly created ones. The goal of the algorithm is to develop a memory set of antibodies that represents a solution to an engineering problem.

**Table 2** Mapping between Immune System and Mutation Testing

| Immune System | Mutation testing |
|---|---|
| B-cell | Model under test |
| Antigen | Mutant |
| Antibody | Test data |
| Affinity | Mutation score |
| Clonal Selection | Evolving test data |
| Memory cells | The memory set to save test data which be able to kill mutants |

# 4  The Proposed Approach

## 4.1  Mapping from Immune System to Mutation Testing

The immune system develops a population of antibodies to protect the body from incursion of antigens. In the context of mutation testing, thus, an antigen is an introduced mutant. An antibody is a test data which is generated to kill mutants. In our work, the affinity of an antibody (test data) is measured by mutation score. Antibody evolution occurs through the process of clonal selection, guided by the affinity values. The high affinity antibodies generate more clones than low affinity ones, but mutate less. This process aims at refining antibodies to kill as many mutants as possible. The best antibodies will be added to a memory set to save and return to the tester when testing process finish. Table 2 presents the mapping between immune system and mutation testing.

## 4.2  Clonal Selection Algorithm for Generation of Mutation Based Test Data

The Clonal Selection Algorithm dynamically evolves a memory set with each antibody can kill at least one mutant which is not killed by any antibody else. Thus, we do not need to initialize a number of tests for memory set. In this section, we propose an adoption of the CLONALG with some modifications to automatically evolve a set of initial tests by basing on mutation score as a quality criterion. The proposed algorithm is presented in Algorithm 1. The algorithm works by initializing the population with *popSize* antibodies which are randomly generated using **CreateRandomPopulation** method. Then, the affinity (mutation score) of each antibody in population is computed (line 2). Antibodies in initial population that are able to kill at least one mutant not killed by any antibodies else are added to memory set, $M$ (line 4). The process of clonal selection is performed when stop conditions are not satisfied yet. Stop conditions of algorithm are to reach the mutation score threshold or the maximal number of tests. In this process, firstly, we choose

---

**Algorithm 1** The proposed clonal selection algorithm to generate test data for Simulink model.

---

**Input:**
- *popSize*: size of population
- *selectionSize*: a number of antibodies that is selected to undergo clonal selection.
- *randomPopSize*: a number of randomly generated antibodies
- *cloneRate*: rate of cloning antibodies.

**Output:**
- A set of antibodies is able to kill mutants of Simulink model $M$

**Algorithm:**
1:   *population* = **CreateRandomPopulation**(*popSize*)
2:   **CalAffinity**(*population*)
3:   $M = \emptyset$
4:   **AddToMemory**(*population*, $M$)
5:   **while** (stop conditions are not satisfied) **do**
6:      *populationSelect* = **Select**(*population*, *selectionSize*)
7:      *populationClones* = **Clone**(*populationSelect*, *cloneRate*)
8:      **foreach**(*ad* **in** *populationClones*)
9:        **HyperMutate**(*ad*, 1 - **GetAffinity**(*ad*))
10:     **end foreach**
11:     **CalAffinity**(*populationClones*)
12:     **AddToMemory**(*populationClones*, $M$)
13:     *populationNew* = **Combine**(*population*, *populationClones*)
14:     *population* = **Select**(*populationNew*, *popSize*)
15:     *populationRandom* = **CreateRandomPopulation**(*randomPopSize*)
16:     **CalAffinity**(*populationRandom*)
17:     **AddToMemory**(*populationRandom*, $M$)
18:     **Replace**(*population*, *populationRandom*)
19:   **end while**
20:   **return** $M$

---

*selectionSize* the highest affinity antibodies of current population. After that, these antibodies undergo clonal selection with affinity proportional cloning and inversely proportional hyper-mutation (line 7-10). The number of clones produced for each antibody in selected population, *populationSelect*, is calculated by: *affinity \* cloneRate*, with a minimum of one clone. After clonal selection process, the affinity of each mutated antibody is computed by using **CalAffinity** function. The clone antibodies that can kill at least one mutant not killed by any test data in $M$ are added to memory set. We use **Combine** function to combine the current population and clone population in order to create new population, *populationNew*. Then, we select *popSize* best antibodies from this new population. In order to maitain the diversity of population, we create *randomPopSize* antibodies randomly and replace the *randomPopSize* lowest affinity antibodies in population with these antibodies (line 18). If any antibody, which is randomly generated, can kill at least one mutant not killed, then it will be added to memory set. When the process of evolving test data finishes, memory set $M$, which contains test cases killing mutants, are returned to testers. In the hyper-mutation step, if any variable is mutated, we will add an integer number which is randomly generated in [-5, 5] to that variable. If the generated random value is 0, the mutated variable is assigned by 0. Each mutant model and the original model will be executed with the generated test data. In order to reduce the execution time, we use the parallel computing approach in [17].

## 4.3 Optimizing the Set of Test Data

We found that the final set of all the memorized test data may not be minimal since the algorithm only saves the best antibodies of one generation and it misses information to guide process of minimizing the number of test case in the memory set through the generations. The minimization can be done in a separate phase once the algorithm finishes. We can build a boolean matrix, called $A$, whose rows are test data and whose columns are mutants. $A_{ij} = 1$ means that the $i^{th}$ test data kills the $j^{th}$ mutant and $A_{ij} = 0$ means that it does not. Then, we optimize this matrix in order to create the minimal tests set. Algorithm 2 presents the algorithm optimizing the obtained set of tests.

---

**Algorithm 2** The algorithm optimizing the obtained set of test data.

**Input:**
- $T$: test set in the memory which is not optimal,
- $N$: number of test data in $T$.
- $A$: a boolean matrix presents if the $i^{th}$ test data kills the $j^{th}$ mutant

**Output:**
- $T$: test set in the memory which is optimal.

**Algorithm:**
1: $S = \varnothing$ // $S$ is used to mark the index of the retained test data
2: **for** each(testdata $i^{th}$ in $T$) )
3:   - Let $M$ is the set that contains index of the mutants which are killed by the $i^{th}$ test data.
4:   - Removing the $i^{th}$ test data from the memory set $T$ when one of two following conditions is satisfied $\forall j \in M$ :
   ✓ $\exists k \in S : A_{kj} = 1$
   ✓ $\exists k \in [1, N], k > i : A_{kj} = 1$
5:   - if the $i^{th}$ test data is not removed, we will add $i$ to $S$.
6: **endfor each**
7: **return** $T$

---

## 5 Experimentation

We implemented the approach in the *MuSimulink* tool and used the Tiny model from [5] and the Quadratic_v2 model from [4] in Table 3 to do some assessments of our approach. We conducted our experiments on a PC with a 2.4 GHz Intel Core 2 Quad CPU Q6600 and 2 GB memory, running the Windows 7 operating system. In this experimentation, we generated test data in order to kill the mutants created by a subset of mutation operators in Table 1. We ignore three mutation operators TRO, SRO, DCO because they generate many equivalent mutants and they are less effective for our case studies.

When we design a clever algorithm to solve a specific problem, we not only describe a representation, and a set of components to tackle the problem but also need
to choose parameter values by expecting the algorithm gives the best results. The
AIS depends on the size of population (*popSize*), the number of the best individuals that are selected to proliferate (*numSelectedInd*), the user-defined parameter
about number of clones created for each selected member (*numClones*), the number
of new randomly generated antibodies to replace the lowest affinity antibodies in
population (*randomPopSize*). Table 4 shows the values that we configured for the
AIS. These values are derived from our experiments and we used them for the case
studies.

In this paper, we want to compare the effectiveness between the AIS and random test data generation in terms of mutation score and the number of generated
test data in order to achieve the same mutation score. For each execution, we will
perform the algorithm with the same of input domain description table. We set up
the same mutation score threshold and the maximal number of generated test data,
then comparing the effectiveness of AIS and random approach.

We realize five individual runs for each mutation score threshold. We show the
average results when generating test data for the models by using the AIS and random approaches by the following. Table 5 shows results of the Tiny model. Table 6
presents the results for the Quadratic_v2 model.

Based on experimental results, we found that the proposed approach is stable
and it has also significantly improved mutation score as well as generated less test
data than random test data generation approach in order to kill the same number of
mutants.

During the experimentation, even though four mutants were executed on four-
cores processor in parallel, the process of test data generation and evolvement was
still time-consuming. In the third case of both models, with the same generated
test data, we can state that the execution time of the AIS approach is higher than
the random approach because random test data is executed only on alive mutants
whereas in order to compute the affinity for each antibody of the AIS, test data must
be executed on all mutants.

We can find that the AIS approach cannot reach to 85% mutation score threshold
for the Quadratic_v2 model and 94% mutation score threshold for the Tiny model
because of several reasons. First, the input domain description table plays an important role in our work. If data is chosen from a appropriate input domain, the
number of killed mutants will be high and vice versa. Second, there are a number

**Table 3** Experimental Models

| Model | No of Input Var | No of Mutants |
|---|---|---|
| Tiny | 3 | 144 |
| Qutdraaic_v2 | 3 | 140 |

**Table 4** Parameters of the AIS

| Model | popSize | numSelectedInd | randomPopSize | numClones |
|---|---|---|---|---|
| Tiny | 300 | 40 | 40 | 40 |
| Quadratic_v2 | 600 | 30 | 30 | 30 |

**Table 5** Results[1] of Tiny Model

| MST (%) | MNTD | AIS | | | | Random | | | |
|---|---|---|---|---|---|---|---|---|---|
| | | MS (%) | No. GTD | No. TDKM | Time (s) | MS (%) | No. GTD | No. TDKM | Time (s) |
| 88 | 10000 | 88.89 | 460 | 6 | 1026.07 | 88.89 | 2050 | 13 | 2099.90 |
| 93 | 10000 | 93.06 | 1528 | 6 | 3701.78 | 93.06 | 4821 | 15 | 4516.81 |
| 94 | 10000 | 93.06 | 10000 | 6 | 25240.18 | 93.06 | 10000 | 15 | 9389.68 |

**Table 6** Results[1] of Quadratic_v2 Model

| MST (%) | MNTD | AIS | | | | Random | | | |
|---|---|---|---|---|---|---|---|---|---|
| | | MS (%) | No. GTD | No. TDKM | Time (s) | MS (%) | No. GTD | No. TDKM | Time (s) |
| 80 | 10000 | 80.71 | 924 | 5 | 4836.14 | 71.57 | 10000 | 11 | 20182.85 |
| 82 | 10000 | 82.14 | 1308 | 6 | 6911.42 | 70.29 | 10000 | 12 | 19484.31 |
| 85 | 15000 | 82.14 | 15000 | 7 | 76065.10 | 71.71 | 15000 | 11 | 31877.12 |

of equivalent mutants for every model and none of them was eliminated from the set of candidate mutants. Third, the fitness function is defined by basing on the achieved mutation score. The drawback of this fitness is that it does not guide the search process by measuring the closeness of reaching to target test data, which be able to kill specific mutants, from current test data. Thereby, a model has many the hard-to-kill mutants, its mutation score is not high. For these reasons, we intend to build a more effective fitness function in order to kill a higher number of mutants in further work. Fourth, the AIS algorithm is influenced by different parameters on the evolution of the fitness value. Thus, many trials are necessary to investigate implementation choices and to study alternative combinations of parameter settings in order to determine the most appropriate combination.

---

[1] MST: MS Threshold, MNTD: Maximal Number of Test data, No. GTD: Number of generated test data, No. TDKM: Number of test data kill mutants.

# 6 Conclusion and Future Work

As model-based development is now established in many industrial fields and practitioners have noticed opportunities to test software as earlier as possible, we proposed an approach to support mutation testing for Simulink models by automatically generating and evolving test data.

The process of test data generation based on mutation coverage criterion is time-consuming and complicated because of number of generated mutants from original model. However, test data generated by this criterion is very effective for testing critical systems such as designs in Simulink. In this work, we proposed the use of the AIS approach to improve mutation score of test data for Simulink models. The obtained results show that our approach outperformed the random test generation method in terms of the number of killed mutants and the number of generated test data in order to reach the same mutation score. Besides, the parallel mutant execution approach also reduced the test data generation time significantly. Although these results are not really high, they are very promising so that we are continuing to realize further research. We intend to improve fitness function by using mutation distance in order to increase the effectiveness for the process of test data search. We are also going to carry out more experiments to determine the influence of the AIS configuration parameters on results as well as automatically adjust these parameters in searching process.

# References

[1] Beizer, B.: Software Testing Techniques, 2nd edn. Thomason Computer Press (1990)
[2] DeMillo, R., Lipton, R., Sayward, F.: Hints on Test Data Selection: Help for Practicing for Programmer. IEEE Computer (11), 34–41 (1978)
[3] Brillout, A., He, N., Mazzucchi, M., Kroening, D., Purandare, M., Rümmer, P., Weissenbacher, G.: Mutation-based Test case generation for Simulink models. In: de Boer, F.S., Bonsangue, M.M., Hallerstede, S., Leuschel, M. (eds.) FMCO 2009. LNCS, vol. 6286, pp. 208–227. Springer, Heidelberg (2010)
[4] Zhan, Y.: A Search-Based Framework for Automatic Test-Set Generation for Matlab/Simulink models. University of York. PhD Thesis (2005)
[5] Ghani, K., Clark, J.A., Zhan, Y.: Comparing Algorithms for Search-based Test Data Generation of Matlab Simulink Model. In: 10th IEEE Congress on Evolutionary Computation (CEC 2009), Trondheim, Norway (2007)
[6] He, N., Rummer, P., Kroening, D.: Test-Case Generation for Embedded Simulink via Formal Concept Analysis. In: DAC, San Diego, California, USA (2011)
[7] Li, M., Kumar, R.: Model-Based Automatic Test Generation for Simulink/Stateflow using Extended Finite Automaton (2011)
[8] Godboley, S., Sridhar, A., Kharpuse, B., Mohapatra, D.P., Majhi, B.: Generation of Branch Coverage Test Data for Simulink/Stateflow Models using Crest Tool. International Journal of Advanced Computer Research III(13), 222–229 (2013)
[9] Oh, J., Harman, M., Yoo, S.: Transition Coverage Testing for Simulink/Stateflow Models Using Messy Genetic Algorithms. In: GECCO 2011, Dublin, Ireland, pp. 1851–1858 (2011)

[10] Satpathy, M., Yeolekar, A., Ramesh, S.: Randomized directed testing (redirect) for simulink/stateflow models. In: Proceedings of the 8th ACM International Conference on Embedded Software, New York, USA, pp. 217–226 (2008)

[11] De Castro, L.N., Timmis, J.: Artificial Immune Systems: A New Computational Intelligence Approach. Springer (2002)

[12] May, P., Timmis, J., Mander, K.: Immune and Evolutionary Approaches to Software Mutation Testing. In: de Castro, L.N., Von Zuben, F.J., Knidel, H. (eds.) ICARIS 2007. LNCS, vol. 4628, pp. 336–347. Springer, Heidelberg (2007)

[13] Pachauri, A., Gursaran: Use of Clonal Selection Algorithm as Software Test Data Generation Technique. In: Second International Conference on Advanced Computing & Communication Technologies, pp. 1–5 (2012)

[14] The Matwork Inc., http://www.mathworks.com/products/simulink/

[15] Hanh, L.T.M., Binh, N.T.: Automatic Generation of Mutants for Simulink Models. In: 16th National Conference: Selected Problems About IT And Telecommunication, Danang, Vietnam, pp. 339–346 (2013)

[16] Hanh, L.T.M., Binh, N.T.: Mutation Operators for Simulink Models. In: KSE 2012 - The Fourth International Conference on Knowledge and Systems Engineering, Danang, pp. 54–59 (2012)

[17] Hanh, L.T.M., Tung, K.T., Binh, N.T.: Improving Mutation Execution in Mutation Testing for Simulink Models Using Parallel Computing. Journal of Science and Technology, University of Danang II(1(74)), 9–13 (2014)

[18] Freitas, A.A., Timmis, J.: Revisiting the Foundations of Artificial Immune Systems: A Problem Oriented Perspective. In: Timmis, J., Bentley, P.J., Hart, E. (eds.) ICARIS 2003. LNCS, vol. 2787, pp. 229–241. Springer, Heidelberg (2003)

[19] Burnet, F.M.: The Clonal Selection Theory of Acquired Immunity. Cambridge University Press (1959)

[20] De Castro, L.N., Timmis, J.: Artificial Immune Systems: A Novel Paradigm to Pattern Recognition, vol. 2, pp. 67–84. Springer Verlag, University of Paisley, UK (2002)

[21] de Castro, L.N., Von Zuben, F.J.: Learning and optimization using the clonal selection principle. IEEE Transactions on Evolutionary Computation 6(3), 239–251 (2002)

[22] Offutt, A.J., Voas, J.M.: Subsumption of Condition Coverage Techniques by Mutation Testing. Technical Report ISSE-TR-96-01 (1996)

# A Priority-Based Flow Control Method for Multimedia Data in Multi-hop Wireless Ad hoc Networks

Ngo Hai Anh, Nguyen Tien Lan, and Pham Thanh Giang

**Abstract.** Currently, there are many kinds of applications, from light load e.g. web access to heavy load e.g. multimedia services, which operate over wireless network. The *quality of service* (QoS) for these applications is complex, especially the effectiveness of shared bandwidth between multimedia data streams. IEEE 802.11e was released with some QoS mechanisms, such as EDCA (Enhanced Distributed Channel Access), to provide effective bandwidth sharing between data streams in the network. However, this standard has not yet ensured fairness between flows, especially between short-hop and long-hop flows. In this paper, we examine the contention at both MAC and link layer and present a method that improves scheduling mechanism in order to achieve an appropriate level of fairness between data flows.

**Keywords:** IEEE 802.11, IEEE 802.11e, ad hoc network, asymmetric ad hoc network, fairness, back-off, contention window.

## 1 Introduction

The IEEE 802.11e has been approved as a standard that defines a set of Quality of Service (QoS) mechanisms for WLANs [1]. It differentiates traffic basc on its types and sources and is considered to be of critical importance for delay-sensitive and bandwidth-sensitive applications, such as streaming multimedia and wireless Voice over IP (VoIP). It addresses the QoS issues in the MAC sublayer.

To improve the QoS limitations of DCF and PCF, IEEE 802.11e specifies a new coordination function, the Hybrid Coordination Function (HCF). It provides a hybrid access method approach to achieve a better QoS performance. The purpose of HCF is to combine contention-based and contention-free medium access method.

Ngo Hai Anh · Nguyen Tien Lan · Pham Thanh Giang
Institute of Information Technology,
Vietnam Academy of Science and Technology, Hanoi, Vietnam
e-mail: {ngohaianh,ntlan,ptgiang}@ioit.ac.vn

Within the HCF, there are two methods of channel access, *Enhanced Distributed Channel Access* (EDCA) and *HCF Controlled Channel Access* (HCCA). EDCA, which provides distributed access method, can be viewed as an enhancement of DCF, and used in both infrastructure networks and ad hoc networks. While HCCA provides centralized access method and can be used only in infrastructure networks.

The main drawback of the DCF, regarding QoS provisioning, is that it can only support random access and cannot provide any service differentiation since all stations have the same priority, for example, the same CWmin, CWmax and waiting time before backoff or transmission (DIFS) [1].

EDCA is also a differentiated, distributed and contention-based medium access mechanism, but it enhances the original DCF to provide prioritized QoS. The EDCA mechanism defines four access categories (ACs) that provide support for the delivery of traffic with user priorities (UPs) at the stations. An AC which is based on user priority or frame type, is assigned to each frame before it enters the MAC. Values of ACs are described in the table below:

**Table 1** User Priority and Access Category

| Priority | User Priority (UP) | Access Category (AC) | Designation |
|----------|--------------------|----------------------|-------------|
| lowest   | 1                  | AC_BK                | Background  |
|          | 2                  | AC_BK                | Background  |
|          | 0                  | AC_BE                | Best effort |
|          | 3                  | AC_BE                | Best effort |
|          | 4                  | AC_VI                | Video       |
|          | 5                  | AC_VI                | Video       |
|          | 6                  | AC_VO                | Voice       |
| highest  | 7                  | AC_VO                | Voice       |

There are four traffic types corresponding to the four ACs: AC_BK, AC_BE, AC_VI and AC_VO stands for background, best effort, video, and voice traffic, respectively. For example, management frames should be sent with the highest priority, so AC_VO is selected for these frames.

An enhanced variant of the DCF, Enhanced Distributed Channel Access Function (EDCAF), is assigned to each AC. This enhanced DCF variant is called an EDCA parameter set. It is used by each AC to contend for medium access.

The AC parameter set contains the following parameters:

*Arbitrary interframe space number* (AIFSN): the number of time slots after a SIFS duration that a station has to defer before either invoking a backoff or starting a transmission. AIFSN affects the arbitration interframe space (AIFS), which specifies the duration (in time instead of number of time slots) a station must defer before backoff or transmission:

$$AIFS = SIFS + AIFSN \times SlotTime$$

So an AC with lower value of AIFSN has less AIFS and is thus given a high priority.

*Contention Window* (CW): a random number is drawn from the interval between CWmin and CWmax for calculating the total backoff time:

$$Backoff = AIFS + random[CWmin, CWmax]$$

An AC with lower values of CWmin and CWmax has higher probability to draw a smaller random number, thus it is given higher priority.

*TXOP limit*: the maximum duration in which a station can transmit after obtaining a TXOP. A value of zero means when this AC gets access to the medium, it is allowed to send only one frame from the AC queue. This is to limit the low priority traffic. And a value higher than zero means that an AC may transmit multiple frames only from its AC queue since a TXOP is given to an EDCAF in a specific AC, not to a station. As long as the duration of the transmissions does not exceed the TXOP limit, the station is allow to transmit frames from the specific AC queue. Thus an AC with a higher value of TXOP limit has a higher priority.

The EDCA parameters for each AC are shown in the table below:

**Table 2** Calculation of contention window boundaries

| AC | CWmin | CWmax | AIFSN | TXOP limit (ms) |
|---|---|---|---|---|
| AC_BK | aCWmin | aCWmax | 7 | 0 |
| AC_BE | aCWmin | aCWmax | 3 | 0 |
| AC_VI | (aCWmin+1)/2-1 | aCWmax | 2 | 6.016 |
| AC_VO | (aCWmin+1)/4-1 | (aCWmin+1)/2-1 | 2 | 3.264 |

For a typical of aCWmin=15 and aCWmax=1023, as used. The EDCA parameter set information is only present in infrastructure mode. With proper tuning of AC parameters, traffic performance from different ACs can be optimized and prioritization of traffic can be achieved.

## 2 Related Work

Many simulation and analysis have been done regarding the performance of IEEE 802.11e EDCA. In the research of Ferre et al. [18], the MAC level performance gain of IEEE 802.11e is analyzed. The network level QoS performance supported by 802.11e EDCA is enhanced by using a parameter configuration algorithm to provide guaranteed throughput, and the optimal parameter selection algorithm is analyzed to efficiently utilize the network resources [19]. In the research of Liang et al. [20], the analysis is based on the application level comparison between EDCA and DCF to represent the perceptual quality of the end users. Also, many researches have been done to analyze the performance of IEEE 802.11e in the condition of saturated network load. And in the paper [21], the author proposes an analytical model to evaluate the saturation throughput which is based on the use of mean value analysis, and this scheme accurately demonstrated the effects of the change of contention window size and Arbitration Interframe Space. And in the research of Xiong et al. [22], a novel Markov chain based model with a simple architecture for EDCA performance analysis under the saturated traffic load is proposed. It incorporates more features of EDCA into the analysis and has better performance accuracy.

## 3   Evaluation of Queue Control in Ad hoc Network

The contentions in link layer affect the per-flow fairness between *direct flows* and *forwarding flows*. We will examine these contentions in FIFO and Round Robin (RR) scheduling.

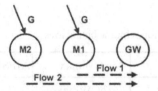

**Fig. 1** A basic multihop wireless network model

There are some kinds of unfairness problem topologies. In this section, a basic multihop wireless network topology will be illustrated as an example. Considering the topology in Fig. 1, there are three stations. Stations M1 and M2 are in one *transmission range*, in which a packet can be transmitted and received successfully. Station M1 and gateway GW are also in another transmission range. Station M2 and GW are out of transmission range but in *carrier-sensing range*, in which a transmission can be detected. The carrier-sensing range is larger than the transmission range, and may be more than two times of the transmission range [12]. It is noted that the sizes for the transmission and carrier-sensing ranges vary according to the power levels. Stations M1 and M2 are assumed to generate the same offered load G to GW. Let $B$ be the maximum medium bandwidth, and $B_1$, $B_2$ be the allocated bandwidth for stations M1 and M2 in the saturation state, respectively. We have $B = B_1 + B_2$. In the communication process, depending on the status of the stations, the offered load $G$ can affect their bandwidth. Normally near stations tend to have higher bandwidth ($B_1 > B_2$), can even get big amount of the bandwidth ($B_1 \simeq 4B_2$) [13]. Therefore, we need adjust the flow's throughput in a proper way.

### 3.1 Contention in FIFO Scheduling

First, if the bandwidth $B$ is sufficiently large compared to the sum of all flow's offered loads, each flow can get its required throughput, hence:

$$Th(flow1) = Th(flow2) = G, \text{if } G < B/3 \qquad (1)$$

Second, if the bandwidth $B$ is not enough for all flows, because $B_1$ is much greater than $B_2$, the flow 1 (direct flow) can get required throughput, and the remaining bandwidth is used for the flow 2 (forwarding flow). Throughputs of the flow 1 and flow 2 are calculated as follows.

$$\begin{cases} Th(flow1) = G \\ Th(flow2) = \frac{B-G}{2} \end{cases} \text{if } B/3 \leq G < B_1 - B_2 \qquad (2)$$

Third, the network is in the saturation state. In FIFO scheduling, a common queue is used for all flows. We assume that buffer size is infinite. The ratio of the buffer allocation $Q_{flow1}$ for the flow 1 to $Q_{flow2}$ for the flow 2 at station M1 is $Q_{flow1}$ : $Q_{flow2} = G : B_2$. Hence, the throughputs of the flow 1 and flow 2 at station M1 are calculated as follows.

$$\begin{cases} Th(flow1) = B_1 \frac{G}{G+B_2} \\ Th(flow2) = B_1 \frac{B_2}{G+B_2} \end{cases} \text{if } G \geq B_1 - B_2 \qquad (3)$$

Equations should be punctuated in the same way as ordinary text but with a small space before the end punctuation mark.

### 3.2 Contention in Round Robin Scheduling

In RR scheduling, we have the same results as in the first and second cases as in (1) and (2) described in Sect. 3.1. In the third case, both flow 1 and flow 2 share the bandwidth, but the throughput of the forwarding flow 2 $Th(flow2)$ is limited by the receiving bandwidth $B_2$, and flow 1 can get all remaining bandwidth. The throughputs for flow 1 and flow 2 are calculated as follows.

$$\begin{cases} Th(flow1) = B_1 - B_2 \\ Th(flow2) = B_2 \end{cases} \text{if } G \geq B_1 - B_2 \qquad (4)$$

## 4 Improving Data Flow Performance by Control

### 4.1 Probabilistic Control on Round Robin Queue Scheduling

The reason why RR mechanism cannot give a satisfactory throughput for forwarding flows is the limited receiving bandwidth at the MAC layer. Only a small number of forwarding packets can reach the relay station, the forwarding flow's queues often become empty and thus RR mechanism misses turns for forwarding flows. It is clearly of great advance to the direct flow. Our idea is to manage RR queues to ensure fair buffer and bandwidth allocation.

We propose *Probabilistic Control on Round robin Queue* (PCRQ) scheduling. In the PCRQ scheduling, RR queues are used with three algorithms: Algorithm 1 controls the number of input packets to queues, Algorithm 2 controls the turn of reading queues, and Algorithm 3 controls the number of output packets from queues. Figure 2 shows our proposed method.

## 4.2 The Control Algorithms for Scheduling

In 802.11e, data traffic is classified into different priority levels to support different QoS requirements, each Access Category (AC) will have separately buffer queue, corresponding to the parameters EDCA. However, the method using in AC queue management is fixed and not adaptive to changes in data transmission, especially for highly distributed network such as multihop ad hoc networks. That leads to the fact that flows of the different ACs becomes unbalanced, can make larger latency and higher packet loss rate, and when the network traffic becomes heavy, the QoS requirements for services that need high priority will not be guaranteed. To overcome these issues, we propose three algorithms to handle each AC queue: Algorithm 1 controls the number of input packets to queues, Algorithm 2 controls the turn of reading queues, and Algorithm 3 controls the number of output packets from queues. As shown in Figure 1, if combined with three types of AC queue in 802.11e, we have a total of nine queues in station M1 to control the input and output packets.

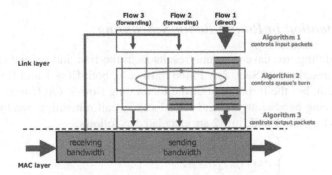

**Fig. 2** Probabilistic control on round robin queue

## Algorithm 1 (Controlling the Number of Input Packets to Queues)

In multihop network, when the offered load $G$ is large, the direct flow's queue tends to occupy completely the buffer space. Algorithm 1 decides to receive or drop an input packet so that not to put too much packets to a queue. An arriving packet from flow $i$ with priority $p$ is put into its queue at the following probability:

$$P_{i\_input}^p = \begin{cases} 1 & \text{if } qlen_i^p \leq ave^p \\ 1 - \alpha \frac{qlen_i^p - ave^p}{(n^p - 1)ave^p} & \text{if } qlen_i^p > ave^p \end{cases} \tag{5}$$

where $\alpha$ is an input weight constant, $ave^p$ is the average of the queue lengths for all flows with the same priority $p$, $n^p$ is the number of flows with priority $p$, $qlen_i^p$ is the queue length of flow $i$ with priority $p$, $p = 0, 1, 2$. Packets from a heavy offered load flow may be dropped with probability in range 0 to $\alpha$. In case the queue of flow $i$ is full while other queues are empty, incoming packet will be dropped with probability $\alpha$. In case the queue length of flow $i$ smaller or equal average queue length, all packets will be enqueued. Algorithm 1 reduces input packets of the flows with heavy offered load, and hence makes the queue length of each flow fairer and smaller.

## Algorithm 2 (Controlling the Turn of Reading Queues)

Generally, the receiving bandwidth at transit node is small, thus the forwarding flow's queues often become empty. In this case, the direct flow will get more turns from RR mechanism, and good per-flow fairness is not ensured. Algorithm 2 keeps the empty queue's turn for an interval time $\delta$ and waits for a new packet. The queue's turn of flow $i$ is hold at the following probability:

$$P_{i\_turn}^p = \begin{cases} \beta \frac{n^p \times ave^p}{qmax} & \text{if } qlen_i^p = 0 \\ 0 & \text{if } qlen_i^p > 0 \end{cases} \tag{6}$$

where $\beta$ is a hold weight constant, $qmax$ is the maximum queue length of all the queues. The turn of the empty queue is kept with probability in the range from $\beta/n$ to $\beta$. In case a queue is full while other queue is empty, the turn of the empty queue is kept with probability $\beta$. In case the queue lengths of all flow are almost equal, this probability is about $\beta/n$. Because a delay time $\delta$ can be used for receiving packets, the receiving bandwidth at the MAC layer will be improved. Thus, Algorithm 2 does not only help RR mechanism working more effectively but also make the receiving and sending bandwidths fairer.

## Algorithm 3 (Controlling the Number of Output Packets from Queues)

The unfairness between the receiving and sending bandwidths is the main reason of per-flow unfairness in RR scheduling. Algorithm 3 prevents heavy offered load flows from sending many packets to the MAC layer, and hence more bandwidth is left for receiving/forwarding flows. As a result, throughputs of forwarding flows are improved. A packet at the head of the queue for flow $i$ is sent from the link layer to the MAC layer at the following probability:

$$P_{i\_output}^p = \begin{cases} 1 & \text{if } qlen_i^p \le ave^p \\ 1 - \gamma \frac{qlen_i^p - ave^p}{(n^p - 1)ave^p} & \text{if } qlen_i^p > ave^p \end{cases} \tag{7}$$

where $\gamma$ is an output weight constant. Packets from a heavy offered load flow may be delay to send with probability in range 0 to $\gamma$. In case the queue is full while other flow is empty, packets may stop sending with probability $\gamma$. In case the queue length smaller or equal average queue length, all packets will be dequeued. If Algorithm 3 decides postpone sending a packet, the packet is delayed for an interval time $\gamma$. Thus Algorithm 3 makes not only the receiving and sending bandwidths fair, but also the throughput for each flow fair.

## 5 Analysis of Simulation Results

We evaluate the performance of PCRQ scheduling by comparing with the original FIFO scheduling in IEEE 801.11e standard, and RR scheduling. We use Network Simulator (NS-2) [14] for evaluation. The simulation parameters are shown in Table 3 and Table 4. In PCRQ scheduling, we set the input weight constant $\alpha = 2.0$, the hold weight constant $\beta = 0.3$, the output weight constant $\gamma = 0.3$ and delay time $\delta = 1$[ms]. We use the fairness index, which is defined by R. Jain [15] as follows:

$$FairnessIndex = \frac{(\sum_{i=1}^{n^p} x_i^p)^2}{n^p \times \sum_{i=1}^{n^p} (x_i^p)^2} \tag{8}$$

where $n^p$ is the number of flows with priority $p$, $x_i^p$ is the throughput of flow $i$ with priority $p$. The result ranges from $1/n^p$ to 1. In the best case, throughput of all flows are equal, the fairness index achieves 1. In the worst case, the network is totally unfair, one flow gets all capacity while other flows get nothing, and fairness index is $1/n$. In this paper, fairness index is evaluated based on goodput at the destination station.

Total throughput: The average of total goodput of all flows during simulation.

**Table 3** Parameters in the simulation

| | |
|---|---|
| Channel data rate | 11 Mbps |
| Antenna type | Omni direction |
| Radio Propagation | Two-ray ground |
| Transmission range | 250 m |
| Carrier Sensing range | 550 m |
| MAC protocol | IEEE 802.11e (EDCA) |
| Connection type | UDP/CBR |
| Buffer size | 100 packet |
| Simulation time | 300 s |

Network topology used in simulation is an ad hoc network that includes three nodes 0, 1, 2; each node generates flows with the corresponding priority of 0, 1, and 2.

**Table 4** 802.11e parameters in the simulation

| Priority | CWmin | CWmax | AIFS | TXOP Limit (ms) |
|----------|-------|-------|------|-----------------|
| 0 | 7 | 15 | 2 | 3.264 |
| 1 | 15 | 31 | 2 | 6.016 |
| 2 | 31 | 1023 | 3 | 0 |

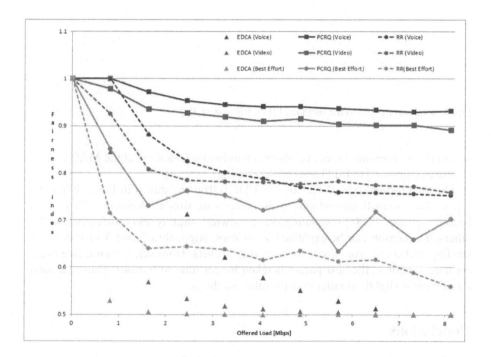

**Fig. 3** Fairness index with Voice, Video, Best effort

Fairness Indexes are shown in Fig.3. When the offered load $G$ is small, all scheduling methods get Fairness Index equal 1. When the offered load G becomes high, in FIFO scheduling, the direct flow gradually occupies completely the buffer space, then Fairness Index becomes very bad. In PCRQ scheduling, input and output packets to RR queues are controlled, then the throughput of each flow with the

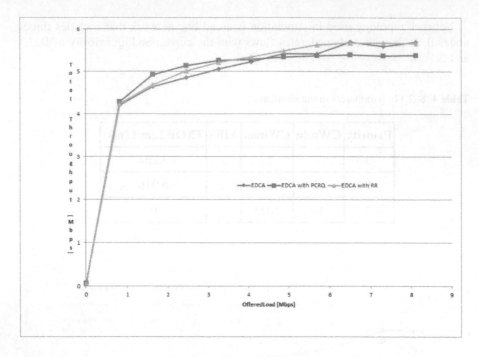

**Fig. 4** Total throughput in the methods FIFO, RR and PCRQ

same priority becomes fairer and also the bandwidth allocation at the MAC layer is improved compared to FIFO and RR scheduling.

The total throughputs of all flows in UDP traffic are shown in Fig.4. When the offered load is small, throughputs in all methods are similar. When the offered load becomes greater, PCRQ scheduling uses bandwidth slightly less efficiently than the others. The reason can be explained as follows. Algorithm 2 and 3 give delay to sending packets to give chance for receiving packets. However, if forwarding packets may not come, the next packet is taken longer time to transfer. Thus, our total throughput is slightly smaller than the other methods.

## Conclusions

In this paper, the problem of ensuring throughput's fairness in multihop wireless ad hoc is examined, mainly focused on why the FIFO and Round Robin scheduling provide low performance. FIFO scheduling provides only one queue, so it cannot resolve the contention at the link layer. Similarly, Round Robin scheduling is not efficient because the bandwidth allocated at MAC layer and not suitable for forwarding flows and direct flow at the link layer.

Our research examined a new proposed scheduling method in the link layer, namely PCRQ to solve the problem. By controlling the number of input packets to different queues, the turn of reading queues, and the number of output packets from queues, scheduling PCRQ will indirectly control shared channels at MAC layer. Therefore, our mechanism ensures the fairness between flows with the same priority in IEEE 802.11e.

# References

[1] Part11: Wireless LAN Medium Access Control (MAC) and Physical Layer (PHY) Specifications: Amendment 8: Medium Access Control (MAC) Quality of Service Enhancements. ANSI/IEEE Std 802.11e (2005)

[2] I.S. Department. IEEE 802.11 wireless lan medium access control (MAC) and physical layer (PHY) specifications. ANSI/IEEE Standard 802.11 (1999)

[3] Karn, P.: Maca: A new channel access method for packet radio. In: ARRL/CRRL Amateur Radio 9th Computer Networking Conference, pp. 134–140 (1990)

[4] Bharghavan, V., Demers, A., Shenker, S., Zhang, L.: Macaw: A media access protocol for wireless lan's. In: SIGCOMM 1994: Proc.Conference on Communications Architectures, Protocols and Applications, pp. 212–225 (1994)

[5] Nandagopal, T., Kim, T.E., Gao, X., Bharghavan, V.: Achieving MAC layer fairness in wireless packet networks. In: MobiCom 2000: Proc. 6th Annual International Conference on Mobile Computing and Networking, pp. 87–98 (2000)

[6] Bensaou, B., Wang, Y., Ko, C.C.: Fair medium access in 802.11 based wireless ad-hoc networks. In: MobiHoc 2000: Proc. 1st ACM International Symposium on Mobile Ad Hoc Networking & Computing, pp. 99–106 (2000)

[7] Jiang, L.B., Liew, S.C.: Improving throughput and fairness by reducing exposed and hidden nodes in 802.11 networks. IEEE Trans. Mobile Computing 7(1), 34–49 (2008)

[8] Li, Z., Nandi, S., Gupta, A.K.: Ecs: An enhanced carrier sensing mechanism for wireless ad-hoc networks. Comput. Commun. 28(17), 1970–1984 (2005)

[9] Jangeun, J., Sichitiu, M.: Fairness and qos in multihop wireless networks. In: IEEE Vehicular Technology Conference, vol. 5, pp. 2936–2940 (2003)

[10] Shagdar, O., Nakagawa, K., Zhang, B.: Achieving per-flow fairness in wireless ad hoc networks. Electron. Commun. Jpn. 1, Commun. 89(8), 37–49 (2006)

[11] Izumikawa, H., Sugiyama, K., Matsumoto, S.: Scheduling algorithm for fairness improvement among subscribers in multihop wireless networks. Electron. Commun. Jpn. 1, Commun. 90(4), 11–22 (2007)

[12] Gossain, H., de, C., Cordeiro, M., Agrawal, D.P.: Energy efficient mac protocol with spatial reusability for wireless ad hoc networks. International Journal of Ad Hoc and Ubiquitous Computing 1(1/2), 13–26 (2005)

[13] He, J., Pung, H.K.: One/zero fairness problem of mac protocols in multi-hop ad hoc networks and its solution. Proc. International Conference on Wireless Networks, 479–485 (2003)

[14] The Network Simulator: ns-2, http://www.isi.edu/nsnam/ns/

[15] Jain, R., Chiu, D.M., Hawe, W.: A quantitative measure of fair-ness and discrimination for resource allocation in shared conputer systems. Technical Report TR-301, DEC Research Report (1984)

[16] Giang, P.T., Nakagawa, K.: Archieving fairness over 802.11 multi-hop wireless ad hoc networks. IEICE Trans. Commun. E92-B(8), 2628–2637 (2009)

[17] Giang, P., Nakagawa, K.: Cross-Layer Scheme to control Contention Window for per-flow Fairness in Asymmetric Multi-hop Networks. IEICE Transactions on Communications E93-B(9), 2326–2335 (2010)

[18] Ferre, P., Doufexi, A., Nix, A., Bull, D.: Throughput analysis of IEEE 802.11 and IEEE 802.11e MAC. In: Wireless Communications and Networking Conference, pp. 783–788 (2004)

[19] Banchs, A., Perez-Costa, X., Qiao, D.: Providing throughput guarantees in IEEE 802.11e wireless LANs. In: 18th International Teletrac Congress (2003)

[20] Liang, H.M., Ke, C.H., Shieh, C.K., Hwang, W.S., Chilamkurti, N.K.: Performance Evaluation of 802.11e EDCF in Infrastructure Mode with Real Audio/Video Traffic. In: International Conference on Networking and Services, pp. 92–92 (2006)

[21] Lin, Y., Wong, V.: Saturation Throughput of IEEE 802.11e EDCA Based on Mean Value Analysis. In: IEEE Wireless Communications and Networking Conference (WCNC) (2006)

[22] Xiong, L., Mao, G.: Saturated throughput analysis of IEEE 802.11e EDCA. Computer Networks, 3047–3068 (2007)

# Using Attribute Relationships for Person Re-Identification

Ngoc-Bao Nguyen, Vu-Hoang Nguyen, Thanh Ngo Duc,
Duy-Dinh, and Duc Anh Duong

**Abstract.** Person re-identification is the problem of matching pedestrian images observed by different cameras in non-overlapping regions. Semantic features, also called attributes, have demonstrated to produce state-of-the-art performances in this problem. In existing works, attributes are detected independently to each other. In this paper, we propose using relationships between attributes to refine the attribute detection result. Experimental results on two datasets VIPeR and PRID prove the effectiveness on performances when our method is applied.

**Keywords:** Person Re-Identification, Attribute Relationship, Re-Score.

## 1 Introduction

Person re-identification is the task of recognizing human images from various cameras in non-overlapping regions at different time. It plays an important role in surveillance systems which are more and more popular in public places such as supermarkets, airports, hospitals. Person re-identification faces challenges such as: different viewpoints, various resolutions, illumination changes, cluttered background, occlusion.

To compare a people to others, we need to represent people by visual features. Semantic features, also known as attributes, are used widely recently and produce good performances. This method allows to describe people by hair-style, clothing-style, gender.

Because attributes are used as features, person re-identification's performance depends on the attribute detection accuracy. Existing works examine attribute

Ngoc-Bao Nguyen · Vu-Hoang Nguyen · Thanh Ngo Duc · Duy-Dinh · Duc Anh Duong
Multimedia Communications Laboratory
University of Information Technology, VNU-HCM
Ho Chi Minh City, Vietnam
e-mail: 10520228@gm.uit.edu.vn,
        {vunh,thanhnd,ledduy,ducda}@uit.edu.vn

© Springer International Publishing Switzerland 2015                                    195
V.-H. Nguyen et al. (eds.), *Knowledge and Systems Engineering*,
Advances in Intelligent Systems and Computing 326, DOI: 10.1007/978-3-319-11680-8_16

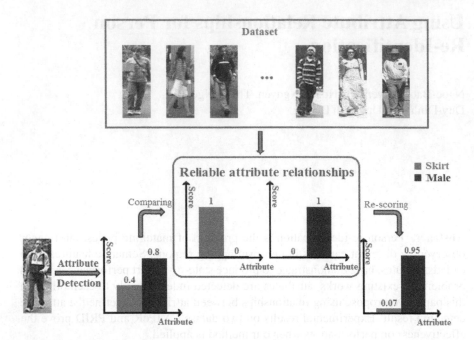

**Fig. 1** Attribute Relationships

detection independently while the relationships between attributes could help identify better. For example: assuming we use two attributes Male, Skirt to describe a person (Figure 1). It may be incorrect if the result of attribute detection is a man and wear skirt (Most of men doesn't wear skirt!). In this paper, we utilize the relationships between attributes to make the attribute detection result more accurate. In particular, attribute relationships are applied to re-score attribute confidence produced by attribute detectors. To prove the effectiveness of our proposed method, we conduct experiments on two datasets VIPeR [5] and PRID [7]. The results show that the person re-id performance is boosted when our method is applied.

The remaining of this paper is organized as follows: In Sec. 2 we review related work, Sec. 3 will discuss details of our proposed method, and in Sec. 4 we describe the experiment on the VIPeR and the PRID datasets, Sec. 5 is conclusions of the paper.

## 2  Related Work

Many existing works focus on visual features to represent human images. In the first place, low level features are used widely. Because of the low resolution of images, simple features like color histograms demonstrate their effectiveness in person re-identification problem [3, 11, 12]. [3] uses histograms of HSV color system. [6, 13]

uses histograms of eight color channels selected from of RGB, YCbCr, and HSV color systems. In addition to color, texture feature is also used in [6, 13, 15]. [6] uses two families of textures, Schmid filter and Gabor filter. The authors convolve each filter with the luminance channel to form a texture channel. In [15], the authors extract texture features by using co-occurrence matrices. Ma et al. [11] combine biologically inspired feature (BIF) with covariance descriptors for the re-identification problem. Beside low level features, semantic features are used commonly recently in [8, 10] and produce promising performances.

Several works are proposed to match human body parts for more accurate re-identification. In [4], Gheissari et al. proposed a model fitting technique to fit a triangulated graph to a human body. In [3], Farenzena et al. proposed a method for dividing a human body into upper part and lower part. Interest point matching also engages attention in [4]. [12] tries to compare detected blobs of pairs of images.

Beside using traditional distances like L1 distance, L2 distance, or Bhattacharyya distance, many works focus on learning a new distance which is more suitable on a specific dataset. This direction attracts attention from many researches [6, 13, 15]. Another direction is learning transformation between two cameras. [1, 16] learn a transformation function from a fixed camera to another fixed camera.

In this paper, we focus on the feature representation approach, more specifically we concentrate on semantic feature which is an emerged topic recently.

## 3   Attribute Re-Scoring (ARS)

A common framework of person re-identification comprises two parts: feature extraction and matching (Figure 2). Feature extraction transforms the input data (images) into feature vectors. The matching component aims to compute similarity of image pairs using their feature vectors.

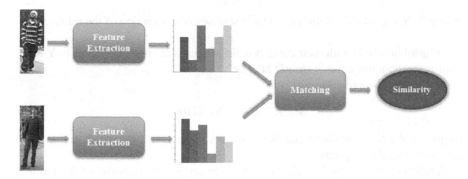

**Fig. 2** Framework of person re-identification

For feature extraction, if we use n attributes $\{A_1, A_2, \ldots, A_n\}$ to describe a person image, then its feature vector is:

$$\vec{f} = (s(A_1), s(A_2), \ldots, (A_n)) \tag{1}$$

where $s(A_i)$ is score which represents the confidence of $A_i$ on the image.

Existing works treat attributes independently in presentation. In our work, we propose a new method using attribute relationships called **Attribute Re-Scoring (ARS)** (Figure 3) for boosting the attribute detection accuracy as well as boosting the person re-identification performance .

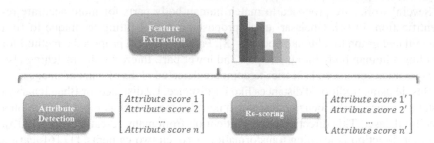

**Fig. 3** Attribute Re-Scoring

More specifically, we define a set of relationships between attributes, each relationship includes at least two attributes and their statuses. Then we use the relationships to re-score attribute detection result.

## 3.1 Attribute Relationships

For an image, each attribute belongs to one of two statuses: either appear or disappear. A relationship between attributes is defined as follows:

$$X \longrightarrow Y \tag{2}$$

where X, Y are sets of attributes and their statuses. The meaning of the relationship is:

"If attributes in X with their corresponding statuses are true, attributes in Y with their corresponding statuses *should be true*".

For instance, the relationship

$$Male, Suit \longrightarrow No\, Skirt$$

implies if the Male attribute and Suit attribute of an image appear then the Skirt attribute should not appear.

A reliable relationship is a relationship which reflects the majority of the dataset. Such reliable relationships should be used to re-score attribute detection results (mentioned in section 3.2). In this paper, we propose a set of relationships manually based on our observation on datasets.

## 3.2 Re-Scoring Attribute Detection Confident

Assuming we have a set of reliable relationships provided. The task is to use them to adjust attribute detection results to make them more accurate. In this section, we describe our method called Attribute Re-Scoring (ARS) for that goal.

### 3.2.1 Re-Scoring Attributes Using a Single Relationship

Because a reliable relationship is correct in the majority of the dataset, an attribute detection result which contradicts a reliable relationship would be highly likely wrong. Therefore attribute detection scores of images should be adjusted base on the relationships to make them more reliable.

For each attribute $A_i$ in the contradicted relationship, let $s'(A_i)$ be the $A_i$'s score before re-scoring and $s'(A_i) = F(s(A_i), st)$ be $A_i$'s score after re-scoring where st is the status of $A_i$ in the relationship (1 for appear or 0 for disappear). Considering two cases of st:

**Case 1: st = 1.** In this case, $s'(A_i)$ should be adjusted to be nearer to 1. The higher $s(A_i)$ is, the higher $s'(A_i)$ should be. Furthermore, the post re-scoring score should be higher than the original score. Specifically:

$$\begin{cases} 0 \leq F(s_1(A_i), 1) < F(s_2(A_i), 1) \leq 1 & for\,all\,s_1(A_i) < s_2(A_i) \\ F(s(A_i), 1) \geq s(A_i) \end{cases} \tag{3}$$

**Case 2: st = 0.** In contrast, $s'(A_i)$ should be adjusted to be nearer to 0. The lower $s(A_i)$ is, the lower $s'(A_i)$ should be. Furthermore, the post re-scoring score should be lower than the original score. Specifically:

$$\begin{cases} 0 \leq F(s_1(A_i), 0) < F(s_2(A_i), 0) \leq 1 & for\,all\,s_1(A_i) < s_2(A_i) \\ F(s(A_i), 0) \leq s(A_i) \end{cases} \tag{4}$$

There are many functions $F(s(A_i), st)$ satisfying (3) and (4). For example, every monotonically increasing functions with their part of graph in the dark regions satisfy the conditions. In this paper, we choose a simple form of $F(s(A_i), st)$ for efficiency. In particular, our method's function is a part of circle above (case 1) and below (case 2) the line y = x (Figure 4). Specifically:

$$F(s(A_i), st) = \begin{cases} 1 - m + \sqrt{m^2 + (m-1)^2 - (s(A_i) - m)^2} & if\,st = 1 \\ m - \sqrt{m^2 + (m-1)^2 - (s(A_i) - 1 + m)^2} & if\,st = 0 \end{cases} \tag{5}$$

where m $\geq$ 1.

### 3.2.2 Re-Scoring Attributes Using Multiple Relationships

With multiple relationships, we filter for all contradicted relationship to apply re-scoring. The filtered relationships will be ordered by their distances to the attribute detection results and then applied to attribute detection results using the procedure

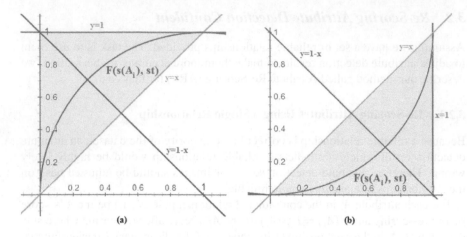

**Fig. 4** Re-scoring curve. (a) $F(s(A_i), 1)$ is the result of $s(A_i)$ after re-scoring when st = 1 and (b) $F(s(A_i), 0)$ is the result of $s(A_i)$ after re-scoring when st = 0.

for single relationship as discussed in the previous section. The detail of our algorithm is described below:

Algorithm stops when **Contradicted Relationship** or **Relationships** is empty. Therefore, the maximum number of iterations is the number of relationship (length of **List_Relationship**).

(a) VIPeR examples                              (b) PRID examples

**Fig. 5** Examples from the VIPeR dataset (a) and single-shot version in PRID dataset (b). Four people captured by one camera (top row) and another camera (second row).

**Algorithm 1.** Re-scoring attribute detection confident

---

Create **List_Relationship**.

For each person **p**

    **Relationships = List_Relationship**.

    **Step 1:** Search **Contradicted Relationship** of **p** in **Relationships**.

    **Step 2:** Compute distance between **p** and each status of relationship in **Contradicted Relationship**.

    **Step 3:** Sorted ascending distance.

    **Step 4:** For each relationship **r** in **Contradicted Relationship**

        Using **r** to re-score **p**.

        If (status changes) *//attribute's status changed after re-scoring*

            Remove **r** in **Relationships**.

            Return **Step 1**.

        End

    End

End

---

# 4 Experiments

## 4.1 Datasets and Evaluation Methods

### 4.1.1 Datasets

We conduct our experiments on VIPeR[5] and PRID[7] datasets (Figure 5).

- VIPeR is one of the most commonly used dataset for evaluating re-identification methods. It contains 632 pedestrian image pairs. Each pair contains two images of the same individual seen from different viewpoint by two cameras. The images are scaled to 128x48 pixels.
- PRID provides two versions, one representing the single-shot scenario and one representing the multi-shot scenario. Following [9], we use single-shot version and use the first 200 shots from each view. The images are scaled to 128x64 pixels.

### 4.1.2 Evaluation Methods

The output of a person re-identification system are lists of persons in camera B (gallery persons) ranked by their similarity to a person in camera A (probe person). To compare performances of various methods, similar to [3] and [9] we use average Cumulative Match Characteristic (CMC) curves, normalized area under the CMC curve (nAUC), Rank i and Expected Rank (ER). Rank i denotes the percentage of true matches found within the first i ranked instance (The higher value of Rank i implies better performance). CMC is the curve presenting all Rank i. ER is the mean rank of true matches (The best ER is 1 and the higher ER implies worse performance).

## 4.2 Experimental Settings

### 4.2.1 Attribute Selection

We use attributes and annotation provided in [9].

- VIPeR: We use 21 attributes: *Redshirt, Blueshirt, Lightshirt, Darkshirt, Greenshirt, Nocoats, Notlightdarkjeans colour, Darkbottoms, Lightbottoms, Hassatchel, Barelegs, Shorts, Jeans, Male, Skirt, Patterned, Midhair, Darkhair, Bald, Hashandbag carrierbag, Hasbackpack.*
- PRID: There are a four attributes without enough data for training classifier. So we choose 17/21 attributes for our experiments, include: *Redshirt, Blueshirt, Lightshirt, Darkshirt, Nocoats, Darkbottoms, Lightbottoms, Hassatchel, Barelegs, Jeans, Male, Skirt, Midhair, Darkhair, Bald, Hashandbagcarrierbag, Hasbackpack.*

### 4.2.2 Attribute Detection

There are two parts in this phase. First is extracting feature representing for each image. Second is using feature to build classifiers.

- Feature: We follow the method in [14] to detect attributes. We first divide an image into 6 equal horizontal parts. In each part, we extract 8 color channels (RGB, HS, and YCbCr) and 21 texture filters (Gabor and Schmid) from the bright channel. Our parameters $\gamma, \theta, \lambda, \sigma^2$ of Gabor filter are (0.3, 0, 4, 2), (0.3, 0, 8, 2), (0.4, 0, 4, 1), (0.4, 0, 8, 1), (0.3, $\pi/2$, 4, 2), (0.3, $\pi/2$, 8, 2), (0.4, $\pi/2$, 4, 1), (0.4, $\pi/2$, 8, 1). The parameters $\tau$ and $\sigma$ of Schmid filter are (2, 1), (4, 1), (4, 2), (6, 1), (6, 2), (6, 3), (8, 1), (8, 2), (8, 3), (10, 1), (10, 2), (10, 3), and (10, 4). The extracted feature on 6 parts are concatenated to form the feature vector of the whole image.

- Classifier: We use Support Vector Machines with RBF kernel for attribute classification. The implement provided by LIBSVM [2]. The output of SVM is changed to scores by Sigmoid function.

### 4.2.3 Train and Test Set

For each dataset (VIPeR and PRID), we divide the available data into training and testing partitions. We train the classifier in training partition. Then we test person re-identification performance on testing partition. Size of testing partition in VIPeR and PRID is 316 and 100 respectively. To avoid bias due to imbalanced data in two datasets VIPeR and PRID, we simply train each attribute detector with the same number of negative training examples and positive training examples.

**Table 1** Proposed relationships for VIPeR and PRID

| VIPeR | PRID |
| --- | --- |
| 1. $Blueshirt \longrightarrow No\,Redshirt$ | 1. $Midhair \longrightarrow No\,Bald$ |
| 2. $Shorts \longrightarrow Barelegs$ | 2. $Lightbottoms \longrightarrow No\,Darkbottoms$ |
| 3. $Skirt \longrightarrow Barelegs$ | 3. $Skirt \longrightarrow No\,Jeans$ |
| 4. $Skirt \longrightarrow No\,Notlightdark\,jeanscolour$ | 4. $Skirt \longrightarrow No\,Male$ |
| 5. $Shorts \longrightarrow No\,Jeans$ | 5. $Blueshirt \longrightarrow No\,Redshirt$ |
| 6. $Blueshirt \longrightarrow No\,Greenshirt$ | 6. $Lightshirt \longrightarrow No\,Darkshirt$ |
| 7. $Bald \longrightarrow No\,Darkhair$ | 7. $Darkshirt \longrightarrow No\,Lightshirt$ |
| 8. $Greenshirt \longrightarrow No\,Redshirt$ | |
| 9. $Skirt \longrightarrow No\,Male$ | |
| 10. $Darkshirt \longrightarrow No\,Lightshirt$ | |

### 4.2.4 Attribute Relationships

As aforementioned, in this paper, we choose attribute relationships manually based on our observation on the whole dataset. Because of the difference between the two datasets, we define two different sets of attribute relationships for them. In particular, we build 10 and 7 relationships in VIPeR and PRID show in Table 1.

#### 4.2.5    Distance Metric

For similarity comparison, we use traditional L1 distance.

## 4.3    Results and Discussions

#### 4.3.1    Attribute Detection Result

Accuracy of attribute detection in VIPeR and PRID is shown in Table 2. From the
results we could see that attributes with high popularity such as "Red Shirt" tends to
obtain higher detection accuracy. This is because they have big numbers of training
data which could cause better generalization in training.

#### 4.3.2    Person Re-Identification Result

The re-identification performance is summarized in Table 3 in terms of ER, Rank i
and nAUC and Figure 6 in terms of CMC. The results in all metrics: rank 5, rank 10,
rank 25, ER and nAUC is better when applying the relationship to re-score attribute
score. In VIPeR, the performance improvements are uniform in all ranks. In PRID
the improvement is largest in rank 35 to 70 of PRID.

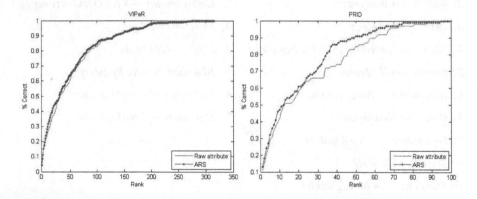

**Fig. 6** Final person re-identification CMC plots for VIPeR gallery sizes p = 316 (left) and
PRID gallery sizes p = 100 (right)

The results show that our proposed method improves performance of attribute
detection on the detected attributes and as a result person re-identification performance is also promoted.

**Table 2** Attribute detection result

|  | VIPeR | PRID |
| --- | --- | --- |
| Redshirt | 75.95 | 23.5 |
| Blueshirt | 59.02 | 54.4 |
| Lightshirt | 84.18 | 70 |
| Darkshirt | 84.02 | 82 |
| Greenshirt | 41.14 | - |
| Nocoats | 64.72 | 68.5 |
| Notlightdarkjeanscolour | 37.5 | - |
| Darkbottoms | 69.46 | 73.5 |
| Lightbottoms | 69.78 | 58.5 |
| Hassatchel | 47.31 | 46 |
| Barelegs | 62.66 | 50 |
| Shorts | 47.94 | - |
| Jeans | 63.29 | 41 |
| Male | 58.23 | 46.5 |
| Skirt | 53.64 | 36.5 |
| Patterned | 52.37 | - |
| Midhair | 58.70 | 59 |
| Darkhair | 63.29 | 74.5 |
| Bald | 23.42 | 37.5 |
| Hashandbagcarrierbag | 34.18 | 48.5 |
| Hasbackpack | 57.75 | 36 |
| **Mean** | **57.55** | **53.29** |

**Table 3** Person re-identification result

|                         | VIPeR  | VIPeR ARS         | PRID   | PRID ARS           |
|-------------------------|--------|-------------------|--------|--------------------|
| Mean attribute detection| 57.55  | 63.96 (**+6.41%**)| 53.29  | 59.94 (**+6.65%**) |
| Rank 5                  | 16.77  | 17.41 (**+0.64%**)| 29     | 32 (**+3%**)       |
| Rank 10                 | 24.05  | 28.16 (**+4.11%**)| 42     | 47 (**+5%**)       |
| Rank 25                 | 42.09  | 45.89 (**+3.8%**) | 64     | 67 (**+3%**)       |
| ER                      | 53.01  | 50.31 (**-2.7 rank**)| 23.86 | 20.1 (**-3.76 rank**)|
| nAUC                    | 0.8337 | 0.8423 (**+0.86%**)| 0.7659 | 0.8034 (**+3.75%**)|

## 5 Conclusions

Relationships between attributes are very important information. In this paper, we proposed a post-processing method using attribute relationships to boost person re-identification performance. Our experiments conducted on two datasets VIPeR and PRID show that performance of person re-identification is improved by applying our method.

**Acknowledgement.** This research is funded by Vietnam National University HoChiMinh City (VNU-HCM) under grant number B2013-26-01.

## References

[1] Avraham, T., Gurvich, I., Lindenbaum, M., Markovitch, S.: Learning Implicit Transfer for Person Re-identification. In: Fusiello, A., Murino, V., Cucchiara, R. (eds.) ECCV 2012 Ws/Demos, Part I. LNCS, vol. 7583, pp. 381–390. Springer, Heidelberg (2012)

[2] Chang, C.-C., Lin, C.-J.: LIBSVM: a library for support vector machines. ACM Transactions on Intelligent Systems and Technology (TIST) 2(3), 27 (2011)

[3] Farenzena, M., Bazzani, L., Perina, A., Murino, V., Cristani, M.: Person re-identification by symmetry-driven accumulation of local features. In: Proceedings of the 2010 IEEE Computer Society Conference on Computer Vision and Pattern Recognition (CVPR 2010), San Francisco, CA, USA. IEEE Computer Society (2010)

[4] Gheissari, N., Sebastian, T.B., Hartley, R.: Person reidentification using spatiotemporal appearance. In: 2006 IEEE Computer Society Conference on Computer Vision and Pattern Recognition, vol. 2, pp. 1528–1535 (2006)

[5] Gray, D., Brennan, S., Tao, H.: Evaluating appearance models for recognition, reacquisition, and tracking. In: IEEE International Workshop on Performance Evaluation of Tracking and Surveillance. Citeseer (2007)

[6] Gray, D., Tao, H.: Viewpoint invariant pedestrian recognition with an ensemble of localized features. In: Forsyth, D., Torr, P., Zisserman, A. (eds.) ECCV 2008, Part I. LNCS, vol. 5302, pp. 262–275. Springer, Heidelberg (2008)

[7] Hirzer, M., Beleznai, C., Roth, P.M., Bischof, H.: Person re-identification by descriptive and discriminative classification. In: Heyden, A., Kahl, F. (eds.) SCIA 2011. LNCS, vol. 6688, pp. 91–102. Springer, Heidelberg (2011)

[8] Layne, R., Hospedales, T., Gong, S.: Person re-identification by attributes. In: Proceedings of the British Machine Vision Conference, pp. 24.1–24.11. BMVA Press (2012)

[9] Layne, R., Hospedales, T.M., Gong, S.: Attributes-based re-identification. In: Person Re-Identification, pp. 93–117. Springer (2014)

[10] Liu, C., Gong, S., Loy, C.C., Lin, X.: Person re-identification: What features are important? In: Fusiello, A., Murino, V., Cucchiara, R. (eds.) ECCV 2012 Ws/Demos, Part I. LNCS, vol. 7583, pp. 391–401. Springer, Heidelberg (2012)

[11] Ma, B., Su, Y., Jurie, F.: Bicov: a novel image representation for person re-identification and face verification. In: Proceedings of the British Machine Vision Conference, pp. 57.1–57.11. BMVA Press (2012)

[12] Oreifej, O., Mehran, R., Shah, M.: Human identity recognition in aerial images. In: 2010 IEEE Conference on Computer Vision and Pattern Recognition (CVPR), pp. 709–716 (2010)

[13] Prosser, B., Zheng, W.-S., Gong, S., Xiang, T.: Person re-identification by support vector ranking. In: Proceedings of the British Machine Vision Conference, pp. 21.1–21.11. BMVA Press (2010), doi:10.5244/C.24.21

[14] Prosser, B., Zheng, W.-S., Gong, S., Xiang, T., Mary, Q.: Person re-identification by support vector ranking. In: BMVC, vol. 1, p. 5 (2010)

[15] Schwartz, W.R., Davis, L.S.: Learning Discriminative Appearance-Based Models Using Partial Least Squares. In: Proceedings of the XXII Brazilian Symposium on Computer Graphics and Image Processing (2009)

[16] Lindenbaum, M., Brand, Y., Avraham, T.: Transitive re-identification. In: Proceedings of the British Machine Vision Conference. BMVA Press (2013)

# Contingent Information: A Four-Valued Approach

Seiki Akama, Jair Minoro Abe, and Kazumi Nakamatsu

**Abstract.** One of the interesting types of information in information systems is contingent information. Information systems should be able to deal with such information. It is interesting to formalize contingent information, and the task is closely related to the so-called future contingents. We try to model contingent information by means of Łukasiewicz's four-valued logic. Although modalities in his logic are non-standard, the non-modal fragment is useful to express contingent information in information systems.

**Keywords:** Contingent information, future contingents, information system, three-valued logic, four-valued logic, Łukasiewicz.

## 1 Introduction

*Information system* is a computer system capable of handling information in our world. In information systems, we face various types of information. One of the interesting types of information is *contingent information*. In fact, some information can be viewed as contingent in the sense that the interpretation is open at certain time point.

We believe that information systems, in particular, those handle temporal information, should thus be able to deal with such information. It is interesting to

Seiki Akama
C-Republic, 1-20-1, Higashi-Yurigaoka, Asao-ku, Kawasaki 215-0012, Japan
e-mail: akama@jcom.home.ne.jp

Jair Minoro Abe
Paulista University, R. Dr. Bacelar, 1212, 04026-002, São Paulo, SP, Brazil
e-mail: jairabe@uol.com.br

Kazumi Nakamatsu
University of Hyogo, 1-1-12, Shinzaike-Honmachi, Himeji, 670-0092, Japan
e-mail: nakamatsu@shse.u-hyogo.ac.jp

© Springer International Publishing Switzerland 2015                                    209
V.-H. Nguyen et al. (eds.), *Knowledge and Systems Engineering,*
Advances in Intelligent Systems and Computing 326, DOI: 10.1007/978-3-319-11680-8_17

formalize contingent information, and the task is closely related to the so-called *future contingents*.

By future contingents, we mean statements about future events. To give a logical interpretation of future contingents, i.e., the problem of future contingents, has challenged logicians for many years. There are several approaches to the problem in the literature.

One of the most intuitive interpretations was found in Aristotle's *De Interpretatione IX*; see Aristotle [6]. He thought that only propositions about the future which are either necessarily true, or necessarily false, or something determined have a determinate truth-value.

The interpretation suggests that a future contingent has no truth-values, i.e., its truth-value is a *gap*. To make his interpretation more formal, we need two logical concepts, i.e., *the law of excluded middle* (LEM) and *the principle of bivalence* (PB).

(LEM) is a syntactic thesis of the form $A \lor \neg A$ (*A* or not-*A*). (PB) is a semantic thesis saying that every proposition is either true or false. Although they are equivalent in classical logic (in which (PB) holds), they should be distinguished if we admit gaps. Aristotle accepts (LEM), but rejects (PB) for future contingents.

Łukasiewicz introduced famous three-valued logic and regarded future contingents as "possible"; see Łukasiewicz [8]. Unfortunately, his approach was not successful and has been criticized by many people; see Prior [10] and Urquhart [12]. Now, a question arises. Was Łukasiewicz wrong with future contingents? Intuitively speaking, his interpretation seems correct. But it has some technical difficulties.

The aim of this paper is to provide an alternative interpretation for contingent information in many-valued setting. In this paper, we propose a four-valued approach to future contingents using Łukasiewicz's four-valued logic in Łukasiewicz [9]. Although modalities in his logic are non-standard, the non-modal fragment is useful to express contingent information in information systems.

The paper is organized as follows. In section 2, we survey Łukasiewicz's idea of formalizing future contingents in his three-valued logic. In section 3, we review Łukasiewicz's four-valued logic. In section 4, we propose a four-valued approach to future contingents using Łukasiewicz's four-valued logic. We also discuss the interpretations of modal operators in the four-valued logic. Section 5 concludes the paper with some conclusions.

## 2   Łukasiewicz on Future Contingents

It is well known that Łukasiewicz proposed a three-valued logic to formalize future contingents in Lukasiewicz [8]. We denote the logic by Ł₃, which uses the truth-values, $T$ (true), $F$ (false), and $U$ (possible), obeying the following truth-value tables for negation ($\neg$), conjunction ($\land$), disjunction ($\lor$) and implication ($\rightarrow$).

| → | $T$ | $U$ | $F$ | ¬ |     | ∧ | $T$ | $U$ | $F$ |     | ∨ | $T$ | $U$ | $F$ |
|---|-----|-----|-----|---|-----|---|-----|-----|-----|-----|---|-----|-----|-----|
| $T$ | $T$ | $U$ | $F$ | $F$ | | $T$ | $T$ | $U$ | $F$ | | $T$ | $T$ | $T$ | $T$ |
| $U$ | $T$ | $T$ | $U$ | $U$ | | $U$ | $U$ | $U$ | $F$ | | $U$ | $T$ | $U$ | $U$ |
| $F$ | $T$ | $T$ | $T$ | $T$ | | $F$ | $F$ | $F$ | $F$ | | $F$ | $T$ | $U$ | $F$ |

We write $\models_3 A$ to mean that $A$ is valid in Ł$_3$. Unfortunately, $\not\models_3 A \vee \neg A$ and $\not\models_3 \neg(A \wedge \neg A)$. These facts mean that typical laws in classical logic do not hold in Ł$_3$. In addition, when $A = U$ the following two hold.

$$A \rightarrow A = T$$
$$A \rightarrow \neg A = T$$

The former is defensible, but the latter is not. Łukasiewicz's starting point is Aristotle's idea, but his formulation cannot implement it faithfully. In fact, gaps are allowed but the excluded middle does not hold.

Some people criticized Łukasiewicz's approach; see Prior [10]. Prior was dissatisfied with Ł$_3$ and felt that there are no finite-valued logics to model future contingents. Urquhart also says in [12]:

> The logic of the 'possible' in Łukasiewicz's sense is just not truth-functional (an observation first made by Gonseth [1941]). This is no more surprising than the fact that the probability calculus is not truth-functional, and it holds for the same reasons. However, it throws in doubt Łuksiewicz's claim to provide a serious logical and philosophical alternative to Aristotelian logic.

His comments reveal that Ł$_3$ is not satisfactory logically as well as philosophically to obtain Łukasiewicz's goal. From such criticisms, we are inclined to say that Łukasiewicz (three-valued) approach to future contingents fails.

## 3 Łukasiewicz's Four-Valued Logic

Even if Łukasiewicz's three-valued approach is not successful, we believe, his intuition cannot be denied. It is thus interesting to explore an improvement or extension of his formulation. Our idea is to adopt a four-valued approach. We will discuss whether it can give an improved approach.

Łukasiewicz also proposed a four-valued logic and its modal extension in Łukasiewicz [9]. We here rehearse his four-valued logic which is denoted by Ł$_4$. It has four truth-values, i.e., 1 (true), 0 (false) and two indeterminacies 2 and 3. For the interpretation of logical symbols Ł$_4$ is enough to provide the following matrix:

| → | 1 | 2 | 3 | 0 | ¬ |
|---|---|---|---|---|---|
| 1 | 1 | 2 | 3 | 0 | 0 |
| 2 | 1 | 1 | 3 | 3 | 3 |
| 3 | 1 | 2 | 1 | 2 | 2 |
| 0 | 1 | 1 | 1 | 1 | 1 |

By implication and negation, we can define conjunction and disjunction.

$A \wedge B =_{def} \neg(A \to \neg B)$
$A \vee B =_{def} \neg A \to B$

We write $\models_4 A$ to mean that $A$ is valid in Ł$_4$. Then, we have that $\models_4 A \vee \neg A$. Ł$_4$ also has the nice property, namely all classical tautologies are tautologies of Ł$_4$, and vice versa.

Further, Łukasiewicz added necessity operator L and possibility operators M to obtain four-valued modal logic. The following matrix provides the interpretations of the necessity operator L and the possibility operator M.

| A | M | L |
|---|---|---|
| 1 | 1 | 2 |
| 2 | 1 | 2 |
| 3 | 3 | 0 |
| 0 | 3 | 0 |

## 4  The Proposed Interpretation

As described in the previous section, the semantics for Ł$_4$ can be presented by the truth-value table. It is far from *intuitive* semantics in the sense that we cannot naturally capture the meanings of 2 and 3. A matrix formulation is often usually taken for the formalization of non-classical logics including many-valued logics, but we need an intuitive semantics in the context of our study.

We now suggest the following interpretations for them. 2 reads "positively possible" and 3 "negatively possible", respectively. Based on these interpretations, one can formalize the following semantics for Ł$_4$. First, the truth-values of Ł$_4$ can be represented as pairs of classical truth-values. We denote 1 by classical truth and 0 by classical falsity, respectively. Then, four truth-values in Ł$_4$ are defined as follows:

$$1 = (1,1), 2 = (1,0), 3 = (0,1), 0 = (0,0)$$

We use the numbers $0, 1$ both for classical truth-values and the truth-values for Ł$_4$, and they can be distinguished by the context. Let $v$ be the valuation mapping from the set of propositional variables $\mathcal{P} = \{a, b, c, ...\}$ to $\{0, 1\}^2$.

Then, $v$ can be extended for negation and implication using unary operator $\sim$: $\{0,1\}^2 \to \{0,1\}^2$ and binary operator $\Rightarrow: \{0,1\}^2 \times \{0,1\}^2 \to \{0,1\}^2$ such that $\sim (a,b) = (1-a,1-b)$ and $(a,b) \Rightarrow (c,d) = (\max(1-a,c),\max(1-b,d))$. One can then define $v(\neg A) = \sim v(A)$ and $v(A \to B) = v(A) \Rightarrow v(B)$ for any valuation. It is easy to check that the given semantics satisfies the truth-value tables above.

An intuition behind the semantics is that there are two sorts of possibility which evaluate as 2 and 3. These correspond to the truth-values of future contingent sentences (1) and (2):

(1) There will be a sea-battle.
(2) There will not be a sea-battle.

Let (1) be $A$. Then, (2) is written as $\neg A$. We here neglect the problems related to internal and external negation. Both sentences are in fact future contingents, but these truth-values should be distinguished, since they express different future contingents. It is also obvious that the excluded middle $A \lor \neg A$ is valid from the truth-value tables for Ł$_4$.

Assuming that future contingents read 'possible', (1) can be interpreted as 'positively possible' by positive future contingent sentence and (2) as 'negatively possible' by negative future contingent sentence. And the negation of (1) must be (2) and the negation of (2) must be (1), respectively.

The source of difficulties in Łukasiewicz's three-valued approach lies in the fact that two kinds of possibility are simply interpreted as 'possible'. We list the interpretations (truth-values) of some future contingent sentences as follows:

(1) 2
(2) 3
(3) $(1) \to (1) = 2 \to 2 = 1$
(4) $(1) \to (2) = 2 \to 3 = 3$
(5) $(2) \to (2) = 3 \to 3 = 1$
(6) $(2) \to (1) = 3 \to 2 = 2$

Here, we should notice (4) and (6). In fact, (4) corresponds to a sentence asserting possibility of $\neg A$ and (6) to a sentence asserting of possibility of $A$ in classical sense, respectively. It appears that we can solve several defects of Łukasiewicz's three-valued approach by means of Ł$_4$.

However, we must be careful to draw a conclusion. For we face at least two fundamental questions. Without appropriate replies to them, our approach is not defensible. The first question is: are four truth-values enough to handle the problem of future contingents? The second question is: are the interpretations of modal operators are compatible with those of truth-values?

Now, we discuss the first question. There seem to be two difficulties related to it. The opponents might be able to attack our approach on the basis of these difficulties. The first difficulty is concerned with implicational future contingent of the form $A \to B$, where $A$ and $B$ are about completely different future events.

The situation obviously differs from the ones for $(3)-(6)$. Consider future contingent sentences (7) and (8):

(7) There will be a concert.
(8) There will not be a concert.

which are expressed as $B$ and $\neg B$, respectively. If we assume that $B$ receives the truth-value 2, then the truth-values of the following sentences are as follows:

(9) $(1) \rightarrow (7) = 2 \rightarrow 2 = 1$
(10) $(1) \rightarrow (8) = 2 \rightarrow 3 = 3$
(11) $(7) \rightarrow (1) = 2 \rightarrow 2 = 1$
(12) $(8) \rightarrow (1) = 3 \rightarrow 2 = 2$

Here, the truth-values of $(9) - (12)$ are the same as those of $(3) - (6)$, but some of which, in particular, (9) and (11) might be claimed to be problematic. But, we can reply to the objection. We can claim that (9) and (11) are acceptable. Let us consider $A \rightarrow B$. When both $A$ and $B$ are possibly true, i.e., $v(A) = v(B) = 2$, it means that there are (future) possibilities where $A$ is true and $B$ is false. This implies that there are also possibilities where $A \rightarrow B$ is false. Hence, Hence, $A \rightarrow B$ is possibly true (i.e., $v(A \rightarrow B) = 2$), but not necessarily true (i.e., $v(A \rightarrow B) = 1$). Consequently, it seems counter-intuitive to require that $v(A \rightarrow B) = 1$ when $v(A) = v(B) = 2$.

The other problem is that in our four-valued approach to future contingents there is room for using the fifth truth-values for 'neither positively possible nor negatively possible'. The need of extra truth-values could be applied to any finite-valued logic. Presumably, this is the reason some people object to many-valued (or truth-functional) approach.

One way out is to use infinite-valued logic, and it leads to probabilistic (or statistical) interpretation of future contingents. Another reply is that *pragmatically* four is enough number for truth-values for future contingents. For instance, it is not clear that the above mentioned fifth truth-value is needed for normal discourse.

Some of the opponents to our approach may be able to support the alternative three-valued approach based on *supervaluation* due to van Fraassen [13] who introduced the concept for different context. Supervaluation is a promising technique, keeping the excluded middle but allowing gaps.

Of course, Łukasiewicz's interpretation can be utilized in a three-valued setting by means of supervaluation. Thomason detailed supervaluational semantics for temporal logic in [11], which is pioneer work in this direction. Most workers in the area seem to believe that supervaluational semantics is one of the most attractive approaches.

However, one can point out some shortcoming. Indeed supervaluational semantics can naturally capture Aristotle's interpretation of future contingents in three-valued logic, but we give up some classical laws. In a similar way, we can also propound the approach using dual of supervaluation allowing gluts. But, it has also difficulties similar to those in supervaluational semantics.

We turn to the second question. Since Łukasiewicz proposed $Ł_4$ as a modal logic, we should comment on its modalities. There are also some difficulties. $Ł_4$ can be axiomatized by classical propositional logic plus the following axioms:

$$L(A \to B) \to (LA \to LB)$$
$$LA \to A$$
$$LA \to (B \to LB).$$

Note that Ł$_4$ does not have the rule of necessitation (i.e., $\vdash A \Rightarrow \vdash LA$). This means that Ł$_4$ is non-standard modal logic, since it is a rule for normal modal logic.

We can also point out that modal operators in Ł$_4$ could capture the meaning of future contingents. But, unfortunately there are currently no intuitive semantics, i.e., Kripke semantics.

Ł$_4$ cannot offer intuitive interpretation for future contingents. If future contingent sentences are read as "possible" ones, then they should be interpreted using a possibility operator. However, this would conflict with Łukasiewicz's own possibility operator, according to which a sentence asserting possibility would never take the truth-value 2. Thus, in fact, the interpretation which reads "possible" as "future" conflicts with Łukasiewicz'sown ideas.

It would be possible to give an alternative interpretation of modal operators. We denote by $L'$ and $M'$ based on the interpretation. They are introduced by means of the following definitions:

$$L'A = def \; \neg(A \to \neg A)$$
$$M'A = def \; \neg A \to A$$

These definitions may be found in the ones for a modal extension of Ł$_3$, and the ones in conditional logics. The matrices of $L'$ and $M'$ are as follows:

| $A$ | $M'$ | $L'$ |
|---|---|---|
| 1 | 1 | 1 |
| 2 | 2 | 2 |
| 3 | 3 | 3 |
| 0 | 0 | 0 |

From the matrices, we see that the alternative definitions of modal operators do not work. $L'$ and $M'$ cannot be interpreted as "modal" operators. If $M'$ is a possibility operator, then $M'A$ should be true when $A$ is possibly true, i.e., $v(M'A) = 1$ if $v(A) = 2$.

However, we can see that $v(M'A) = v(L'A) = v(A)$ for any four-valued truth assignment. In fact, $M'A$ and $L'A$ do not make sense since they are both equivalent to $A$ in the given four-valued semantics. It may not be a surprising fact because the non-modal fragment of Ł$_4$ is exactly the classical propositional logic.

From the discussion above, we can conclude that Łukasiewicz's project of formalizing future contingents as possibility fails even if four-valued logic Ł$_4$ is employed. We believe that there are no appropriate formal and philosophical justifications of his ideas in many-valued setting.

# 5   Conclusions

We considered a logical foundation for contingent information by means of Łukasiewicz's four-valued logic Ł4. Our approach uses his four-valued logic instead of three-valued logic to model contingent information, and the subject is closely connected with the problem of future contingents. We showed that there are some problems with the modal operators in Ł4 and claimed that Łukasiewicz's original ideas of formalizing future contingents cannot be justified.

However, the use of four truth-values in Ł4 seems to be promising to represent contingent information for practical applications. They can in fact express two kinds of contingent information. In addition, the non-modal fragment of Ł4 is equivalent to the classical propositional logic, and it is easy to use the fragment as a representation language for information systems.

There appear to be other approaches to contingent information. For instance, three-valued approaches are also attractive. In fact, various three-valued logics were extensively used to solve the problems in computer science. For example, Kleene's three-valued logic is well-known for computer scientists; see Kleene [7]. However, his logic has no logically justified implications.

We think that we should work out non-classical temporal logics for contingent information for various types of information. We investigated several approaches to future contingents; see [1, 2, 3, 4, 5], and they could be adapted to the formalization of contingent information.

**Acknowledgments.** We are grateful to three referees for valuable comments and suggestions.

# References

[1] Akama, S., Murai, T., Kudo, Y.: Uncertainty in future: A paraconsistent approach. In: Huynh, V.-N., Nakamori, Y., Lawry, J., Inuiguchi, M. (eds.) Integrated Management and Applications. AISC, vol. 68, pp. 335–342. Springer, Heidelberg (2010)

[2] Akama, S., Murai, T., Kudo, Y.: Partial and paraconsistent approaches to future contingents. To appear in Proc. of Arthur Prior Centenary Conference

[3] Akama, S., Murai, T., Miyamoto, S.: A three-valued modal tense logic for the Master Argument. Logique et Analyse 213, 19–30 (2011)

[4] Akama, S., Nagata, Y., Yamada, C.: A three-valued temporal logic for future contingents. Logique et Analyse 198, 99–111 (2007)

[5] Akama, S., Nagata, Y., Yamada, C.: Three-valued temporal logic $Q_t$ and future contingents. Studia Logica 88, 215–231 (2008)

[6] Aristotle: De Interpretatione. In: Edghill, E.M., Ross, W.D. (eds.) The Works of Aristotle. Oxford University Press, Oxford (1963)

[7] Kleene, S.: Introduction to Metamathematics. North-Holland, Amsterdam (1952)

[8] Łukasiewicz, J.: On 3-valued logic. In: McCall, S. (ed.) Polish Logic, pp. 16–18. Oxford University Press, Oxford (1920, 1967)

[9] Łukasiewicz, J.: A system of modal logic. In: Borkowski, L., Łukasiewicz, J. (eds.) Selected Work, pp. 352–390. North-Holland, Amsterdam (1970)

[10] Prior, A.N.: Three-valued logic and future contingents. Philosophical Quarterly 3, 317–326 (1953)

[11]  Thomason, R.H.: Indeterminist time and truth-value gaps. Theoria 36, 264–811 (1970)
[12]  Urquhart, A.: Basic many-valued logic. In: Gabbay, D., Gunthner, F. (eds.) Handbook of Philosophical Logic, 2nd edn., vol. 2, pp. 249–295. Springer, Berlin (2001)
[13]  van Fraassen, B.: Singular terms, truth-value gaps, and free logic. Journal of Philosophy 63, 481–495 (1966)

# On Automating Inference of OCL Constraints from Counterexamples and Examples

Duc-Hanh Dang and Jordi Cabot

**Abstract.** Within model-based approaches, defining domains and domain restrictions for conceptual models or metamodels is significant. Recently, a domain is often presented as a class diagram, and domain restrictions are expressed using the *Object Constraint Language* (OCL). An effective method to define a domain is based on a description of the domain at the instance and example level. So far such a method has often focused on the generation of structure aspects, but have omitted the inference of OCL restrictions that could complement the domain structure and improve the precision of the domain. This paper proposes an approach to automating the inference of OCL restrictions from a domain description in terms of counter- and examples. Candidates are generated by a problem solving, and irrelevant ones are eliminated using the user feedback on generated counter- and examples. Our approach is realized with the support tool InferOCL.

## 1 Introduction

To capture a domain corresponding to a conceptual model of a system or a metamodel is a significant step to manipulate models within model-driven approaches. An effective approach to define domains is based on a description of the domain at the instance and example level [1]. So far such a method has often focused on the generation of structure aspects captured by a class diagram, but have omitted the inference of restrictions that could complement the domain structure and improve the precision of the domain. The restrictions are often expressed in the Object Constraint Language (OCL) [2]. It puts forward a need to infer OCL restrictions from the instance level information of the domain.

Duc-Hanh Dang
VNU - University of Engineering and Technology, Hanoi, Vietnam
e-mail: hanhdd@vnu.edu.vn

Jordi Cabot
AtlanMod, cole des Mines de Nantes - INRIA, LINA, Nantes, France
e-mail: jordi.cabot@inria.fr

© Springer International Publishing Switzerland 2015                                    219
V.-H. Nguyen et al. (eds.), *Knowledge and Systems Engineering,*
Advances in Intelligent Systems and Computing 326, DOI: 10.1007/978-3-319-11680-8_18

The essence of such an inference is to consider the relationship between snapshots (object diagrams) and OCL invariants. Several authors offer methods to check if a snapshot is valid [3]. Current works often translate OCL specifications into other specification environments such as CSP [4], Alloy [5], and relational logic [6] in order to effectively validate snapshots or to find valid (invalid) snapshots or to check properties. The work in [7] proposes a genetic algorithm in order to generate OCL well-formedness rules from counter- and examples. However, the work lacks of considering the relevance of generated candidates.

This paper introduces an approach to automating the inference of OCL invariants. At the first step, from input valid and invalid snapshots, OCL invariant candidates are generated using patterns [8]. This paper enhances this step at two points: (1) input snapshots are preprocessed in order to increase the relevance of generated candidates, and (2) the algorithm is improved in order to remove the duplication of candidates. At the second step, *model finding* is employed in order to generate counter- and examples. By getting *user feedback* on generated snapshots, irrelevant cases could be eliminated. We realize the approach with the tool InferOCL, based on the tool EMFtoCSP [4], and the solver ECL$^i$PS$^e$ [9]. We have applied the approach on the domain of Role-Based Access Control (RBAC) models [10]. The experiment shows a possibility to apply the tool InferOCL in practice as well as threads to validity of our approach.

The remainder of this paper is organized as follows. Section 2 motivates the work with a running example. Section 3 introduces an approach to automating the inference of OCL invariants. Section 4 explains our realization for the approach with the tool InferOCL. We present experimental results in Sect. 5 and discusses threats to validity of our approach in Sect. 6. Section 7 surveys related work. This paper is closed with conclusions and an outlook on future work.

## 2 Running Example

Role-based access control (RBAC) [10] is a popular approach to restricting system access to authorized users. In order to build and integrate RBAC concrete policies in the system, it is necessary to define a conceptual model for it. Figure 1 presents such a conceptual model in form of a class diagram. In this way a RBAC concrete policy would be presented by an object diagram (a snapshot) as depicted in Fig. 2.

Domains like RBAC often require to add certain restrictions on the class model in order to obtain a more precise specification of them. For example, the example RBAC policy shown in Fig. 2 is actually an invalid snapshot since the number of roles that the `ada` could play is greater than the value of the `maxRoles` attribute. Such restrictions are often expressed in OCL and referred to as OCL invariants. We might find the OCL invariants for the simplified RBAC domain [11] as depicted in Fig. 2: (1) The maximum number of roles to which a user is assigned must be less than its attribute `maxRoles`; (2) The assignment of a dependent role with a user postulates the assignment of required roles with the user; (3) A user must not be assigned to both of the mutually exclusive roles.

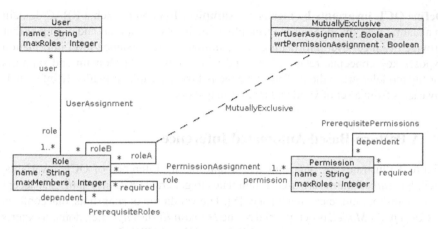

**context** User **inv** maxCard_Role: self.role —>size() ≤
self.maxRoles

**context** User **inv** requiredInclusion_role: self.role —>forAll(
d | d.required —>forAll(r | self.role —>includes(r)))

**context** MutuallyExclusive **inv** retrictedAssoc_user:
self.wrtUserAssignment implies   self.roleA.user
—>excludesAll(self.roleB.user)

**Fig. 1** Simplified RBAC conceptual model

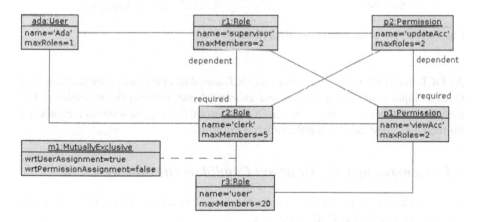

**Fig. 2** The instance-level information with snapshots in form of object diagrams

**Define OCL invariants by means of examples.** To capture such a RBAC domain in a natural way, the modeler often employs snapshots as counter- and examples that provide instance-level information of the domain. Such instance-level information exposes key concepts, relationships, and restrictions of the domain and supports for the modeler grasp them. The challenge is how we can automatically infer OCL invariants from a set of valid and invalid snapshots.

## 3   A Pattern-Based Automated Inference

Figure 3 illustrates for our approach to automating inference of OCL invariants: *OCL Invariant Patterns* are used as templates to generate OCL invariants that accept valid snapshots and reject invalid ones [8]. The validation of snapshots, that conform to *Conceptual Model*, is determined by the *Domain Knowledge*, i.e., domain experts or model-generating tools such as UML2CSP [4] and USE [3].

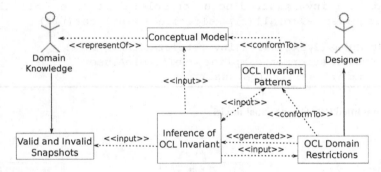

**Fig. 3** Overview of the pattern-based approach

An OCL invariant pattern consists of *an OCL template* and *a matching function* that takes as input a valid snapshot set and an invalid one, binding the variables of the template to values in order to form an OCL invariant [8]. Figure 4 shows the patterns for the OCL invariants within the running example.

### 3.1   Generating OCL Invariant Candidates as Solving a CSP

We view the generation of OCL invariants as solving a CSP, where OCL invariant patterns are encoded as CSP constraints.

**Fetching Patterns from a Catalog.** The basic idea is that each OCL invariant set candidate corresponds to a partition of the invalid set $SNOK = \bigcup SNOK_i$, where each invariant is obtained by applying a pattern such that it accepts $SOK$ and rejects $SNOK_i$. We encode the algorithm in prolog as follows:

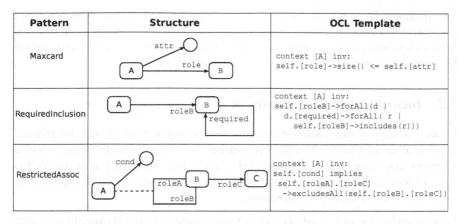

**Fig. 4** OCL invariant patterns for the running example

```
apply_all(SOK, SNOK, PATTERN, INV) :-
  sort(SNOK, SNOKo),
  %---------- Get all partitions of the invalid snapshot set SNOK -------
  partition(SNOKo, SnokGroups),
  %--------------------------------------------------------------------
  ( foreach(SnokGroup, SnokGroups),
    fromto([], InINV, OutINV, INV),
    param(SOK, PATTERN) do
      member(Pattern, PATTERN),
      %-- Each SnokGroup is rejected by a pattern-generated invariant ---
      applyPattern(Pattern, SOK, SnokGroup, Para),
      %----- Each invariant is captured by a pair (Pattern, Para) -------
      OutINV = [ [ Pattern, Para ] | InINV ]
  ).
```

The aim of the constraint `applyPattern(Pattern, SOK, SnokGroup, Para)`
is to define the matching value `Para` as we apply the `Pattern` on the valid snapshots `SOK` and the invalid ones `SnokGroup`. Note that the generated invariant
could be formed by the pair `[Pattern, Para]`. We encode the `applyPat-`
`tern(_, _, _, _)` based on the encoding of each pattern as mentioned below.

**Encoding OCL Invariant Patterns.** We map each pattern to a CSP constraint so
that generated invariants could accept valid snapshots and reject invalid ones. For
example, the `maxCard` pattern could be encoded as follows:

```
apply_maxCard(SOK, SNOK, Para):-
  %----------------Ensuring the invariant accepts SOK-----------------
  ( foreach(SnapshotOk, SOK),
    param(Para) do
      maxCard(SnapshotOk, Para, 1)
  ),
  %----------------Ensuring the invariant rejects SNOK-----------------
  ( foreach(SnapshotNok, SNOK),
    param(Para) do
      maxCard(SnapshotNok, Para, 0)
  ).
```

The parameterized OCL invariant `maxCard` is translated into the CSP constraint `maxCard(Snapshot,Para,Ret)`, where the `Ret`, that may be 0 or 1, determines the validation of the input snapshot. Such a CSP predicate could be obtained by a translation of OCL invariants into CSP as explained in [4].

## 3.2 Incorporating User Feedback

In order to capture user's knowledge of the underlying domain in OCL we propose incorporating user feedback at the two aspects: (1) The user could locate part of a snapshot that makes it invalid, and (2) she could determine whether a snapshot is valid or invalid. The first one is to preprocess input snapshots, and the second one is to eliminate irrelevant inferred results.

**Preprocessing Input Snapshots.** Let us focus on the snapshot shown in Fig. 2, that could be presented as an instance of the following snapshot structure:

```
SnapshotStructure = [ user(oid,name,maxRoles), role(oid,name,maxMembers),
  mutuallyexclusive(oid,roleA,roleB,id,wrtUserAssignment,wrtPermissionAssignment),
  permission(oid,name,maxRoles), userassignment(user,role),
  prerequisiteroles(dependent,required),
  assoccls_mutuallyexclusive(roleA,roleB),
  prerequisitepermissions(dependent,required),
  permissionassignment(role,permission) ]
```

Since this is an invalid snapshot w.r.t the invariant `maxCard_Role` as shown in Fig. 1, the user could highlight relevant part of the snapshot. Technically, we could present preprocessed snapshots in this way:

```
PreprocessedSnapshot = [ [user(1,_,1)],[role(1,_,_),role(2,_,_))],[],[],
  [userassignment(1,1),userassignment(1,2)],[],[],[],[] ]
```

**Eliminating Irrelevant Results.** We aim to get user feed back in order to eliminate irrelevant generated sets of OCL invariants. The basic idea of this method is that we consider each pair of generated invariant sets, $INV_1 = \{inv_{11}, inv_{12}, ..., inv_{1m}\}$ and $INV_2 = \{inv_{21}, inv_{22}, ..., inv_{2n}\}$, and generating a valid snapshot for the new invariant set $INV_1 \cup \{not\ inv_{2i}\}$, w.r.t each invariant $inv_{2i} \in INV_2$. We then capture user feedback for the generated snapshot example: (1) If the user gives a positive feedback, the result $INV_2$ as well as the other generated results that reject the snapshot example should be irrelevant; (2) In the other case the result $INV_1$ and the other generated results that accept the snapshot example should also be irrelevant. In case no valid snapshot w.r.t the new invariant set could be found, we would have $INV_1 \Rightarrow inv_{2i}$, and then if that is true for all $inv_{2i} \in INV_2$, we would have $INV_1 \Rightarrow INV_2$. As we would also have $INV_2 \Rightarrow INV_1$, they are logically equivalent.

## 4   Tool Support

This section first overviews the support tool InferOCL based on the tool EMFtoCSP [4], and the solver ECL$^i$PS$^e$ [9]. Then, it illustrates the tool and our inference method by applying them for the running example.

## 4.1 Overview of the InferOCL Tool

Figure 5 outlines the InferOCL tool. First, a model is loaded with an xmi file and encoded in prolog using the EMFtoCSP tool. The control module analyzes input snapshots provided by the user, and sending a query to the $ECL^iPS^e$ solver to obtain OCL invariant candidates. In order to generate counter- and examples, the user needs to provide a domain restriction for snapshots.

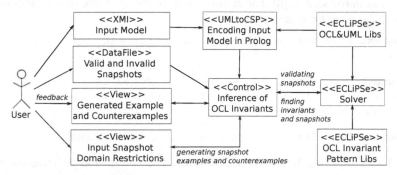

**Fig. 5** Overview of the InferOCL tool

The elimination step finishes as no example could be found, and the InferOCL tool brings out final candidates, each of which includes OCL invariants. At this point the candidates are logically equivalent to each other. The user may continue validating the result by examining as much as possible on counter- and examples.

## 4.2 Applying the Method to the Running Example

We could apply the method to the running example with the following steps.

**Step1 - Preprocessing Input Snapshots.** The user provides as input 1 valid snapshot and 3 invalid ones that have been preprocessed. The first invalid snapshot corresponds to the `maxCard_Role` invariant, mentioned in Fig. 1. The second and third one corresponds to the `retrictedAssoc_user` and `requiredInclusion_user` invariant, respectively. Note that for sake of clarity, we employ the type Integer instead of the String in this example.

```
SOK = [[[user(1,100,3)], [role(1,100,5),role(2,200,5),role(3,300,5)],
  [mutuallyexclusive(1,1,2,1,0),mutuallyexclusive(2,2,3,1,0)],
  [permission(1,100,2),permission(2,200,2)],
  [userassignment(1,1),userassignment(1,3)], [prerequisiteroles(1,3)],
  [assoccls_mutuallyexclusive(1,2),assoccls_mutuallyexclusive(2,3)],
  [prerequisitepermissions(2,1)],
  [permissionassignment(1,1),permissionassignment(2,1),
   permissionassignment(3,1),permissionassignment(3,2)]]],
SNOK = [[[user(1,_,2)], [role(1,_,_),role(2,_,_),role(3,_,_)],[],[],
  [userassignment(1,1),userassignment(1,2),userassignment(1,3)],[],[],[],[]],
  [[user(1,_,_)],[role(1,_,_),role(2,_,_),role(3,_,_)],
```

```
[mutuallyexclusive(1,1,2,1,_),mutuallyexclusive(2,1,3,1,_),
 mutuallyexclusive(3,2,3,1,_)],[],[userassignment(1,1),userassignment(1,3)],[],
[assoccls_mutuallyexclusive(1,2),assoccls_mutuallyexclusive(1,3),
 assoccls_mutuallyexclusive(2,3)],[],[]],
[[user(1,_,_)],[role(1,_,_),role(2,_,_)],[],[],[userassignment(1,1)],
 [prerequisiteroles(1,2)],[],[],[]]])
```

**Step 2 - Generating Candidates.** With the input data from Step 1, the InferOCL tool could generate 18 candidates by an inference on our experimental catalog of OCL invariant patterns, including the 3 patterns as depicted in Fig. 4 and the 5 patterns as introduced in [8]. For example, the following is one of the candidates, that includes three OCL invariants.

```
context Role inv  cardInv_roleB: self.roleB -> size()
< 2

context User inv  requiredInclusion_role: self.role
-> forAll(d | d.required -> forAll(r | self.role
-> includes(r)))

context User inv  intv_maxRoles: 2 < self.maxRoles
```

**Step 3 - Eliminating Irrelevant Candidates.** The elimination of irrelevant candidates for the running example is summarized as in Table 1. At the first step the user provides a domain restriction for snapshots, that includes bounds for the size of classes and associations and attribute values:

```
User :: 0..10; name: Integer :: 1..10; maxRoles: Integer :: 1..10;
Role :: 0..10; name: Integer :: 1..10; maxMembers: Integer :: 1..10;
MutuallyExclusive :: 0..10; wrtUserAssignment: Boolean :: 0..1;
                          wrtPermissionAssignment: Boolean :: 0..1;
Permission :: 0..10; name: Integer :: 1..10; maxRoles: Integer :: 1..10;
UserAssignment :: 0..10; PrerequisiteRoles :: 0..10;
PrerequisitePermission :: 0..10; PermissionAssignment :: 0..10;
```

The generated example at step 2 in Table 1 is irrelevant w.r.t the `retrictedAssoc_user` restriction. The irrelevant snapshots at step 3 and 5 correspond to the `requiredInclusion_user` and `maxCard_Role` restriction.

**Step 4 - Examining on Equivalent Candidates.** At this step the InferOCL tool brings out the final candidate including 3 OCL invariants as mentioned in Fig. 1. The user could continue to validate this result and each equivalent candidate in general by examining on generated counter- and examples.

## 5  Experimental Results

This experiment is performed on a 64-bit computer with 2.70GHz Intel Core i7 processor and 8Gb RAM. We take as input 05 arbitrary snapshots for the valid set *SOK* and from 3 to 11 other ones for the invalid set *SNOK*. In the first case the input *SNOK* is preprocessed and the other case without preprocessing. The result of generating candidates and the corresponding performance time is presented as in Table 2. We could realize that the number of generated candidates decreases as

**Table 1** Eliminating irrelevant candidates for the running example

| Step | Generated Example Snapshot | Feedback | Candidate |
|---|---|---|---|
| 1 | [[],[role(1,1,1),role(2,1,1)],[mutuallyexclusive(1,1,1,0,0), mutuallyexclusive(2,2,1,0,0)],[permission(1,1,1)],[],[], [assoccls_mutuallyexclusive(1,1), assoccls_mutuallyexclusive(2,1)],[], [permissionassignment(1,1),permissionassignment(2,1)]] | Yes | 12 |
| 2 | [[user(1,10,3)],[role(1,1,1),role(2,1,1)], [mutuallyexclusive(1,1,1,1,0)],[permission(1,1,1)], [userassignment(1,2),userassignment(1,1)],[], [assoccls_mutuallyexclusive(1,1)],[], [permissionassignment(1,1),permissionassignment(2,1)]] | No | 6 |
| 3 | [[user(1,10,3)],[role(1,1,1),role(2,1,1),role(3,1,1)],[], [permission(1,1,1)],[userassignment(1,2), userassignment(1,1)],[prerequisiteroles(1,3)],[],[], [permissionassignment(1,1),permissionassignment(2,1), permissionassignment(3,1)]] | No | 3 |
| 4 | [[user(1,10,3)],[role(1,1,1),role(2,1,1),role(3,1,1)],[], [permission(1,1,1)],[userassignment(1,3), userassignment(1,2),userassignment(1,1)],[],[],[], [permissionassignment(1,1),permissionassignment(2,1), permissionassignment(3,1)]] | Yes | 2 |
| 5 | [[user(1,10,3)],[role(1,1,1),role(2,1,1),role(3,1,1), role(4,1,1)],[],[permission(1,1,1)],[uscrassignment(1,4), userassignment(1,3),userassignment(1,2), userassignment(1,1)],[],[],[],[permissionassignment(1,1), permissionassignmcnt(2,1),permissionassignment(3,1), permissionassignment(4,1)]] | No | 1 |

the input is preprocessed, i.e., the preprocessing highlights invalid part of snapshots so that irrelevant cases are eliminated. The generation of candidates is also faster performed as the input is preprocessed.

The experiment points out a certain restriction on the size of input data: The tool works well on 10 invalid snapshots of a conceptual model including 4 classes and 5 associations. It could accept a large domain restriction in order to generate counter- and examples. Specifically, with a quite large domain and applicable in practice like 0..500 for attribute values and a simple domain for the number of classes and associations like 1..10, the finding is performed just in few seconds.

**Table 2**  The number of generated candidates and the performance time in two cases, (1) with preprocessing and (2) without preprocessing

| SNOK Size | Candidates 1 | Candidates 2 | Time 1 | Time 2 |
|:---:|:---:|:---:|:---:|:---:|
| 3 | 1 | 2 | 0.31s | 0.29s |
| 4 | 1 | 6 | 0.78s | 0.92s |
| 5 | 1 | 6 | 2.09s | 2.55s |
| 6 | 2 | 18 | 8.05s | 10.55s |
| 7 | 4 | 54 | 34.62s | 46.32s |
| 8 | 4 | 54 | 152.72s | 210.77s |
| 9 | 4 | 108 | 738.04s | 996.49s |
| 10 | 8 | 324 | 4077.09s | 5244.91s |
| 11 | 8 | 324 | 23030.21s | 30252.04s |

# 6  Threats to Validity

There are certain threats to validity of our inference approach. First, the approach is just evaluated on simple examples. Further explorations on larger case studies need to be performed. Second, the method to generate invariant candidates currently retains limitations in performance. We would need a better preprocessing solution, e.g., a web-based solution, to effectively get user feedback on a large number of input snapshots. Last but not least, the current method to eliminate irrelevant candidates puts forward challenges: (1) How we can always confirm whether a counter- or an example exists on a given domain, and (2) how a user-given domain restriction could be defined with less effort from the user.

# 7  Related Work

**OCL Inference.** The need to infer OCL restrictions from instance-level information has been considered in the work [12], where a technique to infer metamodels from instances is introduced. The work in [7] proposes an approach based on genetic programming. They encode sets of invariants as populations, and genetic operators is provided in order to evolve the population and to obtain the final generation that accepts examples and rejects counterexamples. However, the work does not consider the relevance of generated candidates.

**OCL and Snapshot.** There is significant work concentrating on the relationship between OCL constraints and snapshots. In [3] the USE tool allows us to check OCL restrictions on class diagrams. The paper in [13] proposes to validate UML and OCL models by generating snapshots from a declarative description. Other works often translate the OCL specification into specification environments such as CSP [4], Alloy [5], and relational logic [6] in order to effectively validate snapshots or to find valid (invalid) snapshots or to check properties. We here focus on how OCL constraints are defined by analyzing snapshots.

**OCL Pattern.** Our method to specify OCL invariant pattern is related to the paper in [14], in which a method to represent OCL expressions in a SBVR form in order to generate natural language explanations for business rules is proposed. The paper in [15] introduces a method to generate semantically equivalent alternatives for the initially defined OCL constraints. It assists the designer to better define OCL constraints.

**OCL Learning.** The essence of this work is a concept learning, an issue in machine learning. The paper in [16] introduces the L* algorithm in order to generate the best assumption in form of deterministic final-state automata from examples. The paper in [17] proposes a SAT-based method to acquire binary constraint networks. The paper in [18] discusses a method to generate OCL constraints from natural language. Our work focuses on learning OCL constraints.

This work continues our previous work in [8]. In this work a new algorithm to fetch patterns is introduced so that we could get over the duplication of inferred results. Moreover, this work supports for multiple restrictions instead of only one as explained in the previous work. The proposal in this work to incorporate user feedback could help decreasing the number of generated candidates. This work also offers a more effective strategy to eliminate irrelevant cases.

# 8  Conclusions

This paper has introduced an approach to automating the inference of domain restrictions as OCL invariants from the instance-level information captured by valid and invalid snapshots. In the language engineering context the underlying domain classifies models, while within the software engineering context the domain corresponds to a conceptual model, representing for system snapshots. The basic idea of this approach is to employ OCL invariants patterns as inference patterns in order to infer invariant candidates from a set of valid and invalid snapshots. The user could help to preprocess input snapshots so that less irrelevant candidates are generated. Getting user feedback on generated snapshots also help to remove irrelevant candidates. The approach is realized with the InferOCL tool, based on the model finding tool EMFtoCSP and the ECL$^i$PS$^e$ solver. The experiment on the simplified RBAC model shows a possibility of the InferOCL tool in practice as well as threads to validity of the approach.

In future, we aim to apply the approach on larger case studies in order to get detailed feedback on it. Such a task could be to validate and to infer invariants for the UML metamodel. It would require us to enrich and maintain the catalog of patterns as well as to extend the method to generate invariant candidates and to eliminate irrelevant ones. To enhance the InferOCL tool with such new features is also on the focus of our future work. Our enhancements in model finding would be further assessed in order to make contribution to that issue.

**Acknowledgement.** This work was supported by the research project QG.14.06, Vietnam National University, Hanoi. We also thank anonymous reviewers for their comments on the earlier version of this paper.

# References

[1] Sutcliffe, A.G., Maiden, N.A.M., Minocha, S., Manuel, D.: Supporting Scenario-Based Requirements Engineering. IEEE Trans. Software Eng. 24(12), 1072–1088 (1998)

[2] Warmer, J., Kleppe, A.: The Object Constraint Language: Getting Your Models Ready for MDA, 2nd edn. Addison-Wesley Professional (2003)

[3] Gogolla, M., Büttner, F., Richters, M.: USE: A UML-Based Specification Environment for Validating UML and OCL. Science of Computer Programming 69(1-3), 27–34 (2007)

[4] Cabot, J., Claris, R., Riera, D.: UMLtoCSP: A Tool for the Formal Verification of UML/OCL Models Using Constraint Programming. In: Kurt Stirewalt, R.E., Alexander Egyed, B.F. (eds.) Proc. 22th Int. Conf. Automated Software Engineering (ASE), pp. 547–548. ACM, New York (2007)

[5] Anastasakis, K., Bordbar, B., Georg, G., Ray, I.: UML2Alloy: A Challenging Model Transformation. In: Engels, G., Opdyke, B., Schmidt, D.C., Weil, F. (eds.) MODELS 2007. LNCS, vol. 4735, pp. 436–450. Springer, Heidelberg (2007)

[6] Kuhlmann, M., Gogolla, M.: From UML and OCL to Relational Logic and Back. In: France, R.B., Kazmeier, J., Breu, R., Atkinson, C. (eds.) MODELS 2012. LNCS, vol. 7590, pp. 415–431. Springer, Heidelberg (2012)

[7] Faunes, M., Cadavid, J.J., Baudry, B., Sahraoui, H.A., Combemale, B.: Automatically searching for metamodel well-formedness rules in examples and counter-examples. In: Moreira, A., Schätz, B., Gray, J., Vallecillo, A., Clarke, P. (eds.) MODELS 2013. LNCS, vol. 8107, pp. 187–202. Springer, Heidelberg (2013)

[8] Dang, D.H., Cabot, J.: Automating Inference of OCL Business Rules from User Scenarios. In: Proc. 20th Asia-Pacific Conf. Software Engineering (APSEC), pp. 156–163. IEEE (2013)

[9] ECLiPSe: The ECLiPSe Constraint Programming System. Version 6.1 (June 2013)

[10] Ferraiolo, D., Kuhn, D.: Role-Based Access Control. In: Proc. 15th National Computer Security Conf., pp. 554–563 (1992)

[11] Kuhlmann, M., Sohr, K., Gogolla, M.: Comprehensive Two-Level Analysis of Static and Dynamic RBAC Constraints with UML and OCL. In: Baik, J., Massacci, F., Zulkernine, M. (eds.) Proc. 5th Int. Conf. Secure Software Integration and Reliability Improvement (SSIRI), pp. 108–117. IEEE (2011)

[12] Javed, F., Mernik, M., Gray, J., Bryant, B.R.: MARS: A Metamodel Recovery System Using Grammar Inference. Information & Software Technology 50(9-10), 948–968 (2008)

[13] Gogolla, M., Bohling, J., Richters, M.: Validating UML and OCL Models in USE by Automatic Snapshot Generation. Software and System Modeling 4(4), 386–398 (2005)

[14] Pau, R., Cabot, J.: Paraphrasing OCL Expressions with SBVR. In: Kapetanios, E., Sugumaran, V., Spiliopoulou, M. (eds.) NLDB 2008. LNCS, vol. 5039, pp. 311–316. Springer, Heidelberg (2008)

[15] Cabot, J., Teniente, E.: Transformation Techniques for OCL Constraints. Science of Computer Programming 68(3), 152–168 (2007)
[16] Angluin, D.: Learning Regular Sets from Queries and Counterexamples. Information and Computation 75(2), 87–106 (1987)
[17] Bessière, C., Coletta, R., Koriche, F., O'Sullivan, B.: A SAT-Based Version Space Algorithm for Acquiring Constraint Satisfaction Problems. In: Gama, J., Camacho, R., Brazdil, P.B., Jorge, A.M., Torgo, L. (eds.) ECML 2005. LNCS (LNAI), vol. 3720, pp. 23–34. Springer, Heidelberg (2005)
[18] Bajwa, I., Bordbar, B., Lee, M.: OCL Constraints Generation from Natural Language Specification. In: Proc. 14th Int. Conf. Enterprise Distributed Object Computing Conference (EDOC), pp. 204–213. IEEE (2010)

# Improving Text-Based Image Search with Textual and Visual Features Combination

Xuan-Son Vu, Thanh Vu, Huong Nguyen, and Quang-Thuy Ha

**Abstract.** With the huge number of available images on the web, an effective image retrieval system has been more and more needed. Improving the performance is one of crucial tasks in modern text-based image retrieval systems such as Google Image Search, Frickr, etc. In this paper, we propose a unified framework to cluster and re-rank returned images with respect to an input query. However, owning to a difference to previous methods of using only either textual or visual features of an image, we combine the textual and visual features to improve search performance. The experimental results show that our proposed model can significantly improve the performance of a text-based image search system (i.e. Flickr). Moreover, the performance of the system with the combination of textual and visual features outperforms the performance of both the textual-based system and the visual-based system.

**Keywords:** Image search system, Meta-search engine, Textual features, Visual features, Re-ranking, Clustering.

## 1 Introduction

With the development of photo sharing sites (e.g. Flickr, Photobucket) and social networks (e.g. Facebook), the internet user can easily upload and share their images

Xuan-Son Vu
School of Computer Science and Engineering,
Kyungpook National University, Korea
e-mail: sonvx@sejong.knu.ac.kr

Thanh Vu
The Open University, Milton Keynes, UK
e-mail: thanh.vu@open.ac.uk

Huong Nguyen · Quang-Thuy Ha
KTLab, University of Engineering and Technology (VNU-UET),
Vietnam National University, Hanoi (VNUH), Ha Noi, Viet Nam
e-mail: {huongnguyen,thuyhq}@vnu.edu.vn

© Springer International Publishing Switzerland 2015                                      233
V.-H. Nguyen et al. (eds.), *Knowledge and Systems Engineering,*
Advances in Intelligent Systems and Computing 326, DOI: 10.1007/978-3-319-11680-8_19

with other users. It leads to a fact that the number of images on those systems has been growing exponentially. As mentioned in [17], in August 2011, there was six billion images were stored in Flickr and 300 million images are daily uploaded to Facebook [23]. The large number of available images has been raising a need of an effective image retrieval system.

In general, an image search system can be characterized into (1) a text-based retrieval method [21, 18], (2) an image content-based retrieval method [8, 4] or a mixed method [16](that is the combination of both text-based and content-based methods). The main difference between a text-based system and an image content-based system is that the input query issued in the text-based system is the textual query (i.e. terms/words representing the user's needs) as opposed to the visual query (i.e. an image indicating the user's needs). Text-based image retrieval systems are more popular in real systems because searchers can easily express their needs in a textual query instead of in an image. Therefore, in this research, we aim at improving the performance of a text-based image retrieval system.

Modern text-based image search engine (TISE) such as Google Image Search, Alta Vista Image Search and Flickr are widely used. The TISE, however, faces two problems. Firstly, with respect to an input query, the TISE returns an uncategorized list of images based on keyword matching algorithms. For example, Figure 1 shows the search results returned from Flickr with the query *"Pluto"*. The returned result list is a mixture of *animated images* and *dwarf planets*. Therefore, the searcher needs to spend extra time and effort surfing irrelevant images before getting to their relevant images. Secondly, because a search query is usually short and ambiguous, it is difficult to return suitable results that match the user interests. For example, the searching intent of the query *"Pluto"* is either planets or cartoons. However, as shown in Figure 1, beside planets or cartoons, garbage images are also returned.

As a solution for the problems, this research focuses on improving the performance of a TISE by clustering and re-ranking the returned images. In our proposed model, we leverage not only the textual features [20] but also the visual features [10, 26, 22, 18] of an image to make a final feature set of the image. To our best knowledge, there have been only few research [25, 14, 24] focusing on classifying the image search results. By clustering images to different number of highly related group of images, user can quickly surf to find their interest images. Moreover, the ranking process allows promoting the more relevant images in a higher rank. For example, in the figure 2 is real returned images of our proposed system for query "Pluto". The main contributions of our research are listed as follows:

- Build a unified framework for clustering and re-ranking the returned images from a TISE.
- Propose a method to use RankComplete for a combined feature approach.
- Evaluate the frame work on a dataset from a major TISE (i.e. Flickr) and show the improvement on the search performance.

The remainder of this paper is constructed as follows. In the next section, we review related work of using textual or/and visual features on TISE. We then describe

**Fig. 1** Flickr Search Results for Query *Pluto*, (Search on May 13rd, 2014)

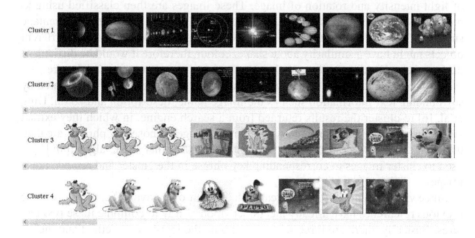

**Fig. 2** Our Proposed System Results for Query *Pluto*

our proposed framework in Section 3. Section 4 shows the experimental results. Conclusions and future work are demonstrated in Section 5.

## 2  Related Work

Using colour feature is a common and the most widely used method in content-based image processing system [13, 5]. This is a simplest and fastest method [13]. Its performance is, however, limited compared with other methods. One of main descriptors of colour feature is a colour histogram. By using some of typical colour space such as RGB, HSV, an enhanced HSV, a colour histogram of an image is drawn, in which RGB is the most common one. Yet, there is a problem when generating a histogram based on RGB system. Because the histogram descriptor in RGB does not take into account the color spatial distribution of the image, therefore different images may have a same histogram descriptor [10]. Subsequently, this problem will cause garbage images in image search results. Thus, Liu et al. [10] proposed a method allowing cluster image based on HSV colour space for top n returned images from a image search engine. In another way, Wang et al. [25] proposed a method to cluster images by extracting features of different regions inside images. In their approach, they assume that each image is composed from two regions: (1) key region representing the main semantic content (e.g. animals, humans) and (2) environment region representing the context (e.g. earth, grasslands). Because the returned images by Web search engines are assumed to have the same semantic content, it is the context of one image that determines to which cluster it should be assigned. Another approach to cluster image based on image content is taking into account consistent feature extracted from image [3, 22]. These features are invariant against a change of light intensity and rotation of image. These images are then classified using k-means/KNN with feature selection [3]. Even visual feature has a strong advantage since it reflects truly what the picture is about. However, in some case, two different objects might have a similarity about shape, colour, therefore it would be difficult to distinct them. For that reason, we also need to take into account textual feature.

For images on the web, along with visual content, image is also represented by surrounding text information such as the image name, description and so on. Ding et al. [6] re-cluster the results returned from a search engine, in which they extract key phrases in the text associated with returned images. Those key phrases are then used to cluster based on semantic similarity [7]. After that, BBC algorithm [1] is used to cluster images to corresponding key phrase in the cluster and remove noisy images.

Since visual feature and textual feature have their own advantage, therefore some previous researches have already mentioned about a way to combine these two features. Basically, there are three ways to combine the text-based method and the visual content-based method: (1) use the text-based method first; (2) use the visual content-based method first; (3) use the two methods at the same time. With the first approach, Luo et al. [12], initially, use a text-based image meta-search engine to retrieve images from the Web based on the text information on the image host pages so as to provide an initial image set. Hence, the image content based processing is employed to produce a high relevant output. However, the pipeline solution from Luo et al. has a limitation is that they cannot flexibly combine those two features since in some cases visual feature is more important than textual feature, and vice versa. Hence, lines on the third way, our work proposes a method to combine the features, in which we put a tradeoff factor to give a flexible combination between the features.

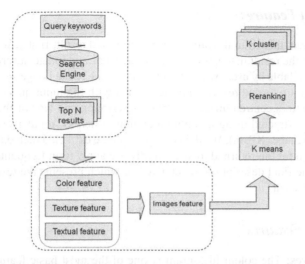

**Fig. 3** System workflow of our proposed model

**Fig. 4** A result returned from the image search engine with query "Bletchley park"

## 3 Proposed Model

Figure 3 describes the pipeline of our proposed model. With an input query, in the first step, we download the top N results returned by an image search engine (i.e. Flickr system[1]). After that, in the second step, we extract visual and textural features for each returned image. The images along with their extracted features are finally passed to a machine learning framework for clustering and re-ranking. In the framework, we apply K-Means clustering algorithm to group images into different clusters and re-rank the images within each cluster using Rank Compete method [2].

---

[1] https://www.flickr.com/

## 3.1  Textual Features

In order to extract the textural features from a returned result, firstly, we download
the metadata of the result that is image title, image tags and image description. This
data is made available by users who uploaded the image to the image search system.
Figure 4 shows a returned result from the system with the input query "Bletchley
Park". Along with a centred image, the title, description and tags of the image are
available to be extracted using a simple technique, such as, regular expression or
XML Path Language (XPath). We then combine all the extracted metadata to form a
final description for the returned image. Finally, we apply term frequent-based vec-
torisation on the final description to get a vector which presents the textual feature
of the returned result.

## 3.2  Visual Features

**Colour Features:** The colour histogram is one of the most basic features used in
content-based image search system [5]. In this paper, we use the colour histogram
features as the colour features of an image. We extract the histogram features of
an image using the RGB colour space. By applying method described in [5], we
calculate the histogram features using Eq. 1.

$$H_{R,G,B}(r,g,b) = N.Prob(R = r, G = g, B = b) \tag{1}$$

Here, $R$, $G$ and $B$ are the three colour channels of the RGB colour space, and $N$ is
the number of pixels in the image. However, to reduce the complexity of our system,
we quantise an image with 256 colour bins in RGB colour space (256 x 256 x 256)
into a 16 colour bins to form a 16 x 16 x 16 (= 4096) image. After that, we get a
vector of the histogram features with 4096 dimensions.

**Textural Features:** Texture is one of the prominent features, which has been widely
used in identifying objects in an image or in image classification, with application
in image search systems. As mentioned in [13], texture can be defined informally
as the repetition of a pattern or patterns over a region in an image. It expresses
valuable information regarding the underlying structure arrangement of the image's
surfaces such as desks, chairs or wardrobes, etc. We combine two feature sets of
textures to form the textural features of an image. The former one contains four
features extracted from a co-occurrence matrix of gray levels of an image, that is,
*entropy*, *energy*, *contrast* and *homogeneity*. Mitra et al. [13] defined the gray level
co-occurrence matrix $C(i, j)$ as the number of all pairs of pixels classified by a
displacement vector $d_{i,j}$ and having gray levels $i$ and $j$. By using the co-occurrence
matrix, the textural features can be defined as follows:

$$Entropy = -\sum_i \sum_j C(i,j) log C(i,j) \tag{2}$$

$$Energy = -\sum_i \sum_j C^2(i,j) \tag{3}$$

$$Contrast = -\sum_i \sum_j (i-j)^2 C(i,j) \tag{4}$$

$$Homogeneity = -\sum_i \sum_j \frac{C^2(i,j)}{1+|i-j|} \tag{5}$$

The latter set contains *coarseness, contrast, directionality, linelikeness, regularity,* and *roughness* features proposed in [20].

## 3.3 Feature Combination

After getting the textual and visual vectors which represents the textual and visual contents of an image respectively, we combine the two vectors to form a final vector $F$ representing the image as follows:

$$F = [\alpha X \frac{L_t}{L_v} X F_t \bigcup F_v] \tag{6}$$

where, $F_t$ is the textual vector, $F_v$ is the visual vector. $L_t$ and $L_v$ are the sizes of $F_t$ and $F_v$ respectively. $\alpha$ is a controlled weight used to adjust the contributions of $F_t$ and $F_v$ to the final vector $F$.

## 3.4 Machine Learning Framework

The final vector, then, is passed into a machine learning framework (i.e. K-Means clustering algorithm [9]) to get K clusters, in which each cluster contains a disjointed image set. Figure 5 shows an example of 5 clusters returned by K-means algorithm.

## 3.5 Re-ranking Images

For each cluster extracted in the previous section, we re-rank the images in the cluster using the method proposed in [19]. First, interest points on the images are extracted using SIFT operator [11]. Second, the similarity between two images is calculated by matching two interest point sets of the images. Then, we get a fully connected graph, in which a note indicates an image, and the weight of an edge denotes the similarity between two images. After that, the graph is converted into a binary graph using suitable threshold (e.g. we only keep edges with a weight greater than the threshold). Finally, the rank of an image is calculated as the number of images that connect to the current image. Figure 6 indicates an example of a binary graph. After re-ranking, the center image (K playing card) will be re-ranked in the top because it connects with the largest number(5) of other image.

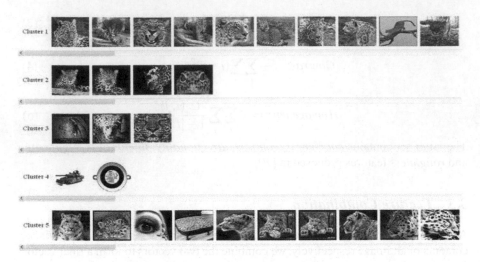

**Fig. 5** An output of K-means clustering algorithm

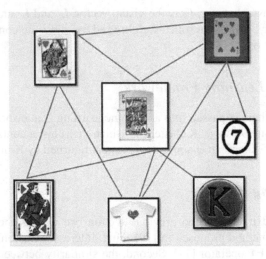

**Fig. 6** An output of K-means clustering algorithm

# 4 Experiments

In this section, we will prove the effectiveness of our approach through two experiments. By these experiments, we will answer two questions: (1) Can the method solve ambiguous queries? and (2) Can the proposed method using combined feature (e.g. textual and visual features) be better than using each feature separately? (e.g. either textual or visual features).

**Table 1** Confusion Metric to calculate Precision, Recall, and F-score

|  |  | Gold standard | |
|---|---|---|---|
|  |  | $\in$ cluster 1 | $\notin$ cluster 1 |
| System prediction | $\in$ cluster 1 | $tp$ (true positive) | $fp$ (false positive) |
|  | $\notin$ cluster 1 | $fn$ (false negative) |  |

To collect data for the experiments, we perform 14 text-based queries on Flickr. For each query, we get top 50 returned images from Flickr in accordance with related text information (e.g., tag, title, file name). By the fact that user often uses object or subject name to find their needs. Thus, we collect returned images from TISE with some queries about object name, animal, or landscape. Moreover, we collect returned images for some ambiguous queries such as *"apple"* (a fruit or a logo of Apple .Inc), *"pluto"* (a cartoon or a dwarf planet) .etc. The crawling data process is done by using FlickrSearcher of Cam-Tu Nguyen [15].

## 4.1 Feature Vector Extraction

We extract represented feature vector for each image as follows:

- Colour Feature Vector: by using RGB colour space with three colour channel R, G, and B, we read every image pixel to fill up into the three channels. Finally, a vector of 512 dimensions is formed.
- Texture Feature Vector: this vector is composed by 3 particles including (1) Tamura feature regarding to coarseness, directionality, and contrast information; (2) Entropy information; (3) A co-occurrence matrix of 64 fields with distance of 1; (4) A statistical on co-occurrence matrix with 4 fields. Eventually, texture feature vector has 72 fields (or 72 dimensions).
- Textual Feature Vector: we extract tag, title of each image from a corresponding file *info.txt* with image, which is stored by the FlickrSearcher when crawling. From all texts, we state a textual vector for each image following Bag-of-Words model with Term-Frequency weighting scheme.

## 4.2 Evaluation Methodology

We use Precision, Recall, and F-score to evaluate clustering performance. To calculate those scores, we need to calculate $tp$ (true positive), $fp$ (false positive), $fn$ (false negative) values for each cluster. The final evaluation result of the system is the average of all cluster values. Value of *tp, pf, fn* is calculated by using the confusion matrix 1.

For example: from gold standard, cluster 1 has picture $a,b,c,d,e$. Meanwhile, a system predicts that cluster 1 has picture $a,b,d,e,h,k,l$. Compared to gold standard,

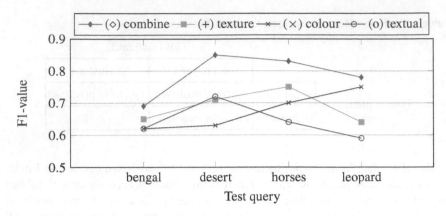

**Fig. 7** F1-value on average for 4 queries

system has predicted truly 4 images $(a,b,d,e)$, therefore $tp = 4$. There are 3 wrongly predicted images $(h,k,l)$, therefore $fp = 3$. From the returned images, image $c$ is absence, therefore $fn = 1$.

From those values of $tp$, $fp$, $fn$ we can calculate the three measurement coefficients as following equations:

$$Precision = \frac{tp}{tp + fp} \qquad (7)$$

$$Recall = \frac{tp}{tp + fn} \qquad (8)$$

$$F_\beta = (1 + \beta^2) \cdot \frac{Precision \cdot Recall}{(\beta^2 \cdot Precision)) + Recall} \qquad (9)$$

## 4.3 Results

The table 2 and the figure 7 are the answers to the two mentioned questions. Firstly, the table 2 indicated performance of our method in terms of clustering. By using combined feature as the equation 6 with a heuristic $\alpha$ of 750, we start to cluster all returned images of 10 different queries. The table 2 illustrated that on average, our method achieved a high F1 with 0.82. Especially, with an ambiguous query such as *"Pluto"*, the F1 value still remains high with 0.89. However, as in case of the query *"cup"* (e.g. a "world cup" or a "tea cup"), the clustering performance is limited because the object's shape and colour are similar (e.g. round shape objects with main yellow colour), therefore there is no distinct visual differences. Meanwhile, textual information is not rich enough for enhancing their different features. Secondly, the figure 7 shows that the proposed method is always higher than the others from 3% to 4%. Especially, for the query *"desert"* the proposed method achieves a significant performance that is higher than the others at least 13%.

**Table 2** F1-value on average for 10 queries

|  | Precision | Recall | F1 |
|---|---|---|---|
| Apple | 0.76 | 0.79 | 0.77 |
| Cup | 0.54 | 0.45 | 0.49 |
| Kids | 0.86 | 0.87 | 0.86 |
| Lake | 0.86 | 0.83 | 0.84 |
| Lemon | 0.88 | 0.91 | 0.89 |
| Sun | 0.92 | 0.90 | 0.91 |
| Train | 0.85 | 0.87 | 0.86 |
| Tree | 0.76 | 0.79 | 0.77 |
| Pluto | 0.90 | 0.88 | 0.89 |
| Wave | 0.85 | 0.86 | 0.85 |
| **On average** | 0.82 | 0.82 | 0.82 |

## 5 Conclusions and Future Work

The main purpose of this paper is to provide an efficient approach for improving results of returned search images for satisfying user requirement. We first perform a text-based meta-search to obtain an initial image set. Then we combine visual feature and textual feature to be a unique feature supporting for clustering images. Additionally, we remove unrelated resultant images by ranking images in each cluster. Afterward, we could increase accuracy of our system.

The main contribution of this work is three-fold including: (1) build a unified framework for clustering and re-ranking the returned images from a TISE; (2) propose a method to use RankComplete for a combined feature to improve system performance; and (3) we prove that our approach is effective by showing high performance in experiment results. Based on both theory and experiment, it shows that, our approach is efficient model for building a meta-search image for clustering image.

For future plan, we manage to improve our combined feature method in order to handle images with high-similarity in shape and colour.

**Acknowledgments.** This work was partially supported by the project HN01-C08.xx *"Building cholera predictive models in Hanoi"*.

# References

[1] Banerjee, A., Merugu, S., Dhillon, I., Ghosh, J.: Clustering with bregman divergences. Journal of Machine Learning Reserach, 1705–1749 (2005)

[2] Cao, L., Pozo, A.D., Jin, X., Luo, J., Han, J., Huang, T.S.: Rankcompete: Simultaneous ranking and clustering of web photos. In: Proceedings of the 19th International Conference on World Wide Web, WWW 2010, pp. 1071–1072. ACM, New York (2010)

[3] Chang, R., Lin, S., Ho, J., Fann, C., Wang, Y.: A novel content based image retrieval system using k-means/knn with feature extraction. Computer Science and Information Systems 9, 1645–1662 (2012)

[4] Chang, S.K., Hsu, A.: Image information systems: Where do we go from here? IEEE Trans. on Knowl. and Data Eng. 4(5), 431–442 (1992)

[5] Deselaers, T., Keysers, D., Ney, H.: Features for image retrieval: An experimental comparison. Inf. Retr. 11(2), 77–107 (2008)

[6] Ding, H., Liu, J., Lu, H.: Hierarchical clustering-based navigation of image search results. In: Proceedings of the 16th ACM International Conference on Multimedia, MM 2008, pp. 741–744. ACM, New York (2008)

[7] Fischer, I.: An algorithm for vector quantization design. Report No.IDSIA-12-04 (2004)

[8] Gupta, A., Jain, R.: Visual information retrieval. Commun. ACM 40(5), 70–79 (1997)

[9] Kanungo, T., Mount, D.M., Netanyahu, N.S., Piatko, C.D., Silverman, R., Wu, A.Y.: An efficient k-means clustering algorithm: Analysis and implementation. IEEE Transactions on Pattern Analysis and Machine Intelligence 24, 881–892 (2002)

[10] Liu, G., Lee, B.: A color-based clustering approach for web image search results. In: Proceedings of the 2009 International Conference on Hybrid Information Technology, ICHIT 2009, pp. 481–484. ACM, New York (2009)

[11] Lowe, D.G.: Distinctive image features from scale-invariant keypoints. Int. J. Comput. Vision 60(2), 91–110 (2004)

[12] Luo, B., Wang, X., Tang, X.: A world wide web based image search engine using text and image content features. In: Proc. of IS-T/SPIE Electronic Imaging 2003, Internet Imaging IV (2003)

[13] Mitra, S., Acharya, T.: Data mining - multimedia, soft computing, and bioinformatics. Wiley (2003)

[14] Mukherjea, S., Hirata, K., Hara, Y.: Using clustering and visualization for refining the results of a www image search engine. In: Proceedings of the 1998 Workshop on New Paradigms in Information Visualization and Manipulation, NPIV 1998, pp. 29–35. ACM, New York (1998)

[15] Nguyen, C.-T.: Flickrsearcher: A java implementation for image retrieval with flickr API, http://www.hori.ecei.tohoku.ac.jp/~ncamtu/

[16] Niblack, C.W., Barber, R., Equitz, W., Flickner, M.D., Glasman, E.H., Petkovic, D., Yanker, P., Faloutsos, C., Taubin, G.: Qbic project: querying images by content, using color, texture, and shape. In: Proc. SPIE, vol. 1908, pp. 173–187 (1993)

[17] Olivarez-Giles, N.: Flickr reaches 6 billion photos uploaded, http://latimesblogs.latimes.com/technology/2011/08/flickr-reaches-6-billion-photos-uploaded.html

[18] Pentland, A., Picard, R.W., Sclaroff, S.: Photobook: Content-based manipulation of image databases. Int. J. Comput. Vision 18(3), 233–254 (1996)

[19] Sevil, S., Zitouni, H., Ikizler, N., Ozkan, D., Duygulu, P.: Re-ranking of image search results using a graph algorithm. In: IEEE 16th Signal Processing, Communication and Applications Conference, SIU 2008, pp. 1–4 (April 2008)

[20] Tamura, H., Mori, S., Yamawaki, T.: Texture features corresponding to visual perception. IEEE Transactions on Systems, Man and Cybernetics 8(6) (1978)

[21] Tamura, H., Yokoya, N.: Image database systems: A survey. Pattern Recognition 17(1), 29–43 (1984), Knowledge Based Image Analysis

[22] Thomas, D., Daniel, K., Hermann, N.: Clustering visually similar images to improve image search engines. Informatiktage der Gesellschaft für Informatik

[23] Thomas, O.: Facebook: Users upload 300 million images a day, http://www.businessinsider.com/facebook-images-a-day-instagram-acquisition-2012-7

[24] Upstill, T., Nagappan, R., Craswell, N.: Visual clustering of image search results. In: SPIE Visual Data Exploration and Analysis VIII, San Jose, California USA (2001), http://citeseer.ist.psu.edu/upstill01visual.html

[25] Wang, X.-J., Ma, W.-Y., He, Q.-C., Li, X.: Grouping web image search result. In: Proceedings of the 12th Annual ACM International Conference on Multimedia, MULTIMEDIA 2004, pp. 436–439. ACM, New York (2004)

[26] Ding, Z.Y.: An image retrieval algorithm based on hue entropy. Taiyuan Normal Uni: Nat. Sci. Ed., 10–11

# Cloud Detection Algorithm for LandSat 8 Image Using Multispectral Rules and Spatial Variability

Duc Chuc Man, Viet Hung Luu, Van Thang Hoang,
Quang Hung Bui, and Thi Nhat Thanh Nguyen

**Abstract.** Since 1972, LandSat program has experienced six successful missions that have contributed to nearly 40 years record of Earth Observations for monitoring the land cover and change dynamics. LandSat images have provided new dataset for field monitoring with geospatial information at high spatial resolution and become one of the most widely used sources of satellite images in various domains. The LandSat 8 generation launched in February 2013 continues the mission of collecting images of the Earth with an open and free data policy. However, clouds are often obscure the detection of land surface and restrict the analysis. The paper proposes improvements for an effective method of cloud detection that widely used in LandSat 5-7 in order to obtain better results for the new satellite generation LandSat 8's images. The validation demonstrates that cloud and cloud contaminated pixels were almost detected with overall recall, precision, accuracy and error of over $98.59, 99.94, 99.89$ and $0.11\%$, respectively on testing datasets. In comparison with the original method, our approach is able to detect $13.64\%$ more cloud pixels with a lower error of $1.03\%$ and $1.04\%$ higher accuracy rate.

**Keywords:** Cloud detection, satellite image, LandSat 8, fuzzy set, multispectral rules, spatial variability, standard deviation.

## 1 Introduction

Remote sensing is the science (and to some extent, art) of acquiring information about the Earth's surface without actually being in contact with it [1]. The electromagnetic radiation is normally used as an information carrier in remote sensing. This is done by sensing and recording reflected or emitted energy, processing, analyzing, and applying that information. The products of a remote sensing system are usually an remote sensing image representing the scene being observed.

Duc Chuc Man · Viet Hung Luu · Van Thang Hoang ·
Quang Hung Bui · Thi Nhat Thanh Nguyen
University of Engineering and Technology, VNU, 144 Xuan Thuy, Hanoi, Vietnam
e-mail: {chucmd_55,hunglv_540,thanghv_55,hungbq,
thanhntn}@vnu.edu.vn

© Springer International Publishing Switzerland 2015     247
V.-H. Nguyen et al. (eds.), *Knowledge and Systems Engineering,*
Advances in Intelligent Systems and Computing 326, DOI: 10.1007/978-3-319-11680-8_20

One of the most popular and valued remote sensing images are obtained from LandSat satellites (LandSat 5, LandSat 7 and LandSat 8) which are a joint venture between NASA and U.S. Geological Survey. Each LandSat generation has different sensing devices recording radiation from surface of the Earth at different wavelengths. Since the beginning of the LandSat program in 1972, data have been available to all countries around the world for monitoring and managing natural resources (agriculture and forestry, mapping, land use planning, water resource planning,...), environment quality, public health and human well-being. LandSat observations have found increasingly wide acceptance within the science and applications communities over the program's lifetime [4].

Cloud detection from multispectral remote sensing images is a well-known problem in the literature. The presence of clouds can alter the energy budget of the Earth atmosphere system through scattering and absorption of short way radiation and absorption and re-emission of infrared radiation of longer wavelength [2]. As a result, clouds are considered as a contaminant whose presence serves as source of errors when sensing other phenomena for many remote sensing applications. Therefore, detecting the cloud cover region of over a region in satellite images is an important pre-processing process. For satellite images which contain clouds, in order to derive the correct radiance, top of atmosphere reflectance and other parameters values, scientist have to isolate cloud pixels from images using cloud detection algorithm. The output of cloud detection algorithm is often called a cloud mask, an image of the same size as the input images of interest, whose a pixel can have only two (or a few) values [3]. Since the scattering and absorption characteristics of clouds vary with the microphysical properties of clouds [2], hence the clouds can be classified into different types such as thin cloud, thick cloud, un-classified pixels...

Different cloud detection algorithms have been developed for various kinds of satellite images. The high accuracy for cloud detection under a variety of conditions is challenging although in some recent studies, the probability of detecting clouds correctly has been reported to exceed 90% [5][6][7]. Cloud is recognized by applying a set of logical rules at different spectral bands with defined thresholds on each satellite image's pixel value, which is called pixel-based methods. Otherwise, the context-based methods focus on extracting cloud using their spatial variability characteristic. Some approach combine both strategies in order to obtain more accurate results. The key to the success of these algorithms relies on the selection of the threshold [2].

A simple pixel-based method using threshold of visible and infrared channels was developed to detect the cloud contaminated pixels from NOAA-17/AVHRR visible and thermal infrared day time images [8]. For each pixel in NOAA-17/AVHRR image, the Top Of Atmosphere (TOA) reflectance and brightness temperature are calculated from channel 1 and 4, respectively. Then, all of pixels whose TOA reflectance at the channel 1 is greater than a predefined threshold will be defined as cloudy pixels [9]. Moreover, pixels with brightness temperature less than a predefined value are considered as cloudy pixels. Finally, the combination of two above tests is applied to detect shadow of cloud on the surface and on another cloud [8].

The Automatic Cloud Cover Assessment (ACCA) was developed for LandSat 7 data [10] using reflectance value of bands 2, 3, 4, 5 at the wavelengths($\mu$m) 0.519-0.601, 0.631-0.692, 0.772-0.898, 1.547-1.749 respectively and at-sensor temperature value of band 6 (10.31-12.36 $\mu$m). In the first test, cloud is detected by applying a set of 8 different spectral rules on band 2, 3, 4 and 5's reflectance values. At the end of the process, each pixel in the image has been assigned to one of four labels (i.e.: non-cloud, ambiguous, warm cloud, or cold cloud). Then, ambiguous pixels are resolved during the second test using the band 6 statistic parameters such as mean, standard deviation,... The value that is larger than the 95% of the numbers in dataset is considered as a threshold. Pixels that below that threshold and pass 3 another conditions are classified as clouds. In additional, a window of 3x3 pixels around each non-cloud pixel is considered ans reclassified as a cloud pixel to fill the cloud holes if 5 out of 8 adjancy pixels are clouds.

Due to the new acquisition (first product image is acquired in March 2013), there were few research on cloud detection that specifies for LandSat 8 images. However, as mention above, most these cloud detection approaches use thermal emission variations at short (3.5-4.0 $\mu$m) and long thermal wavelengths (10.0-12.0 $\mu$m) to identify clouds using different spectral thresholds. Thus, in theory one could apply the same algorithm on LandSat 8 images using input bands at the same wavelength [3]. In this article, we investigate application of detecting clouds for LandSat 8 images using the combination of pixel-based method and context-based method for detecting clouds over land and ocean.

The first method is based on spectral rules per-pixel on reflectance and at-sensor temperature values of six bands from new Operational Land Imager (OLI) and Thermal Infrared Sensor (TIRS) sensors, which is originally developed for LandSat 7 images. Detection algorithm is applied on reflectance of the wavelengths($\mu$m): 0.441-0.514, 0.519 0.601, 0.631-0.692, 0.772-0.898, 1.547-1.749 and 10.31-12.36. They are corresponding to the bands 1, 2, 3, 4, 5 and 6 of LandSat 7 but the bands 2, 3, 4, 5, 6 and 10 of LandSat 8, respectively. Due to the slightly different wavelength dynamic range for spectral bands of LandSat 7 and LandSat 8, the original rule set is redesignated appropriately. Furthermore, in order to overcome weakness of cloud detection algorithm for thin or cirrus cloud.

The article is organized as follows. In Section 2, we present methodologies including data pre-processing, features extraction, multispectral rules and method strategy to conduct new rules. Experiments and results will be described and discussed in Section 3. Finally, conclusion is given in Section 4, together with future work.

# 2 Methodologies

## 2.1 Preprocessing

The standard LandSat 8 products use quantized and calibrated scaled Digital Number (DN) representing multispectral image data. To work on LandSat 8 data, DNs firstly need to be converted to Radiance (Eq. 1) and Top Of Atmosphere (TOA) reflectance (Eq. 2) following formulas described in [13].

### 2.1.1 Conversion to Radiance

$$L_\lambda = M_L Q_{CAL} + A_L \tag{1}$$

where $L_\lambda$ is radiance value, $M_L$ is band specific multiplicative rescaling factor from metadata, $A_L$ is band specific additive rescaling factor from metadata and $Q_{CAL}$ is digital number (DN).

### 2.1.2 Conversion to TOA Reflectance

$$p_{\lambda'} = M_P Q_{CAL} + A_P \tag{2}$$

where $p_{\lambda'}$ is TOA reflectance without correction for solar angle, $M_P$ is band specific multiplicative rescaling factor from metadata, $A_P$ is band specific additive rescaling factor from metadata and $Q_{CAL}$ is digital number (DN). TOA reflectance with a correction for the sun angle is then:

$$p_\lambda = \frac{p_{\lambda'}}{\cos \Theta_{SZ}} \tag{3}$$

where $p_\lambda$ is TOA reflectance with correction for solar angle, $p_{\lambda'}$ is TOA reflectance without correction for solar angle calculated above and $\Theta_{SZ}$ is local solar zenith angle.

## 2.2 Fuzzy Features Extraction and Basic Multispectral Rules

The original methodology was tested and applied for LandSat 5 TM and LandSat 7 ETM+ and achieved high accuracy in cloud detection as reported in [12]. Although, LandSat 8 is next generation of LandSat 7 with some extensions and improvements, application of cloud detection method still need to investigate, adapt and improve.

Input data of the original rule-based RS mapping system for LandSat 7 imagery are seven bands from TM1 to TM7 (TM1 - TM7) calibrated into planetary reflectance values. The extracted features capable of capturing image independent characteristics of different land cover classes are proposed: Brightness (Bright), Visible (Vis), Near-infrared reflectance (NIR), Middle-infrared reflectance (MIR1 and MIR2), Thermal-infrared reflectance (IR), Normalized difference vegetation index (NDVI), Normalized difference bare soil index (NDBSI), Normalized difference snow index (NDSI), MIR/TIR composite (MIRTIR), Normalized difference index for blue band in built-up area and barren land (NDBBBI). Spectral data primitives TM1-TM7 are employed to extract these eleven features. Features characterized for each object are organized into three fuzzy sets (low, medium and high sets) with experimental thresholds as mentioned in [12].

Each kernel spectral rule represented by a logical expression of LandSat TM and EMT+ bands TM1-TM5 and TM7. These rules are generated from well-known spectral signatures adopted as fuzzy templates in different portions of the electromagnetic spectrum. Fourteen kernel spectral rules are: Thick Clouds Spectral Rule (TKCL_SR), Thin Clouds Spectral Rule (TNCL_SR), Snow Or Ice Spectral Rule (SNIC_SR), Water Or Shadow Spectral Rule (WASH_SR), Pitbog Or Greenhouse Spectral Rule (PBGH_SR), Dominant Blue Spectral Rule (DB_SR), Vegetation Spectral Rule (V_SR), Rangeland Spectral Rule (R_SR), Barren Land Or Built Up Or Cloud Spectral Rule (BBC_SR), Flat Response Barren Land Or Built Up Or Cloud Spectral Rule (FBB_SR), Shadow With Barren Land Spectral Rule (SHV_-SR), Shadow Cloud Or Snow Spectral Rule (SHCLSN_SR), Wetland Spectral Rule (WE_SR) (see their definition in detail in [12]).

For LandSat 8 imagery the same feature extraction and multispectral rules was applied with bands: 2, 3, 4, 5, 6, 10, 7 in corresponding with band: 1, 2, 3, 4, 5, 6, 7 in Landsat 7 imagery.

## 2.3   Multispectral Rules for Cloud Classification

In order to detect cloud, the original method combined basic multispectral rules and fuzzy features into two rules based on expert's knowledge as follows:

**Rule 1:** *(TKCL_SR or TNCL_SR) and not (LBright or LVis or LNIR or HNDSI or LMIR1 or LMIR2 or HTIR or HMIRTIR)*
**Rule 2:** *(DB_SR and SHCLSN_SR) and not (HNDSI or LNIR or LBright or LVis or HNDBSI or HTIR)*

In order to improve cloud detection quality, the article follows a practical method in which we selected a training data set of cloud and non-cloud pixels extracted from LandSat 8 images. All eleven features at three level (high, medium, low) and fourteen basic spectral rules were calculated for both cloud and non-cloud pixels and then we identified percentage of pixels in total satisfying those features and rules. Fig. 1 and Fig. 2 present percentages of cloud and non-cloud pixels satisfying basic multispectral rules for cloud (i.e.: TKCL_SR or TNCL_SR). As shown in Fig. 1, a large number of non-cloud pixels also meet these conditions. In addition, the usage of some feature such as LBright, LMIR1 and LMIR2 will reject about

20% of cloud pixels (see Fig 2.). Therefore, we proposed to add multispectral rule FBB_SR that is able to reject nearly 70% of non-cloud pixels but still keep all cloud information. The following features including LNIR, HMIRTIR, HTIR and LNDBBBI are considered to in the new rule 1 to enhance cloud recognitions but still keep appropriate accuracy. The selection of rule's parameter is based on identifying fuzzy features having low cloud rejection rate and high non-cloud acceptance rate (see Fig. 2).The new rule 1 is proposed as:

*(TKCL_SR or TNCL_SR) and not (LNIR or HMIRTIR or HTIR or LNDBBBI) and FBB_SR)*

Fig. 3 and Fig. 4 present percentages of cloud and non-cloud pixels satisfying a part condition of original rule (i.e.: DB_SR and SHCLSN_SR). As show in Fig. 3, all cloud and non-cloud pixels in training datasets satisfy two basic rules. The use of other basic multispectral rules other than DB_SR and SHCLSN_SR would not reject non-cloud pixels without affecting on cloud pixels, so no basic rule is added. Besides, the use of LNIR, LVIS and HTIR just reject about 4.5% of cloud pixels but almost 100% of non-cloud pixels (see Fig. 4). Therefore, they should be consider as impact factors and propose to use in the new rule 2:

*(DB_SR and SHCLSN_SR) and not (LNIR or LVIS or HTIR)*

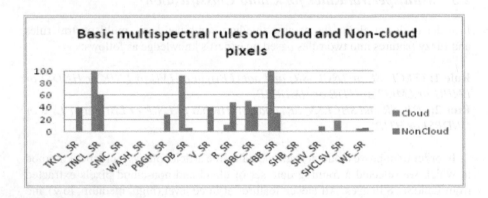

**Fig. 1** Percentages of cloud and non-cloud pixels satisfying basic multispectral rules in the dataset selected by cloud detection rule 1 ((TKCL_SR or TNCL_SR))

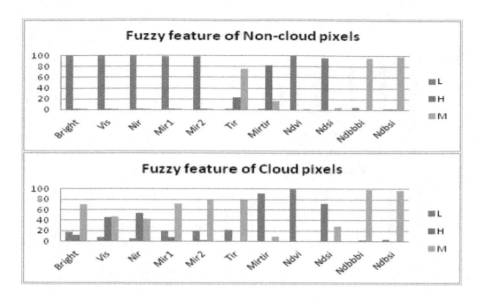

**Fig. 2** Percentages of cloud and non-cloud pixels satisfying fuzzy features at three levels (Low, High, Medium) in the dataset selected by cloud detection rule 1 ( (TKCL_SR or TNCL_SR))

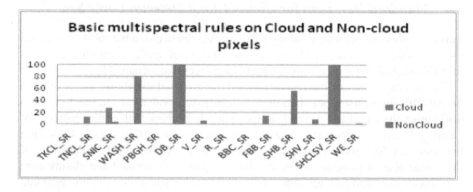

**Fig. 3** Percentages of cloud and non-cloud pixels satisfying basic multispectral rules in the dataset selected by cloud detection rule 2 (DB_SR and SHCLSN_SR)

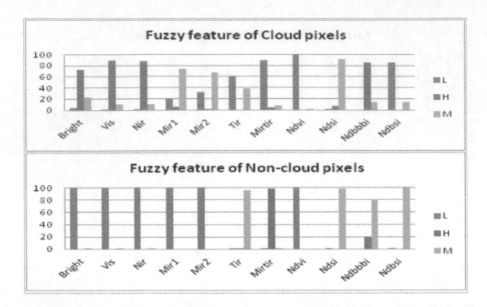

**Fig. 4** Percentages of cloud and non-cloud pixels satisfying fuzzy features at three levels (Low, High, Medium) in the dataset selected by cloud detection rule 2 (DB_SR and SHCLSN_SR)

## 2.4  Standard Deviation for Cirrus Cloud Detection over Sea

Two new proposed rules have strong ability to recognize cloud but thin/cirrus cloud over sea is difficult to detect. We investigated the use of spatial variability, a powerful method applying successfully for MODIS images over ocean [11].

The context-based cloud detection method using spatial variability was developed for MODIS image using standard deviation of pixel's values at the band green to distinguish the cloud and aerosol over oceans. It is show that spatial patterns of cloud and aerosol reflectance observed from satellite are different [11]. Following the approach, 3x3 pixels are grouped together to determine standard deviation of the center pixel. Histogram of the standard deviation for aerosol and clouds were constructed over studied areas shows that cloud and aerosol are completely separated with very distinct 3x3-STD. Based on these histogram, a threshold value of 3x3-STD = 0.0025 was defined as the separator between aerosols and cloud for band green.

Due to the correlation between similar spectral bands of MODIS and LandSat imageries have studied and reported with good results [14]. In the paper, two types of brightness temperature are calculated. The first type is calculated from band 6 (10.31-12.36 $\mu$m) of Landsat-7/ETM+ and the second are from bands 31(10.78-11.28 $\mu$m), 32(11.77-12.27 $\mu$m) of Terra/MODIS. After that, correlation between the two types of values were calculated and the results show that there are high correlation of brightness temperature values calculated from bands having similar wavelength. Thus, we could apply the same algorithm on LanSat 8 images using input bands that are the same wavelength with MODIS [3].

In this paper, we follow the same approach ([11]) to identify thin cloud over sea using band green of LandSat 8, which is sensitive to thin cloud and successfully isolated cloud from aerosol over oceans in MODIS images, to calculate the standard deviation of pixels in a window of 3x3 pixels (3 x 3-STD). However, there are differences between different imageries such as spatial resolution, spectral resolution, different sensors... then, threshold should be recalculated to adapt with new purposes and LandSat 8 data. Fig. 5 show the histogram of the 3 x 3-STD showing the statistical separation between sea cloud, cloud shadow and sea water, respectively. Based on this histogram, a threshold value of 3 x 3-STD was proposed as 0.018 in order to separate effectively cloud, cloud shadow and sea water. The threshold allows for a significant fraction of cloud shadow but allow only for a very small fraction of cloud contamination (1-2% of the studied cases).

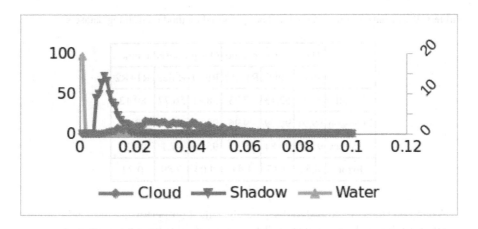

**Fig. 5** Histogram of 3 x 3-STD at band green of cloud, cloud shadow, and water over ocean

## 3 Experiments and Results

Experiments are conducted using data of Vietnam region in 2013 in order to assess the proposed algorithm's performance in comparison with the original one. Datasets built from 13 LandSat 8 images are separated into training and testing datasets including 7 and 6 images, respectively. In the training dataset, we collected 584802 cloud pixels and 8278193 non-cloud pixels. The testing dataset contains 668973 cloud pixels and 8092876 non-cloud pixels. In the article, we used training datasets for multispectral rule analysis and threshold determination as presented in Section 3. The original approach and our proposed methodology are applied on datasets to investigate their performance based on recall, precision, accuracy and error rate [15].

Since we use analysis techniques, experiments were conducted on both training and testing dataset to see their behaviors. For each method on training or testing datasets, each of two cloud classification rules is applied frequently on datasets of cloud and non-cloud pixels. After that, the final assessment on combination of both rules is considered. At each test, we calculated four measurement factors including recall, precision, accuracy and error. Tab.1 and Tab.2 shows results on training and testing datasets, respectively. As shown in Tab. 1, the original method is able to recognize 77.73% of cloud pixels (i.e: recall rate) with very high precision and accuracy (99.99% and 98.85%, respectively). The most achievement of our proposed method is capacity of recognizing cloud pixels in LandSat 8 images much more than the original one but still having similar accuracy. The new method's recall rate is 89.12% (i.e.: increasing 11.39% in total) with comparable accuracy (99.27%). Overall accuracy and error show that our proposed method slightly outperforms the original one. Tab. 2 shows performance of two methods on testing datasets. The recall rate increases from 84.95 to 98.59 meanwhile precision rates are almost the same. Accuracy is improved 1%.

**Table 1** Performance of the original and our proposed method on training datasets

|           | The original method | | | Our proposed method | | |
|-----------|-------|-------|-------|-------|-------|-------|
|           | Rule  | Rule2 | R1+R2 | Rule1 | Rule2 | R1+R2 |
| Recall    | 31.16 | 52.13 | 77.73 | 38.81 | 56.77 | 89.12 |
| Precision | 99.91 | 99.99 | 99.96 | 99.81 | 99.99 | 99.91 |
| Accuracy  | 95.45 | 96.84 | 98.52 | 95.95 | 97.1  | 99.27 |
| Error     | 4.54  | 3.15  | 1.47  | 4.04  | 2.89  | 0.72  |

When two rules are considered separately, rule 2 outperforms rule 1 with higher recall, precision and accuracy rates. Two proposed rules have better results in comparison with old ones in both training and testing datasets, in which the most improvement can be seen at the second rule. In addition, the F-measure values (Tab. 3) also show about 5% higher performance of the new proposed rules on the two datasets in comparison with the old rules.

**Table 2** Performance of the original and our proposed method on testing datasets

|           | Theoriginalmethod | | | Ourproposed method | | |
|-----------|-------|-------|-------|-------|-------|-------|
|           | Rule1 | Rule2 | R1+R2 | Rule1 | Rule2 | R1+R2 |
| Recall    | 48.98 | 66.95 | 84.95 | 49.52 | 80.57 | 98.59 |
| Precision | 99.99 | 100   | 99.99 | 99.88 | 99.99 | 99.94 |
| Accuracy  | 96.1  | 97.47 | 98.85 | 96.14 | 98.5  | 99.89 |
| Error     | 3.89  | 2.52  | 1.14  | 3.85  | 1.48  | 0.11  |

**Table 3** F-measure values of the original and our proposed method on the two datasets

|                | F-measure | |
|----------------|------------------|-----------------|
|                | Training dataset | Testing dataset |
| Original rules | 87.45            | 94.2            |
| Proposed rules | 91.86            | 99.26           |

Fig. 6 visualizes of two algorithm's results of the LandSat image recorded on 7 Jun 2013. The cloud mask of proposed method is obviously filled by cloud better than that of original one.

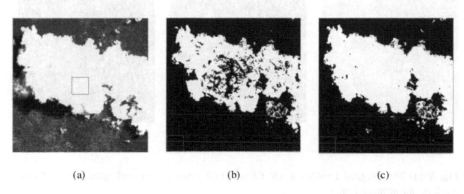

(a)                              (b)                              (c)

**Fig. 6** (a) The original LandSat 8 recorded on 7 June 2013, (b) Cloud mask made by original rules and (c) Cloud mask of our proposed rules

Rule 3 is proposed to solve problem of cloud detection over ocean. The original cloud detection method works almost perfectly over land but has limitation over sea. Based on STD techniques with a new threshold for LandSat 8, the rule 3 is used over sea and ocean after rule 1 and 2 are applied. Fig. 7 illustrates cloud detection over sea taken by LandSat 8 on $28^{th}$ May 2013. We can see that the old rule is able to detect bright, thick cloud but thin/cirrus cloud over see. Fig. 8 is an example that thin cloud is totally unrecognized by the old method but our application of the new rule 3.

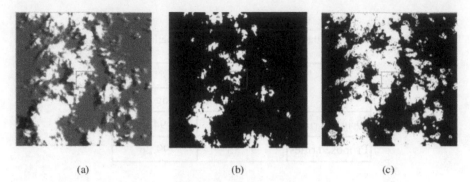

(a)    (b)    (c)

**Fig. 7** (a) The original LandSat 8 taken on $28^{th}$ May 2013, (b) Cloud mask made by original rules and (c) Cloud mask of our proposed rules

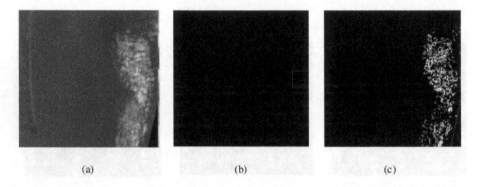

(a)    (b)    (c)

**Fig. 8** (a) The Original LandSat 8, (b) Cloud mask made by original rules and (c) Cloud mask of our proposed rules

## 4  Conclusion

In the article, a hybrid technique which take advantage of both pixel- and context-based strategies for cloud detection strategy has been developed to detect cloud over land and ocean for LandSat 8 images. The pixel-based approach uses multispectral rule sets for fuzzy features, which were largely applied for the LandSat 5-7's images, with specific adaptations to the new LandSat 8 datasets. The context-based method relies on spatial variability based on standard deviation to improve thin/cirrus cloud detection over ocean. The experiments results show effectiveness of improvements in cloud detection capacity and accuracy for LandSat 8 data. In the future, the band 9 specially designed for thin cloud detection of LandSat 8 images will be investigated.

**Acknowledgements.** The authors would like to thank the VNU research project QGTD.13.27 "Air pollution monitoring and warning system" and the Asia Research Center, VNU research project "Air pollution modelling using multi-source, multi-date, and multi-spatial resolution satellite imagery and ground-based measurements" for supports.

# References

[1] A Canada Centre for Remote Sensing Remote Sensing Tutorial: Fundamentals of Remote Sensing, Natural Resources Canada, p. 5
[2] Jedlovec, G.: Automated detection of cloud in satellite imagery
[3] Memarsadegh, N.: A Review of LandSat-7 and EO-1 Cloud Detection Algorithms
[4] http://LandSat.gsfc.nasa.gov/?page_id=2298
[5] Saunders, R.W., Kriebel, K.T.: An improved method for detecting clear sky and cloudy radainces from AVHRR data. Int. J. Remote Sensing 9, 123–150 (1988)
[6] Jedlovec, G.J., Haines, S.L., LaFontaine, F.J.: Spatial and temporal varying thresholds for cloud detection in GOES imagery. IEEE Trans. Geosci. Rem. Sens. 46(6), 1705–1717 (2008)
[7] Reuter, M., et al.: The CM-SAF and FUB Ccloud detection schemes for SEVIRI: Validation with synoptic data and initial comparison with MODIS and Calipso. J. Appl. Meteor. And Climate 48, 301–316 (2009)
[8] Ghosh, R.R., et al.: A Simple Cloud Detection Algorithm Using NOAA-AVHRR Satellite Data. International Journal of Scientific & Engineering Research 3(6) (June 2012)
[9] Kubota, M.: A new cloud detection Algorithm for Nighttime AVHRR/HRPT. Journal of Oceanography 50, 31–41 (1994)
[10] El-Araby, E., El-Ghazawi, T., Le Moigne, J., Irish, R.: Reconfigurable Processing for Satellite On-Board Automatic Cloud Cover Assessment(ACCA). Journal of Real Time Image Processing 4(3), 245–259 (2009)
[11] Martins, J.V., Tanré, D., Remer, L., Kaufman, Y., Mattoo, S., Levy, R.: MODIS cloud screening for remote sensing of aerosol over oceans using spatial variability (2002)
[12] Baraldi, A., Puzzolo, V., Blonda, P., Bruzzone, L., Taratino, C.: Automatic Spectral Rule-Based Preliminary Mapping of Calibrated LandSat TM and ETM+ Images. IEEE Transactions on Geoscience and Remote Sensing 44(9) (September 2006)
[13] http://LandSat.usgs.gov/LandSat8_Using_Product.php
[14] Oguro, Y., Ito, S., Tsuchiya, K.: Comparisons of Brightness Temperatures of LandSat-7/ETM+ and Terra/MODIS around Hotien Oasis in the Taklimakan Desert. Applied and Environmental Soil Science 2011 (2011)
[15] Han, J., Kamber, M., Pei, J.: Data Mining Concepts and Techniques, p. 365

# Binary Hybrid Particle Swarm Optimization with Wavelet Mutation

Quang-Anh Tran, Quan Dang Dinh, and Frank Jiang

**Abstract.** Particle swarm optimization (PSO) is an evolutionary algorithm in which individuals, called particles, move around a multi-dimensional problem space at different directions (trajectories) and speeds (velocities) to find the best solution. A particle movement is based on its previous best result and the previous best result of the entire population. In one of PSO variants – the HPSOWM [4], a mutation process based on wavelet theory was added to the original PSO to prevent premature conclusion of the best solution. This hybrid PSO has improved solution stability and quality over the original algorithm as well as many other hybrid PSO algorithms. However, it is limited to work on a continuous problem space. In this paper, we propose Binary Hybrid Particle Swarm Optimization with Wavelet Mutation (BHPSPWM) – a reworked version of such algorithm which operates on binary-based problem space. The movement mechanisms of particles as well as the mutation process have been transformed. The new algorithm was applied in training block-based neural network (BBNN) as well as finding solutions for several mathematical functions. The results showed significant improvement over genetic algorithms.

**Keywords:** Particle swarm optimization, binary, wavelet mutation.

## 1 Introduction

Genetic Algorithm (GA) [1] was the first algorithm which utilizes the power of biological mechanisms: the crossover and mutation of individuals' chromosome in

Quang-Anh Tran · Quan Dang Dinh
Faculty of Information Technology, Hanoi University, Vietnam
e-mail: anhtq@hanu.edu.vn, quandd.vnip@gmail.com

Frank Jiang
School of Engineering and IT, University of New South Wales, Canberra, Australia
e-mail: F.Jiang@adfa.edu.au

© Springer International Publishing Switzerland 2015                                261
V.-H. Nguyen et al. (eds.), *Knowledge and Systems Engineering,*
Advances in Intelligent Systems and Computing 326, DOI: 10.1007/978-3-319-11680-8_21

order to create new, better individuals. It started a new field of research, which is often called "evolutionary algorithms". GA has been successfully applied to various problems including optimization of artificial networks, scheduling, solving engineering problems, rule discovery, etc [3].

In 1995, Kennedy and Eberhart [5] introduced a completely new evolutionary algorithm: particle swarm optimization. They first intended to use their method for simulating the movement of birds. They tried to simulate the mechanisms which bird flocks use to find food in nature. Interestingly, the method can conveniently be used to find solution in a problem space, where "solution" can be referred as "food location" and a real-number-based problem space can be referred as the actual "area of land" in which the bird flock travels. Following is a description of the original PSO algorithm:

Firstly, a search space is defined by a real number space with a specific number of dimensions. There are boundaries in each dimension which are denoted as $para^{max}$ and $para^{min}$. The value of each dimension cannot exceed these boundaries. Then a number of particles are randomly generated with random velocities. A particle is essentially a position in the search space. It contains a set of coordinates, for instance:

$$x^p = [x_0^p, x_1^p, ..., x_k^p]$$

Velocity is a vector which indicates the direction and speed at which a particle travels.

$$v^p = [v_0^p, v_1^p, ..., v_k^p]$$

In PSO, a particle "moves" by updating its magnitudes $x_j^p$ using its velocity vector. The velocity vector of each particle changes at each iteration based on its own best previous position $pbest^p$ and the best previous position of all particles $gbest$. Basically, $gbest$ is the $pbest$ value of the particle which has achieved the lowest cost value. Following is the formula to update the velocity vector:

$$v_j^p = v_j^p + 2 \cdot rand() \cdot (pbest_j^p - x_j^p) + 2 \cdot rand() \cdot (gbest_j - x_j^p) \qquad (1)$$

The velocity is also limited by a $v^{max}$ value to prevent particles from moving too fast. Then, a particle is updated by adding the velocity vector to its current position:

$$x_j^p = x_j^p + v_j^p \qquad (2)$$

Since the introduction of PSO, many variations of this algorithm have been proposed. In 1998, Shi and Eberhart [6] added an inertia weight $\omega$ to create a balance between global search and local search. With this modification, formula (1) changes to:

$$v_j^p = \omega \cdot v_j^p + 2 \cdot rand() \cdot (pbest_j^p - x_j^p) + 2 \cdot rand() \cdot (gbest_j - x_j^p)$$

If $\omega$ is greater than or 1, the velocity update favors local search, which enables the particle to try its own direction. Otherwise, the particle tends to move towards the global optima.

In 2002, Clerc and Kennedy [7] proposed a method to control the constriction coefficients so that the need of $v^{max}$ is eliminated. The new formula is as follows:

$$v_j^p = \chi(v_j^p + \phi_1 \cdot rand() \cdot (pbest_j^p - x_j^p) + \phi_2 \cdot rand() \cdot (gbest_j - x_j^p))$$

where $\phi = \phi_1 + \phi_2 > 4$ and

$$\chi = \frac{2}{\phi - 2 + \sqrt{\phi^2 - 4\phi}}$$

In 1997, Kennedy and Eberhart [8] modified PSO into a version which supports binary number space. In this version, the terms position, velocity, trajectory, etc are redefined in ways that allow smooth transition of the algorithm from the point where it only works with real numbers into a point where it operates on problems defined as bit strings. Some of their ideas were adapted in this paper which will be described later on.

## 2 From BBNN Optimization Problem to Binary HPSOWM

In another work [2], the authors had to optimize the BBNN structures and weights for an IDS implementation which follows the anomaly detection approach. Originally, the neural network was trained using the original GA described in [1] and the author noticed that GA took a lot of time to train the model. At that moment, we thought of PSO as an alternative since PSO has proved to be more suitable in training neural networks than GA [8]. However, the original version of PSO only works in a continuous problem space (with real numbers) while the training of a BBNN is done in a binary problem space. There had been a reworked version of PSO which is able to operate in discrete binary space [8]. However, the authors of this paper did not show the improvements of their binary version of PSO over GA or any other evolutionary algorithm.

After studying variations of PSO, we found a hybrid PSO algorithm which has superior performance over standard PSO and other PSO variations [4]. Unfortunately, this hybrid version of PSO is not able to solve binary-based problems. In order to use this algorithm to train our BBNN, we reworked it into a version which works in binary space instead of continuous space to replace GA in training BBNN.

## 3 From Real Numbers to Probabilities

A binary-based particle consists of a bit string in which each element can take a value of one or zero. Such particle "moves" by flipping a number of its bits from 0 to 1 and vice versa. A velocity vector composed of bits could be used to modify a binary particle. However, applying formula (1) in a binary space by replacing $v^p$, *pbest*$^p$ and *gbest* with bit strings would actually make the particle move randomly around the problem space, not following any direction. The reason for this is that flipping a bit from 0 to 1 means taking a full jump from the lower bound to the upper bound.

What would be the velocity for a binary-based particle? How trajectory should be understood in a binary problem space? These two questions can be answered by defining the values in the velocity vector as *"probabilities that a bit will take value 1 or 0"*. Such probability value is used in the following way:

```
if (rand() < probability) then
     x = value1;
else
     x = value2;
```

Note that the *rand*() function returns a real number $x$ such that $0 \leq x < 1$.

Thus, the velocity vector $v^p$ must be scaled down to real numbers in the range $[0.0, 1.0]$ and $x_j^p$ will take either 0 or 1 as value. Specifically, the velocity value, which is between 0.0 and 1.0, will be compared to a random number in the same range. If it is higher than the random value, $x_j^p$ will take value 1, otherwise $x_j^p$ will take value 0. Note that in the original version of PSO, the magnitudes on the dimensions of velocity vector are limited by their dynamic range, which can be quite large. The rate at which the velocity increases or decreases is also high. To address this problem, a logistic function – the sigmoid function – was used to scale down the velocity values. The function is as follow:

$$S(t) = \frac{1}{(1 + e^{-t})} \tag{3}$$

With the sigmoid function, we can keep the old velocity update formula (1). This function makes sure it generates a number between 0.0 and 1.0 from any real number $t$. This property eliminates the need of a $v^{max}$ value which was used in the original version to limit the particle speed to about one tenth of the search space's length. With the new definition of velocity and the sigmoid function, now particles are updated using the following logic instead of formula (2):

$$x_j^p = \begin{cases} 1, & rand() < S(v_j^p) \\ 0, & rand() > S(v_j^p) \end{cases} \tag{4}$$

This transformation makes the particles move more slowly on each dimension towards the two best values. As a result, one bit would change from 0 to 1 before another does, whereas previously, many bits would flip at the same time due to using a bit string as a velocity vector. This approach allows the particles to scan the problem space much more carefully, reducing the risk of "stepping over" a good solution.

Trajectory in a binary space can now be explained. In a continuous space, the direction of the velocity vector decides the particle's trajectory. As a particle moves closer to the global best position, its coordinates get closer to the coordinates of *gbest*. In a binary space, the "position" of a particle is actually defined by the velocity vector, which contains probabilities. A particle is said to be close to another particle when the probabilities of corresponding bits are close to each other. One interesting thing about this algorithm is that, even when two particles are at the same position they could result in different bit strings thanks to the random value in the particle update process described in (4). Note that with a probability value of 0.9, there is still a 10% chance that a bit would be 0.

## 4 The Reworked Mutation Process

Mutation is a feature of hybrid PSO which was adapted from GA in order to avoid premature convergence. In the HPSOM algorithm in [9], a particle $x^p$ is selected randomly at each stage for the mutation process which is described using the following rule:

$$x_j^p = \begin{cases} x_j^p + \omega, & r \geq 0 \\ x_j^p - \omega, & r < 0 \end{cases}$$

where $x_j^p$ is a randomly chosen element of the particle $x^p$, $\omega$ is a random number which ranges from 0 to $0.1 \times (para^{max} - para^{min})$ – one tenth of the dimension's dynamic range – and $r$ is a random number between -1.0 and 1.0. This approach helps reduce the likelihood of premature convergence (also called stagnation). Nevertheless, the mutating space in HPSOM is limited by $\omega$ and thus the full potential of the mutation process may not have been utilized. In HPSOWM [4], the mutating space is controlled by wavelet theory. In short, a "wavelet" or "mother wavelet" is any continuous function $\Psi(x)$ which satisfies the following two properties:

$$\int_{-\infty}^{+\infty} \Psi(x)dx = 0$$

$$\int_{-\infty}^{+\infty} |\Psi(x)|^2 dx = 0$$

In HPSOWM, the authors used the Morlet wavelet taken from [10]. The Morlet wavelet is defined as follows:

$$\Psi(x) = e^{\frac{-x^2}{2}} \cos(5x)$$

How its graph looks is presented in Fig. 1.

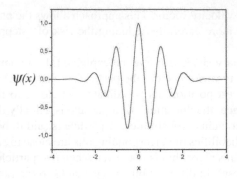

**Fig. 1** The original form of the Morlet wavelet

In a non-modified version of the Morlet wavelet, over 99% of energy concentrates around a short range of $x$ where $-2.5 \leq x \leq 2.5$. To extend the energy distribution range, we use the following function:

$$\Psi_{(a,b)}(x) = \frac{1}{\sqrt{a}} \Psi(\frac{x-b}{a})$$

In this function, $a$ is the dilation parameter and $b$ is called the translation parameter. The dilation parameter $a$ makes the function energy spread over a wider range while the amplitude is decreased. The dilation parameter can be used in various ways to fine-tune the mutation process.

The mutation in HPSOWM is described by the following formula:

$$\bar{x}_j^p = \begin{cases} x_j^p + \sigma \times (para^{max} - x_j^p), & (\sigma > 0) \\ x_j^p + \sigma \times (x_j^p - para^{min}), & (\sigma \leq 0) \end{cases} \tag{5}$$

In (5), $x_j^p$ is a randomly selected element from a randomly selected particle and $\sigma$ is calculated as follows:

$$\sigma = \Psi_{(a,0)}(\varphi) \tag{6}$$

where $\varphi$ is a random number within $[-2.5a, 2.5a]$.

To transform this mutation process into the binary algorithm, we considered several points. Firstly, in formula (5), the $para^{max}$ value is 1 and $para^{min}$ is 0 in a binary space. Secondly, the value $\sigma \times (para^{max} - x_j^p)$ and $\sigma \times (x_j^p - para^{min})$ are amounts used to mutate particle elements. They should be transformed into the probabilities for the mutation to happen in the binary version. Thirdly, since these probability values are used to compare to a random number between 0.0 and 1.0, a sigmoid function should be applied to scale them down. Overall, the mutation process in a binary space is described as follows:

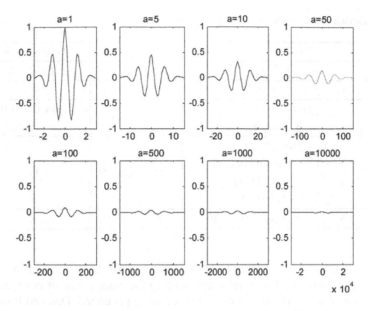

**Fig. 2** The Morlet wavelet with different dilation parameter values (adapted from [4])

$$x_j^p = \begin{cases} 1, & (rand() < S(wm_j^p)) \\ 0, & (rand() \le S(wm_j^p)) \end{cases} \tag{7}$$

The function $S$ is the sigmoid function as described in (3) and $wm_j^p$ is calculated using the following formula:

$$wm_j^p = \begin{cases} \sigma \times (1 - x_j^p), & (\sigma > 0) \\ \sigma \times x_j^p, & (\sigma \le 0) \end{cases} \tag{8}$$

where $\sigma$ is defined in (6).

## 5 Comparison between BHPSOWM and GA (Testing Functions)

The reason we chose GA for comparison is that GA originally works with binary numbers, unlike PSO which originally works in continuous space. Although GA is not the current best method to compete, it suits our goal to prove that our transformed algorithm works properly and it is even better than a decent method such as GA. We used the benchmark testing functions $f_1$, $f_6$, $f_{12}$, $f_{14}$, $f_{15}$ in [4] to compare the performance of our Binary HPSOWM with that of GA. In this paper, we renamed the functions to $f_1$, $f_2$, $f_3$, $f_4$, $f_5$ for easier reference. Our BHPSOWM outperforms GA in solving all of these functions. Table 1 contains the definition of these functions.

**Table 1** Definition of benchmark testing functions

| Function | Domain range | Optimal point |
|---|---|---|
| $f_1(x) = \sum_{i=1}^{30} x_i^2$ | $-100 \le x_i \le 100$ | $f_1(0) = 0$ |
| $f_2(x) = \sum_{i=1}^{30} |x_i| + \prod_{i=1}^{30} |x_i|$ | $-10 \le x_i \le 10$ | $f_2(0) = 0$ |
| $f_3(x) =$ $-\sum_{i=1}^{30} c_i \exp[-\sum_{j=1}^{3} a_{ij}(x_j - p_{ij})^2]$ | $0 \le x_i \le 1$ | $f_3(0.114, 0.556, 0.852)$ $= -3.8628$ |
| $f_4(x) = \{sin^2(\pi 3x_1)$ $+\sum_{i=1}^{29}(x_i - 1)^2 \cdot [1 + sin^2(3\pi x_{i+1})]$ $+(x_{30} - 1)^2 [1 + sin^2(2\pi x_{30})]\}$ $+\sum_{i=1}^{30} u(x_i, 5, 100, 4)$ | $-50 \le x_i \le 50$ | $f_4(1) = 0$ |
| $f_5(x) = \sum_{i=1}^{30} [x_i^2 - 10\cos(2\pi x_i) + 10]$ | $-50 \le x_i \le 50$ | $f_5(0) = 0$ |

The results from using BHPSOWM and GA to solve these functions were analyzed and the following line graphs representing the comparison of our method to
GA in terms of achieved cost value and speed were generated. Detailed results are
given in Table 3.

## 6 Comparison between BHPSOWM and GA (Industrial Data)

We used the Fold-1 of DARPA IDS Netflow dataset (in previous report [2]) to test
SPSOWM and GA performance. Two tests were performed: (1) We fixed the training
time and compare the objectives (genuine detection rate while $FAR < 0.2$); and (2)
We fixed the objectives (genuine detection rate while $FAR < 0.2$) and compare the
training times required to obtain these objectives. The results were produced after
50 runs of the each algorithm.

The following formula was used to compute how much percent BHPSOWM is
better than GA in term of objective (Detection Rate) obtained:

$$\Delta = Average\left(\frac{DR_{BHPSOWM} - DR_{GA}}{DR_{GA}}\right)$$

In average, the objective obtained by BHPSOWM is 5.5% higher than the objective obtained by GA. The following formula was used to compute how much faster
BHPSOWM is compared to GA in terms of training time:

$$\Delta_t = Average\left(\frac{T_{GA} - T_{BHPSOWM}}{T_{GA}}\right)$$

In average, the training time by BHPSOWM is 31.8% less than the training time
by GA. Following are line graphs representing the improvement of our method over
GA. Detailed data are shown in Table 4 and Table 5.

**Table 2** Comparison between BHPSOWM and GA using 5 benchmark testing functions

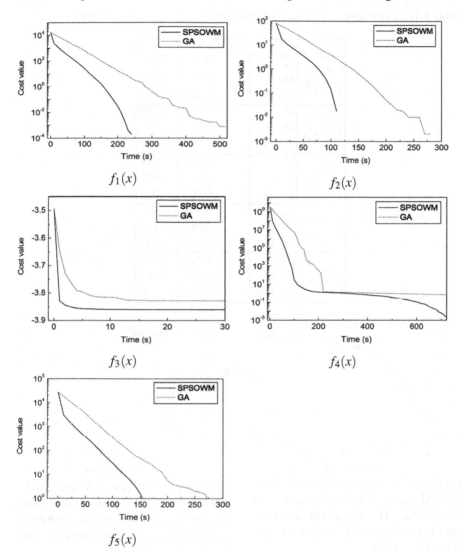

**Table 3** Detailed comparison between BHPSOWM and GA for testing functions

|  |  | BHPSOWM | GA |
|---|---|---|---|
| $f_1(x)$<br>$T = 250$ | Best | 0.000 | 3.005 |
|  | Mean | 0.000 | 0.410 |
|  | Std Dev | 0.000 | 6.082 |
| $f_2(x)$<br>$T = 120$ | Best | 0.000 | 1.908 |
|  | Mean | 0.000 | 0.400 |
|  | Std Dev | 0.000 | 0.752 |
| $f_3(x)$<br>$T = 100$ | Best | -3.861 | -3.828 |
|  | Mean | -3.862 | -3.862 |
|  | Std Dev | 0.001 | 0.031 |
| $f_4(x)$<br>$T = 600$ | Best | 0.110 | 0.816 |
|  | Mean | 0.000 | 0.100 |
|  | Std Dev | 0.125 | 0.237 |
| $f_5(x)$<br>$T = 200$ | Best | 0.000 | 5.540 |
|  | Mean | 0.000 | 0.000 |
|  | Std Dev | 0.000 | 11.270 |

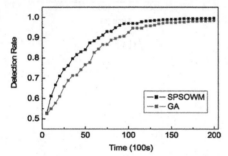

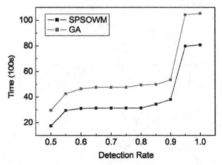

Comparison between GA and BHP-SOWM using industrial dataset with fixed time. The objectives are compared. More detailed results are presented in Table 4.

Comparison between GA and BHP-SOWM using industrial dataset with fixed objectives. The training time measurements are compared. Detailed data are displayed in Table 5.

**Table 4**

|         |         | BHPSOWM | GA    |
|---------|---------|---------|-------|
|         | Best    | 0.841   | 0.767 |
| $T = 50$  | Mean    | 1.000   | 1.000 |
|         | Std Dev | 0.212   | 0.224 |
|         | Best    | 0.971   | 0.926 |
| $T = 100$ | Mean    | 1.000   | 1.000 |
|         | Std Dev | 0.072   | 0.139 |
|         | Best    | 0.991   | 0.975 |
| $T = 150$ | Mean    | 1.000   | 1.000 |
|         | Std Dev | 0.022   | 0.070 |

**Table 5** Comparison between BHPSOWM and GA for training time (seconds) vs. expected detection rate. The results are the mean of detection rate over 50 runs.

|           |         | BHPSOWM | GA      |
|-----------|---------|---------|---------|
|           | Best    | 3114.0  | 4645.8  |
| $DR = 0.6$  | Mean    | 110.0   | 238.0   |
|           | Std Dev | 2824.3  | 4018.9  |
|           | Best    | 3138.7  | 4936.8  |
| $DR = 0.8$  | Mean    | 110.0   | 238.0   |
|           | Std Dev | 2812.3  | 4207.1  |
|           | Best    | 8075.5  | 10538.4 |
| $DR = 1.0$  | Mean    | 588.0   | 914.0   |
|           | Std Dev | 6445.0  | 7285.4  |

## 7 Conclusion and Further Research

This paper describes the binary rework of a powerful hybrid PSO algorithm. As this new PSO variant operate in a binary space, it should only be compared to other algorithms which work with binary-based problems. In this work, the author compared BHPSOWM to GA – a proven, powerful revolutionary algorithm. The authors succeeded in transforming the algorithm so that it not only preserves the ability to find high-quality solutions but also achieves better performance than GA does.

After testing our new method against benchmark test functions derived from [4], we see that (1) BHPSOWM outperforms GA in all 5 functions and (2) except for $f_2(x)$, BHPSOWM found better solutions than GA does. As can be seen from the graphs in Table 2, in some cases, BHPSOWM suffered premature convergence while GA continued to generate better solutions. However, the result that BHPSOWM achieved in Table 2 is quite acceptable. This problem inspires the author to discover more mechanisms to improve our algorithm so that premature convergence will be reduced in future works.

From reading the experiment results, it can be seen that BHPSOWM can optimize block-based neural networks much faster than GA but still achieves a significant increase in solution quality. Therefore, the authors suggest that BHPSOWM should replace GA in such task. In future research, we plan to apply our BHPSOWM and its modified versions to more industrial applications in order to obtain more results for analysis. We will also work on increasing our method's performance and compare it with other state-of-the-art methods.

**Acknowledgments.** This research was fully supported by the Vietnam National Foundation for Science and Technology Development (NAFOSTED) under project number 102.01-2012.04.

# References

[1] Holland, J.: Genetic Algorithms, computer programs that evolve in ways that even their creators do not fully understand. Scientific American, 66–72 (1975)

[2] Tran, Q.A., Jiang, F., Hu, J.: A real-time netflow-based intrusion detection system with improved BBNN and high-frequency field programmable gate arrays. In: Proceedings of the 11th IEEE International Conference on Trust, Security and Privacy in Computing and Communications, Liverpool, UK, pp. 201–208 (2012)

[3] Ross, P., Corne, D.: Applications of genetic algorithms. AISB Quaterly on Evolutionary Computation 89, 23–30 (1994)

[4] Ling, S.H., Iu, H.H., Chan, K.Y., Lam, H.K., Yeung, B.C., Leung, F.H.: Hybrid particle swarm optimization with wavelet mutation and its industrial applications. IEEE Transactions on Systems, Man, and Cybernetics, Part B: Cybernetics 38(3), 743–763 (2008)

[5] Kennedy, J., Eberhart, R.: Particle swarm optimization. Proceedings of IEEE International Conference on Neural Networks 4(2), 1942–1948 (1995)

[6] Shi, Y., Eberhart, R.: A modified particle swarm optimizer. In: The 1998 IEEE International Conference on Evolutionary Computation Proceedings. IEEE World Congress on Computational Intelligence, pp. 69–73. IEEE (May 1998)

[7] Clerc, M., Kennedy, J.: The particle swarm-explosion, stability, and convergence in a multidimensional complex space. IEEE Transactions on Evolutionary Computation 6(1), 58–73 (2002)

[8] Kennedy, J., Eberhart, R.C.: A discrete binary version of the particle swarm algorithm. In: International Conference on Systems, Man, and Cybernetics. Computational Cybernetics and Simulation, vol. 5, pp. 4104–4108. IEEE (October 1997)

[9] Esmin, A.A., Lambert-Torres, G., de Souza, A.Z.: A hybrid particle swarm optimization applied to loss power minimization. IEEE Transactions on Power Systems 20(2), 859–866 (2005)

[10] Daubechies, I.: Ten lectures on wavelets, vol. 61, pp. 198–202. Society for industrial and applied mathematics, Philadelphia (1992)

# Collaborative Filtering by Co-Training Method

Tran Nhat Quang, Do Thi Lien, and Nguyen Duy Phuong

**Abstract.** Collaborative filtering is a technique to predict the utility of items for a particular user by exploiting the behavior patterns of a group of users with similar preferences. This method has been widely used in e-commerce systems. In this paper, we propose a collaborative filtering method based on co-training– a semi-supervised technique that iteratively expands the training set by switching between two different feature sets. In the collaborative filtering settings, our co-training based method uses users and items as two different feature sets. Each feature set is used to infer the most reliable predictions which are then added to the new labeled set. This procedure leads to improved prediction accuracy and reduces the negative influence of data sparsity – a main obstacle to the application of collaborative filtering. The experimental results on real data sets show that the proposed method achieves superior performance compared to baselines.

**Keywords:** Collaborative Filtering, Co-Training, Recommendation System, Item-Based Recommendation, User-Based Recommendation.

## 1 Collaborative Filtering

One of the challenges to the Internet users and E-Commercial systems is that there are many options. In order to get approach to useful information, users usually have to process, extrude unnecessary data. The recommender systems (RS) provide a solution that aims at reducing the amount of information by predicting and providing a short list of the items (websites, news, movies, videos, and so on) that the users may pay attention to. There are two techniques that are mainly implemented in the RS: content-based filtering and Collaborative Filtering [1, 2]. Comparing to content-based filtering, Collaborative Filtering is implemented more widely for its simplicity

Tran Nhat Quang · Do Thi Lien · Nguyen Duy Phuong
Posts and Telecommunications Institute of Technology,
Hanoi, Vietnam
e-mail: {quangtn,liendt,phuongnd}@ptit.edu.vn

© Springer International Publishing Switzerland 2015                    273
V.-H. Nguyen et al. (eds.), *Knowledge and Systems Engineering,*
Advances in Intelligent Systems and Computing 326, DOI: 10.1007/978-3-319-11680-8_22

in use and installation, and its accuracy [1, 2]. In this paper, we concentrate on transfer-learning for CF. The problem of CF for the RS is stated as follows:

Given the indefinite set $U = \{u_1, u_2, \ldots, u_N\}$ that includes $N$ users, $P = \{p_1, p_2, \ldots, p_M\}$ that includes $M$ products. Each product $p_x \in P$ is either goods, movies, images, magazines, documents, books, newspapers, services, or any other kind of information that the users need. For convenience, we write $p_x \in P$ as $x \in P$; and $u_i \in U$ as $i \in U$. The relation between user set $U$ and product set $P$ is expressed by the evaluation matrix $R = \{r_{ix}\}, i = 1..N, x = 1..M$. Each $r_{ix}$ indicates the evaluation of the user $i \in U$ for each product $x \in P$. The value of $r_{ix}$ is collected by either directly asking or indirectly through the feedback of the user. $r_{ix} = 0$ means the user $i$ has never evaluated or known of product $x$.

Next, we label $P_i \subseteq P$ as the set of products $x \in P$ reviewed by user $i \in U$ and $U_x \subseteq U$ as the set of the users $i \in U$ who give feedback for $x \in P$. To a user who needs advice $a \in U$ (current user, positive user), the CF task is to predict the feedback of user $a$ for products $x \in (P \backslash P_a)$, so that the system is able to suggest a highly-evaluated products.

Table 1 presents an example with the evaluation matrix $R = (r_{ij})$, in which there are 5 users and 7 products in set $U = \{u_1, u_2, u_3, u_4, u_5\}$ and $P = \{p_1, p_2, p_3, p_4, p_5, p_6, p_7\}$ respectively. Each user gives his/her feedback about the products according to the level of measurement: $\{0, 1, 2, 3, 4, 5\}$. As $r_{ij} = 0$, it means that the user $u_i$ has not given his/her feedback or never known of the product $p_j$. Values of $r_{5,1} = ?$ are the products that the system has to give prediction for user $u_5$.

**Table 1** Evaluation matrix of CF

| Users | Products | | | | | | |
|---|---|---|---|---|---|---|---|
| | $p_1$ | $p_2$ | $p_3$ | $p_4$ | $p_5$ | $p_6$ | $p_7$ |
| $u_1$ | 4 | 2 | 5 | 0 | 3 | 0 | 3 |
| $u_2$ | 5 | 0 | 5 | 5 | 4 | 0 | 0 |
| $u_3$ | 4 | 0 | 0 | 4 | 3 | 4 | 3 |
| $u_4$ | 0 | 3 | 5 | 5 | 0 | 5 | 0 |
| $u_5$ | ? | 5 | ? | ? | ? | 4 | 4 |

The most typically applied CF is k-nearest-neighbors. This method is also called memory-based filtering [9, 10, 11], which is different from model-based filtering [4, 5, 6, 7, 8, 12, 14, 15]. To each user, the system decides whether $k$ users have the same interest to him/her based on the products they have chosen or evaluated in the past and suggests to the user those products that $k$ users have chosen. Similarly, instead of finding nearest $k$ users, we can look for $k$ nearest neighbors for each of the products based on whether the users cared about these neighbors to choose or not choose a product. In the first situation, CF is called User-Based collaborative filtering meanwhile it is called Item-Based in the second situation.

In order to do CF in memory to provide accurate results, we need to identify how identical the users are based on the evaluation matrix of the users and products (or the similarity among the items, depending on the method being used). Generally, the scale for similarity used is the similarity between two vectors such as: cosine or Pearson correlation. However, the results from these methods are not accurate with sparse data, which, in this case, happens when each of the users only evaluates some of the products. This is a typical situation in the systems that use CF.

To limit the negative influence of sparse data on CF, many cotraining methods have been proposed, such as cluster-based smoothing [7, 8, 13] or User-Based and Item-Based filtering combining [15]. Some other methods are based on SVD (Singular Value Decomposition) [4, 5, 6, 7] or common-feature sharing [14].

In this paper, we propose another Co-Training method that directly expands other CF methods based on memory. To focus on the proposal, we will present about CF based on memory in section 2. Section 3 is about cotraining CF. The last section aims explain experimental methods and results evaluation.

## 2 Memory-Based CF

There are two main approaches for memory-based CF: User-Based and Item-Based. Each approach has its advantages that make use of users' or products' attributes. The common characteristic of these two is that they both use the whole evaluation dataset about predicting users' feedback about the products that they need reviews but have never known about.

## 2.1 User-Based and Item-Based Approach

User-Based approach estimates the similarity between the pair of users based on some similariy scale to return the prediction of new appropriate products for users. Item-Based approach estimates the similarity betwwen pair of products based on some similariy scale to return predictions of new appropriate products for users. Each of the two methods is implemented with 3 steps:

**Step 1:** *Calculate the similarity between users' pair or products' pair.* We use correlation or similarity scale to calculate the similarity between users' pair or products' pair [1, 12, 13]. Let $u_{ij}$ be the similarity between user $i$ and user $j$, $p_{xy}$ be the similarity between product $x$ and $y$. Pearson correlation between user $i \in U$ and user $j \in U$ is defined as (1). Similarity of product $x \in P$ and product $y \in P$ is defined as (2) [1, 12, 13]:

$$u_{ij} = \frac{\sum_{x \in P_i \cap P_j} (r_{ix} - \overline{r_i})(r_{jx} - \overline{r_j})}{\sqrt{\sum_{x \in P_i \cap P_j} (r_{ix} - \overline{r_i})^2} \sqrt{\sum_{x \in P_i \cap P_j} (r_{jx} - \overline{r_j})^2}} \quad (1)$$

$$p_{xy} = \frac{\sum\limits_{i \in U_x \cap U_y} (r_{ix} - \overline{r_x})(r_{iy} - \overline{r_y})}{\sqrt{\sum\limits_{i \in U_x \cap U_y} (r_{ix} - \overline{r_x})^2} \sqrt{\sum\limits_{i \in U_x \cap U_y} (r_{iy} - \overline{r_y})^2}} \tag{2}$$

**Step 2:** *Define the neighbor sets for users who need advice.* In this step, we need to sort the values of $u_{ij}$ or $p_{xy}$ in descending order, in which $i \in U$ are the users who need advice about the products $x \in P$. Next, we need to choose $K$ first users for the set of user $i$'s neighbors, or $K$ first users for the set of product $x$'s neighbors [12, 13].

**Step 3:** *Generate predictions for users.* The most familiar method to generate predictions of user $i$ for product $x$ is with (3) [12] and of products with (4) [13].

$$r_{ix} = \overline{r_i} + \frac{\sum\limits_{j \in K_i} (r_{jx} - \overline{r_j}) u_{ij}}{\sum\limits_{j \in K_i} |u_{ij}|} \tag{3}$$

$$r_{ix} = \frac{\sum\limits_{y \in K_x} p_{xy} r_{iy}}{\sum\limits_{y \in K_x} |p_{xy}|} \tag{4}$$

$K_i$ and $K_x$ are the neighbors sets of the users $i \in U$ and products $x \in P$, respectively.

## 2.2 Limitations of User-Based and Item-Based Methods

Although User-based and Item-based methods have been applied in many e-commerce systems, there are still some limitations:

* *Sparse data.* To the CF systems, the number of unranked are much greater than the number of ranked products. This influences directly on the process of calculating similarity between pairs of users or products. In case there are two users $i, j \in U$ that have the evaluation sets $P_i \in P_j = \emptyset$, the similarity between two users is indefinable (user $u_5$ and $u_2$ in Table 1). When two products $x, y \in P$ have the user sets $U_x \cap U_y = \emptyset$, the similarity between these two products is also indefinable. This is the biggest limitation when implementing Step 1 of the algorithm.
* *Sparse data makes defining the neighbor sets in Step 2 less credible.* When calculating the values $s_{ij}$ and $p_{xy}$, the scales are usable only on the sets $P_i \cap P_j \neq \emptyset$ and $U_x \cap U_y \neq \emptyset$. The fact that the sets of values cannot be used in calculation process leaves out many classifying labels that have already been known of. This leads to the result that many similar users or products cannot be matched. On the other hand, many pairs of users that are not very close to each other are still capable of defining the neighbor sets. The improperly defined neighbor sets result in the quality of predictions in each method.
* *New users and new products problem.* For a new user who has no evaluation for the products, both User-based and Item-based are not able to give

prediction about appropriate products. CF defines this as New User and New Product problem [11].

## 2.3 Proposed Method

To restrain these limitations, we consider CF under the perspective of Co-Training [8]. The process of observing the users or products is added with credible classifying labels to help calculating the similarity between pair of products or users, corresponding. To implement, for each user or product, the training instances that recognize the classifying labels are $r_{ij} \neq \emptyset$. The mission of Co-Training is to build a training algorithm which allows adding classifying labels to the set of labels $r_{ij} = \emptyset$.

The process of observing and training based on user $i \in U$ with available classifying labels ($r_{ix} \neq \emptyset$) predicts new products $y \in P$ that are appropriate to this user. The result of this process creates additional credible classifying labels ($r_{iy}$) to calculate the similarity between two products $x \in P$ and $y \in P$. The main focus of this process is to define the credible classifying labels ($r_{iy}$). The process of observing and training each product $x \in P$ includes set the of available classifying labels $r_{ix} \neq \emptyset$ with labels ($r_{iy}$) that are predicted with users. Watching over the products gives us the ability to predict a good match between users $i \in U$ and products $z \in P$ with the values of ($r_{iz}$). The set of values ($r_{iz}$) is then added to the set of available classifying labels to continue the training process in the next step. The procedure ends when we are not able to add credible prediction values. The method that gives different observation methods, provides and adds training results is Co-Training's aim in Machine Learning [7, 8]. Detailed content of Co-Training method is in the next section.

## 3 Collaborative Filtering by Co-Training

To solve the problem using CF by Co-Training method, we need to define two types of observation on the training data: User-based and Product-based observation using CF. Next we need to build the contemporaneous training algorithm between these observations to generate predictions for users.

## 3.1 User-Based Observation

Presented in section 2, User-Based method calculates the similarity between user $i \in U$ and other users in the training dataset [1, 12]. This leads to two drawbacks:

First, if two users $i, j$ have $|P_i \cap P_j|$ small but $r_{ix} = r_{jx}$ with every $x \in P_i \cap P_j$, they are considered to have absolutely the same interest. In Table 1, user $u_2$ is considered to be absolutely the same as user $u_4$ as both of them have the same feedback for $p_3, p_4 (r_{23} = r_{43} = 5; r_{24} = r_{44} = 5)$. The result is $u_4$, which is always the neighbor of $u_2$ when giving the predictions about products for $u_2$.

Second, if two users $i, j$ having $|P_i \cap P_j| = \emptyset$ are considered to be separate in interest, as we observe, they have some interests, then those products $x \in |P_i \cap P_j| = \emptyset$ will not be included in the training and predicting process.

In this proposal, calculating similarity between pair of users $i \in U$ is not used to define neighbor sets $K_i$ as in [1, 12]. Instead, it is used to define credible classifying labels $r_{iy}$ for user $i$. To implement this, we define the generation set for user $i \in U$ as:

**Definition 1.** Generation set $S_i$ for user $i \in U$ is the set that includes all users $j \in U$ whose feedback sets intersect by at least $\gamma$ products, in which $\gamma$ is a positive constant.

$$S_i = \{ j \in U : |P_i \cap P_j| \geq \gamma \} \tag{5}$$

For instance, let $\gamma = 3$, user $u_1$ in Table 1 will have $S_1 = u_2, u_3$ as both $u_2$ and $u_3$ have 3 feedbacks like $u_1$. Similarly, we can see that $S_2 = u_1, u_3, S_3 = u_1, u_2, S_4 = \emptyset, S_5 = \emptyset$.

In the traditional User-Based method, the similarity between user $i \in U$ and user $j \in U$ is calculated in the evaluation matrix to define the set of $K$ neighbors of user $i$. The method to generate prediction about new products uses $K$ neighbors of user $i$ to calculate. This means that $K$ neighbors of user $i$ is a constant set throughout the training process [12].

In our proposal, the similarity of user $i$ and user $j$ is only calculated on generation set $S_i \subseteq U$ with formula (6). This allows us to obstruct the pairs of users that have small $|P_i \cap P_j|$ but are rated with high similarity according to User-Based method. The generation set of user $i \in U$ changes after every step of the Co-Training process.

$$u_{ij} = \begin{cases} 0 & \text{if} \quad j \notin S_i \\ \dfrac{\sum\limits_{x \in P_i \cap P_j} (r_{ix} - \overline{r_i})(r_{jx} - \overline{r_j})}{\sqrt{\sum\limits_{x \in P_i \cap P_j} (r_{ix} - \overline{r_i})^2} \sqrt{\sum\limits_{x \in P_i \cap P_j} (r_{jx} - \overline{r_j})^2}}, & otherwise \end{cases} \tag{6}$$

The similarity of user $i \in U$ and user $j \in S_i$ in (6) is used to find user $i$'s neighbors set. In the previous methods, the set of $K$ neighbors of user $i \in U$ is fixed during the entire evaluation process [1, 12]. In this paper, we consider the neighbors set of user $i$ to be always changing, depending on generation set $S_i$ of that user. In detail, the change of the neighbors set is defined as:

**Definition 2.** The neighbors set of user $i \in U$, denoted as $K_i$, is the set that includes users $j$ in $S_i$ that have the similarity $u_{ij}$ calculated with (6) greater than $\beta$ in which $\beta \in [0, 1]$.

$$K_i = \{ j \in S_i : u_{ij} > \beta \} \tag{7}$$

Based on the neighbors set $K_i$ of user $i \in U$, credible classifying labels are predicted with (8):

**Table 2** Estimation Matrix for Users

| Users | Products | | | | | | |
|-------|-------|-------|-------|-------|-------|-------|-------|
|       | $p_1$ | $p_2$ | $p_3$ | $p_4$ | $p_5$ | $p_6$ | $p_7$ |
| $u_1$ | 4 | 2 | 5 | 4 | 3 | 4 | 3 |
| $u_2$ | 5 | 2 | 5 | 5 | 4 | 0 | 3 |
| $u_3$ | 4 | 2 | 5 | 4 | 3 | 4 | 3 |
| $u_4$ | 0 | 3 | 5 | 5 | 0 | 5 | 0 |
| $u_5$ | ? | 5 | ? | ? | ? | 4 | 4 |

$$r_{ix} = \overline{r_i} + \frac{\sum\limits_{j \in K_i} (r_{jx} - \overline{r_j}) u_{ij}}{\sum\limits_{j \in K_i} |u_{ij}|} \tag{8}$$

For the set of users in Table 1, we have $K_1 = \{u_3\}, K_2 = \{u_1\}, K_3 = \{u_1\}$. Consequentially, the credible prediction values for $u_1$, are $r_{14} = 4, r_{16} = 4$; the credible prediction values for $u_2$ are $r_{22} = 2, r_{27} = 3$; the credible prediction values for $u_3$ are $r_{32} = 2, r_{33} = 5$.

We can see that the set of classifying labels $r_{ix}$ calculated from (8) is less than that calculated from (3). However, this will be improved as we observe the classifying labels in items.

## 3.2 Item-Based Observation

The similarity between pair of products $x \in P$ is not used to define the neighbor set $K_i$ as in [1, 13] but to clarify how appropriate the new product $x \in P$ is to user $i \in U$. To implement, we define generation set for product $x \in P$:

**Definition 3.** Generation set $C_x$ for products $x \in P$ includes all products whose feedback sets intersect by at least $\gamma$ users, in which $\gamma$ is a positive constant.

$$C_x = \{y \in P : |U_x \cap U_y| \geq \gamma\} \tag{9}$$

Let $\gamma = 3$, with the set of users in Table 1, we are able to find that $C_1 = \{0\}, C_2 = \{0\}, C_3 = \{0\}, C_4 = \{0\}, C_5 = \{p_1\}, C_6 = \{0\}, C_7 = \{0\}$. However, if the process of item-based observation is implemented after the process of User-Based observation as in Table 2, we have the result: $C_1 = \{p_2, p_3, p_4, p_5, p_7\}, C_2 = \{p_1, p_3, p_4, p_5, p_7\}$, $C_3 = \{p_1, p_2, p_4, p_5, p_7\}, C_4 = \{p_1, p_2, p_3, p_5, p_6, p_7\}, C_5 = \{p_1, p_2, p_3, p_4, p_7\}$, $C_6 = \{p_2, p_3, p_4, p_7\}, C_7 = \{p_1, p_2, p_3, p_4, p_5, p_6\}$.

In the approaches used before, the calculation of how appropriate a product $x \in P$ to a user $i \in U$ is performed on the evaluation matrix to find set of $K$ neighbors of product $x$. The method to predict the views of user $i \in U$ is implemented based on $K$ neighbors of product $x \in P$. Thus, $K$ neighbors of product $x \in P$ is a fixed set during the training process [1, 13].

In this proposal, the similarity of each product $x \in P$ and product $y \in P$ is calculated using generation set $C_x \subseteq P$ with formula (10). This can prevent the products whose $|U_x \cap U_y|$ is small but have high estimated similarity according to Item-based method. The generation set $C_x$ of product $x \in P$ changes in every step of the Co-Training process.

$$r_{41} = r_{44} = 5, r_{55} = r_{57} = 4, and r_{26} = r_{24} = 5 \tag{10}$$

The similarity of product $x \in P$ and product $y \in C_x$, according to (10), is used to find the neighbors set for product $x$. In the traditional approaches, neighbors set $K$ of product $x \in P$ stays unchanged throughout the evaluation process. In this proposal, we consider the neighbors set of product $x \in P$ changes based on its $C_x$. In detail, the neighbors set of $x \in P$ is defined as:

**Definition 4.** The neighbors set of $x \in P$, denoted as $K_x$, comprises of products $y$ in $C_x$ that have the similarity $p_{xy}$ calculated by (10) greater than some threshold $\beta$, in which $\beta \in [0, 1]$.

$$K_x = \{y \in C_x : p_{xy} > \beta\} \tag{11}$$

Based on the neighbors set $K_x$ of product $x \in P$, credible classifying labels for user $i \in U$ are predicted using formula (12).

$$r_{ix} = \frac{\sum\limits_{y \in K_x} p_{xy} r_{iy}}{\sum\limits_{y \in K_x} |p_{xy}|} \tag{12}$$

For the information given in Table 2, we can find that $K_1 = \{p_4\}, K_2 = \{p_7\}, K_3 = \{p_4\}, K_4 = \{p_1\}, K_5 = \{p_7\}, K_6 = \{p_4\}, K_7 = \{p_5\}$. Therefore, the credible prediction values for $p_1, p_5$, and $p_6$ are $r_{41} = r_{44} = 5, r_{55} = r_{57} = 4$, and $r_{26} = r_{24} = 5$ respectively. The results are shown in Table 3.

**Table 3** Item-based Estimation Matrix

| Users | Products | | | | | | |
|---|---|---|---|---|---|---|---|
| | $p_1$ | $p_2$ | $p_3$ | $p_4$ | $p_5$ | $p_6$ | $p_7$ |
| $u_1$ | 4 | 2 | 5 | 4 | 3 | 4 | 3 |
| $u_2$ | 5 | 2 | 5 | 5 | 4 | 5 | 3 |
| $u_3$ | 4 | 2 | 5 | 4 | 3 | 4 | 3 |
| $u_4$ | 5 | 3 | 5 | 5 | 0 | 5 | 0 |
| $u_5$ | ? | 5 | ? | ? | 0 | 4 | 4 |

Two mechanisms of observation: Item-based and User-based add credible classifying labels in every step of the Co-Training method and maximizes the number of values $rix = \emptyset$ in the evaluation matrix. In order to anticipate more classifying labels, we observe the products. This is the last important step of the method Co-Training for CF being presented in the next section.

## 3.3   Combining Two Types of Observation

In this section, we present two Co-Training methods: User-based Co-Training (denoted as Co-UserBased) and Item-based Co-Training (denoted as Co-ItemBased). The difference in these two methods is which process runs first.

**Co-UserBased Method**

Co-UserBased is detailed in Figure 1, implemented with loops $t$. In the initializing step $t = 0$, prediction matrix $R^{(0)} = (r_{ij}^{(0)})$ is set equal to the initial evaluation matrix $R = (r_{ij})$. In each loop, the User-based training process is implemented through (2.1.a-c). In step (2.1.a), we need to define the generation set $S_i^{(t)}$ for user $i$ at loop $t$ th with (5) and define the similarity $u_{ij}^{(t)}$ of user $i \in U$ and user $j \in S_i^{(t)}$ with (6).

In step (2.1.b), using $S_i^{(t)}, u_{ij}^{(t)}$ defined in step (2.1.a), we get $K_i^{(t)}$ as the neighbors set of user $i$ at loop $t$th with formula (7). In step (2.1.c), using $K_i^{(t)}$ from previous step, we can predict $r_{ix}^{(t)}$ as the credible views of user $i$ for products $x$ at loop $t$th with formula (8). Values of $r_{ix}^{(t)}$ predicted by User-based approach at loop $t$th are added to the Item-based training process in step 2.2. In step (2.2.a), we need to find generation set $C_x^{(t)}$ of product $xinP$ using formula (9) and find the similarity $p_{xy}^{(t)}$ of products $x \in P$ and $y \in C_x^{(t)}$ using (10). In step (2.2.b), using $C_x^{(t)}, p_{xy}^{(t)}$ from (2.2.a), we are able to find $K_x^{(t)}$ as the neighbors set of product $x \in C_x^{(t)}$ at step $t$th with (10). In step (2.2.c), using $K_x^{(t)}$ from (2.2.b), we can predict $r_{ix}^{(t)}$ as the accurate views of user $i$ for products $x$ at step $t$th with (11).

In step 2.3, $t$ increments by 1 loop cycle and implements the next Co-Training process. Co-UserBased converges in loop $t$th that has $r_{ix}^{(t)} = r_{ix}^{(t-1)}$.

**Co-ItemBased Method**

This method, presented in detail in Figure 2, is implemented as follows:

At the initializing step $t = 0$, the prediction matrix $R^{(0)} = (r_{ij}^{(0)})$ is set equal to the initial evaluation matrix $R = (r_{ij})$. The Co-ItemBased process is implemented in prior. In step (2.1.a) we define generation set $C_x^{(t)}$ of product $x \in P$ with (9) and $p_{xy}^{(t)}$ with (10). Using $C_x^{(t)}, p_{xy}^{(t)}$, we find $K_x^{(t)}$ with (11) in step (2.1.b). With $K_x^{(t)}$, we are able to predict $r_{ix}^{(t)}$ as the accurate views of user $i$ for products $x$ in loop cycle $t$th using (12). The values $r_{ix}^{(t)}$ that predict based on products are added to the User-based training in step 2.2.

In step (2.2.a), we need to find sets $S_i^{(t)}$ using (5) and $u_{ij}^{(t)}$ using (6). With $S_i^{(t)}, u_{ij}^{(t)}$ found in (2.2.a), we find $K_x^{(t)}$ using (7). Then, the accurate view of user $i$ for products $x$ $r_{ix}^{(t)}$ are predicted using $K_x^{(t)}$ at loop cycle $t$th in (8). In step (2.3), loop cycle

**Input**: Evaluation matrix $R^0 = (r_{ij}^{(0)}) = (r_{ij})$.

**Output**: Prediction matrix $R^{(t)} = \left( r_{ij}^{(t)} \right)$.

**Steps**:

    1. Initialize number of loops: $t \leftarrow 0$;

    2. Loops:

  **Repeat**

        2.1. User-based training:

            a) Find $S_i^{(t)}, u_{ij}^{(t)}$ with (5),(6).

            b) Find $K_i^{(t)}$ with (7).

            c) Predict $r_{ix}^{(t)}$ with (8).

        2.2. Item-based training:

            a) Find $C_x^{(t)}, p_{xy}^{(t)}$ with (9),(10).

            b) Find $K_x^{(t)}$ with (11).

            c) Predict $r_{ix}^{(t)}$ with (12).

        2.3. Increment the loop cycle: $t \leftarrow t + 1$;

**Until** Converges.

**Fig. 1** Co-UserBased algorithm

$t$th increments by 1 and implements the next Co-Training process. Co-ItemBased algorithm converges at loop cycle$t$th when $r_{ix}^{(t)} = r_{ix}^{(t-1)}$.

**Input**: Evaluation matrix $R^0 = (r_{ij}^{(0)}) = (r_{ij})$.

**Output**: Prediction matrix $R^{(t)} = \left( r_{ij}^{(t)} \right)$.

**Steps**:

    1. Initialize number of loops: $t \leftarrow 0$;

    2. Loops:

  **Repeat**

        2.1. Item-based training:

            a) Find $C_x^{(t)}, p_{xy}^{(t)}$ with (9),(10).

            b) Find $K_x^{(t)}$ using (12).

            c) Predict $r_{ix}^{(t)}$ using (11).

        2.2. User-based training:

            a) Find $S_i^{(t)}, u_{ij}^{(t)}$ using (5),(6).

            b) Find $K_i^{(t)}$ using (8).

            c) Find $r_{ix}^{(t)}$ using (7).

        2.3. Increment the loop cycle: $t \leftarrow t + 1$;

**Until** Converges.

**Fig. 2** Co-ItemBased algorithm

# 4 Experiments and Results

The efficiency of Co-Training is determined based on the capability to generate accurate users' accurate predictions. In this section, we present the referendum used to evaluate and compare the method presented with different methods.

First, the data for experiment is divided into two parts: $U_{tr}$ is the training data, $U_{te}$ is the testing data. Set $U_{tr}$ has 75% of evaluation meanwhile $U_{te}$ has 25% of evaluation. Training data is used to build a model that the algorithm presented above. To each user $i$ in the testing data set, the evaluations (including pre-available evaluations) of users are divided into two parts $O_i$ and $P_i$. $O_i$ is considered to be known. $P_i$, on the other hand is the set of evaluations that need to be predicted based on testing data and $O_i$. We calculate the Mean Absolute Error [11] for each user $u$ in the testing data.

Co-Training algorithm is tested with the data MovieLens [21]. The dataset ml-100K comprises of 100,000 evaluations of 943 users for 1682 movies. The evaluation values range from 1 to 5. The sparsity of the data is 98.7%.

Randomly step by step choose 200, 400, and 600 users from the dataset to be the training data. 200 users are randomly chosen from the rest to be the set of testing data. To test the capability of the method presented in comparison with other methods, having little data, we changed the number of evaluations of each user in the testing set so that the number of known evaluations are: 5, 10, and 20. The rest are evaluations that need predicting.

Choose $\beta = 0.8, \gamma = 4, 7, 14$ for the training data set to implement with the Co-Training model. The neighbors sets of the users or products are taken only from generation set that have similarity of 0.8, which means the two users are very similar. Based on this calculation, the prediction of accurate classifying labels for the next User-based or Item-based training process is implemented.

The MAE values in Table 4 are estimated from the average of the results in 10 random tests. The results show that both Co-Training CF methods give better result in comparison with the User-based and Item-based filtering methods. MAE values of both Co-UserBased and Co-ItemBased method are smaller in all sizes of training data sets and the number of users' evaluations. This states that the presented method improves sharply the accuracy of the results for CF.

In case of having complete data (many evaluations of users in the testing data are known), Co-UserBased and Co-ItemBased methods give similar results. However, as the size decreases, for instance when 5 or 10 users' evaluations are known, in most cases, Co-ItemBased has a smaller value of MAE, in comparison to Co-UserBased method. The main reason is the number of elements in set $C_x$ are more than that of $S_i$. This makes predicting the additional classifying labels in the User-based training process better.

**Table 4** MAE values on set ml-100K

| Size of training data | Method | Known evaluations | | |
|---|---|---|---|---|
| | | 5 | 10 | 20 |
| 200 users | User-Based | 0.732 | 0.711 | 0.645 |
| | Item-Based | 0.742 | 0.722 | 0.673 |
| | Co-UserBased | **0.621** | **0.594** | **0.512** |
| | Co-ItemBased | **0.598** | **0.572** | **0.507** |
| 400 users | User-Based | 0.694 | 0.675 | 0.644 |
| | Item-Based | 0.711 | 0.697 | 0.653 |
| | Co-UserBased | **0.615** | **0.615** | **0.587** |
| | Co-ItemBased | **0.607** | **0.607** | **0.517** |
| 600 users | User-Based | 0.693 | 0.686 | 0.686 |
| | Item-Based | 0.697 | 0.687 | 0.687 |
| | Co-UserBased | **0.548** | **0.519** | **0.511** |
| | Co-ItemBased | **0.534** | **0.524** | **0.514** |

## 5   Conclusion

The paper presents a CF algorithm using Co-Training. In the algorithm, the User-based training process adds accurate classifying labels for the Item-based training process. On the other hand, Item-based training adds accurate classifying labels for the User-based training process. Implementing these two processes instantaneously allows the adding and sharing of common classifying labels. The main advantage of this method is that the immediate classification of both users and products can use the information from the similar users or products so that the accuracy of classification for relatively small amount of data is improved. The results on the datasets of MovieLens show that the presented method gives a better result than the other two methods when data is full or sparse.

The aim for the next researches is to apply Co-Training for Content-based Filtering so that a model based on Collaborative Filtering and Content-based Filtering can be built.

## References

[1] Su, X., Khoshgoftaar, T.M.: A Survey of Collaborative Filtering Techniques. Advances in Artificial Intelligence, 1–20 (2009)
[2] Adomavicius, G., Tuzhilin, A.: Toward the Next Generation of Recommender Systems. A Survey of the State-of-the-Art and Possible Extensions. IEEE Transactions On Knowledge And Data Engineering (2005)
[3] Pan, W., Yang, Q.: Transfer learning in heterogeneous collaborative filtering domains. Artification Intelligence 197, 39–55 (2013)
[4] Pan, W., Xiang, E., Yang, N.L.: Transfer Learning in Collaborative Filtering for Sparsity Reduction. In: AAAI 2010, pp. 230–235 (2010)

[5] Pan, W., Xiang, E., Yang, Q.: Transfer Learning in CollaborativeFiltering with Uncertain Ratings. In: AAAI 2012, pp. 662–668 (2012)

[6] Wang, W., Zhou, Z.: A New Analysis of Co-Training. In: Proceedings of International Conference on Machine Learning, pp. 1135–1142 (2010)

[7] Amatriain, X., Torrens, M., Resnick, P., Zanker, M.: Incremental collaborative filtering via evolutionary co-clustering. In: RecSys. ACM (2010)

[8] Blum, A., Mitchell, T.: Combining labeled and unlabeled data with co-training. In: COLT, pp. 92–100 (1998)

[9] Xue, G., Lin, C., Yang, Q., Xi, W., Zeng, H., Yu, Y., Chen, Z.: Scalable collaborative filtering using cluster-based smoothing. In: SIGIR 2005, New York, USA, pp. 114–121 (2005)

[10] Wang, J., de Vries, A.P., Reinders, M.J.T.: Unifying user-based and item-based collaborative filtering approaches by similarity fusion. In: SIGIR 2006, pp. 501–508. ACM (2006)

[11] Herlocker, J.L., et al.: Evaluating Collaborative Filtering Recommender Systems. ACM Trans. Information Systems 22, 5–53 (2004)

[12] Breese, J.S., Heckerman, D., Kadie, C.: Empirical analysis of Predictive Algorithms for Collaborative Filtering. In: UAI 1998 (1998)

[13] Sarwar, B., Karypis, G., Konstan, J., Riedl, J.: Item-Based Collaborative Filtering Recommendation Algorithms. In: Proc. 10th Int'l WWW Conf. (2001)

[14] Phuong, N.D., Phuong, T.M.: Collaborative Filtering by Multi-Task Learning. In: RIVF 2008, pp. 227–232 (2008)

[15] Phuong, N.D., Thang, L.Q., Phuong, T.M.: A Graph-Based Method for Combining Collaborative and Content-Based Filtering. In: Ho, T.-B., Zhou, Z.-H. (eds.) PRICAI 2008. LNCS (LNAI), vol. 5351, pp. 859–869. Springer, Heidelberg (2008)

[16] GroupLens, http://www.grouplens.org/

# Fast K-Means Clustering for Very Large Datasets Based on MapReduce Combined with a New Cutting Method

Duong Van Hieu and Phayung Meesad

**Abstract.** Clustering very large datasets is a challenging problem for data mining and processing. MapReduce is considered as a powerful programming framework which significantly reduces executing time by dividing a job into several tasks and executes them in a distributed environment. K-Means which is one of the most used clustering methods and K-Means based on MapReduce is considered as an advanced solution for very large dataset clustering. However, the executing time is still an obstacle due to the increasing number of iterations when there is an increase of dataset size and number of clusters. This paper presents a new approach for reducing the number of iterations of K-Means algorithm which can be applied to very large dataset clustering. This new method can reduce up to 30 percent of iterations while maintaining up to 98 percent accuracy when tested with several very large datasets with real data type attributes. Based on the significant results from the experiments, this paper proposes a new fast K-Means clustering method for very large datasets based on MapReduce combined with a new cutting method (abbreviated to FMR.K-Means).

**Keywords:** MapReduce, K-Means, Fast K-Means for very large datasets.

## 1 Introduction

In the big data era, collected datasets become very large and complex which not only bring potentially and highly undiscovered values to business, management, scientific researches but also lead to a challenging problem for mining and processing tasks [1]. One of the most concerned issues of partitioning clustering methods is

Duong Van Hieu · Phayung Meesad
Faculty of Information Technology
King Mongkut's University of Technology North Bangkok
Bangkok 10800, Thailand
e-mail: duongvanhieu@tgu.edu.vn, pym@kmutnb.ac.th

© Springer International Publishing Switzerland 2015                    287
V.-H. Nguyen et al. (eds.), *Knowledge and Systems Engineering,*
Advances in Intelligent Systems and Computing 326, DOI: 10.1007/978-3-319-11680-8_23

coping with very large datasets. Various methods were proposed to use for dealing with very large dataset clustering such as dataset size reduction, using representative samples, parallelization, and better initial center selection [2, 3, 4]. However, these methods can not completely solve the problem "process masses of heterogeneous data within a limited time" [5]. Among the data clustering techniques, which are statistical classification techniques to discover the nature of set groupings of data which fall into different groups, K-Means has gained great interest from researchers because of its advantages in data analyzing but has limitations when working with very large datasets nowadays [6]

The K-Means algorithm has strength and performance in terms of working capability to deal with large datasets by utilizing advanced resources and network infrastructure including multicore machines, computer clusters. Parallel K-Means utilizes the available resources of multicore machines [7], and parallel K-Means based on MapReduce [8] is considered as an advanced model for boosting large dataset processing capability on distributed environments, and gained some successes in large dataset clustering [9, 10, 11]. Although parallel K-Means based on MapReduce has made a new trend referring to large dataset clustering, it is believed that this model needs to be improved to defeat the challenging problems of data intensive applications [1].

To reduce executing time of the K-Means algorithms when working with large datasets, Christopher [12] proposed to apply two termination conditions separately or combine each with a fixed number of iterations, $\alpha$, to ensure that the algorithm will be terminated in an acceptable period of time. These terminative criteria are when RSS falls below a threshold $\sigma$ and when decrease in RSS falls below a threshold $\theta$. However, which values the $\alpha$, $\sigma$ and $\theta$ should take is still a question. On the other hand, some sampling methods have been proposed including clustering using representative (CURE) [13], coresets for K-Means and K-Median [14]. These sampling methods are efficient in terms of reducing executing time. However, it is believed that the accuracy is still not significantly high due to working on proportions instead of the whole collected datasets.

Based on the problems mentioned above, this paper proposes a new approach for reducing executing time of the K-Means clustering by cutting off a number of iterations in clustering very large datasets. To be more adequate to the increasing in size of datasets, this paper also proposed a fast K-Means clustering algorithm for very large dataset clustering based on the MapReduce combined with a new iteration cutting method, called FMR.K-Means.

The contents of this paper are organized into five sections. Related work is presented in Section 2. Section 3 explains the proposed approach for cutting off iterations of the K-Means algorithm for very large dataset clustering. Experimental analysis is shown in Section 4. Finally, conclusion and future work will are covered in the last section.

# 2 Related Work

This section presents the related literature review which will be applied in this study including the K-Means algorithm, parallel K-Means based on the MapReduce programming model, and cluster validity based on external information.

## 2.1 K-Means Algorithm

The K-Means clustering is one of the most well-known clustering techniques in the machine learning discipline. This method tries to cluster a provided dataset $X$ having $N$ samples into $K$ clusters in such a way that intra-cluster similarity is high whereas inter-cluster similarity is low. Firstly, $K$ samples from the provided dataset will be selected to be initial cluster centers. Secondly, each remaining sample is assigned to the closest cluster based on the distance between the sample and the cluster center. Thirdly, the new center for each cluster is then calculated using information of samples, which belong to that cluster. This process iterates assigning samples to clusters and recalculating center of clusters until the terminate criterion is satisfied [12, 15].

The K-Means clustering is quite easy to deploy and works well with numerical data types. However, it is inadequate to very large datasets due to intensive calculation. It needs $NxKxI$ distance calculations to calculate distances between data samples to cluster centers where $N$ is number of samples in the provided dataset, $K$ is number of clusters, and $I$ is number of iterations. There will be a significantly high increase in executing time when there is an increase in dataset size or number of clusters, especially when $N$ increases highly.

## 2.2 Parallel K-Means Based on MapReduce

Parallel K-Means based on the MapReduce was proposed in [9, 16] including three functions that are (1) a map function which takes care of calculating distance from each data sample to clusters and assigns this data sample to the closest cluster, (2) a combine function which calculates local centers before sending them to the reducing function, and, (3) a reduce function which obtains local centers and calculates global centers of each cluster.

Theoretically, regardless of data transferring time, the executing time will be reduced $W$ times if this model runs on a machine with $W$ cores or a cluster having $W$ workers. However, this model can only solve the challenge in terms of "divide and conquer". The challenging problem is still there when the data size increases exponentially and the pressure from processing very large dataset within a limited time increases [5].

## 2.3 Cluster Validity

Evaluation of clustering results is one of the most important issues in cluster analysis. Cluster validity is a technical term for a procedure of evaluating the results of a clustering algorithm. Three most used approaches to investigate cluster validity are external criteria based, internal criteria based, and relative criteria based. This section focuses on the first approach because it is the best choice to evaluate results of the FMR.K-Means algorithm. Cluster validity based on external criteria which evaluates the results of a clustering algorithm based on a pre-specified structure [17, 18]. The common use of external criteria is to compare clustering structure $C$ to an independent partition of data $P$. Let

- $C=\{C_1, C_2,...,C_k\}$ be a clustering structure of a provided dataset $X$.
- $P=\{P_1, P_2, ...,P_s\}$ be a predefined partition of $X$.
- $SS$ be points belong to the same cluster of $C$ and to the same group of $P$.
- $SD$ be points belong to the same cluster of $C$ and to different group of $P$.
- $DS$ be points belong to different cluster of $C$ and to the same group of $P$.
- $DD$ be points belong to different cluster of $C$ and to different group of $P$.
- $a, b, c, d$ is value of $SS, SD, DS, DD$, respectively.
- $M$ be the maximum number of pairs in the dataset.

The degree of similarity between $C$ and $P$ can be measured by:

$$M = a+b+c+d. \tag{1}$$

1. Rand Statistic

$$R = (a+d)/M. \tag{2}$$

2. Jaccard Coefficient

$$J = a/(a+b+c). \tag{3}$$

Value of $R, J$ can be either 0 or 1 corresponding to completely different or similar. The larger value $R, J$ gets, the higher similar between $C$ and $P$.

# 3  Fast K-Means Clustering for Very Large Datasets Based on MapReduce Combined with a New Cutting Method

This section firstly states the obstacles of very large dataset clustering tasks in terms of executing time. Based on these barriers and preliminary experiment results, a new approach for cutting off last iterations will be proposed to use for high dimensional very large datasets with real data type attributes. Lastly, a fast K-Means algorithm based on the MapReduce combined with a new cutting method is proposed to use for clustering very large datasets.

## 3.1  The Obstacles of Very Large Datasets Clustering Using K-Means

It is clearly that the K-Means clustering algorithm is a well-known clustering method but two obstacles exist for very large datasets clustering. The first obstacle is computational complexity of distance calculations, which calculates distances between data samples to clusters. This difficulty can be overcome by applying the

MapReduce model to distribute computations to multiple workers in a distributed environment. However, this obstacle is still there when data size increases exponentially.

The second obstacle is the number of iterations which significantly increases when the number of sample data increases. This problem may be solved by using two-stages K-Means algorithm or K-Means++ algorithm [19]. K-Means++ consists of two steps: the first step is to select better initial centroids and the second step is the K-Means. This K-Means variant may reduce the number of iterations of K-Means by applying a new technique for selecting the better initial centroids in the initial stage. However, it also needs more time for this first step due to high computational complexity. In case the best initial centers selected, it also needs more time to execute a large number of iterations at the second stage.

To solve that problem, a new cutting method will be proposed to cut off last iterations of the K-Means clustering, which is presented in the next section.

## 3.2 A New Proposed Method for Reducing a Number of Iterations

This research deals with a problem of how to cut off a number of iterations of the K-Means clustering in order to reduce executing time while maintaining significantly high accuracy. The preliminary experiments show that the last iterations have the least contribution to the percent of correctly clustered objects and the number of iterations will significantly increases if there is a big increase in data size [12].

This section aims to propose a new method called cutting off the last iterations based on differences between centers of each cluster of two adjacent iterations. This proposed method can be explained as follows.

- Let

  - $x_i=(x_{i1}, x_{i2}, ..., x_{iP}) \in R^P$ be a sample having $P$ attributes.
  - $X-\{x_1, x_2, .., x_N\}$ be a provided dataset having $N$ samples, which need to be clustered into $K$ groups.
  - $c_i^{(j-1)}$, $c_i^{(j)} \in R^P$ be center of cluster $i$ at the $(j-1)^{th}$ and the $j^{th}$ iteration.
  - $v = (v_1, v_2, ..., v_P) \in R^P$ be a vector can be driven from $X$.

- Define a new measure called vector Norm of $X$

  - $Norm(X)=(v_1, v_2, ..., v_P)$

where

$$v_j = max(x_{ij}) - min(x_{ij}) \text{ with } j = 1,2,..P; i = 1,2,..N . \tag{4}$$

- Define a new terminative condition

  - $\Delta \leq \varepsilon$.

where

$$\Delta = \frac{|c_i^{(j)} - c_i^{(j-1)}|}{Norm(X)}. \tag{5}$$

$$\varepsilon = (\varepsilon_1, \varepsilon_2, ..., \varepsilon_P) \in R^P \ with \varepsilon_i = 0.01x2^{(-m)}. \tag{6}$$

- $m$ is a fault tolerance degree, it can be from 0 to positive infinitive. The higher the tolerance degree is used, the higher accuracy is obtained.

The K-means algorithm with the new cutting off last iteration method is depicted as Algorithm 1.

---

**Algorithm 1.** K-Means $(X, K, m)$

---

**Input:** A given dataset $X$, number of clusters $K$, fault tolerance degree $m$
**Output:** A set of centers, index of clusters that each object belongs to
  1.   Calculate fault tolerance vector $\varepsilon$
  2.   Calculate $Norm\ (X)$
  3.   Initialize $K$ centers
  4.   Assign each object to the closest cluster
  5.   Calculate new centers
  6.   Calculate vector $\Delta = \frac{|c_i^{(j)} - c_i^{(j-1)}|}{Norm(X)}$
  7.   Check stop condition $\Delta \leq \varepsilon$
  -   If the stop condition is not satisfied, update centers and repeat from step 4
  -   Otherwise, return a set of centers, index and stop

---

## 3.3 A Fast K-Means Algorithm Based on MapReduce Combined with a New Cutting Method (FMR.K-Means)

The proposed method for cutting off a number of last iterations will be integrated into the Parallel K-Means algorithm based on the MapReduce framework which was proposed in [9] to tackle very large datasets, depicted as Algorithm 5. Algorithm 2 is used to calculate distances from each object to all centers and assigns that object to the closest cluster. Algorithm 3 is devoted to local centroid calculations. It obtains inputs from map function instances and calculates local centers of clusters associated with the number of objects assigned to each cluster. Algorithm 4 is devoted to global centroid calculations. It obtains inputs from combine functions and produces global centroids. Algorithm 5 is a fast K-means based on MapReduce combined with the proposed method for cutting off a number of last iterations (abbreviated to FMR.K-Means).

## 4 Experimental Analysis

This section describes the experiment framework which is carried out by the proposed algorithm. The experiment results will be evaluated in terms of four criteria including percent of reduced iterations, percent of correctly clustered objects called accuracy, Rand statistic and Jaccard Coefficient measures.

**Algorithm 2.** Map (*<key1, value1>, global centers*)

**Input:** A list of *<key1, value1>* pairs, a list of *K global centers*. Where *key1* is position and *value1* is content of object.

**Output:** A list of *<key2, value2>*. Where *key2* is composition of *key1* and index of clusters that each object assigned to, *key2* is content of object.

1. Initialize a list of *<key2, value2>* pairs.
2. Get a pair *<key2, value2>* that has not been updated, called $x_i$
3. Calculate distances between $x_i$ to *global centers*
4. Update $x_i$
5. Check stop condition
- If exist a pair *<key2, value2>* has not been updated, repeat from step 2
- Otherwise, return a list of *<key2, value2>*, and stop.

**Algorithm 3.** Combine (*<key2, value2>*)

**Input:** A list of *<key2, value2>*.

**Output:** A list of *<key3, value3>*. Where *key3* is index of clusters, *value3* is local center associated with number of objects belong to that cluster.

1. Initialize a list of *<key3, value3>* pairs.
2. Get a pair *<key3, value3>* that has not been updated, called $x_i$
3. Calculate *value3*
4. Update $x_i$
5. Check stop condition
- If exist a pair *<key3, value3>* has not been updated, repeat from step 2
- Otherwise, return a list of *<key3, value3>*, and stop.

**Algorithm 4.** Reduce (*<key3, value3>*)

**Input:** A list of *<key3, value3>*.

**Output:** A list of *<key4, value4>*. Where *key4* is index of clusters, *value4* is global center of clusters.

1. Initialize a list of *<key4, value4>* pairs.
2. Get a pair *<key4, value4>* that has not been updated, called $x_i$
3. Calculate *value4*
4. Update $x_i$
5. Check stop condition
- If exist a pair *<key4, value4>* has not been updated, repeat from step 2
- Otherwise, return a list of *<key4, value4>*, and stop.

**Algorithm 5.** FMR.K-Means ($X$, $K$, $m$, $W$)

**Input:** A provided dataset $X$, number of clusters $K$, fault tolerance degree $m$, number of workers $W$
**Output:** A list of global centers and indexes of clusters that each object assigned to.
1.  (at master worker)
-   Calculate fault tolerance vector $\varepsilon$
-   Calculate vector $Norm(X)$
-   Initialize $K$ centers
-   Form a list of $<key1, value1>$ from $X$, called $List1$
-   Divide $List1$ into $W$ block and distribute them to $W$ workers
2.  (at all workers)
-   Run map function to produce $<key2, value2>$ pairs
-   Run combine function to produce $<key3, value3>$ pairs
3.  (at master worker)
-   Get results from all workers
-   Form a list of indexes
-   Form a list of $<key3, value3>$ pairs
-   Run reduce function to produce $<key4, value4>$ pairs are global centers
4.  (at master worker)
-   Calculate vector $\Delta = \frac{|c_i^{(j)} - c_i^{(j-1)}|}{Norm(X)}$
-   Update *global centers*
-   If $\Delta \leq \varepsilon$, return a list of indexes, a list of centers, and stop
-   Otherwise, repeat from step 2.

## 4.1 Experiment Framework

The experimentations were performed on two core and four core machines. All code was written in Matlab. Datasets used in these experiments were datasets extracted from YouTube Multiview Video Games datasets [20] and Daily and Sports Activities datasets [21] obtained from the UCI Machine Learning Repository. The basic information of datasets is illustrated in Table 1 below. Each dataset was tested at least 100 times with 9 values of fault tolerance degree m from 0 to 8 with the FMR.K-Means. Average values will be used to analyze efficiency of the new method.

## 4.2 Results Analysis

In this section, the term accuracy is defined as the ratio of number of samples that are correctly clustered compared to the results without using fault tolerant variable and total number of samples of the provided dataset. Firstly, experimental results from datasets extracted from the Daily and Sport Activity database, named Dataset1, Dataset2, Dataset3, Dataset4 which are listed in Table 1 show that the average number of iterations increases considerably when there is an increase in the number of data samples, illustrated in Fig 1. The increasing in number of iterations will lead to increasing in executing time when there is an increase in number of samples in

**Table 1** Information about experiment datasets

| Dataset | Size | No. of Samples | No. of Attributes | No. of Clusters |
|---|---|---|---|---|
| vision_misc | 706 MB | 97,935 | 838 | 31 |
| vision_hist_motion_estimate | 706 MB | 97,935 | 838 | 31 |
| vision_hog_features | 706 MB | 97,935 | 838 | 31 |
| Dataset1 | 128 MB | 285,000 | 45 | 3 |
| Dataset2 | 257 MB | 570,000 | 45 | 3 |
| Dataset3 | 467 MB | 1,140,000 | 45 | 3 |
| Dataset4 | 934 MB | 2,280,000 | 45 | 3 |

datasets. Moreover, when applying this new proposed cutting method to those four datasets, results also show that the higher fault tolerant degree gained the higher accuracy with lower reduction of iterations which is illustrated in Fig 2.

Secondly, experiment results obtained from the testing of three datasets referring to YouTube Multiview Video Game show when the larger size dataset has the smaller fault tolerant degree this should be selected, as illustrated in Fig 3.

For cluster validity, Rand Statistic and Jaccard Coefficient measures are used to show that the results of the proposed method are highly similar to the results of K-means clustering without using fault tolerant variable. In this case, the predefined partition $P=\{P_1, P_2, \ldots, P_s\}$ is a clustered result without using fault tolerant degree m, or the algorithm will stop iterating when there is no change in centroids. The clustering structure $C=\{C_1, C_2, \ldots, C_k\}$ which is used to compared P is the clustered result with using m. Tables 2 and 3 show percentage of reduced iterations, accuracy; Rand Statistic and Jaccard Coefficient measure values.

**Table 2** Results tested with Daily and Sport Activities databases

| Dataset | No. of Clusters | Fault tolerance degree (m) | Reduced Iterations | Accuracy | Rand Statistic Measure | Jaccard Coefficient Measure |
|---|---|---|---|---|---|---|
| Dataset1 | 3 | 7 | 42.49% | 99.09% | 0.9912 | 0.9909 |
|  |  | 8 | 28.48% | 99.77% | 0.9977 | 0.9977 |
| Dataset2 | 3 | 7 | 42.42% | 93.44% | 0.9452 | 0.9344 |
|  |  | 8 | 25.59% | 99.79% | 0.9978 | 0.9978 |
| Dataset3 | 3 | 7 | 63.29% | 95.79% | 0.9637 | 0.9579 |
|  |  | 8 | 47.78% | 99.43% | 0.9943 | 0.9942 |
| Dataset4 | 3 | 7 | 67.04% | 99.06% | 0.9907 | 0.9906 |
|  |  | 8 | 63.07% | 99.43% | 0.9944 | 0.9944 |

Note: m is fault tolerance degree, R measure is Rand statistic measure, and J measure is Jaccard coefficient measure.

**Table 3** Results tested with YouTube Multiview Video Games database

| Dataset | No. of Clusters | m | Reduced Iterations | Accuracy | R Measure | J Measure |
|---|---|---|---|---|---|---|
| vision_misc | 31 | 4 | 12.27% | 99.29% | 0.9983 | 0.9983 |
| | | 5 | 3.47% | 99.93% | 0.9998 | 0.9998 |
| vision_hist_motion_estimate | 31 | 4 | 22.50% | 97.70% | 0.9962 | 0.9953 |
| | | 5 | 8.94% | 99.83% | 0.9996 | 0.9996 |
| vision_hog_features | 31 | 4 | 60.94% | 88.66% | 0.9865 | 0.9838 |
| | | 5 | 41.49% | 96.19% | 0.9957 | 0.9954 |

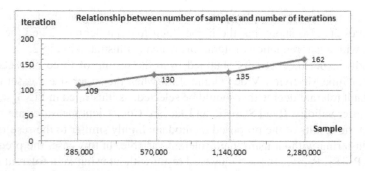

**Fig. 1** Relationship between number of samples and number of iterations

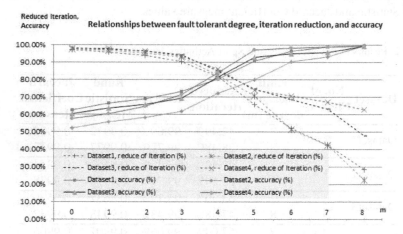

**Fig. 2** Relationship between fault tolerant degree, reduced iteration, and accuracy tested with four datasets extracted from Daily and Sport Activities datasets

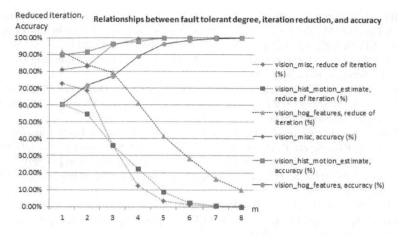

**Fig. 3** Relationship between fault tolerant degree, reduced iteration, accuracy tested with three datasets from YouTube Multiview Video Games database

## 5 Conclusion and Future Work

The experiment results tested with four datasets extracted from Daily and Sport Activities and results tested from three datasets extracted from YouTube Muliview Video Game show that when m=8 and m=4 are selected, respectively, the proposed method can reduce up to 30% of iterations, which is equivalent to 30% of executing time, while maintaining up to 98% accuracy. It can be concluded that this proposed method for cutting off iterations can reduce considerably the number of iterations of K-Means algorithm while maintaining high accuracy. This method of integration with the MapReduce programming model is an appropriate choice for very large dataset clustering jobs. Compared to the previous methods such as clustering using representative (CURE)[8], coresets for K-means and K-median[9], this new method can be considered as more significant because it can work with entire datasets, significantly reduce executing time, and provide high accuracy. For future work, to estimate fault tolerant degrees automatically based on the size of datasets such as number of samples, number of attributes, more experiments need to be carried out with numerous datasets.

## References

[1] Philip Chen, C.L., Zhang, C.-Y.: Data-intensive applications, challenges, techniques and technologies: A survey on Big Data. Information Sciences (in press, 2014)

[2] Barioni, M.C.N., Razente, H., Marcelino, A.M.R., Traina, A.J.M., Traina, C.: Open issues for partitioning clustering methods: an overview. Wiley Interdisciplinary Reviews: Data Mining and Knowledge Discovery 4, 161–177 (2014)

[3] Hadian, A., Shahrivari, S.: High performance parallel k-means clustering for disk-resident datasets on multi-core CPUs. The Journal of Supercomputing, 1–19 (2014)

[4] Bharill, N., Tiwari, A.: Handling Big Data with Fuzzy Based Classification Approach. In: Jamshidi, M., Kreinovich, V., Kacprzyk, J. (eds.) Advance Trends in Soft Computing. STUDFUZZ, vol. 312, pp. 219–227. Springer, Heidelberg (2014)

[5] Chen, M., Mao, S., Zhang, Y., Leung, V.M.: Chapter 1. Introduction. In: Big Data, pp. 1–10. Springer, Heidelberg (2014)

[6] Jain, A.K.: Data Clustering: 50 Years Beyond K-means. In: Daelemans, W., Goethals, B., Morik, K. (eds.) ECML PKDD 2008, Part I. LNCS (LNAI), vol. 5211, pp. 3–4. Springer, Heidelberg (2008)

[7] Stoffel, K., Belkoniene, A.: Parallel k/h-Means Clustering for Large Data Sets. In: Amestoy, P.R., Berger, P., Daydé, M., Duff, I.S., Frayssé, V., Giraud, L., Ruiz, D. (eds.) Euro-Par 1999. LNCS, vol. 1685, pp. 1451–1454. Springer, Heidelberg (1999)

[8] Dean, J., Ghemawat, S.: MapReduce: simplified data processing on large clusters. Commun. ACM 51, 107–113 (2008)

[9] Zhao, W., Ma, H., He, Q.: Parallel K-Means Clustering Based on MapReduce. In: Jaatun, M.G., Zhao, G., Rong, C. (eds.) Cloud Computing. LNCS, vol. 5931, pp. 674–679. Springer, Heidelberg (2009)

[10] Lin, C., Yang, Y., Rutayisire, T.: A Parallel Cop-Kmeans Clustering Algorithm Based on MapReduce Framework. In: Wang, Y., Li, T. (eds.) Knowledge Engineering and Management. AISC, vol. 123, pp. 93–102. Springer, Heidelberg (2011)

[11] Lv, Z., Hu, Y., Zhong, H., Wu, J., Li, B., Zhao, H.: Parallel K-means clustering of remote sensing images based on mapReduce. In: Wang, F.L., Gong, Z., Luo, X., Lei, J. (eds.) Web Information Systems and Mining. LNCS, vol. 6318, pp. 162–170. Springer, Heidelberg (2010)

[12] Manning, C.D., Raghavan, P., Schütze, H.: K-Means. In: An Introduction to Information Retrieval. Cambridge University Press (2009)

[13] Guha, S., Rastogi, R., Shim, K.: CURE: an efficient clustering algorithm for large databases. SIGMOD Rec. 27, 73–84 (1998)

[14] Har-Peled, S., Mazumdar, S.: On coresets for k-means and k-median clustering. Presented at the Proceedings of the Thirty-Sixth Annual ACM Symposium on Theory of Computing, Chicago, IL, USA (2004)

[15] Jain, A.K., Dubes, R.C.: Chapter 3. Clustering Methods and Algorithms. In: Algorithms for Data Clustering, vol. Computer Science. Prentice Hall (1988)

[16] Anchalia, P.P., Koundinya, A.K., Srinath, N.K.: MapReduce Design of K-Means Clustering Algorithm. In: 2013 International Conference on Information Science and Applications (ICISA), pp. 1–5 (2013)

[17] Dom, B.E.: An Information-Theoretic External Cluster-Validity Measure. In: The Eighteenth Conference on Uncertainty in Artificial Intelligence (UAI 2002), Alberta, Canada, pp. 137–145 (2012)

[18] Wagner, S., Wagner, D.: Comparing Clusterings - An Overview. Institute of Theoretical Informatics (2007)

[19] Xu, Y., Qu, W., Li, Z., Min, G., Li, K., Liu, Z.: Efficient k-means++ Approximation with MapReduce. IEEE Transactions on Parallel and Distributed Systems PP, 1–10 (2014)

[20] UCI. YouTube Multiview Video Games Dataset, http://archive.ics.uci.edu/ml/datasets/YouTube+Multiview+Video+Games+Dataset

[21] UCI. Daily and Sports Activities, http://archive.ics.uci.edu/ml/datasets/Daily+and+Sports+Activities

# Neural Networks with Hidden Markov Models in Skeleton-Based Gesture Recognition

Hai-Son Le, Ngoc-Quan Pham, and Duc-Dung Nguyen

**Abstract.** In Gesture Recognition (GR) tasks, a system with a traditional use of Hidden Markov Models (HMMs) usually serves as a baseline. Their performance is often not so good and therefore somehow overlooked. However, in recent years, especially in Automatic Speech Recognition (ASR), there are advanced methods proposed for this type of model which have been shown to improve significantly recognition results. Among them, the use of Neural Networks (NNs) instead of Gaussian Mixture Models (GMMs) for estimating emission probabilities of HMMs has been considered as one of biggest advances [1, 2, 3]. This fact implies that the performance of HMM-based models on GR need to be revised. For this reason, in this study, we show that by carefully tailoring NNs to a traditional HMM-based GR system, we can improve significantly the performance, hence, achieving very competitive results on a skeleton-based GR task which is defined by using Microsoft Research Cambridge 12 (MSRC-12) data [4]. It should be pointed out that, it is straightforward to apply our proposed techniques to more complicated GR tasks such as Sign Language Recognition [5], where basically a sequence of sign gestures need to be transcribed.

**Keywords:** Gesture Recognition, Skeleton-based, Kinect, Hidden Markov Model, Neural Network, Hybrid system, Linear Discriminant Analysis.

## 1 Introduction

Touchless interaction plays an important role in Human Machine Communication (HMC). Thanks to the ability to provide a friendly user experience, touchless movement interface was applied in various fields, such as ambient environment,

Hai-Son Le · Ngoc-Quan Pham · Duc-Dung Nguyen
Institute of Information Technology
Vietnam Academy of Science and Technology,
Hanoi, Vietnam
e-mail: {lehaison,quanpn,nddung}@ioit.ac.vn

© Springer International Publishing Switzerland 2015          299
V.-H. Nguyen et al. (eds.), *Knowledge and Systems Engineering*,
Advances in Intelligent Systems and Computing 326, DOI: 10.1007/978-3-319-11680-8_24

or interactive games. The appearance of Kinect in 2010, a gaming device that utilize touchless movement, is considered as an important breakthrough in HMC as Kinect provides mutli-channel information (audio, RGB image, depth image and skeleton joints). Among them, the skeleton joints (points of a human body) seems to have a more compact representation of a human body movement, hence while being taken into account, usually improving significantly the performance of the system. Since its appearance, the number of related studies to Kinect for Gesture Recognition (GR) is constantly increased. The approaches in GR are often induced from various studies in Computer Vision, Automatic Speech Recognition (ASR), Machine Learning and Statistical Methods. They are many but can be grouped into two families: Template Matching-based and State-Space-based approaches. Hidden Markov Models (HMMs), a subject of our study, follow the second approach. Regardless of the differences between types of features, it is quite easy to adapt already available methods which use traditional features induced from a RGB image to a set of skeleton joints. However, one problem arises as it is evident that different features should be processed differently.

In the literature for GR, the performance of standard HMM-based systems is often subpar, hence, it only serves as a baseline. However, in recent years, notably in ASR, many advanced methods have been shown to have significant achievement. One of biggest advances is the use of Neural Networks (NNs) in place of Gaussian Mixture Models (GMMs) [1, 2, 3]. In the conventional system, GMMs are used to represent the relationship between HMM states and the input feature. With enough number of mixtures, they can model probability distributions to any required level of accuracy. But GMMs have serious drawbacks. The first one is that they are statistically inefficient for modeling data that lie on or near a nonlinear manifold in the data space [3]. The second one is that a small increase in the size of feature would drastically increase the number of parameters. This problem makes GMMs subject to the estimation problem, which may cause performance degradation when the dimensionality is high. In both cases, NNs are proved to work better. However, we were not able to prove it practically due to their expensive computational cost. Only recently, thanks to the significant progresses in both training algorithms and computer hardware, many studies have shown that NNs with many layers can outperform GMMs for ASR on a variety of data sets including large data sets with large vocabularies. Therefore, our goal in this article is to show that NNs can actually improve significantly the performance, hence, achieving state-of-the-art results on a skeleton-based GR task defined by using MSRC-12 data [4].

The remainder of this paper is organized as follows. Section 2 is for related works. Section 3 briefly shows the conventional HMM framework. Sections 4 explains the application of NNs to this framework. Section 5 is for the experimental setup, results and the following evaluations. Section 6 concludes this paper.

## 2   Related Work

For the Template Matching-based approach, feature extraction plays an important role. The common way is to construct a fix length feature from all of the individual features of all frames as for example, in [4]. In [6], Fourier Temporal Pyramids was used. Recently, in [7], features based on covariance of human body points were shown to achieve the best performance. The most notable shortcoming of the first family is that it is not straightforward to extend their discriminative power for using in a more general framework, where sequences of gestures need to be recognized. The reason is that they need a priori a segmentation (a boundary) for each gesture, which must be identified earlier from a Gesture Spotting (GS) framework.

HMM, Conditional Random Field (CRF) follow the State-Space-based approach. Because of their natural capacity to deal with the temporal variability problem, this problem seems to be less important. For instance, in [8], the structure of HMMs learned to do GR was modified to construct filler and gesture completion HMM models in order to do GS and GR at the same time. The dominating use of HMMs in ASR, where sentences (sequence of words) need to be recognized, make this difference so clear. Therefore, they are more suitable for GR and have been progressively studied. In [9], for an isolated 10 gesture recognition task, HMMs were used with a feature of 13 points of a human body. To reduce the complexity, data-dependent statistics and relative entropy were applied to decrease the number of states [9, 10]. To improve the performance, in [11], a global parametric variation was included in the output probabilities of the states of HMM. The model can handle arbitrary smooth dependencies because the output densities were defined as a function of the parameter vector.

In [12], with a large set of experiments, it is shown that CRFs typically outperform HMMs. In [10], the extension of CRFs with the increase of the weights of self-transition feature functions was proposed, achieving a significant improvement. A new model, called Hidden CRF (HCRF), was proposed in [13]. This type of model was constructed based on the introduction of hidden states into CRF. From the experimental result, HCRF is the best while HMM and CRF achieve comparable performance. To tackle the complexity problem, the authors in [14] proposed the Semi-Markov CRF. Compared to HMMs, the advantage of CRFs is that they allow arbitrary dependencies on the observation sequence. However, due to this characteristic, the class of CRFs is much more expensive, leading to generally lower recognition accuracy.

## 3   Hidden Markov Model for Gesture Recognition

HMM is a graphical model for objects that have the temporal variability structure. Supposing that an object (a gesture) is first considered as a sequence of observations (or frames) with constant interval between them. As the duration of a gesture is variable, each sequence has in general a different number of observations. Each observation, in its turn, is represented as a feature (a vector). For example, in Fig. 1,

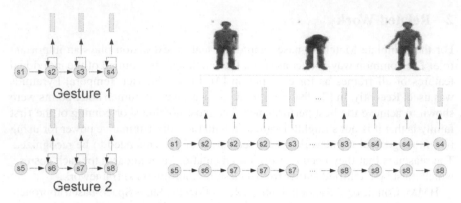

**Fig. 1** Examples of HMMs for GR. Note that, the human gesture is taken from [4]

observations are represented as blue rectangles. Each feature is computed from raw data signal in a way depending on the nature and the particular characteristics of data signal. For Skeleton-based GR, it could be based on the coordinates of 3D skeleton joints, the angles between these joints.

Structure

We use $O_1^T = O_1, O_2, \ldots, O_T$ to denote a sequence of observations. An observation is usually a continuous vector that can have any values in a high dimensional continuous space $O_t \in \mathcal{R}^d$. A Markov process generates this sequence. It means that for each observation $O_t$, a HMM assigns a hidden state $q_t$ whose value is in a set of $N$ states: $S = \{s_1, s_2, \ldots, s_N\}$. We say that $q_t$ generates $O_t$ according to a distribution. In a simple case, it is a multivariate Gaussian (or normal) distribution:

$$p(O_t = o | q_t = s_n) = P(O_t = o | \mu_n, \Sigma_n^2) = N(o; \mu_n, \Sigma_n)$$
$$= \frac{1}{(2\pi)^{d/2} |\Sigma_n|^2} \exp\left(-\frac{1}{2}(o - \mu_n)^T \Sigma_n^{-1}(o - \mu_n)\right) \quad (1)$$

In fact, emission probabilities of HMMs are often estimated by using more complicated models, called Gaussian Mixtures Models (GMM) [15]. If we use $M$ mixtures of Gaussians, the emission probabilities are computed as follows:

$$P(O_t = o | q_t = s_n) = \sum_{j=1}^{M} W_k(j) N(o; \mu_{n,j}, \Sigma_{n,j}) \quad (2)$$

So each state $s_n$ has a set of parameters: $\{W_n(j), \mu_{n,j}, \Sigma_{n,j}, 1 \leq j \leq M\}$. We use $B$ to represent a set of all parameters used to estimate emission probabilities. Now, a HMM can be represented as a set of parameters $\lambda = (A, B, \pi)$:

1. The initial state distribution $\pi = \{\pi_n\}$. $\pi_n = P(q_1 = s_n)$ is the probability that a Markov process starts at state $s_i$.

2. The state transition probabilities $A = \{a_{mn}\}$. $a_{mn} = P(q_{t+1} = s_n | q_t = s_m)$ is the probability for the transition from state $s_i$ to state $s_j$.
3. The emission probabilities: $P(O_t = o | q_t = s_n) = \sum_{j=1}^{M} W_k(j)N(o; \mu_{n,j}, \Sigma_{n,j})$: the probability that state $s_n$ generates an observation $o$.

There are three main problems for HMMs:

1. Evaluation problem: How to compute the probability of an observation sequence given the model.
2. Decoding problem: How to determine the best sequence of states that generates an observation sequence.
3. Estimation problem: How to learn parameters of models from the training data.

To show how HMMs work, we present briefly how to solve these problems. For the first one, we first need to define forward and backward variables $\alpha, \beta$:

$$\alpha_t(n) = P(O_1, O_2, \ldots, O_t, s_n | \lambda) \tag{3}$$

$$\beta_t(n) = P(O_{t+1}, O_{t+2}, \ldots, O_T, s_n | \lambda) \tag{4}$$

Note that, these forward and backward variables can be efficiently computed by using recursive equations with dynamic programming. We find that:

$$\alpha_t(n).\beta_t(n) = P(O, s_n | \lambda) \tag{5}$$

Now, it turns out that for the first problem, mathematically, we need to calculate $P(O|\lambda)$ which can be done as follows:

$$P(O|\lambda) = \sum_{n=1}^{N} \alpha_t(n).\beta_t(n), \text{any } t \tag{6}$$

The second problem aims to explain the best state sequence generating the observations $O_1^T$ through a model $\lambda = (A, B, \pi)$ with maximum likelihood. To achieve this goal, Viterbi algorithm is employed [16, 17]. We need to compute $\delta$ using the following recursive equation:

$$\delta_{t+1}(n) = \max_m [\delta_t(m).A_{mn}].P(O_{t+1}|s_n) \tag{7}$$

In addition, we use $\phi$s with argmax operator to keep track of the best path:

$$\phi_{t+1}(n) = \arg\max_m [\delta_t(m).A_{mn}] \tag{8}$$

Having all $\phi$ values, the best path $q_1^*, \ldots, q_T^*$ is reconstructed from $T$:

$$q_t^* = \phi_{t+1}(q_{t+1}^*) \tag{9}$$

To train a HMM (solve the estimation problem), a traditional way is to adjust its model parameters to optimize with maximum likelihood $P(O|\lambda)$. For this purpose, the Baum-Welch (BW), also known as Forward-Backward, an instance of a generalized Expectation Maximization (EM) algorithm, is used. Basically, it is an iterative algorithm, alternating between an Expectation (E) step and a Maximization (M) step. At E-step, using the current model parameters, the distribution of the latent variables (related to hidden states) is determined, the expectation of the log-likelihood is evaluated. At M-step, model parameters are reestimated so as to maximize the expected log-likelihood computed from the previous E-step. For further detail on HMMs, see, for example, [18, 19].

Application to Isolated Gesture Recognition

Now, supposing that for each gesture, we define a structure (or topology) of HMMs as on the left of Fig. 1. It is called left-to-right structure as we accept only transitions from a state to itself or to any state which is on its right. This topology is usually used in GR and ASR. Each gesture has one initial state and a (can be varied) number of emission states (3 in this example). To this end, we have a set of HMMs, one for each gesture. For the decoding problem, i.e., to find the most likely gesture given an observation sequence (the top right part of Fig. 1), we carry out a Viterbi algorithm for each HMM to obtain their best paths (the bottom right part of Fig. 1) and their scores. A gesture whose model gives a best score will be considered as a solution. For the training problem, supposing that we have a training data which contains gestures instances and their labels, it is straightforward to apply BW algorithm to estimate parameters for each model.

## 4  Neural Network with Hidden Markov Model

Instead of using GMMs, we can use Neural Networks (NNs) to estimate the emission probabilities. This type of system is often called a hybrid NN-HMM system. A standard HMM system is constructed, then GMMs are replaced by NNs while all other parameters are kept the same. More concretely, we first carry out a Viterbi alignment algorithm presented just earlier with a baseline system to obtain a label (a HMM state index) for each observation. After that, we consider a problem of multi-classification, i.e., how to classify observations into $N$ classes, which are states of HMMs. Therefore, NNs are used to estimate $P(s_i|o)$, a probability of a state given an observation. However, in the HMM framework, we require $P(o|s_i)$. Fortunately, thanks to a Bayes equation:

$$P(o|s_i) = \frac{P(s_i|o).P(o)}{P(s_i)}, \qquad (10)$$

the output of the neural network can be converted into the scaled likelihood by dividing them by the frequencies of the HMM states in the training data [20]. In this way, all of the likelihoods are scaled by the same unknown factor of $P(o)$, but this has no effect on the alignment and on the decoding algorithm.

Structure

A neural network is normally feed-forward with one or more hidden layers between its inputs and outputs. Here, a neural network takes as input an observation vector and output a probability for each state $P(s_i|o)$.

The structure of a neural network with one hidden layer consists of three layers. The first one is called the input layer which is $o$, a feature used to represent an observation. The second one is a hidden layer that introduces a nonlinear transform, where the output layer activation values are defined by:

$$h = \text{sigm} \left( W^h o + b^h \right), \tag{11}$$

where $o$ is the input, $H$ is the hidden layer size, $W^h \in \mathbb{R}^{H \times d}$ and $b^h \in \mathbb{R}^H$ are the weights and biases of this layer. We use here a sigmoid function: $\text{sigm}(x) = \frac{1}{1+\exp(-x)}$. In general, it could be any function with a well-behaved derivative.

The last one is an output layer that consists of $N$ nodes, each node is associated with one state of HMMs. Its activation values are computed as follows:

$$c = W^c h + b^c, \tag{12}$$

where $W^c \in \mathbb{R}^{N \times H}$ and $b^c \in \mathbb{R}^N$ are the weights and the biases of this layer.

Then we can estimate the desired probability, thanks to the *softmax* function:

$$p_i = \frac{\exp(c_i)}{\sum_j \exp(c_j)} \tag{13}$$

The $i^{\text{th}}$ component of $p$ corresponds to the estimated probability of the $i^{\text{th}}$ state of HMMs given an observation $o$.

Training Algorithm

NNs can be discriminatively trained by backpropagating derivatives of a cost function that measures the discrepancy between the target outputs and the actual outputs produced for each training case. The natural cost function $C$ is the cross entropy between the target probabilities $t$ and the outputs of the softmax, $p$.

$$C = -\sum_i t_i \log p_i, \tag{14}$$

where the target probabilities, taking values of one (for a desired output class) or zero (for the others), are the supervised information provided to train the NN. For large training set, it is typically more efficient to use a Stochastic Gradient Descent method, i.e., computing the derivatives on a small "minibatch" of training cases, rather than the whole training set, before updating the parameters. To prevent overfitting, validation data should be used, i.e., we stop training while the performance on the validation data degrades. Detailed techniques for training NNs can be found in [21].

Feature Processing

The feature extracted or the representation of an observation at time $t$ is not only a raw data extracted at this time. To achieve better accuracy, we need to take into account context information. Therefore, it is necessary to concatenate raw data from time $t - w$ to time $t + w$ to form a feature[1]. For this reason, the dimensionality of a feature is often large, e.g., 1170 in our experiment. Two problems arise. The first one is the estimation problem. In the case with HMMs using GMMs, the number of parameters would drastically increased with even a small increase in the dimensionality of a feature, which may cause performance degradation. So with GMMs, the best choice for input dimension in ASR is widely believed to be about 40. For the hybrid system with NNs, this problem seems to be less important as the number of parameters doesn't change much [22]. The second one concerns the correlation between elements of features. One remedy is to use Linear Discriminant Analysis (LDA) in a similar way as, for example, in [23] for ASR. LDA is a technique in statistical pattern classification, which is based on a linear transformation of a feature vector for improving discrimination (doing decorrelation) and compressing the information contents (doing dimension reduction). It acts as follows.

Again, supposing that we have good forced alignment labels from a baseline system and considering a problem of how to classify observations into $N$ classes (HMM states). LDA aims to find a linear transformation of feature vectors $o$ in a $d$-dimensional space to vectors $o'$ in an $e$-dimensional space ($e < d$) such that the class separability is maximum [24]. To formulate the problem, we need to define the within-class scatter matrix $M_w$ and between-class scatter matrix $M_b$:

$$M_w = \sum_{i=1}^{S} M_i = \sum_{i=1}^{S} \sum_{o \in s_i} (o - \mu_i)(o - \mu_i)^T, \tag{15}$$

where $\mu_i = \frac{1}{|s_i|} \sum_{o \in s_i} o$, $|s_i|$ stands for the cardinality of $s_i$; and

$$M_b = \sum_{i=1}^{S} |s_i|(\mu_i - \mu)(\mu_i - \mu)^T, \tag{16}$$

where $\mu = \frac{1}{\sum_{i=1}^{N} |s_i|} \sum_{\forall o} o$.

Now, in a similar way, we have two matrices $M'_w, M'_b$ for $o'$ which is the projection of $o$ onto the subspace using a projection matrix $P$ ($o' = P^T o$). It is easy to prove that:

$$M'_b = P^T M_b P; M'_w = P^T M_w P \tag{17}$$

Note that, we are looking for a projection that after projecting, maximizes the class separability. One way to do that is to maximize the ratio of between-class to within-class scatter:

$$J(P) = \frac{|M'_b|}{|M'_w|} = \frac{|P^T M_b P|}{|P^T M_w P|} \tag{18}$$

---

[1] $w$ is called a window size.

It turns out that the optimal projection matrix $P^*$ contains columns, which are the eigenvectors $w_i$ corresponding to the largest eigenvalues $\varepsilon_i$ of the generalized eigenvalue problem: $(M_b - \varepsilon_i M_w)w_i = 0$.

After having this matrix, we project feature vectors of an original system, obtaining new vectors, which can be considered as observations for a new HMM-based system. Then we can use GMMs or NNs for emission probabilities as usual. Now, features have a dimension reduced and are decorrelated.

## 5  Experiment

Experimental Setup

The dataset used in our paper is MSRC-12 [4]. This dataset contains twelve actions which are performed by 30 people, with a total of 6244 samples. Each sequence in the dataset is an action repetitively performed by an actor. Due to some errors in the data, we had to remove 40 samples, hence only 6204 samples were used in the experiments. To have the boundary for each gesture, we exploit the manual annotation from [7].

The Matlab source codes provided with the dataset enable us to extract a raw feature for each frame: 60 coordinates for 20 skeleton joints; 60 velocities at the skeleton joints: the differences of joints between the adjacent frames; 35 angles: the angle between the segments on either side of the joints; 35 joint angular velocities: the difference of angles between the adjacent frames. Similar to [4], we use joint velocities, angles and angle velocities to have feature of 130 dimension for each frame. In the baseline, the representation of an observation at time $t$ is a concatenation of data taken from time $t - 4$ to time $t + 4$. HMM topology is a left-to-right models with 6 emission states for each gesture. For all experiments, we follow leave-person-out approach, which means that we carry out 30 experiments. For experiment $i$, we choose Person $i$ as a test data, Person $i + 1^2$ as a validation data and the remaining part as a training data. Note that, we use validation data not only for selecting a best system for each configuration but also for optimizing hyper-parameters such as number of emission states, number of Gaussians, dimension of LDA or to do early stopping training strategy for NNs. Test results are only used for final comparisons between configurations. All result numbers below are the average of the F-measure[3] over 30 leave-person-out experiments. For implementation, we adapt a powerful ASR Kaldi toolkit [25].

---

[2] The only exception is for Person 30 where we use Person 1.

[3] Calculate F-measure for each type of gesture, and find their average, weighted by the number of true instances for each type of gesture.

Results

For the first experiment, our goal is to build a baseline system. The total number of Gaussians is optimized. From Table 1, we find that the best number is 500 with the result on test data is 79.3%.

**Table 1**   Results with numbers of Gaussians, in F-measure (%)

| Number of Gaussians | Valid | Test |
|---|---|---|
| 200 | 78.0 | 78.5 |
| 500 | **78.6** | **79.3** |
| 1000 | 77.9 | 78.0 |
| 5000 | 74.6 | 74.4 |

For the second experiment, after having a forced alignment from the best baseline system, we train LDA parameters. The number of classes is $72(12 \times 6)$ because we have for each gesture model, 6 emission states. In Table 2, two columns 30-LDA and 60-LDA are for two cases with different choices of a reduction dimension after LDA. We find that with the optimum number of Gaussians, F-measure is increased roundly 5% in both cases and 30-LDA is slightly better.

**Table 2**   Results with LDA, in F-measure (%)

| Number of Gaussians | 30-LDA | | 60-LDA | |
|---|---|---|---|---|
| | Valid | Test | Valid | Test |
| 200 | 82.9 | 83.1 | 80.8 | 80.2 |
| 500 | 84.4 | 85.3 | 82.2 | 81.8 |
| 1000 | **85.9** | **85.4** | 83.1 | 83.8 |
| 5000 | 84.9 | **85.4** | 83.6 | **84.3** |

For the third experiment, after having a forced alignment from the best LDA system, we train NNs. The number of output classes is 72. We use NNs with 2 hidden layers of 1200. The input is 1170 (Raw) or 72 (LDA[4]). The first two lines of Table 3 shows that in both cases, the performance increases. The best achievement is with LDA, which is around 4% as compared to the previous LDA system (85.4% and 89.1%) or 9% as compared to the baseline (79.3% and 89.1%).

For the final experiment, for each gesture instance, the raw data is first normalized by subtracting their mean value. Note that, this normalization, called per-utterance normalization, is often used in ASR. After preprocessing raw feature, we rebuild new systems following the same scheme as above: constructing a baseline, applying

---

[4] After LDA dimension reduction.

**Table 3** Results with Neural Network, in F-measure (%)

| Input feature | Normalization | Valid | Test |
|---------------|---------------|-------|------|
| Raw | no | 86.2 | 87.2 |
| LDA | no | 88.3 | 89.1 |
| Raw | yes | 92.0 | 92.1 |
| LDA | yes | **92.3** | **92.5** |

LDA, training NNs. Comparing two top lines and two bottom lines of Table 3, we find that the F-measure is increased further, from 89.1% to 92.5% when LDA is used. However, the difference between the use of Raw or LDA feature is smaller than in the previous experiment.

**Table 4** Summary results, in F-measure (%)

| System | Baseline | | + Normalization | |
|--------|----------|------|-------|------|
| | Valid | Test | Valid | Test |
| Baseline | 78.6 | 79.3 | 87.8 | 87.8 |
| Baseline + LDA | 85.9 | 85.4 | 90.2 | 89.9 |
| Baseline + LDA + NNs | 88.3 | 89.1 | **92.3** | **92.5** |

In total, Table 4 indicates that by using recent advanced techniques induced from ASR, we increase the performance 13.1% (from 78.6% to 92.5%), achieving 16% relative improvement. It is a very competitive result as applying the state-of-the-art methods based on co-variance feature described in [7] to our experimental setup, the best result is 92.9%.

# 6 Conclusion

In this article, we have investigated deeply the performance of a HMM-based system for GR. We found that by using NNs instead of GMMs to estimate emission probabilities and by using LDA to reduce the dimension of an observation representation, we can achieve very competitive results on an isolated GR task well defined on MSRC-12 data [4]. Compared to other approaches for GR, the main advantage of the HMM-based approach is its scale up capacity which has been proved for Sign Language Recognition [5]. More concretely, it is the capacity to deal with the increase in the number of gestures, in the complexity of gesture, the necessity of sequence gesture recognitions, the necessity of introducing other models such as a language model. Verifying the improvement of our proposal methods on more complicated tasks as such is our work in the future.

**Acknowledgments.** This work was supported in part by the National Key Lab for Networking, Technology and Multimedia, Institute of Information Technology, Vietnamese Academy of Science and Technology.

# References

[1] Seide, F., Li, G., Yu, D.: Conversational speech transcription using context-dependent deep neural networks. In: INTERSPEECH 2011, pp. 437–440 (2011)

[2] Mohamed, A., Dahl, G., Hinton, G.: Acoustic modeling using deep belief networks. IEEE Transactions on Audio, Speech, and Language Processing 20(1), 14–22 (2012)

[3] Hinton, G.E., Deng, L., Yu, D., Dahl, G.E., Rahman Mohamed, A., Jaitly, N., Senior, A., Vanhoucke, V., Nguyen, P., Sainath, T.N., Kingsbury, B.: Deep neural networks for acoustic modeling in speech recognition: The shared views of four research groups. IEEE Signal Process. Mag. 29(6), 82–97 (2012)

[4] Fothergill, S., Mentis, H.M., Kohli, P., Nowozin, S.: Instructing people for training gestural interactive systems. In: Konstan, C.J.A., Chi, E.H., Höök, K. (eds.) CHI, pp. 1737–1746. ACM (2012)

[5] Forster, J., Koller, O., Oberdörfer, C., Gweth, Y., Ney, H.: Improving continuous sign language recognition: Speech recognition techniques and system design. In: Proceedings of the Fourth Workshop on Speech and Language Processing for Assistive Technologies, pp. 41–46. Association for Computational Linguistics (2013)

[6] Wang, J., Liu, Z., Wu, Y., Yuan, J.: Mining actionlet ensemble for action recognition with depth cameras. In: 2012 IEEE Conference on Computer Vision and Pattern Recognition (CVPR), pp. 1290–1297 (June 2012)

[7] Hussein, M.E., Torki, M., Gowayyed, M.A., El-Saban, M.: Human action recognition using a temporal hierarchy of covariance descriptors on 3d joint locations. In: Proceedings of the Twenty-Third IJCAI, pp. 2466–2472. AAAI Press (2013)

[8] Malgireddy, M., Corso, J., Setlur, S., Govindaraju, V., Mandalapu, D.: A framework for hand gesture recognition and spotting using sub-gesture modeling. In: 20th International Conference on Pattern Recognition, pp. 3780–3783 (2010)

[9] Yang, H.-D., Park, A.-Y., Lee, S.-W.: Gesture spotting and recognition for human ndash; robot interaction. IEEE Transactions on Robotics 23(2), 256–270 (2007)

[10] Elmezain, M., Al-Hamadi, A., Sadek, S., Michaelis, B.: Robust methods for hand gesture spotting and recognition using hidden markov models and conditional random fields. In: 2010 IEEE International Symposium on Signal Processing and Information Technology (ISSPIT), pp. 131–136 (December 2010)

[11] Wilson, A., Bobick, A.: Parametric hidden markov models for gesture recognition. IEEE Transactions on Pattern Analysis and Machine Intelligence 21(9), 884–900 (1999)

[12] Sminchisescu, C., Kanaujia, A., Li, Z., Metaxas, D.: Conditional models for contextual human motion recognition. In: Tenth IEEE International Conference on Computer Vision, ICCV 2005, vol. 2, pp. 1808–1815 (October 2005)

[13] Wang, S.B., Quattoni, A., Morency, L., Demirdjian, D., Darrell, T.: Hidden conditional random fields for gesture recognition. In: IEEE Computer Society Conference on Computer Vision and Pattern Recognition, vol. 2, pp. 1521–1527 (2006)

[14] Vinh, L., Lee, S., Le, H., Ngo, H., Kim, H., Han, M., Lee, Y.-K.: Semi-markov conditional random fields for accelerometer-based activity recognition. Applied Intelligence 35(2), 226–241 (2011)

[15] Juang, B.-H., Levinson, S., Sondhi, M.: Maximum likelihood estimation for multivariate mixture observations of markov chains (corresp.). IEEE Transactions on Information Theory 32(2), 307–309 (1986)

[16] Viterbi, A.: Error bounds for convolutional codes and an asymptotically optimum decoding algorithm. IEEE Transactions on Information Theory 13(2), 260–269 (1967)

[17] Forney Jr., G.D.: The viterbi algorithm. Proceedings of the IEEE 61(3), 268–278 (1973)

[18] Baum, L.E., Petrie, T., Soules, G., Weiss, N.: A maximization technique occurring in the statistical analysis of probabilistic functions of markov chains. The Annals of Mathematical Statistics 41(1), 164–171 (1970)

[19] Rabiner, L.: A tutorial on hidden markov models and selected applications in speech recognition. Proceedings of the IEEE 77(2), 257–286 (1989)

[20] Bourlard, H.A., Morgan, N.: Connectionist Speech Recognition: A Hybrid Approach. Kluwer Academic Publishers, Norwell (1993)

[21] LeCun, Y.A., Bottou, L., Orr, G.B., Müller, K.-R.: Efficient backprop. In: Orr, G.B., Müller, K.-R. (eds.) Neural Networks: Tricks of the Trade. LNCS, vol. 1524, pp. 9–50. Springer, Heidelberg (1998)

[22] Rath, P.S., Povey, D., Veselý, K., Černocký, J.: Improved feature processing for deep neural networks. In: Proceedings of Interspeech 2013. International Speech Communication Association, vol. 8, pp. 109–113 (2013)

[23] Haeb-Umbach, R., Ney, H.: Linear discriminant analysis for improved large vocabulary continuous speech recognition. In: IEEE ICASSP, vol. 1, pp. 13–16 (1992)

[24] Fukunaga, K.: Introduction to Statistical Pattern Recognition, 2nd edn. Academic Press Professional, Inc., San Diego (1990)

[25] Povey, D., Ghoshal, A., Boulianne, G., Burget, L., Glembek, O., Goel, N., Hannemann, M., Motlicek, P., Qian, Y., Schwarz, P., Silovsky, J., Stemmer, G., Vesely, K.: The kaldi speech recognition toolkit. In: IEEE 2011 Workshop on Automatic Speech Recognition and Understanding. IEEE Signal Processing Society (December 2011)

# Semantic Regions Recognition in UAV Images Sequence

Stéphane Lathuilière, Hai Vu, Thi-Lan Le,
Thanh-Hai Tran, and Dinh Tan Hung

**Abstract.** In this work, we describe a framework to analyze UAV videos content. A multi-class image segmentation approach is proposed considering UAV videos specific characteristics. A static image segmentation is applied on each frame. After a preprocessing step on resulting segments, a SVM classifier is used to recognize regions. A Markov model is introduced to combine the results from the previous frames in order to improve the accuracy. The framework has been designed to be as flexible as possible with an eye to allow to insert holistic information into the model.

## 1 Introduction

It is commonly accepted that there is a growing need for efficient video content analysis tools. In addition UAVs have been used more and more lately and gained popularity both in the general public and the engineering world. Image sequences captured from a camera attached to a small drone usually contain background such as trees, constructions (buildings and roads), grass and sky. The goal is to partition each frame in multiple segments and to label them with these basic categories. Detecting more special objects as cars is another problem and requires using other techniques. Nevertheless, our model has to be flexible to be able to integrate this kind of methods.

Stéphane Lathuilière · Hai Vu · Thi-Lan Le · Thanh-Hai Tran
Research Institute MICA,
Hanoi University of Science and Technology, Hanoi, Vietnam
e-mail: {stephane.lathuiliere,hai.vu,thi-lan.le,
        thanh-hai.tran}@mica.edu.vn

Stéphane Lathuilière
Ensimag - Grenoble INP, France

Dinh Tan Hung
Department of Aeronautical & Space Engineering
Hanoi University of Science and Technology, Hanoi, Vietnam
e-mail: dinhtanhung@gmail.com

© Springer International Publishing Switzerland 2015
V.-H. Nguyen et al. (eds.), *Knowledge and Systems Engineering,*
Advances in Intelligent Systems and Computing 326, DOI: 10.1007/978-3-319-11680-8_25

Videos taken from such devices have their own specific characteristics: moving camera, high distance from objects, view from up high, non constant angle between camera axis and the ground. As a consequence, specific techniques and software have to be developed to analyze their content. Research on image segmentation for the last 30 years has led to various algorithms and methods. Choosing an efficient tool in a particular case is not an easy task. In addition temporal information has to be included at best whereas algorithms on static image are not necessary easily adaptable. In the presented work, a framework is proposed to analyze UAV videos content. In the proposed method, the frames are analyzed one after the other but a temporal model is used to combine results through time and increase the precision on every single frame.

The paper is organized as follows. Section 2 describes previous related work and gives a first overview of our method. Section 3 goes through the segmentation algorithm which makes up the first step of our procedure. Section 4 focuses on the classification step. The temporal model is described in Section 5. Then the results are presented in Section 6. A discussion about future work concludes the paper in Section 7.

## 2   Background and Overview of the Proposed Algorithm

### 2.1   Previous Work on Video Analysis and Image Segmentation

UAVs are widely used for different applications. The main one is high quality images generation for applied fields as glaciology [14] or soil surface modeling [8]. UAV videos content has been analyzed to detect and count vehicles [4] but scene taken from UAV understanding is still a domain which needs to be more deeply explored.

Two classical computer vision issues can be compared with our issue: classical scene understanding and aerial images analysis.

First, scene understanding is considered as one of the next great challenges of computer vision. Many approaches are studied: objects detection [7], multiclass segmentation [11, 22] or reasoning about the 3D scenes[15]. Some algorithms try to obtain an holistic scene understanding combining the different approaches [13]. While such methods provide good results on single image, it seems difficult to apply it by now on video because of high time cost [12]. In addition, hypothesis are done about the camera height in most of these techniques [12, 13, 15]. Some methods [13] also compute the distances to the horizon for each pixel whereas such a measure cannot be used in our case. In fact the angle between the camera axis and the ground is nonconstant. Then retrieving the horizon position would be more difficult [2] and not very helpful.

Secondly, the particular camera height makes the studied problem similar to aerial images analysis. Even though, objects are further than in our case, color and texture features have a role as much crutial. Several techniques have been used for this problems: color, texture and structure features [9], 3-D reasoning [18], morphological operators [1].

## 2.2 Proposed Algorithm Overview

Our algorithm tries to use spatial information and temporal dimension to get a satisfying segmentation. However, a substantial part of our algorithm is applied on each frame ignoring the temporal dimension.

- Static step: this step is illustrated in Fig.1. On each frame a two parts algorithm is performed:
  - First, a segmentation of the frame is computed.
  - Then each segment of the image is labeled thanks to a recognition algorithm. Statistical descriptors of each component are computed and a Support Vector Machine (SVM) algorithm is used to predict the class of each component of the frame. More details about the segmentation step are given in Section 3.

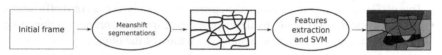

**Fig. 1** Outline of the static algorithm step applied on every single frame

- Temporal step: to combine results frame after frame, a Markov chain is used pixel by pixel (Fig.2).

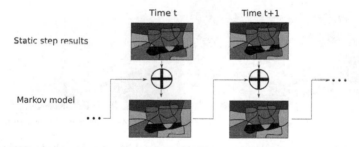

**Fig. 2** Outline of the temporal model

More details about the temporal model are given in Section 5.

## 3 Proposed Approach for Segmentation Step

### 3.1 Segmentation Algorithm

A really precise segmentation cannot be obtained with a high frame rate. For instance, computing time of sophisticated algorithms such as [12] can reach up to 10 minutes per frame.

As a consequence, a mean-shift algorithm has been chosen. In a first time, the segmentation can be only geometric and not semantic. Mean-shift method has already been widely studied [5, 6, 10].

Defining the best parameters for the segmentation step is not an easy task. In fact, segmentation parameters affect the size, the homogeneity of components and border of components. A solution is proposed in Section 5.

## 3.2  Segmentation Results Pruning

After applying the mean-shift algorithm, the small components are automatically removed. Two reasons for that:

- If an object is made of many parts with different colors, all the parts must be in the same component to be efficiently recognized. For example, a building can consist of many different colors and a similar variety of colors cannot occur in *Tree* or *Sky*. If each color is isolated, segments cannot be recognized efficiently.
- If a component is too small, its descriptors have irregular properties due more by the short size of this sample than by the nature of the object.

However after this step, there is a remaining problem: some components are bigger than the minimal requested size but contain meaningless long branches. This situation is shown in Fig.3.

(a) Original image            (b) Zoom on a meaningless segment

**Fig. 3** Example of a meaningless segmentation: original image and resulted segmentation

To solve this problem, morphological operators have been used. An opening and a closing algorithms [21] are applied on the matrix containing all the labels. Here morphological operators are applied whereas the classical order over integers has no meaning. So opening and a closing have been applied not to favor components with high or low labels. However, after these operations, a component can be divided in multiple non connected components. The resulting segmentation must be analyzed again to separate not connected components.

## 3.3  Over-segmentation of Vast Segments

Initially, applying the algorithm on vast and uniform components as sky led to bad results. It could be explained by the presence of some objects in the middle of the zone or a not clear boundary with neighboring segments. An example is shown in Fig.4.

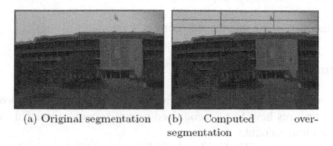

(a) Original segmentation      (b)     Computed     over-
                                        segmentation

**Fig. 4** Example of object in a vast and uniform segment: computed over-segmentation

To solve this problem the sky must be divided into many rectangular parts as shown in Fig.4. Most of the parts have a uniform color and can easily be recognized as belonging to the sky.

## 4  Classification Scheme

### 4.1  Learning from Segmented Images

The learning algorithm follows the scheme below (Fig.5):

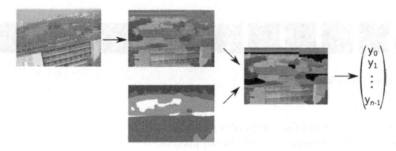

**Fig. 5** Scheme of the 3 steps features extraction algorithm: segmentation, comparison with mask resulting of handmade and feature vectors extraction

More precisely, the 3 steps are:

- The segmentation from each training image are computed (see Section 3).
- The result of the segmentation is compared with the handmade labeled image. Only the segments containing $p\%$ of pixels belonging to only one class are kept. As a consequence, the segments which contain pixels which should belong to several classes are ignored. The parameter $p$ allows to increase or decrease the number of descriptors depending on the precision of the segments used for training.
- The features are computed for each correct segment.

## 4.2   Features Extraction

Statistical descriptors which recognize efficiently textures have to be chosen. In the studied case, textures have to be recognized on not square zones. This specificity has to be taken into account.

Classical descriptors as HOG [7] and SIFT features [19] have not been used for two reasons. They are more suited for close objects recognition than for texture classification and they are quite time-consuming. Local Binary Patterns [20] have not been used as they have not led to good results.

Some of used descriptors have been inspired by [9]. Even though the aim of this article is quite different, the data is similar. Following descriptors have been chosen:

- Histogram of "bicolor" representation: 4 bits of one color space and 4 bits from another base. Best results have been obtained with 4 bits from grayscale and 4 bits from Hue.
- Mean and variance over the all zone of R, G and B dimensions
- Gabor filter: mean over the all zone for different $\sigma$ (standard deviation of the Gaussian envelope) and $\theta$ (orientation of the filter). The chosen kernels are shown below (see Fig.6).

**Fig. 6** Used Gabor kernels

- The y coordinate of the component geometric center.
- The size of the component (number of pixels).

The last two descriptors have been added because a large component on the top of the image is more likely to belong to the sky.

## 4.3 Class Representation

For each segment of the frame, the prediction method of the SVM is performed. It gives us a label corresponding to one of the classes. However labeling segments with a single integer is not flexible. Another class representation has been chosen.

Each pixel of the image matches with a class (*Tree, Building*...). Vectors of probabilities are used for each pixel:

$$Y_t = \begin{pmatrix} y_{t,1} \\ y_{t,2} \\ \vdots \\ y_{t,n} \end{pmatrix}. \tag{1}$$

The variable $n$ is the number of considered classes. $y_{t,i}$ denotes the probability at the time t of belonging to the class $C_i$.

This representation has two advantages:

- More information than only the index of the maximum of the probability are stored. As a consequence, this information can be used to get more stable result through the time.
- Labeling matrix can be seen as a field. If single index representation were used, an interesting norm to compare to pixels could not be defined. The integer order relation is not meaningful in this case. Thanks to this model, the vectors belong to a subset of $[0,1]^4$. This property could allow to use more powerful mathematical tools in the future.

More specifically, two variables are considered in the following paragraphs:

- $X_t$: real class the pixel belongs to.
- $Y_t$: observations, resulting from the recognition algorithm.

The model defining the link between $X_t$ and $Y_t$ is described in 5

## 4.4 Specific Multiclass SVM Use

The basic SVM algorithm can be used only in binary problems. Many techniques have been used to extend SVM to multiclass segmentation.

### 4.4.1 "One-against-one" Approach

"One-against-one" approach [17] is a classical method to build a multiclass SVM based on several binary SVM. The comparisons of methods between multiclass SVM have shown that is a competitive method [16]. That is the method used in the most widely employed SVM library "libsvm" [3].

As carefully described in [17], the algorithm is divided in two parts:

- First, the procedure tries to find classes linearly separable from all others.
- Then each pair of classes is separated.

In our case, the first step has been ignored for convenience. As four classes are considered (*Construction*, *Tree*, *Grass/field* and *Sky*), 6 classifiers have been used.

### 4.4.2 From Classifiers Results to Vectors of Probabilities

After applying the 6 classifiers, votes are computed to get the best class. A class can get up to 3 votes. If a class gets 3 votes, a probability $\alpha$ is assigned to this class. Then the other classes get a probability proportional to their vote count. However a minimal threshold probability $\varepsilon$ is assigned to class without any vote. The parameters $\alpha$ and $\varepsilon$ enable to configure how much the result of our classification can be trusted. If $\alpha$ is close to 1, the prediction gives a really high probability to the class which reaches 3 votes. Conversely, if $\alpha$ is low, probabilities belonging to different classes will be more homogeneous.

## 5 Temporal Model

### 5.1 State-Space Model

In this Section, the probabilistic framework is defined and in this way, the notations defined in 4.3 are used. We assume that the model is first-order Markov. Similarly, the observations are modeled as a first-order Markov model. As a consequence only the following probabilities have to be defined:

- $P(X_t|X_{t-1})$: state-transition function. It is the probability of a pixel changing from one class to another between two frames.
- $P(Y_t|X_t)$: observation function, the result of our static recognition algorithm.

Thanks to the first order Markov assumptions, the probability of each state can be written as follows:

$$P(X_t|Y_{1:t} = y_{1:t}) \propto P(X_t|y_t) \left( \sum_{x_{t-1}} P(X_t|x_{t-1})P(x_{t-1}|y_{1:t-1}) \right) \quad (2)$$

Thus the probability is recursively computed and only the result of the previous frame needs to be memorized. This model is really similar to a hidden Markov model. However, the output $Y_i$ is not known. The observation of the model is a probability vector and not $Y_i$ itself.

In this article, only the following state-transition function is considered:

$$P(X_t = x_{t,i}|X_{t-1} = x_{t-1,j}) = \delta_{i,j}.\alpha + (1 - \delta_{i,j}).\frac{1-\alpha}{n-1} \quad (3)$$

In other words, a pixel has a probability $\alpha$ not to change its class and $(1 - \delta_{i,j}).\frac{1-\alpha}{n-1}$ to change to other classes. The parameter $\alpha$ allows to configure the inertia of the model. The more $\alpha$ is close to 1 the more the model has a high inertia. In future work, more sophisticated state-transition functions will be used. The main idea is to use holistic analysis of the current segmentation to favor some classes.

# 6 Experimental Results

In order to test the algorithm, some frames are selected in testing videos. They are labeled by hand. The algorithm is applied on the testing videos and the results are compared with the labeled masks of selected frames.

The result and the mask are compared pixel by pixel. If the pixel of the mask is white, the pixel is ignored. In fact, some pixels cannot be labeled by hand with certainty because of the high distance of the camera from the objects or because this class of object is not supported. Then, the precision and recall of the recognition algorithm can be computed.

Training data contains 66 images extracted from UAV videos which have been labeled by hand. The tests were performed using an intel Core i5 3.20GHz. The training step took 48s.

Testing experiments have been conducted on 6 other videos. They have been selected to test the algorithm on varied context: urban, countryside and mountains. Videos have been captured in HD($1920 \times 1080$) but have been resized to ($480 \times 270$). The images sequences total duration is 1m22s at 30 frames per second. 47 frames have been labeled by hand. The algorithm reached 0.84 frames per second. Fig.7 shows examples of returned segmentation.

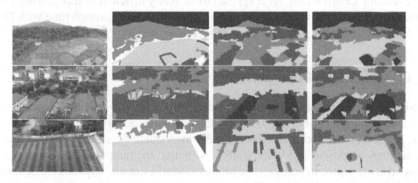

**Fig. 7** Examples of result on test frames: original images, expected segmentations, returned segmentations without temporal model and with temporal model: green=*Tree*, red=*Construction*, blue=*Sky* and yellow=*Grass/field*

Table 1 shows precisions and recalls. In order to illustrate the part played by the proposed Markov model, the output of the static step is compared with the output of the temporal model. Table 2 details more precisely errors made by the proposed model. The results are satisfying for all classes. The good results on sky can be noticed. The sky color and geometric characteristics make it easier to detect.

The main difficulty is recognizing *Construction*. It can be explained by the fact that boundaries between trees and buildings are not very sharp. As a consequence, the segmentation is not precised enough. Moreover, buildings have varied colors and geometric characteristics. The temporal model does not improve significantly

**Table 1** Precision and recall on testing data

|  | Without temporal model | | With temporal model | |
| --- | --- | --- | --- | --- |
|  | Precision | Recall | Precision | Recall |
| *Tree* | 0.56 | 0.39 | 0.94 | 0.54 |
| *Construction* | 0.51 | 0.75 | 0.56 | 0.76 |
| *Sky* | 0.96 | 0.99 | 0.96 | 0.99 |
| *Grass/field* | 0.51 | 0.48 | 0.65 | 0.79 |

**Table 2** Confusion matrix of test frames, expected label on rows and returned label on columns. The numbers of pixels have been divided by the image size.

|  | *Tree* | *Construction* | *Sky* | *Grass/field* |
| --- | --- | --- | --- | --- |
| *Tree* | 5.66 | 3.51 | 0.04 | 1.33 |
| *Construction* | 1.12 | 5.89 | 0.21 | 1.48 |
| *Sky* | 0.00 | 0.00 | 6.04 | 0.01 |
| *Grass/field* | 0.27 | 1.18 | 0.00 | 5.36 |

the results for *Construction* but it appears to be really efficient at recognizing *Tree* and *Grass/field*. As the results of the static step are unstable through time with *Tree* and *Grass/field*, the proposed temporal model helps results.

## 7   Conclusion and Perspectives

In this paper, we have described an algorithm for segmentation of UAV videos. We got satisfying results on our test videos. In further work, the proposed framework has to be tested on more varied data sets in order to evaluate the sensitivity to lighting, illumination conditions and video quality. Although the temporal dimension is not included in the segmentation step, the proposed Markov model expands the possibilities of better temporal analysis. The temporal dimension can be used to converge to a geometrically and semantically consistent segmentation. First tries using holistic approaches to define the state-transition function ensure a promising development.

**Acknowledgements.** This research is funded by Hanoi University of Science and Technology under grant number T2014-130.

## References

[1] Benediktsson, J.A., Pesaresi, M., Amason, K.: Classification and feature extraction for remote sensing images from urban areas based on morphological transformations. IEEE Transactions on Geoscience and Remote Sensing 41(9), 1940–1949 (2003)

[2] Boroujeni, N.S., Etemad, S.A., White, A.: Robust horizon detection using segmentation for uav applications. In: 2012 Ninth Conference on Computer and Robot Vision (CRV), pp. 346–352. IEEE (2012)

[3] Chang, C.-C., Lin, C.-J.: Libsvm: a library for support vector machines. ACM Transactions on Intelligent Systems and Technology (TIST) 2(3), 27 (2011)

[4] Cheng, P., Zhou, G., Zheng, Z.: Detecting and counting vehicles from small low-cost uav images. In: ASPRS 2009 Annual Conference, Baltimore, vol. 3, pp. 9–13 (2009)

[5] Comaniciu, D., Meer, P.: Mean shift analysis and applications. In: The Proceedings of the Seventh IEEE International Conference on Computer Vision, vol. 2, pp. 1197–1203. IEEE (1999)

[6] Comaniciu, D., Meer, P.: Mean shift: A robust approach toward feature space analysis. IEEE Transactions on Pattern Analysis and Machine Intelligence 24(5), 603–619 (2002)

[7] Dalal, N., Triggs, B.: Histograms of oriented gradients for human detection. In: IEEE Computer Society Conference on Computer Vision and Pattern Recognition, CVPR 2005, vol. 1, pp. 886–893. IEEE (2005)

[8] Eltner, A., Mulsow, C., Maas, H.G.: Quantitative measurement of soil erosion from tls and uav data. In: ISPRS-International Archives of the Photogrammetry, Remote Sensing and Spatial Information Sciences, vol. 1(2), pp. 119–124 (2013)

[9] Fauqueur, J., Kingsbury, N., Anderson, R.: Semantic discriminant mapping for classification and browsing of remote sensing textures and objects. In: IEEE International Conference on Image Processing, ICIP 2005, vol. 2, pp. II–846. IEEE (2005)

[10] Fukunaga, K., Hostetler, L.: The estimation of the gradient of a density function, with applications in pattern recognition. IEEE Transactions on Information Theory 21(1), 32–40 (1975)

[11] Fulkerson, B., Vedaldi, A., Soatto, S.: Class segmentation and object localization with superpixel neighborhoods. In: 2009 IEEE 12th International Conference on Computer Vision, pp. 670–677. IEEE (2009)

[12] Gould, S., Fulton, R., Koller, D.: Decomposing a scene into geometric and semantically consistent regions. In: 2009 IEEE 12th International Conference on Computer Vision, pp. 1–8. IEEE (2009)

[13] Gould, S., Gao, T., Koller, D.: Region-based segmentation and object detection. In: NIPS, vol. 1, p. 2 (2009)

[14] Hodson, A., Anesio, A.M., Ng, F., Watson, R., Quirk, J., Irvine-Fynn, T., Dye, A., Clark, C., McCloy, P., Kohler, J., et al.: A glacier respires: quantifying the distribution and respiration co2 flux of cryoconite across an entire arctic supraglacial ecosystem. Journal of Geophysical Research: Biogeosciences (2005–2012), 112(G4) (2007)

[15] Hoiem, D., Efros, A.A., Hebert, M.: Putting objects in perspective. International Journal of Computer Vision 80(1), 3–15 (2008)

[16] Hsu, C.-W., Lin, C.-J.: A comparison of methods for multiclass support vector machines. IEEE Transactions on Neural Networks 13(2), 415–425 (2002)

[17] Knerr, S., Personnaz, L., Dreyfus, G.: Single-layer learning revisited: a stepwise procedure for building and training a neural network. In: Neurocomputing, pp. 41–50. Springer (1990)

[18] Lin, C., Nevatia, R.: Building detection and description from a single intensity image. Computer vision and image understanding 72(2), 101–121 (1998)

[19] Lowe, D.G.: Distinctive image features from scale-invariant keypoints. International Journal of Computer Vision 60(2), 91–110 (2004)

[20] Ojala, T., Pietikainen, M., Maenpaa, T.: Multiresolution gray-scale and rotation invariant texture classification with local binary patterns. IEEE Transactions on Pattern Analysis and Machine Intelligence 24(7), 971–987 (2002)

[21] Serra, J., Vincent, L.: An overview of morphological filtering. Circuits, Systems and Signal Processing 11(1), 47–108 (1992)

[22] Shotton, J., Winn, J., Rother, C., Criminisi, A.: Textonboost: Joint appearance, shape and context modeling for multi-class object recognition and segmentation. In: Leonardis, A., Bischof, H., Pinz, A. (eds.) ECCV 2006, Part 1. LNCS, vol. 3951, pp. 1–15. Springer, Heidelberg (2006)

# Fast Optimization of the Pattern Shapes in Board Games with Simulated Annealing

Huy Nguyen, Simon Viennot, and Kokolo Ikeda

**Abstract.** Monte-Carlo Tree Search is a popular method to implement computer programs for board games, and its performance can be significantly improved by including static knowledge about the game, for example in the form of patterns learned from game records. Finding the right pattern shapes is still an open problem, and we propose in this paper an evolutionary-like method to optimize the pattern shapes. We avoid direct optimization through the heavy Monte-Carlo framework by using instead the performance of a machine-learning algorithm as an early indicator of the quality of the pattern shapes. We have implemented this general method on the specific case of the game of Othello. The final pattern shapes obtained after optimization would be hard to find manually, and they greatly improve the strength of our Othello program.

## 1   Introduction

Monte-Carlo Tree Search (MCTS) is a sampling-based search algorithm [1], which received considerable attention due to its successful application first to the game of Go, and then to many other games. The search is based on random simulations, and is usually much more efficient when combined with a probability model containing static knowledge about the game, for example to make the simulations realistic, or to bias the search towards the promising moves.

Such probability models to encode some knowledge about the game usually relies on the local patterns appearing on the board. The patterns can appear in different shapes, which depend on the game. Historically, both the pattern shapes and their parameters (weights) were designed manually by the programmers, but it was time-consuming and often far from the optimal. Nowadays, the parameters can be

Huy Nguyen · Simon Viennot · Kokolo Ikeda
Japan Advanced Institute of Science and Technology, Nomi, Japan
e-mail: nqhuy@sgu.edu.vn,
        {sviennot,kokolo}@jaist.ac.jp

© Springer International Publishing Switzerland 2015
V.-H. Nguyen et al. (eds.), *Knowledge and Systems Engineering,*
Advances in Intelligent Systems and Computing 326, DOI: 10.1007/978-3-319-11680-8_26

optimized automatically from game records by machine learning methods such as neural networks. However, it is still an open problem to find the right pattern shapes, and they are still usually designed manually.

In this paper, we propose an optimization method to find automatically pattern shapes that improve the performance of MCTS. The originality of our method is that it does not test the pattern shapes directly with MCTS, which would require much computer power. Instead, we use the learning performance of a machine-learning algorithm as an early indicator of the quality of the pattern shapes. The pattern shapes obtained by our method in the case of the game of Othello greatly improve the performance of MCTS. Interestingly, these pattern shapes would be hard to design manually, because of their unexpected shape.

The rest of the paper is organized as follows. Section 2 gives an overview of our research. Section 3 and section 4 presents the background about Monte-Carlo Tree Search and then about Patterns and Machine Learning. Section 5 describes the problem and our method in details, section 6 presents the experimental results of extracting pattern shapes and evaluating them in the case of the game of Othello, and section 7 presents our conclusions and future work.

## 2    Research Overview

The research presented in this paper involves several different techniques, like game tree search, machine-learning and feature selection. To help the reader understand the context, we present in figure 1 an overview of these techniques and their relationship.

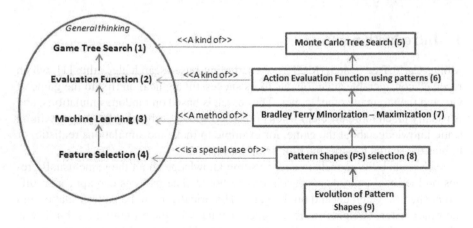

**Fig. 1** Overview of our research

First, game tree search (1) is frequently used for implementing a computer game player.

Then state evaluation functions are often necessary to evaluate a node, either a state evaluation function to evaluate the current state of the board, or an action evaluation function to evaluate the current legal moves (2). Compared to manual design, machine learning is quite effective to learn good evaluation functions automatically (3), and it is known that the selection of the features used in the machine learning is very important for the performance (4).

In our specific case, we focus on the case of MCTS applied to the game of Othello (5). We need a good action evaluation function using some patterns (6), and we employ Bradley-Terry Minorization-Maximization as the machine learning algorithm because it is effective in the game of Go and has attracted attention in the recent years (7). In our case, only patterns are used as the features, but since there are a lot of possible patterns, we make the pattern shapes evolve automatically (8).

There are many ways to select good pattern shapes, especially many possible evaluation measurements to optimize. Our goal is to optimize the performance of MCTS, then it would be straightforward to use to select the pattern shapes that give the highest MCTS performance. However, this optimization would be too expensive, so in this paper, we propose to evaluate and select the pattern shapes with the fast result of BTMM machine learning, run on a reduced set of learning data (9).

## 3 Monte-Carlo Tree Search and Evaluation Functions

This section presents an overview of Monte-Carlo Tree Search (MCTS) and how knowledge can be included in the algorithm to improve its performance.

### 3.1 General Description of Monte-Carlo Tree Search

Monte-Carlo Tree Search (MCTS) is a sampling-based search algorithm [4]. Contrary to classic tree-search methods such as alpha-beta pruning and $A^*$, MCTS does not rely on a heuristic state evaluation function. Instead, the sampling is guided by the result of Monte-Carlo simulations. MCTS is suitable for domains where it is difficult to set up an efficient state evaluation function, such as the game of Go where it was introduced first.

In MCTS, a search tree is expanded gradually. At each step, a leaf node of the expanded tree is selected with a "tree policy", from which a simulation of the game with random moves is performed. The most popular variant of MCTS uses UCB (Upper Confidence Bound) as the tree policy. It is based on a balance of exploitation (explore more the nodes with many won simulations) and exploration (explore more the nodes not well explored yet).

### 3.2 Using Knowledge to Improve MCTS

The performance of MCTS can be improved by incorporating knowledge about the target game. For example, such knowledge can be encoded in a "prediction model" that predicts the probability for a given move to be played by a strong player. Before explaining in the next section how to obtain such a model, we describe here how it can be included in MCTS.

First, the prediction model can be used as a simulation policy, to obtain realistic simulations instead of pure random ones. Secondly, the prediction model can also be used in the tree policy, to limit the exploration to a reduced set of promising moves (*progressive widening*) and to explore more the promising nodes with a bias (*knowledge bias*).

In our Othello program, we use the following formula as the tree policy, as presented for the game of Go in [6] by two of the authors:

$$UCB_{Bias} = \frac{w_j}{n_j} + C \times \sqrt{\frac{\ln n}{n_j}} + \alpha \times \sqrt{\frac{K}{n+K}} \times P(m_j) \qquad (1)$$

The first two terms of equation 5 are the classical exploitation and exploration terms of the UCT algorithm. $n_j$ is the number of simulations for the move $m_j$, $w_j$ the number of won simulations, and $n$ the total number of simulations for the given board. The last term is the progressive bias, with $P(m_j)$ the selection probability of $m_j$ in the prediction model. $C$, $\alpha$ and $K$ are coefficients for tuning the impact of each term. In this paper, we have used $\alpha = 0.5$, $K = 5000$.

## 4 Patterns and Machine-Learning

As mentioned in figure 1, patterns are often used to obtain an evaluation function, and machine learning is then necessary for learning these patterns from game records. In our specific case, the patterns are learned with the Bradley-Terry Minorization-Maximization algorithm, which has attracted a lot of attention for the game of Go. We describe this algorithm below and discuss also some related papers about patterns and machine learning.

### 4.1 Pattern Shapes

A pattern can be defined as a specific configuration of surrounding cells related to a target cell. An important problem is to choose the shape of the patterns, i.e. which surrounding cells should be retained in the pattern or not. For example, figure 2 shows some possible pattern shapes for the target cell B2. Case A uses two shapes {A2, B2, C2, D2, E2, F2, G2, H2} and {B1, B2, B3, B4, B5, B6, B7, B8}, case B two other shapes {A1, B2, C3, D4, E5, F6, G7, H8} and {A3, B2, C1}, and case C a single shape {A1, B1, C1, A2, B2, C2, A3, B3, C3}.

Since the choice of pattern shapes has a direct influence on the performance of the machine learning, we studied methods to evaluate the pattern shapes in [7]. The research presented here continues this previous work by trying to find the best pattern shapes automatically with optimization algorithms.

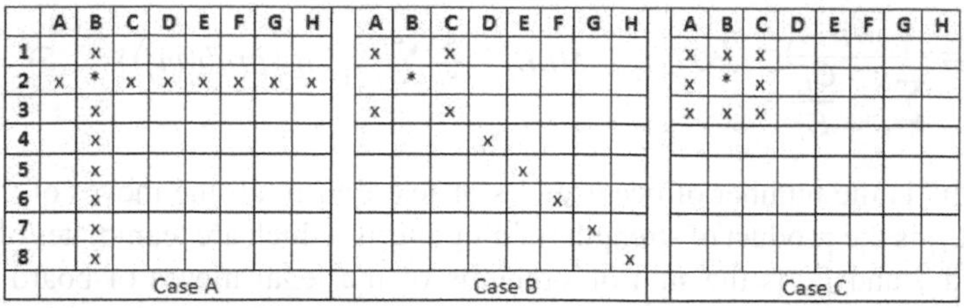

| | A | B | C | D | E | F | G | H |
|---|---|---|---|---|---|---|---|---|
| 1 | | x | | | | | | |
| 2 | x | * | x | x | x | x | x | x |
| 3 | | x | | | | | | |
| 4 | | x | | | | | | |
| 5 | | x | | | | | | |
| 6 | | x | | | | | | |
| 7 | | x | | | | | | |
| 8 | | x | | | | | | |

Case A

| | A | B | C | D | E | F | G | H |
|---|---|---|---|---|---|---|---|---|
| 1 | x | x | | | | | | |
| 2 | | * | | | | | | |
| 3 | x | x | | | | | | |
| 4 | | | | | x | | | |
| 5 | | | | | x | | | |
| 6 | | | | | | x | | |
| 7 | | | | | | | x | |
| 8 | | | | | | | | x |

Case B

| | A | B | C | D | E | F | G | H |
|---|---|---|---|---|---|---|---|---|
| 1 | x | x | x | | | | | |
| 2 | x | * | x | | | | | |
| 3 | x | x | x | | | | | |
| 4 | | | | | | | | |
| 5 | | | | | | | | |
| 6 | | | | | | | | |
| 7 | | | | | | | | |
| 8 | | | | | | | | |

Case C

**Fig. 2** Some possible pattern shapes for the cell B2

Each cell of a pattern shape has three possible states: occupied by Black, occupied by White, or unoccupied. A pattern may be attractive for Black but not for White and vice versa, so we distinguish patterns for Black to play and White to play. Thus, a pattern consists of a target cell, the color Black or White to play, and the status of each cell of the pattern shape.

The number of possible patterns corresponding to a given pattern shape is $2 \times 3^{length-1}$, because the target cell is always empty. Here, the "length" refers to the number of cells of the pattern shape, even if the cells are not necessarily in a line. For example, the number of possible patterns is 18 for the pattern shape $\{A3, B2, C1\}$ and 4374 for $\{A2, B2, C2, D2, E2, F2, G2, H2\}$.

## 4.2 Bradley-Terry Minorization-Maximization

In our case, the patterns are learned with the Bradley-Terry (BT) model and the Minorization-Maximization algorithm (BTMM). This supervised learning algorithm was introduced in 2007 by Coulom for the game of Go [5], and is now used in many top Computer Go programs.

The BT model considers each move of the game as a competition, where the winner is the move actually played, and the losers all the other legal moves. Each move is described as a "team" of features. The parameter of feature $i$, called its strength, is denoted $\gamma_i$. The strength of a move $m$ is the product of strengths of all the features appearing in that move, as in equation 3. The selection probability of a legal move $m$ in a given state of the board is then calculated by equation 2, where $E$ is the sum of strengths of all the other legal moves of the given board.

$$strength(m) = \prod_{i \in m}(\gamma_i) \quad (2) \qquad prob(m) = \frac{strength(m)}{E} \quad (3)$$

We suppose now that we have a collection of $N$ different boards indexed by $j$ from 1 to $N$, and that for each board $j$, the move $m_j$ played by a strong player is known. Then, the goal of the BTMM algorithm is to compute the parameters $\gamma_i$ that will maximize the likelihood estimator $L(\{\gamma_i\}) = \prod_{j=1}^{N}(prob(m_j))$. This maximization can be done with the minorization-maximization method presented in [5], by repeatedly updating the $\gamma_i$ parameters with equation 4.

$$\gamma_i = \frac{W_i}{\sum_{j=1}^{N} \frac{C_{ij}}{E_j}} \quad (4) \qquad MLE = \frac{1}{N} \sum_{j=1}^{N} \log\left(prob(m_j)\right) \quad (5)$$

Here, $W_i$ is the number of occurrences of feature $i$ in all the moves of the game records, $C_{ij}$ is the product of strengths of all patterns which are teammates of pattern $i$ in board $j$ and $E_j$ is the sum of strengths of the legal moves of board $j$, as in equation 2.

The learning and testing performance of the BTMM model is evaluated by the mean-log evidence (MLE), which is the average logarithm of the selection probabilities of all the moves played in game records, as in equation 1.

### 4.3 Other Works Than BTMM

Machine-learning of patterns, or more generally features, has also been studied before the introduction of BTMM. In [2], Bouzy and Chaslot already represent a pattern for the Game of Go by a target intersection and the states of its neighboring intersections. They use a Bayesian approach to compute the pattern probabilities from game records. Stern et al. [9] calculates a probability distribution over the legal moves of a given position in Go by using a Gaussian belief to define a full Bayesian Ranking model. This method uses Go features such as the number of liberties of new chains or the distance to the board edge.

In [3], Buro gives a survey of methods used in strong Othello programs, including the use of machine learning to optimize the parameters of the features used in evaluation functions. In the game of Go, Silver and Muller[8] distinguish location dependent patterns and location independent patterns, and learn the weigths of patterns with temporal difference learning.

## 5 Problem Description and Methods

In this section, we propose an optimization method of pattern shapes (PSs) to improve the performance of MCTS. For designing such an optimization method, it is necessary to consider the following three components: (1) the variables to be optimized, (2) the evaluation functions for the variables, (3) the optimization algorithms.

There are many candidates for each component, so we need to select carefully a set {variables, evaluation function, optimization algorithm} by considering the balance between the optimization cost and the expected performance.

### 5.1 Variables To Be Optimized

The variables to be optimized are the PSs, with an initial value shown on figure 3. The default initial shapes are chosen by hand arbitrarily. We used mainly shapes similar to Logistello [3]. There are three possible ways to work with them:

- The first way is to optimize the pattern shapes one by one in separate optimization steps. For example, we have 25 optimization steps for 25 PSs.
- The second way is to optimize all the PSs of a target cell in one optimization step. For example, PSs of index 1, 2, 3 are optimized for cell B1 in one step, so we have 9 optimization steps in total if working with the PSs of figure 3.
- The third way is to optimize all the 25 pattern shapes in one optimization step.

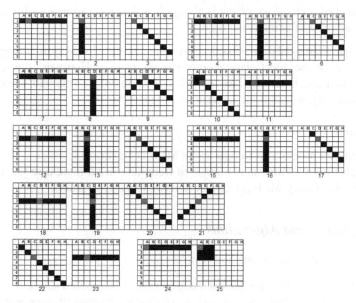

**Fig. 3** The default pattern shapes are selected manually

Usually, cost and difficulty of optimization depends on the number of variables, and the relationship is not linear. If the number of variables is twice, maybe the cost is over twice, possibly 4 or 8 times. But the expected performance is maximum for the third way and minimum for the first one.

## 5.2 Evaluation Function

For the evaluation function, there are the following candidates.

- Battle-performance against a fixed opponent. This evaluation looks like a "trial and error" strategy. The variables are tried directly in the final program, and performance is evaluated with battles against a fixed opponent. Trials are repeated until obtaining good performance, so the cost of this evaluation is huge (5 hours in our case).

- Heavy-MLE, i.e. the performance of a learning algorithm run on a large set of training data with many learning loops. The accuracy of this evaluation is high, but the cost is also high (9 minutes in our case).
- Light-MLE, i.e. the performance of a learning algorithm run on a smaller set of training data with a smaller number of learning loops. The accuracy of this evaluation is not so high (1 minute in our case), but still higher than a measurement without learning.
- A direct measurement, for example the importance measure proposed in [7]. The cost of such measurement is always less than a learning algorithm, but the accuracy of evaluation is not high.

Since the final goal is the battle performance, the first way of using battle-performance is the most natural and simple choice. However, the expected computational cost is too high, so we will use Light-MLE. This will work only is the two following assumptions are verified.

**Assumptions**

- The MCTS program is stronger when MLE is higher.
- A higher Light-MLE (small number of iterations on a limited data set) will lead to a higher Heavy-MLE (more iterations on a bigger data set).

## 5.3   Optimization Algorithms

The following optimization methods can be used.

- Random Search. Given a current solution $x$, if a new random solution $x'$ is better than current solution $x$, then $x = x'$. A solution is obtained after many iterations. The cost of this method is low, but also the expected performance.
- Local Search (LS). Given a current solution $x$, the new solutions are searched in the neighborhood region of the current solution, instead of randomly. The cost of this method is higher than that of random search, and the expected performance is also higher. This method always finds a local optimum.
- Simulated Annealing (SA). This method is able to avoid the trap of local optimums, and to find a global optimum, with a higher cost than LS.
- Genetic Algorithm (GA). It is difficult to compare the performance between SA and GA, but we can see that the cost of SA is lower than GA in generating and evaluating the new solutions.

## 5.4   Our Method

To balance the cost and the performance, the triple {all pattern shapes for a target cell, light-BTMM, Simulated Annealing} is selected. Our method is summarized in figure 4, and described more precisely in Algorithm 1.

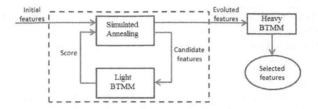

**Fig. 4** Our approach for the evolution of pattern shapes

The evolvedSet contains the evolved PSs that are optimized from the PSs of de-faultSet. In this algorithm, light-BTMM is run on the set (G) of game records, in our case 14,000 Othello game records for learning, 1,000 game records for testing, and the number of loops in BTMM is 5.

## 6 Experiments and Evaluation

In this section, we present the experiments that we have done on the game of Othello with Algorithm 1. The experiments are divided in 4 steps as follows:

1. Game records and MCTS Othello player are given, parameters are fixed.
2. A default set of pattern shapes is defined.
3. Pattern shapes are evolved from the default set with Algorithm 1.
4. Heavy-BTMM ($MLE_{testing}$ value) is used with the evolved set, and the MCTS player using the evolved set is matched to that using the default set.

**[Step 1. Game records and parameters].** Pattern learning was performed on game records played by strong players on a site of Michael Buro [3], au-thor of Logistello program. These game records can be downloaded freely on http://skatgame.net/mburo/ggs/game-archive/Othello. The data set contains 70,000 games, between players sufficiently strong for our purpose of learning Othello pat-terns with machine-learning of patterns.

**[Step 2. Default set of PSs].** With the default PSs of figure 3, we run heavy-BTMM in 65,000 game records for learning, 5,000 game records for testing, and the number of loops in BTMM is 20. Then, the $MLE_{testing}$ is -1.47557. Because of the symmetry of the game board, we need to consider only 9 cells B1, C1, D1, B2, C2, D2, D3, C3 and A1. This set of default PSs is a reasonable basis for comparison with the optimized PSs because they are selected by experts [3], and they lead to good battle-performance of our MCTS program.

**[Step 3. Evolution].** The evolution process is run as in figure 4. BTMM is im-plemented in the evaluation function and in validation step. After some prelimi-nary experiments, the parameters of the complete algorithm are set to ($T_{init} = 0.001$, $T_{stop} = 0.0001$, $\gamma = 0.999$, $n_{trials} = 4$), and the number of SA iterations is set to 2301. One optimization step of 2301 SA iterations needs around 38 hours to complete.

---

Data: Game records (G), a set of default pattern shapes (defaultSet)
Parameters: $0 < T_{stop} < T_{init}$, $\gamma < 1$, $1 \le n_{trials}$
Output: evolvedSet, the set of optimized pattern shapes

baseSet ⟵ defaultSet;
evolvedSet ⟵ ∅;
foreach  *target cell* $c$ ∈ $\{B1, C1, D1, B2, C2, D2, D3, C3, A1\}$ do
    targetSet$_c$ ⟵ pattern shapes of defaultSet associated to the cell $c$;
    baseSet⟵ baseSet \ targetSet$_c$;
    candidateSet ⟵ ∅;
    for $i = 1$ to $n_{trials}$ do                          // can be done in parallel
        currentSet ⟵ targetSet$_c$;
        currentMLE ⟵ lightBTMM(currentSet), MLE value from
          BTMM by using baseSet ∪ evolvedSet ∪ currentSet;
        $T$ ⟵ $T_{init}$;
        while $T > T_{stop}$ do
            newSet is generated by either adding/deleting/changing a cell
              of a PS of currentSet randomly;
            newMLE ⟵ lightBTMM(newSet);
            badness ⟵ currentMLE - newMLE;
            if *badness* ≤ *0* then
              | currentSet ⟵ newSet;
            else
              | currentSet ⟵ newSet with probability $e^{\frac{-badness}{T}}$;
            end
            if *currentSet is replaced* then currentMLE⟵ newMLE;
            $T$ ⟵ $T \times \gamma$;
        end
        candidateSet ⟵ candidateSet ∪ {currentSet}
    end
    bestSet$_c$ is selected from $n_{trials}$ candidates of candidateSet;
    evolvedSet ⟵ evolvedSet ∪ bestSet$_c$;
end

**Algorithm 1:** Pattern shapes optimization with Simulated Annealing

The first evolution step is done with the target cell B1. The final optimized pattern shapes of B1 after 2301 SA iterations are $1'$, $2'$ and $3'$ on figure 5. Then, these optimized PSs are retained for the next step, and in evolution step 2, we optimize the pattern shapes of C1, obtaining $4'$, $5'$ and $6'$ on figure 5. As explained in the algorithm of the previous section, this evolution process is done for the target cells B1, C1, D1, B2, C2, D2, D3, C3, A1.

The MLE column of Table 1 shows the $MLE_{testing}$ after optimizing the pattern shapes of each target cell. As expected, the performance of learning (MLE) increases at each evolution step. Figure 5 shows the cells distribution of the optimized pattern

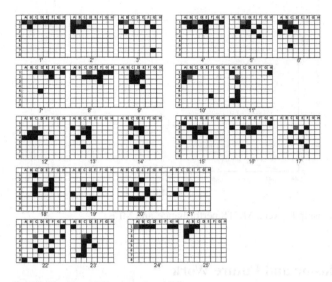

**Fig. 5** The cells distribution of the optimized PSs

**Table 1** Applying the optimized pattern shapes into the MCTS program

| Evolution step | Target PSs | MLE | MCTS(Evo) vs MCTS(Def) | MCTS(Evo) vs Riversi |
|---|---|---|---|---|
| Default PSs | | -1.47557 | 50% | 54.70±3.60% |
| Evolution 1 | B1 | -1.45754 | 52.72±3.66% | 73.72±2.91% |
| Evolution 2 | C1 | -1.44368 | 55.84±2.80% | |
| Evolution 3 | D1 | -1.43955 | 58.85±2.78% | 92.15±1.54% |
| Evolution 4 | B2 | -1.43288 | 62.12±2.65% | |
| Evolution 5 | C2 | -1.42504 | 67.71±2.20% | 94.39±1.32% |
| Evolution 6 | D2 | -1.40159 | 74.48±2.07% | |
| Evolution 7 | D3 | -1.39340 | 79.56±2.47% | 94.70±1.37% |
| Evolution 8 | C3 | -1.38901 | 81.76±2.18% | |
| Evolution 9 | A1 | -1.38994 | 83.45+2.26% | 94.64±1.24% |

shapes. There are many "unusual" shapes compared to the default PSs. It would be hard to find these optimized PSs manually.

[**Step 4. MCTS performance**]. We test the performance of our MCTS program using the optimized PSs against fixed opponents, either MCTS using the default set of PSs, or Riversi, an open source program. Table 1 shows that the program is significantly stronger after each evolution step. The strength improvement is bigger against Riversi than in self-play in the first steps.

Also, Figure 6 shows the performance of MCTS using the optimized PSs in function of MLE. As our initial asssumption, higher MLE implies clearly higher MCTS performance.

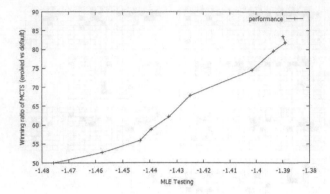

**Fig. 6** Relationship between MCTS performance and MLE

## 7 Conclusion and Future Work

In this research, we propose a method to find automatically the pattern shapes that will give the best performance when included as static knowledge in an MCTS program. The pattern shapes are optimized with a simulated annealing method that relies on the performance of a light machine-learning algorithm for the evaluation. This is much faster than direct testing in the MCTS program, and allows us to perform many steps of the simulated annealing.

The results in the case of the game of Othello are significant. First, the pattern shapes after optimization give a higher learning performance than a default set of natural pattern shapes. Then, when these pattern shapes are included in the final MCTS program, the strength of the program is greatly improved. This confirms that patterns with a higher learning performance lead to a stronger MCTS program. It could be interesting to investigate this relationship in more details in the future, as well as trying other optimization methods of the pattern shapes.

Finally, the optimized pattern shapes are not trivial and would be difficult to design by hand. We can predict that algorithms to find automatically the best pattern shapes (or features) of a game, will have an increasing impact on the strength of the programs in the coming years.

## References

[1] Baier, H., Winands, M.H.M.: Monte-carlo tree search and minimax hybrids. In: Proceedings of the Conference on Computational Intelligence in Games 2013, pp. 129–137. IEEE (2013)

[2] Bouzy, B., Chaslot, G.: Bayesian generation and integration of k-nearest-neighbor patterns for 19x19 go. In: IEEE 2005 Symposium on Computational Intelligence in Games, pp. 1019–1025. IEEE (2005)

[3] Buro, M.: The evolution of strong othello programs. In: Nakatsu, R., Hoshino, J. (eds.) Entertainment Computing. IFIP, vol. 112, pp. 81–88. Springer, Boston (2003)

[4] Cameron, B., Powley, E., Whitehouse, D., Lucas, S.M., Cowling, P.I., Rohlfshagen, P., Tavener, S., Perez, D., Samothrakis, S., Colton, S.: A survey of monte carlo tree search methods. IEEE Transactions on Computational Intelligence and AI in Games 4(1), 1–43 (2012)

[5] Coulom, R.: Computing elo ratings of move patterns in the game of go. Int. Comp. Games Assoc. J. 30(4), 198–208 (2007)

[6] Ikeda, K., Viennot, S.: Efficiency of static knowledge bias in monte-carlo tree search. In: van den Herik, H.J., Iida, H., Plaat, A. (eds.) CG 2013. LNCS, vol. 8427, pp. 26–38. Springer, Heidelberg (2014)

[7] Nguyen, H.Q., Ikeda, K.: Evaluation of pattern shapes in board games before machine learning. International Journal of Electrical Engineering 20(2), 39–49 (2013)

[8] Silver, D., Muller, M.: Reinforment learning of local shape in the game of go. In: Proceedings of the 20th International Joint Conference on Artifical Intelligence, pp. 1053–1058. Springer (2007)

[9] Stern, D., Herbrich, R.: Bayesian pattern ranking for move prediction in the game of go. In: Proceedings of the 23rd International Conference on Machine Learning, pp. 873–880 (2006)

# Modelling Timed Concurrent Systems Using Activity Diagram Patterns*

Étienne André, Christine Choppy, and Thierry Noulamo**

**Abstract.** UML is the *de facto* standard for modelling concurrent systems in the industry. Activity diagrams allow designers to model workflows or business processes. Unfortunately, their informal semantics prevents the use of automated verification techniques. In this paper, we first propose activity diagram patterns for modelling timed concurrent systems; we then devise a modular mechanism to compose timed diagram fragments into a UML activity diagram that also allows for refinement, and we formalise the semantics of our patterns using time Petri nets.

**Keywords:** Unified Modelling Language, formal modelling, timed systems.

## 1 Introduction

UML (Unified Modelling Language) [1] is the *de facto* standard for modelling concurrent systems in the industry. Activity diagrams allow designers to model workflows or business processes. Although UML diagrams are widely used, they suffer from some drawbacks. Indeed, since UML specification is documented in natural language, inconsistencies and ambiguities may arise. First, their rich syntax is quite permissive, and hence favours common mistakes by designers. Second, their informal semantics in natural language prevents the use of automated verification

Étienne André · Christine Choppy
Université Paris 13, Sorbonne Paris Cité, LIPN, CNRS, UMR 7030, F-93430, Villetaneuse, France

Thierry Noulamo
University of Dschang, IUT Fotso Victor, LAIA, Bandjoun, Cameroun

* Work partially supported by STIC Asie project "CATS (Compositional Analysis of Timed Systems)".

** This work is supported by a France-Cameroon cooperation grant through the exchange program funded by the "Service de Coopération et d'Action Culturelle" (SCAC) from the French embassy in Cameroon.

© Springer International Publishing Switzerland 2015     339
V.-H. Nguyen et al. (eds.), *Knowledge and Systems Engineering*,
Advances in Intelligent Systems and Computing 326, DOI: 10.1007/978-3-319-11680-8_27

techniques, that could help detecting errors as early as the modelling phase. In this paper, we propose three contributions.

1. We propose activity diagram patterns for modelling timed concurrent systems using Timed Activity Diagram Components (TADCs),
2. we devise a modular mechanism to compose activity diagram fragments into a UML activity diagram that also allows for refinement, and
3. we formalise the semantics of our patterns using time Petri nets [13].

Our approach guides the modeller task, and allows for automated verification using tools supporting time Petri nets such as TINA (Time Petri Net Analyzer) [4] or Roméo [11].

Our choice of an extension of Petri nets has four advantages. First, the UML specification explicitly mentions Petri nets, and the informal semantics of activity diagrams is given in terms of token flows. Second, they feature a formal semantics [13]. Third, they provide the designer with a graphical notation (close to that of activity diagrams), which helps to be convinced by the "validity" of our translation (although it cannot be formally expressed, due to the lack of formal semantics for activity diagrams). Last, several tools use extensions of Petri nets as input, and can perform efficient verification.

Outline

In Section 2, we discuss related works. In Section 3, we informally recall UML activity diagrams and time Petri nets, and introduce our approach based on the notion of TADCs composed using inductive rules. In Section 4, we provide TADCs with a formal semantics expressed using a translation to time Petri nets. We use as a running example a coffee vending machine. We conclude in Section 5, and provide guidelines for future research.

## 2   Related Work

An important issue is to propose a formal semantics to UML diagrams using a formal notation, which is essential to allow for automated verification. This has been addressed in quite a variety of works using automata, different kinds of Petri nets, etc., so we mention only a few. Instantiable Petri nets are the target of transformation of activity diagrams in [10], and this is supported by tool BCC (Behavioural Consistency Checker). In [7, 3], the issue is performance evaluation, from activity diagrams and others (use case, state diagrams, etc.) to stochastic Petri nets. Also note that [8] proposes an operational semantics of the activity diagrams (for UML 2.2). Börger [5] and Cook et al. [6] present other formalisations of the workflow patterns of [14] using formalisms different from Petri nets, viz. Abstract State Machines and Orc, respectively. In [12], patterns for specifying the system correctness are defined using UML statecharts, and then translated into timed automata. The main differences with our approach are that the patterns of [12] do not seem to be hierarchical:

the "composition" of patterns in [12] refers to the simultaneous verification of different properties in parallel.

In [2], we introduced so-called "precise" activity diagram patterns to model business processes. We inductively defined a set of precise activity diagrams (as a subset of the syntactic constructs of the UML specification [1]), and translated them into coloured Petri nets [9]. Although this work shares with [2] the definition of patterns and their translation into an extension of Petri nets, this work is orthogonal to [2] in the following sense: first, it extends the UML specification with timed constructs, which were not considered at all in [2]. Second, and most importantly, the patterns proposed here are much less restrictive and give more freedom to the designer: the TADCs we define here allow for an arbitrary number of input and output connectors, whereas [2] requires at most one of each. In contrast to [2], we do not restrict the use of the syntactic constructs (in [2], we required that, e.g. each fork is eventually followed by a join). We do not claim that the modular mechanism proposed in this work is better or more useful than [2]: it can be used for a different purpose, when slightly less restrictions are needed. Finally, the scheme that we propose here allows to *refine* TADCs by replacing them with other TADCs, that can be "plugged" into a higher-level one, as long as the connectors are the same. In contrast, due to the presence of at most one input and one output connectors, the patterns of [2] hardly allow this. We give in Table 1 a comparison of the syntactic elements from the specification [1] taken into account in [2] and/or in this work. Note that an element not taken into account in one of these two works does not necessarily mean that that work is less interesting; recall that a main difference between [2] and this work is that [2] is more restrictive in terms of syntax, which may also help to avoid mistakes, whereas this work is more permissive. (We only consider in Table 1 elements taken into account in at least [2] or in this work.)

**Table 1** Summary of the syntactic aspects considered

| Element | [1] | [2] | This work |
|---|---|---|---|
| Activities | Yes | Yes | Yes |
| Data | Limited | Yes | Limited |
| Participants | Limited | Yes | No |
| Initial / final nodes | Yes | Yes | Yes |
| Decision | Yes | Restricted | Yes |
| Merge | Yes | Restricted | Yes |
| Fork | Yes | Restricted | Yes |
| Join | Yes | Restricted | Yes |
| Timed transitions | Limited | No | Yes |

# 3 Modelling Timed Systems Using Activity Diagrams

## 3.1 Preliminaries: Activity Diagrams

We briefly recall here UML activity diagrams [1], and use the coffee vending machine in Fig. 1 to introduce the syntax. First, activity diagrams feature global variables (that are mentioned several times in the UML specification, and used in the examples). In our setting, we require the global variables to be finite domain. This includes Booleans, bounded integers, enumerated types, and possibly more evolved structures such as finite tuples or lists. Such finite domain variables are often met in tools supporting Petri nets and their extensions. Fig. 1 features four variables, viz. Prod (of enumerated type {TEA, COFFEE}), avail (Boolean), state (of enumerated type {on, serving, stand by}), and w_state (of enumerated type {water_ok, water_lack}).

Activity diagrams feature activities (all rounded rectangles in Fig. 1). These activities can involve global variables modifications, either by assigning a value (or the result of a predefined function) to a global variable, or by calling a function with side-effects on several global variables. In our setting, we require this modification to be discrete (i.e. instantaneous). For example, in Fig. 1, the action associated with the Choice activity assigns the result of function P_button() to the global variable Prod.

Activity diagrams also feature an *initial node* (e.g. the upper node in Fig. 1), and two kinds of final nodes, viz. *activity final* that terminates the activity globally, and *flow final* that terminates the local flow (Fig. 1 features no final node). Activity diagrams feature *decision nodes* (e.g. the ChooseProduct node in the middle of Fig. 1), i.e. depending on guards, one path among others is taken; they feature *merge nodes* (e.g. the bottom-most node in Fig. 1), that is the converse. They also feature *fork nodes*, that split the flow into different subactivities executed in parallel, and *join nodes*, the converse operation.

## 3.2 Timed Activity Diagram Components

We define here Timed Activity Diagram Components (TADCs), obtained by composition of basic activity diagrams fragments using inference rules (that will be given in Section 3.3). As shown in Fig. 2(a), a TADC has in general $n \in \mathbb{N}$ input connectors, and $m \in \mathbb{N}$ output connectors. These connectors will be used by the inductive rules to build more complex TADCs, as well as to perform some refinement.

TADCs have three main purposes:

1. define "well-formed" activity diagrams, by both restricting the set of syntactic constructs available in the specification, and augmenting it with some timed constructs,
2. allow a translation to another formalism using an inductive mechanism, and
3. allow a modular specification with possible refinement, or the definition of "black-box" components.

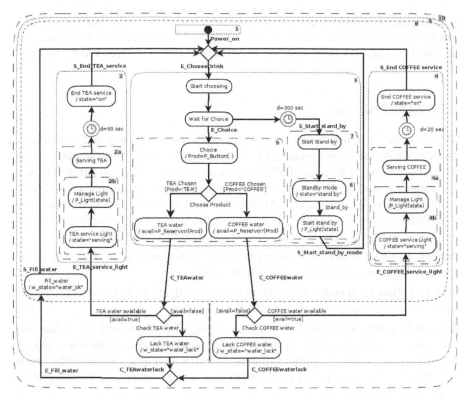

**Fig. 1** Specification of a coffee vending machine using TADCs

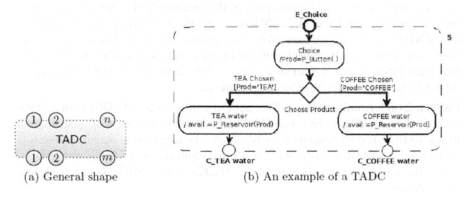

(a) General shape

(b) An example of a TADC

**Fig. 2** TADC: general shape and example

As an example, we give in Fig. 2(b) an example of a TADC responsible for the choice of the drink in a coffee vending machine. This TADC has 1 input connector (i.e. E_Choice) and 2 output connectors (i.e. C_TEA_water and C_COFFEE_-water). This TADC can be refined into another one with a different structure (e.g. a more complex one that could prepare some sugar too), that can then still be plugged into the coffee vending machine, as long as it still has 1 input and 2 output connectors. It is also possible to hide the internal specification of this TADC, by only giving its input and output connectors.

## 3.3 TADC Patterns

We introduce in Table 2 our patterns for inductively creating and composing TADCs. We define 3 basic patterns (1–3), that define TADCs from single activity diagrams syntactic constructs, and 7 composite patterns (4–10), that define TADCs by combining other TADCs together with syntactic constructs (usually transitions or choice nodes). We only depict the connector used in the compositions; recall that each TADC can have an arbitrary number of input and output connectors. (The last column, introduced in Table 2 to save some space, will be used in Section 4.)

**Pattern 1 (initial node).** This pattern is made of the initial node. It has no input connector, and one output connector, which is itself.

**Pattern 2 (flow final node).** This pattern is made of the flow final node.[3] It has one input connector, which is itself, and no output connector.

**Pattern 3 (simple activity).** This pattern is made of a simple activity. It has one input connector, which is itself, and one output connector, which is itself too. Recall that actions can modify the global variables, by either assigning a value (or the result of a predefined function) to a global variable, or using a function with side-effects on several global variables.

**Pattern 4 (sequence).** This pattern combines two TADCs with a sequence transition. The input (resp. output) connectors of the resulting TADC are all input (resp. output) connectors of $TADC_1$ and $TADC_2$, except the output (resp. input) connector of $TADC_1$ (resp. $TADC_2$) connected by the sequence transition. In other words, any free input (resp. output) connector of one of the component TADCs will remain a free input (resp. output) connector of the resulting composite TADC. The mechanism will be the same for all subsequent patterns.

**Pattern 5 (deterministic delay).** This pattern combines two TADCs with a deterministic delay $d \in \mathbb{R}_{\geq 0}$. The lower TADC will start exactly $d$ units of time after the upper TADC has completed its activity. Again, the input (resp. output) connectors of the resulting TADC are all input (resp. output) connectors of $TADC_1$ and $TADC_2$, except the output (resp. input) connector of $TADC_1$ (resp. $TADC_2$) connected by the delay transition.

**Pattern 6 (non-deterministic delay).** This pattern generalises the previous pattern to a non-deterministic delay, i.e. comprised in an interval $[0, d]$.

---

[3] For sake of conciseness, we do not include the activity final node. It could be added to our patterns using the same translation as in [2], i.e. using a "global Boolean variable" that is tested to be true in all guards, and set to false when the activity final node is reached. This mechanism ensures that concurrent activities do not progress any more once an activity final node is reached.

**Pattern 7 (deadline).** In this pattern, after $TADC_1$ completes its execution, activity A is executed. Then, $TADC_2$ starts after at most $d$ units of time; alternatively, $TADC_3$ starts after exactly $d$ units of time.

**Pattern 8 (decision).** This pattern reuses the syntactic "decision" node from the UML specification. After TADC completes its execution, the conditions $cond_1, \ldots, cond_n$ are evaluated, and the TADC with a true guard is executed. The conditions $cond_i$ (for $1 \leq i \leq n$) are Boolean expressions over the activity diagram's variables. If several conditions are true simultaneously, a destination TADC is selected non-deterministically among those with a true condition. In contrast to [2], we do not require that at least one guard is true; this however can result in ill-formed models.

**Pattern 9 (merge).** This pattern reuses the syntactic "merge" node from the UML specification. After some $TADC_i$ (for $1 \leq i \leq n$) completes its execution, then TADC starts executing.

**Pattern 10 (synchronisation).** This pattern merges the syntactic "fork" and "join" nodes from the UML specification. After all $TADC_i$ (for $1 \leq i \leq n$) complete their execution, all $TADC_j$ (for $1 \leq j \leq m$) start executing simultaneously. Note that the classical UML fork (resp. join) from [1] can be obtained from this pattern by setting $n = 1$ (resp. $m = 1$).

## 3.4 Example: A Coffee Vending Machine

The coffee vending machine in Fig. 1 has been built using the inductive rules of Table 2. For example, all activities can be defined by applying Rule 3 ("activity"). The result of the applications of composite rules are outlined using dashed boxes in Fig. 1. For example, the application of Rule 8 ("decision") to the three activities in the centre of Fig. 1 gives the TADC number 5. By applying Rule 7 ("deadline") to this TADC, to its right-hand TADC (number 7), to the `wait for choice` activity and to the `Start choosing` activity considered as a TADC, we get the (large) TADC number 3. The rest of the coffee vending machine is obtained similarly.

## 4 Translation into Time Petri Nets

### 4.1 Time Petri Nets with Global Variables

Time Petri nets (TPNs) [13] are a kind of automaton represented by a bipartite graph with two kinds of nodes, places (e.g. $p_1$ in Fig. 3(a)) and transitions (the rectangle nodes in Fig. 3(a)). In time Petri nets, transitions are associated with a firing interval: for example, in Fig. 3(a), the transition between $p_1$ and $p_2$ (say, $t_2$) can fire between 1 and 3 time units after a token is present in $p_1$, and the transition between $p_1$ and $p_2$ (say, $t_3$) can fire between 2 and 4 time units after a token is present in $p_1$. We use the strong semantics where time elapsing cannot disable a transition: that is, transitions

**Table 2** Activity diagram patterns

| # | Pattern | Syntax | Translation |
|---|---------|--------|-------------|
| 1 | Initial node | | |
| 2 | Final node | | |
| 3 | Simple activity | A /action | action |
| 4 | Sequence | TADC$_1$ TADC$_2$ | $Tr(\text{TADC}_1)$ $Tr(\text{TADC}_2)$ |
| 5 | Deterministic delay | TADC$_1$ $d$ TADC$_2$ | $Tr(\text{TADC}_1)$ $[d, d]$ $Tr(\text{TADC}_2)$ |
| 6 | Non-deterministic delay | TADC$_1$ $[0, d]$ TADC$_2$ | $Tr(\text{TADC}_1)$ $[0, d]$ $Tr(\text{TADC}_2)$ |
| 7 | Deadline | TADC$_1$ A $d$ TADC$_2$ TADC$_3$ | $Tr(\text{TADC}_1)$ $[0, d]$ $[d, d]$ $Tr(\text{TADC}_2)$ $Tr(\text{TADC}_3)$ |
| 8 | Decision | TADC [cond$_1$] [cond$_n$] TADC$_1$ ⋯ TADC$_n$ | $Tr(\text{TADC})$ [cond$_1$] [cond$_n$] $Tr(\text{TADC}_1)$ ⋯ $Tr(\text{TADC}_n)$ |
| 9 | Merge | TADC$_1$ ⋯ TADC$_n$ TADC | $Tr(\text{TADC}_1)$ ⋯ $Tr(\text{TADC}_n)$ $Tr(\text{TADC})$ |
| 10 | Synchronisation | TADC$_1$ ⋯ TADC$_n$ TADC$'_1$ ⋯ TADC$'_m$ | $Tr(\text{TADC}_1)$ ⋯ $Tr(\text{TADC}_n)$ $Tr(\text{TADC}'_1)$ ⋯ $Tr(\text{TADC}'_m)$ |

must fire at the end of their firing interval, unless the transition becomes disabled by another transition in the meanwhile.

We extend the usual TPNs with global variables as follows. We assume a set of finite domain variables over a set of types. For example, Fig. 3(a) makes use of $b$ (of type Boolean) and $i$ (of type integer bounded by some predefined constant). These global variables can be tested in guards (e.g. $\{\neg b\}$ in Fig. 3(a)), and updated when firing transitions (e.g. $i := i + 1$ in Fig. 3(a)). Note that we depict guards within braces (e.g. $\{\neg b\}$ in Fig. 3(a)) to differentiate with firing times depicted within brackets (e.g. $[1, 3]$).

This extension of the usual formalism to finite domain variables is only syntactic, i.e. this formalism does not add expressiveness to usual time Petri nets. Each finite domain variable can be encoded into a finite set of places. For example, a Boolean variable ($b$ in Fig. 3(a)) could be encoded into 2 places $bT$ and $bF$, that encode the fact that $b$ is true (or false, respectively) if it contains a token. Then, testing whether $b$ if false is equivalent to checking the presence of a token in $bF$. For integers, an option in some situations is to encode the value of the integer with a number of tokens in a dedicated place; then adding one to the integer is encoded by adding one token to the dedicated place.

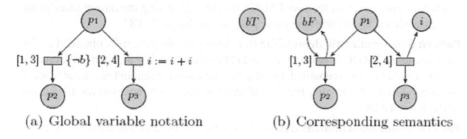

(a) Global variable notation  (b) Corresponding semantics

**Fig. 3** An example of a time Petri net with global variables

*Remark 0.1.* Our definition of time Petri nets with global variables is not very far from coloured Petri nets [9], where types, guards and assignments are also defined. A major difference is that coloured Petri nets feature *coloured* tokens, whereas we use here standard, "null-typed" tokens. A second major difference is that, of course, our definition features time too.

## 4.2 Translation Mechanism

We now explain how to translate the activity diagrams defined in Section 3 into time Petri nets with global variables. The inductive definition of our TADCs (following the rules in Table 2) makes it easy to define an inductive translation.

General Scheme

Recall that each TADC has a set of input and output connectors, that can be used to compose the TADCs in an inductive manner. Here, we translate each TADC into a TPN fragment where the connectors are translated into places. Hence, two TPN fragments can be composed by fusing the corresponding connector places together.

We give the translation of each pattern in the last column of Table 2. In the following, we explain the translation of each pattern.

**Pattern 1 (initial node).** The translation of the initial node is a simple place, that contains a token. At the beginning of the execution of the translated TPN, only these places encoding the initial nodes contain tokens.

**Pattern 2 (final node).** The translation of the final node is a simple place.

**Pattern 3 (simple activity).** The translation of an activity is a TPN transition, preceded and followed by a place, so as to connect in a proper manner with the other translated TADCs. The assignment of variables is easily translated to an assignment on the TPN transition. Note that the functions involving a user input (e.g. P_- Button() in Fig. 1) are translated into a non-deterministic choice in the resulting TPN. Hence, the verification will consider all possible choices.

**Pattern 4 (sequence).** The translation of the sequence pattern is obtained by recursively translating each of the two TADCs, and then by fusing the output place of the upper translated TADC with the input place of the lower TADC.

**Pattern 5 (deterministic delay).** The translation of this pattern is obtained as follows: the upper TADC is connected to a TPN timed transition that can fire exactly $d$ units of time after it was enabled, i.e. after a token was present in the output place of the upper TADC. Then, after the transition fires, the token moves to the translation of the lower TADC.

**Pattern 6 (non-deterministic delay).** This pattern translates the same as the previous pattern, with the exception that the transition can fire any time between 0 to $d$ units of time.

**Pattern 7 (deadline).** First, the upper TADC is translated. Then, it is connected to the TPN transition modelling activity **A**. This transition is then connected to a place. When a token enters this place, both outgoing transitions are enabled. The left-hand transition can fire any time between 0 to $d$ units of time after it is enabled; if it does not, then the right-hand transition must fire exactly $d$ units of time after it is enabled, due to the TPN strong semantics we use.

**Pattern 8 (decision).** The translation of this pattern is straightforward: the output place of the translation of the upper TADC is connected to a set of transitions that have the same guards as the initial TADCs, and then lead to the translation of these TADCs. The semantics of TPNs is the same as for the UML, i.e. if several guards are true, then one is non-deterministically chosen.

**Pattern 9 (merge).** The translation of this pattern is straightforward and is the converse of the previous pattern.

**Pattern 10 (synchronisation).** The translation of this pattern is again straightforward: once the *n* upper TADCs finish their execution, their corresponding token is consumed by the TPN transition; then *m* fresh tokens are created, and the *m* lower TADCs start executing concurrently.

Initial Marking

The initial marking assigns one token to each place corresponding to an initial node. The initial value of the variables can be set to the initial value of the variables in the activity diagram, if any such value is defined, or to a predefined standard value otherwise (e.g. 0 for integers, true for Booleans, etc.).

## 4.3 Application to the Coffee Vending Machine

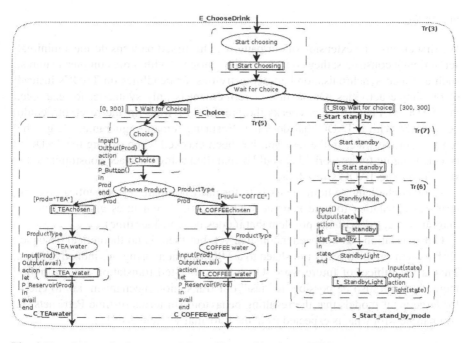

**Fig. 4** Translation of a fragment of the coffee machine into a TPN

We give in Fig. 4 the translation into a TPN of the TADC fragment #3 of Fig. 1 (that corresponds to the application of the deadline pattern to TADCs 5 and 7). We show with dashed lines the patterns, and write in their top-right corner their number. Observe that this TPN is indeed timed (e.g. firing interval $[0, 300]$), and that it features variables: for example, variable Prod is checked in some guards (e.g. Prod = TEA in transition t_TEAchosen), read in side-effect functions (e.g. P_Reservoir(Prod)), and assigned to the result of a function (e.g. P_Button()).

Once the TPN corresponding to the whole TADC of Fig. 1 has been input to a model checker, one can formally verify properties mixing time and value of the global variables, e.g. "if initially w_state = water_ok, then a drink may eventually be delivered within 400 time units". This could be done using tools supporting time Petri nets, such as TINA [4] or Roméo [11].

## 5 Conclusion and Perspectives

In this work, we introduce Timed Activity Diagram Components (TADCs) that help designers to devise complex activity diagrams by the inductive application of predefined patterns. This mechanism guides the designer in the modelling process, and allows one to define black-box components or to replace components by other components. Furthermore, a translation to an extension of time Petri nets allows for formal verification.

Future Work

We first discuss the extension of our patterns. Our timed patterns define a minimal set of timing constructs; they could be further enriched with more complex features, such as timed synchronisation between activities, or deadlines on TADCs instead of on simple activities. Our notion of refinement is only syntactic; for example, the TADC in Fig. 2(b), that checks the presence of water in the reservoir, could be replaced with a TADC that does not. Ensuring semantic guarantees (e.g. "the presence of the water in the reservoir has been checked when exiting the TADC") is a challenging future work, that could be handled using interface constraints to be satisfied on the TADC's variables.

Another challenging future work is to formally compare the semantics given in terms of translation to time Petri nets with a formal semantics given to activity diagrams (e.g. [8], with the problem that it does not consider time).

Our translation has not been automated yet, but there is no theoretical obstacle: each pattern can be directly translated to a TPN fragment using our inductive rules. This is the subject of future work. Once an automated translation tool is implemented, it will also be possible to "test" our translation mechanism, i.e. to check for large input models that the resulting behaviour (in terms of time Petri nets) is compatible with what is expected from the informal semantics of [1].

## References

[1] OMG unified modeling language. version 2.5 beta 2 (September 5, 2013), http://www.omg.org/spec/UML/2.5/Beta2/PDF/

[2] André, É., Choppy, C., Reggio, G.: Activity diagrams patterns for modeling business processes. In: Lee, R. (ed.) SERA 2013. SCI, vol. 496, pp. 197–213. Springer, Heidelberg (2013)

[3] Bernardi, S., Merseguer, J.: Performance evaluation of UML design with stochastic well-formed nets. Journal of Systems and Software 80(11), 1843–1865 (2007)

[4] Berthomieu, B., Vernadat, F.: Time Petri nets analysis with TINA. In: QEST, pp. 123–124. IEEE Computer Society (2006)

[5] Börger, E.: Modeling workflow patterns from first principles. In: Parent, C., Schewe, K.-D., Storey, V.C., Thalheim, B. (eds.) ER 2007. LNCS, vol. 4801, pp. 1–20. Springer, Heidelberg (2007)

[6] Cook, W.R., Patwardhan, S., Misra, J.: Workflow patterns in orc. In: Ciancarini, P., Wiklicky, H. (eds.) COORDINATION 2006. LNCS, vol. 4038, pp. 82–96. Springer, Heidelberg (2006)

[7] Distefano, S., Scarpa, M., Puliafito, A.: From UML to Petri nets: The PCM-based methodology. TOSEM 37(1), 65–79 (2011)

[8] Grönniger, H., Reiß, D., Rumpe, B.: Towards a semantics of activity diagrams with semantic variation points. In: Petriu, D.C., Rouquette, N., Haugen, Ø. (eds.) MODELS 2010, Part I. LNCS, vol. 6394, pp. 331–345. Springer, Heidelberg (2010)

[9] Jensen, K., Kristensen, L.M.: Coloured Petri Nets – Modelling and Validation of Concurrent Systems. Springer (2009)

[10] Kordon, F., Thierry-Mieg, Y.: Experiences in model driven verification of behavior with UML. In: Choppy, C., Sokolsky, O. (eds.) Monterey Workshop 2008. LNCS, vol. 6028, pp. 181–200. Springer, Heidelberg (2010)

[11] Lime, D., Roux, O.H., Seidner, C., Traonouez, L.-M.: Romeo: A parametric model-checker for petri nets with stopwatches. In: Kowalewski, S., Philippou, A. (eds.) TACAS 2009. LNCS, vol. 5505, pp. 54–57. Springer, Heidelberg (2009)

[12] Mekki, A., Ghazel, M., Toguyeni, A.: Validating time-constrained systems using UML statecharts patterns and timed automata observers. In: VECoS, pp. 112–124. British Computer Society (2009)

[13] Merlin, P.M.: A study of the recoverability of computing systems. PhD thesis University of California, Irvine, CA, USA (1974)

[14] Workflow Patterns Initiative. Workflow patterns home page,
http://www.workflowpatterns.com

# SigVer3D: Accelerometer Based Verification of 3-D Signatures on Mobile Devices

Nguyen Ngoc Diep, Cuong Pham, and Tu Minh Phuong

**Abstract.** We present SigVer3D — a convenient authentication method for users of mobile devices with built-in accelerometers. The method works by analyzing streams of signals returned by a mobile device's accelerometer when the user uses the device to draw his (her) signature in 3-D space. We cast authentication as a binary classification problem and train SVM classifiers to identify successful logins. We explore two types of features to represent signal streams, which can be computed very fast even in devices with limited processing power, and demonstrate their effectiveness using gesture data collected from a group of subjects. Experimental results show that the method can differentiate between genuine users and imposters with average EER (equal error rate) of 0.8%. Given the wide availability of accelerometers in mobile devices, the method provides a promising complement to existing mobile authentication systems.

**Keywords:** authentication, accelerometer, SVM.

## 1 Introduction

Nowadays, mobile devices become more pervasive and supportive for people in everyday life activities. It is easy to find a plethora of wearable and mobile devices available on the market such as smart phones, Google Glass [7], Nike+ FuelBand [19], or Pepple Smartwatch [21]. These mobile devices can be taken with people almost everywhere at any time. Among important characteristics of mobile devices are their diversity in sizes, styles, input controls (i.e. displays), and their ability to store substantial volume of data, including sensitive personal data such as bank

Nguyen Ngoc Diep · Cuong Pham · Tu Minh Phuong
Computer Science Department
Posts and Telecommunications Institute of Technology
Hanoi, Vietnam
e-mail: {diepnguyenngoc,cuongpv,phuongtm}@ptit.edu.vn

© Springer International Publishing Switzerland 2015                                  353
V.-H. Nguyen et al. (eds.), *Knowledge and Systems Engineering,*
Advances in Intelligent Systems and Computing 326, DOI: 10.1007/978-3-319-11680-8_28

account or emails. Obviously, if such a mobile device is stolen or lost, its data would potentially be unauthorizedly accessible to thieves or strangers. Therefore a reliable authentication mechanism would be needed for protecting these personal data from strangers. However, traditional authentication schemes for computers (i.e. laptops and desktops) would probably be failed if directly applied for mobile devices as the mobile device's user interfaces are often extremely limited. For example, mobile devices such as tiny wireless sensors [9] have no displays whistle traditional authentication schemes often require the user to type passwords on keyboards for accessible applications on the computers. This dilemma would urge the necessary of a new paradigm which is flexible, comfortable-to use, secured enough but relevant and adaptable to the limitations of mobile device's user interfaces for mobile device authentication.

Emerging authentication methods that overcome the limitations of mobile device's input capacities include biometric authentication systems. These are commonly based on users' physiological and behavioral characteristics such as face [8], fingerprint [17], hand geometries [26], iris [2], keystrokes [4], hand-written signatures [17], voice [32], gait [6], etc. Although significant results (in term of system performance) have been achieved, many biometric authentication systems still encounter the problem of forgeability such as face or fingerprint systems [18]. Another alternative is signature verification. Signature does not comprise of physical properties (i.e. fingerprint or face) but behavioral characteristics. It has been widely accepted for the verification of government, legal and commercial transactions. There are two types of signature-based verification methods: offline (static) signature verification using signature images as input while online (dynamic) signature verification additionally uses signature's dynamic properties (such as direction, stroke, pressure, velocity) often captured by pressure-sensitive tablets or displays. However, one of the drawbacks of online signature-based authentication is its dependence on input devices such as pen or pressure-sensitive tablet while offline signatures seem not highly secured.

In this study, we propose a novel verification method for 3-D signatures, which we call SigVer3D. This method takes the advantages of secured signatures but allowing the users to "login" the mobile devices in a natural and comfortable manner while relieving the dependence of input devices (such as touch displays or keyboards). Our method utilizes the sensing data from accelerometers embedded in mobile devices as follows. A user holds the device and "draws" his (her) signature in the air to login to the mobile device. While the user performs his signature as a private gesture, the acceleration data is captured and used by the authentication system to verify the user's identity. Our method is suitable for mobile authentication as this overcomes the limitations of mobile device displays (i.e. tiny screens on a smart watch [21] or no screen on sensors [23]). The reason for the choice of signature gestures (3-D signatures) as passwords is that the signature gestures are easy to remember and repeat for genuine users while considerably challenging for imposters. The underlying mechanism of the proposed verification method is signature gesture recognition and verification. Two different types of features, one based on global binned distribution on the directions of time points and another utilizes local features on segments of signature patterns, are investigated in this paper. These features are used with support vector machines (SVM) classifiers to recognized signature patterns. The proposed method has been experimentally evaluated in gesture data from 30 genuine users each is forged by 10 different imposters.

## 2 Related Work

Biometric methods are mostly investigated for authentication mobile devices such as face and eye [8], pupil and iris [2], keystroke [4], gait [6], voice [32], fingerprint [17], etc. achieve significant results, they still suffer from forgeability [18] and may require additional equipments such as fingerprint readers. Some of them are inappropriate to mobile devices [8, 25, 26] which are resource-constraint, limited in form factor, with or without tiny screen. Moreover, several individuals raise privacy concerns for some biometric authentication systems like fingerprint, face [11]. Gesture-based authentication approaches would overcome these shortages. Several works utilises accelerometers or gyroscopes embedded inside the mobile phones [3, 5, 13, 20, 28]. Gestures used in these works can be simple or complex. Complex gestures are more secured but more difficult to remember even for genuine users, while simple ones are probably easy to be forged [14]. For example, F. Okumura et. al. [20] proposed an arm sweep gesture but 5% EER is achieved for the best outcomes. uWave [14] can authenticate by gesture recognition with 3% EER in the resilience case but in the case of shoulder surfing attack, the EER increases to 10%. Few works achieved good performance [3] but gestures are too complex with the inclusion of 6 elements: forward, backward, up, down, left, right, tilt left, tilt right, swing left, swing right. This would make the genuine user challenging to remember his password.

In contrast, signatures are memorable to most genuine users. Although 2-D static signatures are probably trivially forged [10, 25], 2-D dynamic signatures are often non-trivially imitated [10, 25]. However, 2-D dynamic signatures are significantly dependent on input devices such as pen or pressure-sensitive displays while not allowing the genuine users to perform his signature in natural, comfortable manner. In addition, the variations of 2-D signatures are often limited to strokes, direction, pressure, and velocity. Minority of studies [1, 12, 33] on mobile device authentication explored 3-D signature gestures. However, their methods are verified on too small datasets (i.e. only 4 users are evaluated in [12]) or achieve unsatisfactory performances. In contrast, in this study, we propose SigVer3D, a 3-D signature verification method that explore 3-dimentional signature gestures for mobile device authentication. The proposed method overcomes the shortages of mobile device's user interfaces (i.e. tiny or no screen on devices) while allowing genuine users to be easily memorable, dynamic and to perform their 3-D signature gestures in a comfortable and natural manner. Our method relies on a simple yet effective feature extraction scheme and is evaluated on the dataset comprises of 1800 3-D signature patterns in which 300 are genuine signatures and 1500 are forged signatures, collected from 30 different subjects.

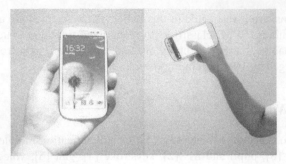

**Fig. 1** Samsung S3 (left) and the way users hold the phone to sign in the air (right)

## 3 Proposed Method

### 3.1 Data Processing

We assume that a mobile device be equipped with an accelerometer. Currently, this is the standard for smart phones [23]. In our study, we use Samsung S3 smart phones (see Fig. 1) with a built-in accelerometer able to sense acceleration along three axes. The accelerometer incorporated inside Samsung S3 is a part of LSM330DLC module which can measure acceleration with a minimum full-scale range of $\pm 2g$ [16] at the sampling rate of 20-120Hz. Each signal sample is a triplet of 3 dimensional values X, Y and Z. In this study, the sampling rate of the accelerometer is set to 100 Hz. Ideally, when running at this frequency, the sensor allows us to capture 100 samples within a second. However, sensor data are often noisy and cannot be used directly. Therefore, some sensing signal pre-processing techniques are inherited from our previous study [24] to remove noise and fill in dropped samples by interpolation. In addition, we truncate data segments at the first and the last seconds of signature patterns as these segments often contain significant noise.

### 3.2 Feature Extraction

We consider the impact of global and local features on the accuracy of the proposed verification system. Two feature sets are extracted from the raw acceleration data. The first set comprises of global features such as mean, standard deviation, energy, entropy, correlation over segments of each signature pattern. The second set is the binned distributions of the directions of time points over the signature pattern.

***Feature Set 1***

Each signature pattern is segmented into $k$ segments of equal size. Each segment contains $n$ samples (each signature pattern comprises of $k*n$ samples after pre-processing). Over $n$ time points on each segment, the following features are extracted:

$$Mean(x) = \frac{\sum_{i=1}^{n} x_i}{n} \tag{1}$$

$$Standard\ deviation(x): \delta_x = \sqrt{\frac{1}{n} \sum_{i=1}^{n} (x_i^2) - [Mean(x)]^2} \tag{2}$$

$$Energy(x) = \frac{\sum_{i=1}^{n} x_i^2}{n} \tag{3}$$

$$Entropy(x) = -\sum_{i=1}^{n} p(x_i) log(p(x_i)) \tag{4}$$

where $x_i$ is an acceleration value; $p(x_i)$, a probability distribution of $x_i$ within the frame, can be estimated as the number of $x_i$ in the frame divided by $n$, and the probability $0*log(0)$ is assumed to be 0.

$$Correlation(X,Y) = \frac{cov(x,y)}{\delta_x \delta_y} \tag{5}$$

in which $cov(x,y)$ is covariance and $\delta_x, \delta_y$ are standard deviations of two series $x$ and $y$.

The computed features over $k$ segments are then concatenated to form the feature vector for each signature pattern.

### Feature Set 2

Each signature pattern is segmented into quanta or slices of a fixed length. The binned distribution is computed over the directions of time points in each quantum. A bin, therefore, is a group of quanta with similar directions. The number of quanta mapped into a relevant bin is a feature. The histogram of quanta over the signature pattern forms a feature vector sized 3x$M$ (from three axis X, Y and Z). Details for the feature extraction process are as follows. The signal stream is a time series represented in $xy$-coordinates where $x$ is in millisecond, $y$ is in $m/s^2$. The direction $d(q)$ of a quantum $q$ is defined as the angle between the vector connecting its start $(x_1, y_1)$ and end $(x_2, y_2)$ points and the x-axis is computed as:

$$d(q) = arctan\left(\frac{y_2 - y_1}{x_2 - x_1}\right) \tag{6}$$

The value of $d(q)$ is in range $[-90^0, 90^0]$.

The direction $d(q)$ is then mapped into one of $M$ bins, where a bin is a group of quanta with similar directions. In other words, $d(q)$ is quantized into $M$ groups. In this paper, we set bins manually by varying the number of bins and stick on the number that maximizes the performance.

The intuition behind using this kind of features is that frames corresponding to similar signature patterns would have similar curves of signal streams and thus have similar number of quanta with close directions. Therefore, these features provide

**Fig. 2** Signatures in dataset

an approximate representation of signal streams, which can be computed efficiently and is suitable for a wide range of classification algorithms.

### 3.3 Pattern Recognition Using SVM

Support vector machines (SVM) is a learning algorithm that can be used to classify unseen data by relying on two techniques: i) mapping input features into a new feature space often of higher dimensions using a so called kernel function; and ii) finding in the new feature space a hyper-plane with max-margin that separate negative examples from positive ones. In this study, we use SMO (Sequential Minimal Optimization) library implemented in Weka [30]. Several kernel functions and parameters are tested using the popular grid search procedure on a subset of collected data. The polynomial kernel with exponent value $E = 3.0$ and SMO's complexity parameter $C = 0.1$ achieves the highest performance and is used for recognizing signature patterns.

## 4 Experiments and Results

### 4.1 Data Collection

30 subjects were asked to draw their 3-D signatures. A logging application developed for this study runs on the Samsung S3 smart phone and records acceleration data. A user starts performing his signature by touching and holding his finger on the phone's screen. While the finger touches the screen, the acceleration values are written into a logging file until the screen is released by the user's finger. The way users hold the phone to sign in the air is described as in Fig. 1 (right). The process is the same for imposters who tried to imitate the signature of the genuine user.

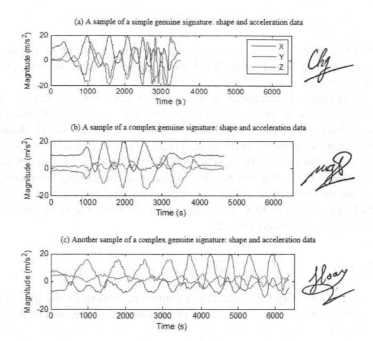

**Fig. 3** Signatures (right) and their acceleration data visualization (left)

Fig. 2 shows signatures of the subjects participating in this experiment as they sign on a paper, and Fig. 3 shows examples of signature patterns that the accelerometer outputs.

Each subject is asked to write down his signature on a piece of paper and perform his signature gesture 10 times. After that, for each subject, 10 other subjects are randomly chosen from the remaining subjects for playing as imposters. Each imposter is given the signature of the tested subject written on the paper and tried 5 times to imitate the signature. Therefore, for each subject, we have 10 genuine logins and 50 forged logins. Consequently, our collected dataset comprises of 1800 signature patterns of which 300 are genuine and 1500 are forged.

## 4.2 Performance Metrics

Evaluation of a verification system requires the analysis of two types of errors: the *false rejection rate* (FRR), indicating the percentage of genuine signatures that are incorrectly rejected by the system, and the *false accepted rate* (FAR), indicating the percentage of incorrectly accepted forgeries. Formally, FAR and FRR are defined as:

$$FAR = \frac{FP}{FP+TN} \qquad (7)$$

$$FRR = \frac{FN}{FN + TP} \tag{8}$$

In which, $FP$ is false positives (incorrectly accepted forgeries), $TP$ is true positives (genuine signatures detected), and $FN$ is false negatives (genuine signatures that are incorrectly rejected).

For a broad range of authentication systems, FAR and FRR are used for measuring the performances. In addition, another performance measure is the error trade-off curve called receiver operating characteristics curve (ROC), which shows how one error changes with respect to the other. The ROC plot is a visual characterization of the trade-off between the FAR and the FRR. A ROC curve is often summarized as area under ROC (AUC). Another metrics, *equal error rate* (EER), commonly used for measuring the performance of the verification systems, is the rate at which FAR equals FRR. In contrast to ROC curve, the lower the EER, the more accurate and reliable the verification system is [15].

## 4.3   Results

For feature set 1, we vary $k$, the number of segments, to evaluate the impact of the number of signature pattern segments on the system performance. The results are plotted in Fig. 4.

In Fig. 4, the best performance is achieved with $k = 7$, at which the EER is below 1% (on the left of the figure) while the area under ROC curve raises up to over 0.99. It is observed that there are about 5 — 8 (7 is dominant) acceleration value peaks on the signature patterns. A signature pattern is partitioned into 7 equally sized segments and each contains a peak value and its neighbor time points. By the experiment presented in our previous study on human activity recognition [24], this partition would exhibit good features from the dataset, and therefore would enhance the performance of the recognition systems [22].

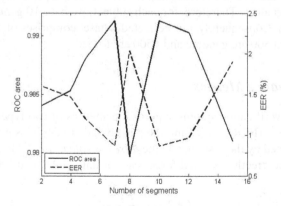

**Fig. 4**  SigVer3D's performance on feature set 1

**Table 1** Result of the verification system on both feature sets

| Subject | Feature set 1 | | Feature set 2 | |
|---|---|---|---|---|
| | ROC area | EER (%) | ROC area | EER (%) |
| 1 | 0.94 | 6 | 0.88 | 12 |
| 2 | 1 | 0 | 0.94 | 6 |
| 3 | 1 | 0 | 0.9 | 10 |
| 4 | 1 | 0 | 0.99 | 1 |
| 5 | 1 | 0 | 0.88 | 12 |
| 6 | 0.95 | 5 | 0.95 | 5 |
| 7 | 1 | 0 | 0.99 | 1 |
| 8 | 1 | 0 | 0.99 | 1 |
| 9 | 1 | 0 | 1 | 0 |
| 10 | 0.95 | 5 | 0.98 | 4 |
| 11 | 1 | 0 | 0.93 | 7 |
| 12 | 0.98 | 2 | 0.97 | 3 |
| 13 | 1 | 0 | 0.83 | 17 |
| 14 | 1 | 0 | 1 | 0 |
| 15 | 1 | 0 | 1 | 0 |
| 16 | 1 | 0 | 1 | 0 |
| 17 | 1 | 0 | 0.99 | 1 |
| 18 | 1 | 0 | 1 | 0 |
| 19 | 0.98 | 2 | 0.97 | 3 |
| 20 | 0.95 | 5 | 0.99 | 10 |
| 21 | 1 | 0 | 0.99 | 10 |
| 22 | 1 | 1 | 0.99 | 1 |
| 23 | 1 | 0 | 0.97 | 3 |
| 24 | 1 | 0 | 0.99 | 1 |
| 25 | 1 | 0 | 1 | 0 |
| 26 | 1 | 0 | 0.99 | 1 |
| 27 | 0.99 | 1 | 0.98 | 2 |
| 28 | 1 | 0 | 0.99 | 1 |
| 29 | 1 | 0 | 0.98 | 2 |
| 30 | 1 | 0 | 1 | 0 |
| **Average** | **0.991** | **0.8** | **0.963** | **3.7** |

The overall performance of our verification system on feature set 2 is illustrated in the Fig. 5. For feature set 2, we investigate how the number of bins $M$ impact on the system accuracy. Therefore, the number of bins is varied from 5 to 10. In Fig. 5, the bottom of EER is obtained at the number of bins $M = 6$ and so is the peak of ROC curve. Overall, feature set 1 yields better performance than feature set 2: the EER values are 0.8% for feature set 1 and 3.7% for feature set 2. A possible explanation

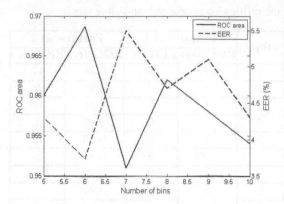

**Fig. 5** SigVer3D's performance on feature set 2

for the superiority of feature set 1 over set 2 is that feature set 2 is proposed to deal with the distinction of short-time patterns such as *fall* or *sit* actions (often occurring in one or two seconds) as the acceleration values changing very quickly. Signatures, however, are mostly performed within 2 to 5 seconds, hence the sensing values can possibly be changed in a moderate fashion over time. It is noticed that feature set 1 often deals well with the patterns in various lengths by segmenting the patterns into sliding windows [22]. The result details of our verification system on both feature sets are represented in the Table 1.

In Table 1, 22 out of 30 subjects have EER of 0% for feature set 1, which means that genuine users performed logins 10 times and all are successful while each imposter tried 50 forges for each signature but all have been failed. By observing the signature images on the Fig. 2, some of these 22 signatures look quite complex, others look simple. It can be concluded that although signatures can probably be easily to be imitated with hand-written [10], it is a significant challenge to forge 3-D signatures even the imposters are given the signatures written on pieces of paper. This can be explained that 3-D signature gestures are natural and intuitive for the genuine users. The implicit of the 3-D signatures in this study is acceleration sensing data that particularly sensitive to the user's gestures, habits, and their intensities. Moreover, the directions of the device and the speed of performing the signature gestures would significantly be sensitive to the data of the 3-D signature patterns.

On feature set 1, 8 other subjects have EER between 1% to 6%. Among them, only one subject has 6% EER, which is high for a verification system. 6% EER indicates that of which 100 times "login", 6 would be failed, and 6 forges are successful from imposters. Looking at the performance of SigVeg3D on feature set 2, the number of subjects with EER of 0% decreases to 7 while 6 subjects have EER excessive 10%, which are substantially less reliable and accurate of SigVer3D's performance on feature set 1. This consolidates our argument that the feature set 1 is appropriate to the recognition of patterns with various time lengths.

## 5 Conclusion

We have presented a novel authentication method for mobile devices equipped with accelerometers. To get access to a device, the user moves the device in 3-D space to imitate his/her signature. The system compares the signal stream generated by accelerometer during this process with the model constructed in the training phase to verify the user's identity. We have shown that, by extracting appropriate features from signal streams, the method can achieve relatively high accuracy. We explored two types of features representing different aspects of signal streams, which can be extracted from accelerometer efficiently. In an experiment with signing gestures provided by 30 subjects, the method achieves an average equal error rate as low as 0.8%. More importantly, the results are robust with respect to the complexity of signatures. Due to its easy usage, the proposed method has potential to be a good complement to existing authentication systems for mobile devices.

## References

[1] Casanova, J.G., Ávila, C.S., de Santos Sierra, A., del Pozo, G.B., Vera, V.J.: A real-time in-air signature biometric technique using a mobile device embedding an accelerometer. In: Zavoral, F., Yaghob, J., Pichappan, P., El-Qawasmeh, E. (eds.) NDT 2010. CCIS, vol. 87, pp. 497–503. Springer, Heidelberg (2010)

[2] Cho, D.H., Park, K.R., Rhee, D.W., Kim, Y., Yang, J.: Pupil and iris localization for iris recognition in mobile phones. In: Seventh ACIS International Conference on Software Engineering, Artificial Intelligence, Networking, and Parallel/Distributed Computing, SNPD 2006, pp. 197–201. IEEE (2006)

[3] Chong, M.K., Marsden, G.: Exploring the use of discrete gestures for authentication. In: Gross, T., Gulliksen, J., Kotzé, P., Oestreicher, L., Palanque, P., Prates, R.O., Winckler, M. (eds.) INTERACT 2009. LNCS, vol. 5727, pp. 205–213. Springer, Heidelberg (2009)

[4] Clarke, N.L., Furnell, S.M.: Authenticating mobile phone users using keystroke analysis. International Journal of Information Security 6(1), 1–14 (2007)

[5] Farella, E., O'Modhrain, S., Benini, L., Riccó, B.: Gesture signature for ambient intelligence applications: A feasibility study. In: Fishkin, K.P., Schiele, B., Nixon, P., Quigley, A. (eds.) PERVASIVE 2006. LNCS, vol. 3968, pp. 288–304. Springer, Heidelberg (2006)

[6] Gafurov, D., Helkala, K., Søndrol, T.: Biometric gait authentication using accelerometer sensor. Journal of Computers 1(7), 51–59 (2006)

[7] Google Glass, http://www.google.com/glass (accessed on May 25, 2015)

[8] Hadid, A., Heikkila, J.Y., Silvén, O., Pietikainen, M.: Face and eye detection for person authentication in mobile phones. In: First ACM/IEEE International Conference on Distributed Smart Cameras, ICDSC 2007. IEEE (2007)

[9] Hooper, J., Preston, A., Balaam, M., Seedhouse, P., Jackson, D., Pham, C., Ladha, C., Ladha, K., Ploetz, T., Olivier, P.: The French Kitchen: Task-Based Learning in an Instrumented Kitchen. In: Proc. of the 14th ACM International Conference on Ubiquitous Computing, Ubicomp 2012, pp. 193–202 (2012)

[10] Jain, A.K., Griess, F.D., Connell, S.D.: On-line signature verification. Pattern Recognition 35(12), 2963–2972 (2002)

[11] Jain, A.K., Ross, A., Prabhakar, S.: An introduction to biometric recognition. IEEE Transactions on Circuits and Systems for Video Technology 14(1), 4–20 (2004)

[12] Ketabdar, H., Yüksel, K.A., Yüksel, K.A., Jahnbekam, A., Roshandel, M., Skirpo, D.: Magisign: User identification/authentication based on 3d around device magnetic signatures. In: The Fourth International Conference on Mobile Ubiquitous Computing, Systems, Services and Technologies, UBICOMM 2010, pp. 31–34 (2010)

[13] Liu, J., Zhong, L., Wickramasuriya, J., Vasudevan, V.: User evaluation of lightweight user authentication with a single tri-axis accelerometer. In: Proceedings of the 11th International Conference on Human-Computer Interaction with Mobile Devices and Services, p. 15. ACM (2009)

[14] Liu, J., Zhong, L., Wickramasuriya, J., Vasudevan, V.: uWave: Accelerometer-based personalized gesture recognition and its applications. Pervasive and Mobile Computing 5(6), 657–675 (2009)

[15] Liu, S., Silverman, M.: A practical guide to biometric security technology. IT Professional 3(1), 27–32 (2001)

[16] LSM330DLC Dataset, `http://www.st.com/st-web-ui/static/active/en/resource/technical/document/datasheet/DM00037200.pdf` (accessed on May 25, 2015)

[17] Maltoni, D., Maio, D., Jain, A.K., Prabhakar, S.: Handbook of fingerprint recognition. Springer (2009)

[18] Matsumoto, T., Matsumoto, H., Yamada, K., Hoshino, S.: Impact of artificial gummy fingers on fingerprint systems. In: Electronic Imaging 2002, pp. 275–289. International Society for Optics and Photonics (2002)

[19] Nike+ FuelBand, `http://www.nike.com/us/en_us/c/nikeplus-fuelband` (accessed on May 25, 2015)

[20] Okumura, F., Kubota, A., Hatori, Y., Matsuo, K., Hashimoto, M., Koike, A.: A study on biometric authentication based on arm sweep action with acceleration sensor. In: International Symposium on Intelligent Signal Processing and Communications, ISPACS 2006, pp. 219–222. IEEE (2006)

[21] Peple Smartwatch, `https://getpebble.com` (accessed on May 25, 2015)

[22] Pham, C., Diep, N.N., Phuong, T.M.: A wearable sensor based approach to real-time fall detection and fine-grained activity recognition. Journal of Mobile Multimedia 9(1-2), 15–26 (2013)

[23] Pham, C., Hooper, C., Lindsay, S., Jackson, D., Shearer, J., Wagner, J., Ladha, C., Ladha, K., Plotz, T., Olivier, P.: The Ambient Kitchen: A Pervasive Sensing Environment for Situated Services. In: Proc. of the ACM Designing Interactive Systems Conference, DIS 2012 (2012)

[24] Pham, C., Phuong, T.M.: Real-time fall detection and activity recognition using low-cost wearable sensors. In: Murgante, B., Misra, S., Carlini, M., Torre, C.M., Nguyen, H.-Q., Taniar, D., Apduhan, B.O., Gervasi, O. (eds.) ICCSA 2013, Part I. LNCS, vol. 7971, pp. 673–682. Springer, Heidelberg (2013)

[25] Plamondon, R., Lorette, G.: Automatic signature verification and writer identification the state of the art. Pattern Recognition 22(2), 107–131 (1989)

[26] Ross, A., Jain, A.K.: A prototype hand geometry-based verification system. In: Proceedings of 2nd Conference on Audio and Video Based Biometric Person Authentication, pp. 166–171 (1999)

[27] Sakoe, H., Chiba, S.: Dynamic programming algorithm optimization for spoken word recognition. IEEE Transactions on Acoustics, Speech and Signal Processing 26(1), 43–49 (1978)

[28] Shu, Y., Gu, Y., Chen, J.: Dynamic authentication with sensory information for the access control systems (2014)

[29] Uludag, U., Pankanti, S., Prabhakar, S., Jain, A.K.: Biometric cryptosystems: issues and challenges. Proceedings of the IEEE 92(6), 948–960 (2004)

[30] Weka, http://www.cs.waikato.ac.nz/ml/weka (accessed on May 25, 2015)

[31] Wiedenbeck, S., Waters, J., Birget, J.C., Brodskiy, A., Memon, N.: Authentication using graphical passwords: effects of tolerance and image choice. In: Proceedings of the 2005 Symposium on Usable Privacy and Security, pp. 1–12. ACM (2005)

[32] Woo, R.H., Park, A., Hazen, T.J.: The MIT mobile device speaker verification corpus: data collection and preliminary experiments. In: Speaker and Language Recognition Workshop, IEEE Odyssey 2006, pp. 1–6. IEEE (2006)

[33] Zaharis, A., Martini, A., Kikiras, P., Stamoulis, G.: "User Authentication Method and Implementation Using a Three-Axis Accelerometer". In: Chatzimisios, P., Verikoukis, C., Santamaría, I., Laddomada, M., Hoffmann, O. (eds.) MOBILIGHT 2010. LNICST, vol. 45, pp. 192–202. Springer, Heidelberg (2010)

# New Mechanism of Combination Crossover Operators in Genetic Algorithm for Solving the Traveling Salesman Problem

Pham Dinh Thanh, Huynh Thi Thanh Binh, and Bui Thu Lam

**Abstract.** Traveling salesman problem (*TSP*) is a well-known in computing field. There are many researches to improve the genetic algorithm for solving *TSP*. In this paper, we propose two new crossover operators and new mechanism of combination crossover operators in genetic algorithm for solving *TSP*. We experimented on *TSP* instances from *TSP*-Lib and compared the results of proposed algorithm with genetic algorithm (*GA*), which used *MSCX*. Experimental results show that, our proposed algorithm is better than the *GA* using *MSCX* on the min, mean cost values.

**Keywords:** Traveling Salesman Problem, Genetic Algorithm, Modified Sequential Constructive Crossover.

## 1 Introduction

The traveling salesman problem is an important problem in computing fields and has many applications in the daily life such as scheduling, vehicle routing, *VLSI* layout design, etc. The problem was first formulated in 1930 and it has been one of the most intensively studied problems in optimization techniques. Until now, researchers have obtained numerous significant results for this problem.

  *TSP* is defined as following: Let 1, 2, ..., $n$ is the labels of the $n$ cities and $C = [c_{i,j}]$ be an $n \times n$ cost matrix where $c_{i,j}$ denotes the cost of traveling from city $i$ to city $j$. *TSP* is the problem of finding the $n$-city closed tour having the minimum cost such that each city is visited exactly once. The total cost A of a tour is.

Pham Dinh Thanh
Tay Bac University, Son La, Vietnam

Huynh Thi Thanh Binh
Hanoi University of Science and Technology, Hanoi, Vietnam

Bui Thu Lam
Le Quy Don Technical University, Hanoi, Vietnam

© Springer International Publishing Switzerland 2015                               367
V.-H. Nguyen et al. (eds.), *Knowledge and Systems Engineering*,
Advances in Intelligent Systems and Computing 326, DOI: 10.1007/978-3-319-11680-8_29

$$A(n) = \sum_{i=1}^{n-1} c_{i,i+1} + c_{n,1} \qquad (1)$$

*TSP* is formulated as finding a permutation of $n$ cities, which has the minimum cost. This problem is known to be *NP*-hard [1, 2] but it can be applied in many real world applications [13] so a good solution would be useful.

Many algorithms have been suggested for solving *TSP*. *GA* is an approximate algorithm based on natural evolution, which applied to many different types of the combinatorial optimization. *GA* can be used to find approximate solutions for *TSP*.

There are a lot of improvements in *GA* that have been developed to increase the performance in solving the *TSP* such as: optimizing creating initial population [3], improving mutation operator [17], creating new crossover operator [12, 20, 21, 22, 23, 24, 25], combining with local search [4, 6, 7, 8, 18].

In this paper, we introduce two new crossover operators: *MSCX_Radius* and *RX*. We propose new mechanism of combination proposed crossover operators and *MSCX* [25] in *GA* to solve *TSP*. This combination is expected to adapt the changing of population. We experimented on *TSP* instances from *TSP*-Lib and compared the results of proposed algorithm with *GA* which used *MSCX*. Experimental results show that, our proposed algorithm is better than the *GA* using *MSCX* on the min, mean cost values.

The rest of this paper is organized as follows. In section 2, we will present related works. Section 3 and 4 contain the description of our new crossovers and the proposed algorithm for solving *TSP* respectively. The details of our experiments and the computational and comparative results are given in section 5. The paper concludes with section 6 with some discussions on the future extension of this work.

## 2   Related Work

*TSP* is *NP*-hard problems. There are two approaches for solving *TSP*: exact and approximate. Exact approaches are based on Dynamic Programming [14], Branch and Bound [2], Integer Linear Programming [21], etc. Exact approaches used to give the optimal solutions for *TSP*. However, these algorithms have exponential running time, therefore they only solved small instances. As M. Held and R. M. Karp [14] pointed out Dynamic Programming takes $O(n^2 \cdot 2^n)$ running time, so that it only solves *TSP* with a small number of the vertices.

In recent years, approximation approaches for solving *TSP* are interested by researchers. These approaches can solve large instances and give approximate solutions near to the optimal solution (sometime optimal). Approximation approaches for solving *TSP* are 2-opt, 3-opt [1], simulated annealing [7, 16], tabu search [7, 16]; nature based optimization algorithms and population based optimization algorithms: genetic algorithm [3, 6, 7, 8, 10, 11, 12, 13, 16, 17, 19, 22, 25], neural networks [15]; swarm optimization algorithms: ant colony optimization [7, 23], bee colony optimization [18].

*GA* is one of computational model inspired by evolution, which has been applied to a large number of real world problems. *GA* can be used to get approximate solutions for *TSP*. High adaptability and the generalizing feature of *GA* help to execute the traveling salesman problem by a simple structure.

M. Yagiura and T. Ibaraki [16] proposed *GA* for three permutation problems including *TSP*; and *GA* solving *TSP* uses *DP* in its crossover operator. The experiments are executed on 15 randomized Euclidean instances (5 instances for each $n = 100$, 200, 500). The proposed algorithm [16] could get better solutions than Multi-Local, Genetic-Local and Or-opt when sufficient computational time was allowed. However, the experimental results have been pointed out that, their proposed algorithm is ineffective to compare some heuristics specially designed to the given *TSP*, such as Lin-Kernighan method [1, 16].

In [5], the authors used local search and *GA* for solving *TSP*. The experiments are executed in kroA100, kroB100 and kroC100 instances. The experiments results show that the combination of two genetic operators, *IVM* and *POS*, and 2-opt have better cost for solving *TSP* problem. However, the algorithm took more time to converge to the global optimum than using 3-opt.

In 1997, Bernd Freisleben, Peter Merz [7] proposed Genetic Local Search for the *TSP*. This algorithm used idea of hill climber to develop local search in *GA*. The experiment shows that the best solutions are better than the one in [24] on running time and better on min cost range from $0.46\% \rightarrow 0.21\%$.

Crossover operator is one of the most important component in *GA*, which generates new individual(s) by combining genetic material from two parents but preserving gene from the parents. The researchers have studied many different optimal crossover operators like creating new crossover operators [22, 24], modifying exist crossover operators [20, 21, 23, 25], and hybridizing crossover operators [10].

Sehrawat, M. et al. [20] modified Order Crossover (*OX*). They selected the first crossover point which is the first node of the minimum edge from second chromosome. The experiment was executed on five sample data. The modifying order crossover (*MOX*) could get better solutions than *OX* on two sample data but number of the best solutions is found by *MOX* more than *OX*.

The new genetic algorithm (called *FRAG_GA*) was developed by Shubhra, Sanghamitra and Sankar [21]. There were two new operators: nearest fragment (*NF*) and modified order crossover (*MOC*). The *NF* is used for optimizing initial population. In the *MOC*, the authors performed two changes: length of a substring for performing order crossover is $y = \max\{2, \alpha\}$, where $n/9 \leq \alpha \leq n/7$ (*n* is the total number of cities) and the length of substring is predefined at any times performing crossover. The experiments are executed in Grtschels24, kroA100, d198, ts225, pcb442 and rat783 instances. The authors compared *FRAG_GA* with *SWAPGATSP* [12] and *OXSIM* (standard *GA* with order crossover and simple inversion mutation) [13]. The experiment results showed that the best result, the average result and computation time of *FRAG_GA* are better than one of *SWAPGATSP*, *OXSIM*.

In [22], the authors proposed an improving *GA* (*IGA*) with a new crossover operator (Swapped Inverted Crossover - *SIC*) and a new operation called Rearrangement. *SIC* creates 12 children from 2 parents then select 10 for applying multi mutation. Finally select 2 best individuals. Rearrangement Operation is applied to all individuals in population. It finds the maximum cost of two adjacent cities then swap one city with three other cities. The experiments are executed 10 times for each instances (KroA100, D198, Pcb442 and Rat783). The experiments show that, performance of *IGA* is better than the three compared *GAs*.

Kusum and Hadush [23] modified the *OX*. In these proposing crossovers, the positions of cut points or the length of the substrings in both parents are different. The experimented on six Euclidean instances derived from *TSP*-lib (eil51, eil76, kroA100, eil101, lin105 and rat195). Crossover rate is 0.9 and mutation rate is 0.01. The experimental results show that results of one modifying crossover are better than *OX* for six *TSP* instances.

In [24], the authors proposed new crossover operator, Sequential Constructive crossover (*SCX*). The main idea of *SCX* is selecting the edges having less value based on maintaining the sequence of cities in the parents. The experiments are performed in 27 *TSPLIB* instances. Results of experiment show that *SCX* is better than the *ERX* and *GNX* on quality of solutions and solution times.

In 2012, Sabry, Abdel-Moetty and Asmaa [25] proposed new crossover operator, Modified Sequential Constructive crossover (*MSCX*), which is an improvement of the *SCX* [24]. The *MSCX* create an offspring and description as follows:

**Step 1:** Start from 'First Node' of the parent 1 (i.e., current node p = parent1(1)).

**Step 2:** Sequentially search both of the parent chromosomes and consider

The first 'legitimate node' (the node that is not yet visited) appeared after 'node p' in each parent. If no 'legitimate node' after node p is present in any of the parent, search sequentially the nodes from parent 1 and parent 2 (the first 'legitimate node' that is not yet visited from parent1 and parent2), and go to Step 3.

**Step 3:** Suppose the 'Node $\alpha$' and the 'Node $\beta$' are found in 1st and 2nd parent respectively, then for selecting the next node go to Step 4.

**Step 4:** If $C_{p\alpha} < C_{p\beta}$, then select 'Node $\alpha$', otherwise, 'Node $\beta$' as the next node and concatenate it to the partially constructed offspring chromosome. If the offspring is a complete chromosome, then stop, otherwise, rename the present node as 'Node p' and go to Step 2.

Although a lot of crossovers were developed for solving *TSP*, but each operator has its property, so, in this paper, we propose two new crossover operators and mechanism of combination them with *MSCX* crossover [25]. This scheme is expected to adapt the changing and convergence of population and improve the effectiveness in terms of cost of tour. The proposed algorithm will be presented in the next section.

## 3 Proposed Crossover Operators

This section introduces two new crossover operators: *MSCX_Radius*, *RX*, which are developed for improving the best solutions and increase the diversity of the population.

## 3.1  MSCX_Radius Crossover

*MSCX_Radius* modify the step two of *MSCX* [25]. In *MSCX_Radius*, if no 'legitimate node' after current node, find sequentially r nodes, which are not visited from the parents. Then select the node having the smallest distance to current node. r is parameter of *MSCX_Radius*.

## 3.2  RX Crossover

This crossover operator is described as following:

**Step 1:** Randomly select *pr%* cities from the first parent to the offspring.

**Step 2:** Copy the remaining unused cities into the offspring in the order they appear in the second parent.

**Step 3:** Create the second offspring in an analogous manner, with the parent roles reverse.

Figure 1 show an example of *RX* crossover operator.

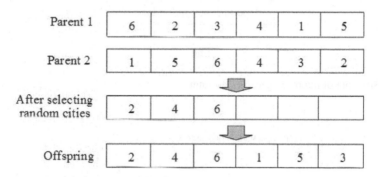

**Fig. 1** Illustration of the RX crossover operator, pr% = 20%

## 4  Proposed Mechanism of Combination Two New Crossovers and MSCX

This section proposes new mechanism of combination two propose crossover operators *MSCX_Radius* and *RX* with *MSCX*. We then use apply this mechanism in an improving genetic algorithm (*CXGA*) for solving *TSP*.

The workflow of *CXGA* is described in Fig.2.

The workflow of *HRX* module is shown in Fig.3.

In the first part, *prx%* of individuals will be chosen for *RX* crossover and the rest for *MSCX_Radius*

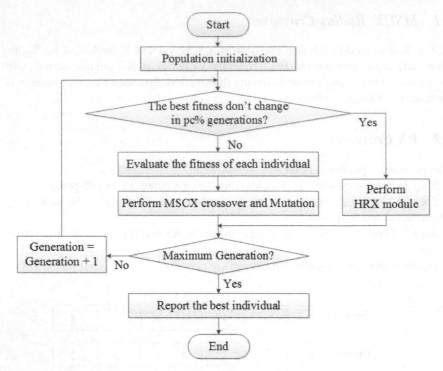

**Fig. 2** Structure of improved genetic algorithm

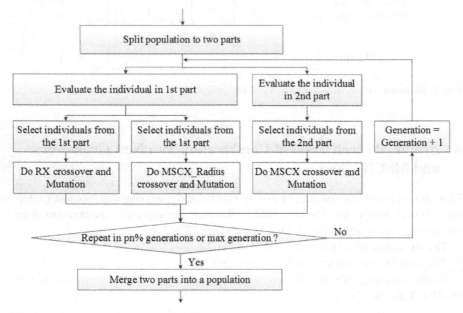

**Fig. 3** Structure of HRX module in CXGA

Sketch of the *HRX* module is presented as below:

```
Procedure: HRX (P, prx, pr, r)
Input:   The population P
         r: parameter of MSCX_Radius
         prx: percent of individuals from first part use RX
         pr: percent of number of cities is gotten random in RX
Output: The optimization population P'

Begin
   Split P into two parts: P1 and P2;
   i ← 0; FPᵢ ← P2; SPᵢ ← P1; numInRX ← (prx * |P1|)/100;
   ng ← number of generations perform HRX module;
   While i < ng do
      For j := 1 to numInRX/2 do
         Select random individuals from SPᵢ;
         Do RX(pr)crossover, mutation;
         Add offsprings to SPᵢ₊₁;
      End for
      For j :=1 to |P1| - numInRX do
         Select random individuals from SPᵢ;
         Do MSCX_Radius(r) crossover, mutation;
         Add offspring to SPᵢ₊₁;
      End for
      For j := 1 to |P2| do
         Select random individuals from FPᵢ;
         Do MSCX crossover, mutation;
         Add offspring to FPᵢ₊₁;
      End for
      i ← i + 1;
   End while
   Merge FPi, SPi into P';
   Return P'
End;
```

# 5 Computational Results

## 5.1 Problem Instances

The results are reported for the symmetric *TSP* by extracting benchmark instances from the *TSP*-Lib [9]. The instances chosen for our experiments are eil51.tsp, Pr76.tsp, Rat99.tsp, KroA100, Lin105.tsp, Bier127.tsp, Ts225.tsp, Gil262.tsp, A280.tsp, Lin318.tsp, Pr439.tsp and Rat575.tsp. The number of vertices: 51, 76, 99, 100, 105, 127, 225, 262, 280, 318, 439, 575. Their weights are Euclidean distance in 2-D.

## 5.2   System Setting

In the experiment, the system was run 10 times for each problem instance. All the programs were run on a machine with Intel Pentium Duo E2180 2.0GHz, 1GB RAM, and were installed by C# language.

## 5.3   Experimental Setup

This paper implemented two sets of experiments. In the first, we run *GA* using *MSCX_Radius* (named *GA1*), *GA* using *RX* (named *GA2*) and compare with *GA* using *MSCX* [25] (named *GA3*). In the second, we compare the performance of *CXGA* with *GA3*.

When execute the *HRX* module, the population is split into two part. The first one includes the best solutions which uses a combination of *MSCX_Radius* and *RX* crossover; the second includes the rest solutions of population, which uses *MSCX* crossover.

The parameters for experiments are:
   Population size: $p_s = 100$
   Number of evaluation: 1000000
   Mutation rate: $p_m = 1/\text{number\_of\_city}$ (chromo length)
   Crossover rate: $p_c = 0.9$

## 5.4   Experimental Resultstitle

The experiments were implemented in order to compare *GA1*, *GA2*, *CXGA* with *GA3* in term of the min, mean, standard deviation values and running times.

For comparing effects of two new crossover operations: *MSCX_Radius* and *RX*. We tested *GA1*, *GA2* with different values of *r*, *pr* parameters. The best results obtain by *GA1*, *GA2* are compared with the ones obtain by *GA3*.

Figure 4 summarizes the mean cost of *GA1* when *r* = 2, 3 and 5 respectively. With *r* = 2, the results found by *GA1* are the best.

The Fig. 5 illustrates the mean cost of *GA2* when *pr%* = 10%, 30% and 50% respectively. The diagrams show that, the mean cost of *GA2* when *pr%* = 10% are better than ones when *pr%* = 30%, 50%.

Experiment results on Fig. 4, Fig. 5 show that *r* = 2 and *pr%* = 10% are the best parameters for *GA1* and *GA2* and they will be selected for comparison with *GA3*.

*HRX* module was implemented in differences parameters to find the best parameter. The size of the first part is 90%, *pc* =15%, *r* = 5, *pr* = 30%, *pn* = 5%, *prx* = 40%.

In order to select the best value of pc in *HRX* module, we analyzed the correlative of the best solution obtaining from *CXGA* with different values of *pc* parameter (*pc%* = 5%, 10%, 15%, 20%, 30% and 50%).

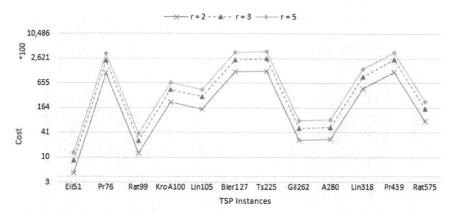

**Fig. 4** The mean cost on TSP instances of GA1 when r = 2, 3 and 5

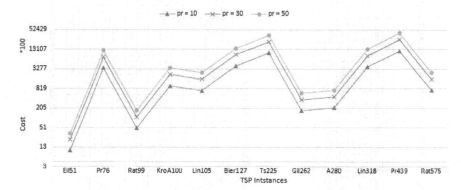

**Fig. 5** The mean cost on TSP instances of GA2 when pr% = 10%, 30% and 50%

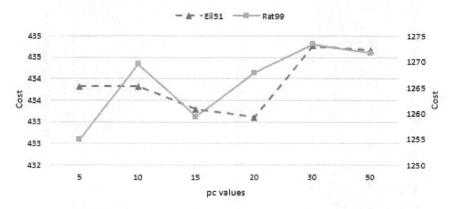

**Fig. 6** The relationship between the pc values, mean cost found by 10 running times of CXGA on Eil51, Rat99 instance

**Table 1** The results of CXGA found by 10 running times on Eil51, Pr76, Rat99, KroA100, Lin105, Bier127, Ts225, Gil262, A280 when r = 2,3,5,7,10

| | Eil51 | | Pr76 | | Rat99 | | KroA100 | | Lin105 | | Bier127 | | Ts225 | | Gil262 | | A280 | |
|---|---|---|---|---|---|---|---|---|---|---|---|---|---|---|---|---|---|---|
| | Min | Mean | Min | Mean | Min | Mean | Min | Mean | Min | Mean | Min | Mean | Min | Mean | Min | Mean | Min | Mean |
| r = 2 | 429.0 | 432.9 | 109695.8 | 111698.7 | 1242.4 | 1256.9 | 21505.5 | 22036.3 | 14614.4 | 14804.5 | 120634.3 | 122241.3 | 127677.2 | 128737.8 | 2527.8 | 2569.2 | 2672.2 | 2732.0 |
| r = 3 | 430.6 | 439.4 | 110165.3 | 111987.0 | 1239.9 | 1261.4 | 21779.4 | 22119.6 | 14645.9 | 14793.3 | 120303.0 | 121845.6 | 127619.0 | 129039.7 | 2534.7 | 2566.8 | 2664.3 | 2726.5 |
| r = 5 | 429.0 | 432.6 | 106651.4 | 111688.5 | 1230.1 | 1256.5 | 21753.5 | 22014.8 | 14593.5 | 14795.9 | 120819.0 | 122073.9 | 128159.5 | 128985.9 | 2484.8 | 2552.4 | 2666.1 | 2731.3 |
| r = 7 | 430.6 | 433.1 | 109695.8 | 111918.3 | 1241.7 | 1259.3 | 21694.8 | 22068.8 | 14612.0 | 14820.2 | 120846.0 | 122113.4 | 127848.3 | 128992.4 | 2484.9 | 2562.7 | 2659.0 | 2732.5 |
| r = 10 | 430.6 | 434.1 | 109695.8 | 111832.7 | 1235.4 | 1256.3 | 21753.5 | 22133.6 | 14673.7 | 14803.7 | 120846.0 | 122211.1 | 128573.6 | 129016.1 | 2473.6 | 2556.8 | 2672.2 | 2728.8 |

Min: Minimum cost; Mean: Mean cost

**Table 2** The Results found by GA1, GA2, GA3 and CXGA algorithm for TSP intstances

| Instances | NCity | GA1 | | | | GA2 | | | | GA USING MSCX | | | | CXGA | | | |
|---|---|---|---|---|---|---|---|---|---|---|---|---|---|---|---|---|---|
| | | Min | Mean | Std | R_Time | Min | Mean | Std | R_Time | Min | Mean | Std | R_Time | Min | Mean | Std | R_Time |
| Eil51 | 51 | 432.1 | 435.1 | 2.9 | 1.0 | 925.6 | 1013.9 | 50.2 | 0.4 | 431.5 | 435.4 | 3.7 | 0.9 | 429.0 | 432.6 | 2.7 | 1.0 |
| Pr76 | 76 | 110919.0 | 115001.6 | 2097.7 | 1.6 | 315495.9 | 349169.1 | 16502.1 | 0.5 | 112022.0 | 113556.0 | 986.5 | 1.5 | 110651.4 | 111688.5 | 804.4 | 1.5 |
| Rat99 | 99 | 1244.3 | 1284.1 | 20.5 | 2.2 | 4555.6 | 4880.2 | 205.4 | 0.6 | 1249.5 | 1271.9 | 20.6 | 2.1 | 1239.1 | 1256.5 | 15.2 | 2.1 |
| KroA100 | 100 | 21978.1 | 22329.4 | 325.8 | 2.2 | 94681.5 | 99235.6 | 3029.9 | 0.6 | 21868.2 | 22388.6 | 421.6 | 2.2 | 21753.5 | 22014.8 | 258.0 | 2.2 |
| Lin105 | 105 | 14751.6 | 15193.1 | 440.9 | 2.4 | 66635.6 | 70254.7 | 2676.8 | 0.6 | 14689.9 | 14966.3 | 215.3 | 2.4 | 14593.5 | 14795.9 | 153.9 | 2.3 |
| Bier127 | 127 | 121263.6 | 122873.1 | 800.8 | 3.2 | 373473.6 | 395735.0 | 14681.4 | 0.6 | 121610.5 | 123742.7 | 985.6 | 3.1 | 120819.0 | 122073.9 | 622.4 | 3.0 |
| Ts225 | 225 | 129210.9 | 129925.2 | 499.4 | 8.1 | 977739.8 | 1025709.4 | 38644.3 | 0.8 | 128282.7 | 129313.1 | 524.1 | 8.1 | 128159.5 | 128985.9 | 501.5 | 7.6 |
| Gil262 | 262 | 2604.3 | 2653.8 | 36.4 | 10.3 | 16309.9 | 17031.1 | 503.5 | 1.0 | 2555.2 | 2615.9 | 28.1 | 10.4 | 2484.8 | 2552.4 | 43.2 | 9.8 |
| A280 | 280 | 2745.4 | 2779.6 | 25.2 | 11.6 | 19353.8 | 21027.9 | 710.1 | 1.0 | 2669.6 | 2737.2 | 45.8 | 11.7 | 2666.1 | 2731.3 | 55.4 | 11.0 |
| Lin318 | 318 | 47113.3 | 48039.3 | 558.1 | 14.3 | 362670.9 | 374143.0 | 7991.4 | 1.1 | 46057.6 | 46927.8 | 704.5 | 14.6 | 45280.5 | 46609.7 | 984.2 | 13.6 |
| Pr439 | 439 | 119595.3 | 123512.6 | 1782.9 | 25.4 | 1120733.8 | 1171935.4 | 44592.6 | 1.4 | 116681.5 | 120078.9 | 2718.7 | 26.4 | 114158.6 | 117905.1 | 3265.9 | 24.3 |
| Rat575 | 575 | 7720.2 | 7858.5 | 100.3 | 41.8 | 70666.9 | 72422.8 | 1120.0 | 1.9 | 7699.8 | 7782.0 | 76.7 | 44.2 | 7527.8 | 7659.5 | 50.5 | 39.9 |

NCity: Number of city; Min: Minimum cost; Mean: Mean cost; Std: Standard Deviation of minimum cost; R_Time: Running time (minutes)

The Fig. 6 shows the dependence between the *pc* values, mean cost values found on 10 running times of *CXGA*. According to the experiments in the Fig. 6, *pc%* = 15% is quite reasonable in our algorithm.

In *MSCX_Radius* crossover, the bigger the *r* parameter is, the more increasing the running times is. In addition, according to the results in the Table 1, the results of *CXGA* when *r* = 5 are better than the ones when *r* = 2, 3, 7 and 10 in most instances on mean and min values (values in bold). So, we chose 5 in all experiments for *r* value.

Table 2 summarizes the results found by *GA3*, *CXGA* and the best results of *GA1*, *GA2* for 12 *TSP* instances of size from 51 to 575.

Mean, min cost value found by *GA1* are worse than *GA3* on 8/12 and 7/12 instances. Standard deviation values found by *GA1* worse than *GA3* on 3 instances. The running time of *GA1* are lower than *GA3* on 3/12 instances. The running time of *GA2* are faster than *GA3* on all instances. Min, mean and standard deviation values found by *GA2* are greater than *GA3* about three times on all instances.

The mean cost values found by *CXGA* algorithm are better than the ones found by *GA3* from 0.2% to 2.4%. The min cost found by *CXGA* are better than the one found by *GA3* from 0.1% to 2.8%. The running time of *CXGA* are faster than the ones found by *GA3* on 11/12 instances. The standard deviation values found by *CXGA* are better than *GA3* on 7/12 instances (values in bold).

# 6  Conclusion

In this paper, we propose two new crossover operators, called *RX* and *MSCX_Radius*, and new mechanism of combination in *GA* to adapt the convergence of the population for solving *TSP*. We experimented on 12 Euclidean instances derived from *TSP*-lib with the number of vertices from 51 to 575. Experiment results show that, the proposed combination crossover operators in *GA* is effective for *TSP*.

In the future, we are planning to apply propose mechanism of combination to another optimization problem.

# References

[1] Lin, S., Kernighan, B.W.: An effective heuristic algorithm for the traveling salesman problem. Operations Research 21, 498–516 (1973)
[2] Eiben, A.E., Smith, J.E.: Introduction to Evolutionary Computing Natural Computing. Series, 1st edn. Springer (2003)
[3] Snyder, L.V., Daskin, M.S.: A random-key genetic algorithm for the generalized traveling salesman problem. European Journal of Operational Research 174, 38–53 (2006)
[4] Paquete, L., Stützle, T.: A two-phase local search for the biobjective traveling salesman problem. In: Fonseca, C.M., Fleming, P.J., Zitzler, E., Deb, K., Thiele, L. (eds.) EMO 2003. LNCS, vol. 2632, pp. 479–493. Springer, Heidelberg (2003)
[5] Neissi, N.A., Mazloom, M.: GLS Optimization Algorithm for Solving Travelling Salesman Problem. In: Second Int. Conf. on Computer and Electrical Engineering, vol. 1, pp. 291–294. IEEE Press (2009)

[6] Bernd, F., Peter, M.: New Genetic Local Search Operators Traveling Salesman Problem. In: Ebeling, W., Rechenberg, I., Voigt, H.-M., Schwefel, H.-P. (eds.) PPSN 1996. LNCS, vol. 1141, pp. 890–899. Springer, Heidelberg (1996)

[7] Freisleben, B., Merz, P.: New Genetic Local Search for the TSP: New Results. In: Int. Conf. on Evolutionary Computation, pp. 159–164. IEEE Press (1997)

[8] Freisleben, B., Merz, P.: A Genetic Local Search Algorithm for Solving Symmetric and Asymmetric Traveling Salesman Problems. In: Int. Conf. on Evolutionary Computation, pp. 616–621. IEEE Press (1996)

[9] TSPLIB,
http://comopt.ifi.uni-heidelberg.de/software/TSPLIB95/

[10] Renders, J.M., Bersini, H.: Hybridizing genetic algorithms with hill-climbing methods for global optimization: two possible ways. In: IEEE World Congress on Computational Intelligence, vol. 1, pp. 312–317. IEEE Press (1994)

[11] Jih, W.-R., Hsu, J.Y.-J.: Dynamic vehicle routing using hybrid genetic algorithms. In: Int. Conf. on Robotics & Automation, vol. 1, pp. 453–458. IEEE Press (1999)

[12] Ray, S.S., Bandyopadhyay, S., Pal, S.K.: New operators of genetic algorithms for traveling salesman problem. In: ICPR 2004, vol. 2, pp. 497–500. Cambridge, UK (2004)

[13] Larranaga, P., Kuijpers, C., Murga, R., Inza, I., Dizdarevic, S.: Genetic algorithms for the traveling salesman problem: A review of representations and operators. In: Artificial Intelligence, vol. 13, pp. 129–170. Kluwer Academic Publishers (1999)

[14] Held, M., Karp, R.M.: A dynamic programming approach to sequencing problems. Journal of the Society for Industrial and Applied Mathematics 10, 196–210 (1962)

[15] Haykin, S.: Neural Networks: A Comprehensive Foundation, 2nd edn. Prentice-Hall (1999)

[16] Yagiura, M., Ibaraki, T.: The Use of Dynamic Programming in Genetic Algorithms for Permutation Problems. European Journal of Operational Research 92, 387–401 (1996)

[17] Murat, A., Novruz, A.: Development a new mutation operator to solve the Traveling Salesman Problem by aid of Genetic Algorithms. In: Expert Systems with Applications, vol. 38, pp. 1313–1320. ScienceDirect (2011)

[18] Mersmann, O., Bischl, B., Bossek, J., Trautmann, H., Wagner, M., Neumann, F.: Local search and the traveling salesman problem: A feature-based characterization of problem hardness. In: Hamadi, Y., Schoenauer, M. (eds.) LION 2012. LNCS, vol. 7219, pp. 115–129. Springer, Heidelberg (2012)

[19] Sourav, S., Anwesha, D., Satrughna, S.: Solution of traveling salesman problem on scx based selection with performance analysis using Genetic Algorithm. International Journal of Engineering Science and Technology (IJEST) 3, 6622–6629 (2011)

[20] Sehrawat, M., Singh, S.: Modified Order Crossover (OX) Operator. International Journal on Computer Science & Engineering 3, 2014–2019 (2011)

[21] Ray, S.S., Bandyopadhyay, S., Pal, S.K.: New genetic operators for solving TSP: Application to microarray gene ordering. In: Pal, S.K., Bandyopadhyay, S., Biswas, S. (eds.) PReMI 2005. LNCS, vol. 3776, pp. 617–622. Springer, Heidelberg (2005)

[22] Sallabi, O.M., El-Haddad, Y.: An Improved Genetic Algorithm to Solve the Traveling Salesman Problem. Proceedings of World Academy of Science: Engineering & Technology 52, 530–533 (2009)

[23] Kusum, D., Hadush, M.: New Variations of Order Crossover for Travelling Salesman Problem. Int. Journal of Combinatorial Optimization Problems and Informatics 2, 2–13 (2011)

[24] Zakir, H.A.: Genetic Algorithm for the Traveling Salesman Problem using Sequential Constructive Crossover Operator. Int. Journal of Biometric and Bioinformatics 3, 96–106 (2010)
[25] Abdel-Moetty, S.M., Heakil, A.O.: Enhanced Traveling Salesman Problem Solving using Genetic Algorithm Technique with modified Sequential Constructive Crossover Operator. Int. Journal of Computer Science and Network Security 12, 134–139 (2012)

# A Hybrid Gravitational Search Algorithm and Back-Propagation for Training Feedforward Neural Networks

Quang Hung Do

**Abstract.** Presenting a satisfactory and efficient training algorithm for artificial neural networks (ANN) has been a challenging task. The Gravitational Search Algorithm (GSA) is a novel heuristic algorithm based on the law of gravity and mass interactions. Like most other heuristic algorithms, this algorithm has a good ability to search for the global optimum, but suffers from slow searching speed. On the contrary, the Back-Propagation (BP) algorithm can achieve a faster convergent speed around the global optimum. In this study, a hybrid of GSA and BP is proposed to make use of the advantage of both the GSA and BP algorithms. The proposed hybrid algorithm is employed as a new training method for feedforward neural networks (FNNs). To investigate the performance of the proposed approach, two benchmark problems are used and the results are compared with those obtained from FNNs trained by original GSA and BP algorithms. The experimental results show that the proposed hybrid algorithm outperforms both GSA and BP in training FNNs.

**Keywords:** Gravitational search algorithm, Back-Propagation algorithm, Feedforward neural networks.

## 1 Introduction

The artificial neural network (ANN), a soft computing technique, has been successfully applied in many manufacturing and engineering areas [1]. Neural networks, which originated in mathematical neurobiology, are used as an alternative to traditional statistical models. Neural networks have the notable ability to derive meaning from complicated or imprecise data and can be used to extract patterns and detect

Quang Hung Do
Department of Electrical and Electronic Engineering,
University of Transport Technology,
No. 54 Trieu Khuc, Hanoi, Vietnam
e-mail: quanghung2110@gmail.com

© Springer International Publishing Switzerland 2015         381
V.-H. Nguyen et al. (eds.), *Knowledge and Systems Engineering,*
Advances in Intelligent Systems and Computing 326, DOI: 10.1007/978-3-319-11680-8_30

trends that are too complicated to be recognized by either humans or traditional computing techniques. This means that neural networks have the ability to identify and respond to patterns that are similar but not identical to the ones with which they have been trained [2]. ANN has become one of the most important data mining techniques, and can be used for both supervised and unsupervised learning. In fact, feedforward neural networks (FNNs) are the most popular neural networks in practical applications. For a given set of data, a multi-layered FNN can provide a good non-linear relationship. Studies have shown that an FNN, even with only one hidden layer, can approximate any continuous function [3]. Therefore, it is the most commonly used technique for classifying nonlinearly separable patterns [4] [5] and approximating functions[6] [7]. The training process is an important aspect of an ANN model when performance of ANNs is mostly dependent on the success of the training process, and therefore the training algorithm. The aim of the training phase is to minimize a cost function defined as a mean squared error (MSE), or a sum of squared error (SSE), between its actual and target outputs by adjusting weights and biases. Two issues are of great importance in ANN training: how to avoid the local minimum, and how to achieve faster convergence. Therefore, presenting a satisfactory and efficient training algorithm has always been a challenging task. Currently, there are a variety of algorithms used to train ANN, such as back-propagation algorithms and heuristic algorithms. A popular approach used in the training phase is the back-propagation (BP) algorithm which includes the standard BP [8] and the improved BP [9] [10] [11]. However, researchers have pointed out that the BP algorithm - a gradient-based algorithm - has its disadvantages [12] [13]. These drawbacks include the tendency to become trapped in local minima [14] [15]. Heuristic algorithms are known for their ability to produce optimal or near optimal solutions for optimization problems. Several heuristic algorithms - including GA [16], PSO [17], and ACO [18] - have been proposed for the purpose of training neural networks to enhance the problems of BP-based algorithms. These algorithms do not use any gradient information, and have a better chance in avoiding local optima by simultaneously sampling multiple regions of the search space; however they still suffer from slow convergence rates around the global optimum [19]. Recently, several heuristic algorithms, inspired by the behaviors of natural phenomena, were developed for solving optimization problems. Gravitational Search Algorithm (GSA), based on the law of gravity and mass interactions was introduced by Rashedi [20]. Actually, the GSA is not truly based on the law of gravity; however, the capabilities of the GSA as an optimization algorithm have still been highly appreciated [21].Through some benchmarking studies, this algorithm has been proven to be an effective one. The comparison of the GSA with other optimization algorithms shows that the GSA performs better in some problems [20] [22]. However, GSA is negatively affected by slow searching speed in the last iterations [13]. Moreover, the well-known No-Free-Lunch theorem (NFL) shows that there are no heuristic algorithms best-suited for all optimization problems [23] [24]. To achieve a better performance, merging different algorithms is a way [25]. In this study, a hybrid method GSABP for training FNN is proposed. The proposed method combines the global search ability of GSA with the local search ability of BP. This method offers a two-stage learning algorithm for optimizing the parameters of the FNN. The remainder of this paper is organized as follows: the GSA is presented in Section 2; a brief description of ANNs is presented in Section 3; the hybrid GSABP algorithm is provided in Section 4; the experimental results are demonstrated in Section 5, and finally, several conclusions are included in Section 6.

## 2  Gravitational Search Algorithm

The GSA can be considered as a system of agents, called masses, that obey the Newtonian laws of gravitation and masses. All masses attract each other by the gravity forces between them. A heavier mass has a bigger force. Consider a system with N masses in which the position of the ith mass is defined as below:

$$X_i = (x_i^1, ..., x_i^d, ..., x_i^n) \quad \text{for} \quad i = 1, 2, ..., N, \tag{1}$$

where $x_i^d$ is the position of the $i$th agent in the $d$th dimension and $n$ presents the dimension of search space. At a specific time, $t$, the force acting on mass $i$ from mass $j$ is defined as follows:

$$F_{ij}^d(t) = G(t) \frac{M_{pi}(t) \times M_{aj}(t)}{R_{ij}(t) + \varepsilon} (x_j^d(t) - x_i^d(t)), \tag{2}$$

where $M_{aj}$ denotes the active gravitational mass of agent $j$; $M_{pi}$ is the passive gravitational mass of agent $i$; $G(t)$ represents the gravitational constant at time $t$; $\varepsilon$ is a small constant; and $R_{ij}(t)$ is the Euclidian distance between agents $i$ and $j$.
The total force acting on agent $i$ in dimension $d$ is as follows:

$$F_i^d(t) = \sum_{j-1, j \neq i}^{N} rand_j F_{ij}^d(t), \tag{3}$$

where $rand_j$ is a random number in $[0,1]$. According to the law of motion, the acceleration of agent $i$ at time $t$ in the $d$th dimension, $a_i^d(t)$, is calculated as follows:

$$a_i^d(t) = \frac{F_i^d(t)}{M_{ii}(t)}, \tag{4}$$

where $M_{ii}(t)$ is the mass of object $i$. The next velocity of an agent is a fraction of its current velocity added to its acceleration. Therefore, the next position and the next velocity can be calculated as:

$$v_i^d(t+1) = rand_i \times v_i^d(t) + a_i^d(t), \tag{5}$$

$$x_i^d(t+1) = x_i^d(t) + v_i^d(t+1). \tag{6}$$

The gravitational constant, $G$, is generated at the beginning and is reduced with time to control the search accuracy. It is a function of the initial value $(G_o)$ and time $(t)$:

$$G(t) = G(G_o, t).$$ (7)

Gravitational and inertia masses are calculated by the fitness evaluation. A heavier mass means a more efficient agent. This means that better agents have higher attractions and move more slowly. The gravitational and inertial masses are updated by the following equations:

$$m_i(t) = \frac{fit_i(t) - worst(t)}{best(t) - worst(t)},$$ (8)

$$M_i(t) = \frac{m_i(t)}{\sum_{j=1}^{N} m_j(t)},$$ (9)

where $fit_i(t)$ denotes the fitness value of agent $i$ at time $t$, and $worst(t)$ and $best(t)$ represents the weakest and strongest agents in the population, respectively.

For a minimization problem, $worst(t)$ and $best(t)$ are as follows:

$$best(t) = \min_{j \in \{1,...,N\}} fit_j(t),$$ (10)

$$worst(t) = \max_{j \in \{1,...,N\}} fit_j(t).$$ (11)

For a maximization problem,

$$best(t) = \max_{j \in \{1,...,N\}} fit_j(t),$$ (12)

$$worst(t) = \min_{j \in \{1,...,N\}} fit_j(t).$$ (13)

The improved GSA algorithm has been proposed to strengthen the standard GSA [13]. The aim of this improvement is to increase information exchange among agents; this, in turn, helps agents to move to a better position. Like the PSO algorithm, in each iteration, the agent positions are updated as follows:

$$v_i(t+1) = w \times v_i(t) + c_1 \times rand \times a_i(t) + c_2' \times rand \times (gbest - x_i(t)),$$ (14)

where $v_i(t)$ is the velocity of agent $i$ at iteration $t$, $c_1'$ and $c_2'$ are acceleration coefficients, $w$ is a weighting function, $rand$ is a random value in the range of $[0,1]$, $a_i(t)$ is the acceleration of agent $i$ at iteration $t$, and $gbest$ is the best solution so far. The positions of agents are then calculated as follows:

$$x_i(t+1) = x_i(t) + v_i(t+1).$$ (15)

In the improved GSA, all agents are first randomly initialized. Each agent is a candidate solution. The force, gravitational constant, and the total force among agents are then calculated by the use of Equations 2, 7, and 3, respectively. The accelerations of agents are obtained as Equation 4. The best solution is updated in each iteration. The velocities of all agents are calculated by Equation 14. Finally, the positions of agents are achieved by Equation 15.

## 3 Artificial Neural Networks

Studies have shown that an FNN even with only one hidden layer can approximate any continuous function [3]. Therefore, FNN is an attractive approach [4]. FNNs have been applied to a wide variety of problems arising from a variety of disciplines, including mathematics, computer science, and engineering [26]. Fig. 1 shows an example of an MLP with one hidden layer. In Fig. 1, $R$, $N$, and $S$ are the numbers of input, hidden neurons, and output, respectively; $iw$ and $hw$ are the input and hidden weights matrices, respectively; $hb$ and $ob$ are the bias vectors of the hidden and output layers, respectively; $x$ is the input vector of the network; $ho$ is the output vector of the hidden layer; and $y$ is the output vector of the network. The neural network in Fig. 1 can be expressed by the following equations:

$$ho_i = f\left(\sum_{j=1}^{R} iw_{ij} \cdot x_j - hb_i\right), \quad \text{for} \quad i = 1,...,N, \tag{16}$$

$$y_i = f\left(\sum_{k=1}^{N} hw_{ik} \cdot ho_k - ob_i\right), \quad \text{for} \quad i = 1,...,S, \tag{17}$$

where $f$ is an activation function, $iw_{ij}$ is the connection weight from the $j$th node in the input layer to the $i$th node in the hidden layer, and $hw_{ik}$ is the connection weight from the $k$th node in the hidden layer to the $i$th node in the output layer.

It is necessary to determine the structure in terms of number of layers and number of neurons in the layers. A network with a structure that is more complicated than necessary may over fit the training data [27]. The fitness function in the training process is given as follows:

$$SSE = \sum_{k=1}^{Q} E_k, \quad E_k = \sum_{i=1}^{S} \left(y_i^k - d_i^k\right)^2, \tag{18}$$

where $Q$ is the number of training samples, $y_i^k$ is the actual output of the $i$th input when the $k$th training sample is used, and $d_i^k$ is the desired output of the $i$th input when the $k$th training sample is used.

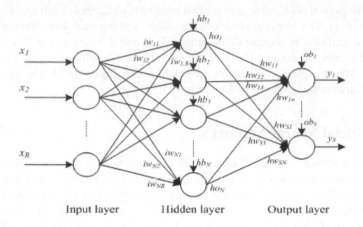

**Fig. 1** A feed-forward network with three layers

## 4 The Hybrid GSABP Algorithm for Training Feedforward Neural Networks

### 4.1 The Hybrid GSABP Algorithm

In a multi-dimensional problem like optimization of FNN architecture, the optimization function has a lot of local minima. The final result reflects the ability of the algorithm in escaping from local minima and achieving a near global optimum. The error of FNN is often large in the initial period of the training process. The GSA algorithm has a strong ability to search for the global optimum. However, when nearing the global optimum the search tends to slow down significantly. The BP algorithm has a strong ability to search local optimum, but its ability to search global optimum is weak. The hybrid GSABP is proposed to combine the global search ability of GSA with the local search ability of BP. This combination takes advantages of both algorithms to optimize the weights and biases of the FNN. The fundamental process of the hybrid algorithm is as follows: at the beginning stage, the GSA is employed to search for the optimum, and find near optimal solutions. When the cost function value has not changed for several iterations, the BP algorithm is applied to find the best solution around global optimum. In this way, the hybrid algorithm may find an optimum quickly. The procedure for the proposed hybrid algorithm can be described as follows:

Step 1: Initialize an FNN structure, as well as the parameters in the GSA. The FNN includes three layers: one input layer, one hidden layer, and one output layer. The parameters are randomly generated the range of [-10, 10].

Step 2: Set up the encoding relationship between the FNN structure and the GSA parameters. In our study, the vector encoding strategy is used. Every agent is encoded for a vector. For FNN, each agent represents all weights and biases of a FNN's structure.

Step 3: Evaluate the fitness of all agents.

Step 4: Update the $G$, $best(t)$, $worst(t)$, and $M_i(t)$ for the population.

Step 5: Calculate $M$, forces, and acceleration of all agents.

Step 6: Update the agents' position and velocity.

Step 7: If the stop condition is satisfied, then go to Step 8; otherwise go to Step 3. When the fitness function has not changed for ten iterations, the searching process is switched to BP algorithm.

Step 8: Decode the near optimal solution of the FNN structure.

Step 9: Use the BP algorithm to search around the near optimal solution. If the search result is better, it can be set as the current search result.

Step 10: Record the global optimum. Output is the current search result.

The number of neurons in the hidden layer is restricted to integer values. The parameters, including weights and biases, are allowed to accept the real values. These parameters are optimized by the hybrid algorithm.

## 4.2 Encoding Strategy

There are three methods of encoding the weights and biases of FNN in evolutionary algorithms: vector, matrix, and binary [15]. In this study, we utilized the vector encoding method. The objective function is to minimize SSE.

## 4.3 Criteria for Evaluating Performance

For the approximation problem, the SSE was employed. For the classification problem, in addition to SSE criterion, an accuracy rate was used. This rate measures the ability of the classifier to produce accurate results and can be computed as follows:

$$Accuracy = \frac{Number\ of\ correctly\ classified\ objects\ by\ the\ classifier}{Number\ of\ objects\ in\ the\ dataset}. \tag{19}$$

A lower SSE and a higher accuracy rate are preferred. In addition, the number of iterations required to reach the final results is also used to evaluate the performances of the training algorithms.

## 5 Experimental Results

In the following experiments, we used two examples to compare the performances of BP, GSA, and GSABP algorithms in training FNNs (hereafter referred to as BP-FNN, GSA-FNN, and GSABP-FNN). The activation function from input to hidden is a sigmoid function, and the activation function from hidden to output is a linear function. Suppose that weights and biases were initially set in the range of [-10, 10]. After several attempts, the ideal parameters were determined as follows: For GSA-FNN, $\alpha$ was set to 20, the initial velocities of particles were randomly generated in the range of [0, 1], the initial values of acceleration and mass were set to 0 for each particle, and the gravitational constant ($G_0$) was set to 1; For BP-FNN, the learning and momentum rates were 0.4 and 0.3; For GSABP-FNN, $\alpha$ was set to 20, the initial velocities of particles were randomly generated in the range of [0, 1], the initial values of acceleration and mass were set to 0 for each particle, the gravitational constant ($G_0$) was set to 1, $c_1'$ was set to 0.5, $c_2'$ was set to 1.5, and the learning and momentum rates were 0.4 and 0.3, respectively. For all problems, GSA-FNN and GSABP-FNN maintained population sizes of 50. In Tables 1 and 2, the best results are signified in bold type.

## 5.1  Experiment 1: Approximation Problem

In this experiment, we used FNNs with the structure of $1 - S_1 - 1$ to approximate the function $f = sin(2x)e^{-x}$, where $S_1$ is the number of hidden nodes with $S_1 = 3, 4, ..., 7$. The training dataset was constructed when x is in the range of [0, $\pi$] with increments of 0.03; while the testing dataset was obtained at an interval of 0.04 in the range of [0, $\pi$]. Therefore, the numbers of samples in the training and testing dataset are 105 and 78, respectively. The search space was [-10, 10]. For every fixed hidden node number, the three algorithms were run five times. Table 1 gives the performance comparisons on the testing dataset for the three algorithms. The results in Table 1 indicate that GSABP -FNN has the best performance in almost all criteria. Table 1 also reveals that the best architecture for this function approximation problem is with $S_1$=6. Fig. 2 shows the convergence rates of BP-FNN, GSA-FNN, and GSABP -FNN based on the average values of SSE in 500 iterations. It is clearly seen that during the initial iterations the convergence of the BP-FNN was very fast; however, in the last iterations, it had almost no improvement and was trapped in the local minima of the parameter space. This figure shows that GSABP-FNN converged to its global optimum in 100-150 iterations, whereas BP-FNN and GSA-FNN were not able to converge to their optimum within 300 iterations. The figure also indicates that GSABP-FNN and GSA-FNN have similar results, but GSABP-FNN is better.

## 5.2  Experiment 2: Classification Problem

A popular benchmark classification problem - the Iris classification problem - was also used to test the performance of the proposed hybrid algorithm in training FNNs. The Iris dataset consists of 150 samples which can be divided into three classes, including Setosa, Versicolor, and Virginica. Each class accounts for 50 samples. All samples have four features: sepal length, sepal width, petal length, and petal width. Therefore, we used FNNs with the structure $4 - S_2 - 3$ to solve this classification problem, where $S_2$ is the number of hidden nodes with $S_2 = 10, 11, ..., 15$. In this study, 100 samples were used for training and the rest were used for testing. The search space was [-50, 50]. Every procedure was run successively for five times, and then the mean values were calculated for these five results and are shown in Table 2. The results show that GSABP-FNN outperforms BP-FNN and GSA-FNN.

**Table 1** Performance statistics in the function approximation problem

| Hidden node ($S_1$) | Algorithm | SSE | | | Standard deviation |
|---|---|---|---|---|---|
| | | Min | Average | Max | |
| 3 | BP | 0.751883 | 1.680321 | 2.289872 | 0.616229 |
| | GSA | 0.350578 | 1.25483 | 1.85848 | 0.605787 |
| | GSABP | **0.344148** | **1.226397** | **1.814407** | **0.601234** |
| 4 | BP | 0.697988 | 1.095909 | 1.541664 | 0.302537 |
| | GSA | 0.286702 | 0.680954 | 1.115703 | 0.297373 |
| | GSABP | **0.284348** | **0.654377** | **1.080321** | **0.285635** |
| 5 | BP | 0.714245 | 1.111784 | 1.394055 | 0.284413 |
| | GSA | 0.191554 | 0.541092 | 0.895821 | 0.291294 |
| | GSABP | **0.169188** | **0.520638** | **0.875466** | **0.289183** |
| 6 | BP | 0.601861 | 0.962664 | 1.336003 | 0.301024 |
| | GSA | 0.303957 | 0.694651 | 0.969553 | 0.27943 |
| | GSABP | **0.29794** | **0.673312** | **0.932586** | **0.267813** |
| 7 | BP | 0.802523 | 0.907582 | 1.054062 | 0.131362 |
| | GSA | 0.362642 | 0.480019 | 0.642164 | 0.138143 |
| | GSABP | **0.331769** | **0.45131** | **0.627703** | **0.151317** |

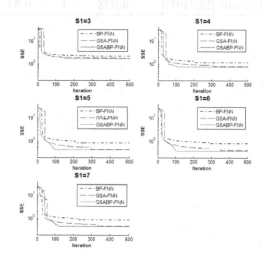

**Fig. 2** Convergence rate in the approximation problem

Fig. 3 shows the convergence rates of BP-FNN, GSA-FNN, and GSABP-FNN based on the average values of SSE with $S_2 = 10, 11, 12, 13, 14,$ *and* 15. This figure confirms that the GSABP -FNN had a trade-off between avoiding premature convergence and exploring the whole search space for all values of hidden numbers. From Table 2 , it can be inferred that GSABP -FNN has a better accuracy rate than BP-FNN and GSA-FNN. For the testing dataset, the best accuracy rates for BP-FNN

**Table 2** Performance statistics in the Iris classification problem

| Hidden node ($S_2$) | Algorithm | SSE | Training accuracy | Testing accuracy |
|---|---|---|---|---|
| 10 | BP | 32.85645 | 0.83 | 0.84 |
| | GSA | 27.2045 | 0.852 | 0.84 |
| | GSABP | **25.25101** | **0.9** | **0.864** |
| 11 | BP | 32.35214 | 0.8 | 0.824 |
| | GSA | 25.86809 | 0.866 | 0.86 |
| | GSABP | **24.57097** | **0.87** | **0.876** |
| 12 | BP | 29.97839 | 0.868 | 0.856 |
| | GSA | 25.12966 | 0.876 | 0.848 |
| | GSABP | **22.57407** | **0.94** | **0.872** |
| 13 | BP | 28.65762 | 0.828 | 0.828 |
| | GSA | 23.88097 | 0.836 | 0.896 |
| | GSABP | **21.38662** | **0.952** | **0.88** |
| 14 | BP | 27.12603 | 0.888 | 0.804 |
| | GSA | 22.53575 | 0.888 | 0.848 |
| | GSABP | **20.21154** | **0.982** | **0.948** |
| 15 | BP | 29.52396 | 0.82 | 0.816 |
| | GSA | 25.30841 | 0.832 | 0.832 |
| | GSABP | **23.31919** | **0.928** | **0.84** |

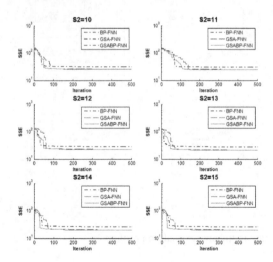

**Fig. 3** Convergence rate in the classification problem

and GSA-FNN were 0.856 and 0.896, respectively; while the best accuracy rate for GSABP-FNN was 0.948. These results prove that GSABP-FNN is capable of solving the Iris classification problem more reliably and accurately than BP-FNN and GSA-FNN.

Based on the obtained results, it can be concluded that GSABP-FNN outper-forms BP-FNN and GSA-FNN due to the capability of the proposed hybrid GSABP algorithm.

# 6 Conclusion

In this study, we propose a hybrid GSABP algorithm based on the GSA and BP algorithms. This algorithm combines the GSA algorithm's strong ability regarding global search and the BP algorithm's strong ability for local search; therefore, it has the superiority of convergent speed and convergent accuracy. We can get better search results using this hybrid algorithm. The GSA, BP, and GSABP were utilized as training algorithms for FNNs. For the two benchmark problems, the comparison results showed that GSABP-FNN outperforms GSA-FNN and BP-FNN in terms of convergence rates and being trapped in local minima. It can be concluded that the proposed hybrid GSABP algorithm is suitable to be used as a training algorithm for FNNs. The results of the present study also show the fact that a comparative analysis of different training algorithms is always supportive in enhancing the performance of a neural network. For future research, we will focus on how to apply this hybrid GSABP algorithm to deal with more optimization problems.

# References

[1] Paliwal, M., Kumar, U.A.: Neural networks and statistical techniques: A review of ap-plications. Expert Systems with Applications 36(1), 2–17 (2009)
[2] Vosniakos, G.C., Benardos, P.G.: Optimizing feedforward Artificial Neural Network Architecture. Engineering Applications of Artificial Intelligence 20(3), 365–382 (2007)
[3] Funahashi, K.: On the approximate realization of continuous mappings by neural net-works. Neural Networks 2(3), 183–192 (1989)
[4] Norgaard, M.R.O., Poulsen, N.K., Hansen, L.K.: Neural networks for modeling and control of dynamic systems. In: A Practitioner's Handbook. Springer, London (2000)
[5] Mat Isa, N.: Clustered-hybrid multilayer perceptron network for pattern recognition ap-plication. Applied Soft Computing 11(1), 1457–1466 (2011)
[6] Homik, K., Stinchcombe, M., White, H.: Multilayer feedforward networks are universal approximators. Neural Networks 2, 359–366 (1989)
[7] Malakooti, B., Zhou, Y.: Approximating polynomial functions by feedforward artifi-cial neural network: capacity analysis and design. Applied Mathematics and Computa-tion 90(1), 27–52 (1998)
[8] Hush, R., Horne, N.G.: Progress in supervised neural networks. IEEE Signal Processing Magazine 10, 8–39 (1993)
[9] Hagar, M.T., Menhaj, M.B.: Training feedforward networks with the Marquardt algo-rithm. IEEE Transactions Neural Networks 5(6), 989–993 (1994)
[10] Adeli, H., Hung, S.L.: An adaptive conjugate gradient learning algorithm for efficient training of neural networks. Applied Mathematics and Computation 62(1), 81–102 (1994)
[11] Zhang, N.: An online gradient method with momentum for two-layer feedforward neural networks. Applied Mathematics and Computation 212(2), 488–498 (2009)

[12] Gupta, J.N.D., Sexton, R.S.: Comparing backpropagation with a genetic algorithm for neural network training. Omega 27(6), 679–684 (1999)

[13] Mirjalili, S.A., Mohd Hashim, S.Z., Sardroudi, H.M.: Training feedforward neural networks using hybrid particle swarm optimization and gravitational search algorithm. Applied Mathematics and Computation 218(22), 11125–11137 (2012)

[14] Gori, M., Tesi, A.: On the problem of local minima in back-propagation. IEEE Transactions on Pattern Analysis and Machine Intelligence 14(1), 76–86 (1992)

[15] Zhang, J.R., Zhang, J., Lock, T.M., Lyu, M.R.: A hybrid particle swarm optimization back-propagation algorithm for feedforward neural network training. Applied Mathematics and Computation 185(2), 1026–1037 (2007)

[16] Goldberg, E.: Genetic algorithms in search, optimization and machine learning. Addison Wesley, Boston (1989)

[17] Kennedy, J., Eberhart, R.C.: Particle swarm optimization. In: Proceedings of IEEE International Conference on Neural Networks, vol. 4, pp. 1942–1948. IEEE (1995)

[18] Dorigo, M., Maniezzo, V., Golomi, A.: Ant system: optimization by a colony of cooperating agents. IEEE Transactions on Systems, Man, and Cybernetics 26(1), 29–41 (1996)

[19] Kiranyaz, S., Ince, T., Yildirim, A., Gabbouj, M.: Evolutionary artificial neural networks by multi-dimensional particle swarm optimization. Neural Networks 22(10), 1448–1462 (2009)

[20] Rashedi, E., Nezamabadi-pour, H., Saryazdi, S.: GSA: A Gravitational Search Algorithm. Information Sciences 179(13), 2232–2248 (2009)

[21] Gauci, M., Dodd, T.J., Grob, R.: Why 'GSA: a gravitational search algorithm' is not genuinely based on the law of gravity. Natural Computing 11(4), 719–720 (2012)

[22] Duman, S., Guvenc, U., Yorukeren, N.: Gravitational Search Algorithm for Economic Dispatch with Valve-Point Effects. International Review of Electrical Engineering 5, 2890–2895 (2010)

[23] BoussaiD, I., Lepagnot, J., Siarry, P.: A survey on optimization metaheuristics. Information Sciences 237, 82–117 (2013)

[24] Ho, Y.C., Pepyne, D.L.: Simple Explanation of the No-Free-Lunch Theorem and Its Implications. Journal of Optimization Theory and Applications 115(3), 549–570 (2002)

[25] Mirjalili, S., Hashim, S.Z.M.: A New Hybrid PSOGSA Algorithm for Function Optimization. In: Proceedings of the International Conference on Computer and Information Application (ICCIA 2010), pp. 374–377 (2010)

[26] Li, L.K., Shao, S., Yiu, K.F.C.: A new optimization algorithm for single hidden layer feedforward neural networks. Applied Soft Computing 13(5), 2857–2862 (2013)

[27] Caruana, R., Lawrence, S., Giles, C.L.: Overfitting in neural networks: backpropagation. In: Proceedings of 13th Conference on Advances Neural Information Processing Systems, USA, pp. 402–408 (2001)

# Efficient Palmprint Search Based on Database Clustering for Personal Identification

Hoang Thien Van and Thai Hoang Le

**Abstract.** This paper proposes an efficient palmprint searching algorithm for personal identification based on database clustering, which reduces the search space of fine matching. A complex filter is applied to double orientation field to detect the symmetry of the palm lines as the main feature at coarse-level search. A K-means clustering technique is applied to partition the symmetry feature space into clusters. A query processing is proposed to facilitate an efficient searching. The experimental results on the public database of Hong Kong Polytechnic University show the effectiveness of the proposed searching algorithm.

**Keywords:** Palmprint identification, Complex filter, Searching, clustering, K-means.

## 1 Introduction

Palmprints are a relatively new biometric modality for personal recognition, specially for deployment for access control at points of entrance, such as airports, and other highly sensitive places [1]. Over the last decade, there has been increased interest in palmprint recognition because of its merits such as distinctiveness, cost-effectiveness, user friendliness, high accuracy, and so on [2]. A palm is defined as the inner surface of a hand between the wrist and the fingers [3]. A palmprint refers to principle lines, wrinkles, and ridges on a palm. Palmprint research employs low

Hoang Thien Van
Department of Computer Sciences, Ho Chi Minh City University of Technology,
Ho Chi Minh City, VietNam
e-mail: lhthai@fit.hcmus.edu.vn

Thai Hoang Le
Department of Computer Sciences, Ho Chi Minh University of Science,
Ho Chi Minh City, VietNam
e-mail: vthoang@hcmhutech.edu.vn

© Springer International Publishing Switzerland 2015      393
V.-H. Nguyen et al. (eds.), *Knowledge and Systems Engineering,*
Advances in Intelligent Systems and Computing 326, DOI: 10.1007/978-3-319-11680-8_31

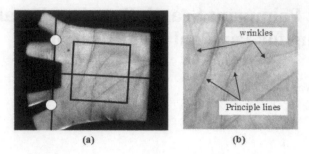

**Fig. 1** (a) Sample palmprint image and (b) its region of interest (ROI)

resolution images (150 dpi or less [2]) for civil and commercial applications (see Fig. 1a). Palmprint recognition uses the persons palm for identifying or verifying who the person is. In verification, the user inputs his palmprint and claims an identity information, then the system verifies whether the input palmprint is consistent with the claimed identity. In identification, the user input his palmprint and the system identifies the potential corresponding one in the database without a claimed identity. Therefore, palmprint identification can be described as follows: given an example palmprint, compare it with all of the possible candidates in the database to determine whether the queried example and the candidates are from the same palm. The search for the best matching is crucial for the performance of the system in terms of accuracy and efficiency. When the size of a palmprint database increases, the one-by-one matching [8], [10], [12], [13] becomes too time-consuming to meet the requirement for on-line personal identification. The problem of real-time identification in large databases has been addressed in two ways: classification and hierarchy. Classification approaches assign a class to each palmprint in a database. Wu et al. [6] define six classes based on the number of principal lines and their intersections. However, the six classes are highly unbalanced (about 80% of palmprints in a category); and the algorithm has high bin errors of 4%. So these classes are not enough for identification. Li et al.[17] proposed dealing with the unbalanced class problem by further dividing the unbalanced class. Hierarchical approaches,also called multi-level matching approaches, employ simple but computationally effective features to retrieve a subset of templates in a given database for further comparison [3], [4], [5], [7]. The coarse level matching (search) is often used to reduce the search space of the time-consuming fine matching and alleviate the accuracy deterioration of identification [5]. You et al. [5] used three-level features at coarse level such as global geometry based key point distance (level-1 feature), global texture energy (level-2 feature) and fuzzy "interest" line (level-3 feature). The use of level-1, level-2, and level-3 features can remove candidates from the database by 9.6%, 7.8%, and 60.6% with an error rate of 4.5%, respectively. Li et al. [3] used texture energy of palm lines in four directions as global features, to guide the fast selection of a small set of similar candidates. The feature vector has four directional components. Only the candidates in the database which are close to the queried palmrint sample with small distance in global feature measurement are selected for fine-level search. With

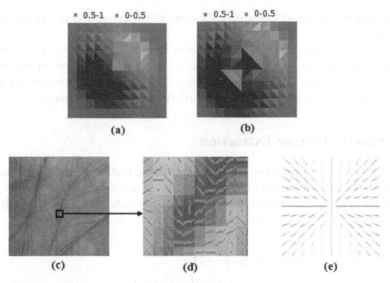

**Fig. 2** (a) Magnitude image of Gaussian filter, (b) magnitude image of filter $h1$, (c) The original palmprint image, (d) the squared orientation field image of a palm line, (e) the orientation field image of filter $h1$

narrowing down the search space to 20% of database, the classification accuracy of the method is 94.5% [3]. Data clustering is a crucial technique used in discovering the underlying structure in a data set by unsupervised grouping of the similar patterns. It accelerates the content based image retrieval by comparing the query image with a few cluster representatives instead of all database templates [14]. This paper proposes an efficient palmprint searching algorithm for personal identification based on database clustering, which reduces the search space of fine matching. A complex filter is applied to orientation field to detect the symmetry of the palm lines. A sequence of local symmetry energies of filtered image is grouped in a one-dimensional (1-D) array, called symmetry energy vector (SEV), for coarse palmprint searching. The symmetry energies captures the local ridge flow pattern, while the ordered enumeration of SEV describes the global relationships among the local patterns. Thus, SEV is robust feature for classification. Our proposed palmprint search algorithm consists of two phases: offline database clustering and online query processing. During the online database clustering, a k-means clustering technique is applied to partition the symmetry feature space into clusters.Based on the offline database clustering, a hierarchical online query processing is proposed to facilitate an efficient searching. In cluster search, each query palmprint is compared with the cluster prototypes to retrieve the closest clusters. Palmprint search is performed on the retrieved clusters to find the templates close to the query palmprint as the candidates at fine-level search for identification. At fine-level search, we use DORIR feature [31] for identification. The experimental results on the public database of Hong Kong Polytechnic University illustrate the effectiveness of the proposed approach.

The rest of the paper is organized as follows. Section 2 presents the global symmetry feature for coarse level matching. Section 3 presents the proposed palmprint search algorithm based on database clustering. The experimental results are presented in section 4. Finally, the paper conclusions are drawn in section 5.

## 2 Symmetry Feature Extraction

Symmetry filters are robust tools for feature extraction, matching and pattern recognition [18], [19]. A symmetry in the image and the associated filter detecting it can be modeled by $\exp(im\varphi)$, where $m$ represents its symmetry order [18]. A polynomial approximation of these fields in Gaussian windows yields:

$$(x+iy)^2 g(x,y) \tag{1}$$

where $g$ is a Gaussian defined as the formula:

$$g(x,y,\sigma) = \exp\left\{-\frac{x^2+y^2}{2\sigma^2}\right\} \tag{2}$$

These filters are applied to the complex valued orientation tensor field image $f(x,y) = (f_x + if_y)^2, i = \sqrt{-1}$. Here, $f_x$ is the derivative of the original image in the x-direction and $f_y$ is the derivative in the y-direction [18]. For palmprint, the orientation of each pixel is the double angle of the gradient vector, in order to avoid an ambiguity of deciding the orientation as either $\theta$ or $\theta + \pi/2$. The squared gradient can be expressed in complex number by

$$f = (G_x + iG_y)^2 = (G_x^2 - G_y^2) + i(2G_x G_y) \tag{3}$$

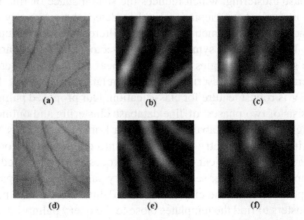

(a)                     (b)                     (c)

(d)                     (e)                     (f)

**Fig. 3** The original images (a), (d), the resulted symmetry images with size $128 \times 128$ (b), (e), and the resulted scaled symmetry images with size $8 \times 8$

$$\bar{f} = \cos(2\theta) + i\sin(2\theta) \tag{4}$$

$G_x$ and $G_y$ denote the gradients of the original palmprint image in the $x$ and $y$ direction, respectively; $f$ is called the complex image. $\bar{f}$ is the normalized complex image. Figure 2b shows the squared orientation field image of a palm line in a palmprint image. We use the parabolic symmetry model of order $m = 1$, since the parabolic pattern is most similar to pattern of palm lines (Figure 2b). The filter of parabolic symmetry [19] are used as

$$h_1(x,y) = z(1)g(x,y) = (x,iy)g(x,y) \tag{5}$$

The filter is implemented by compute the 2-D scalar product for each image point of complex image $\bar{f}$ as the following formulas:

$$sym(u,v) = \frac{1}{M}\sqrt{F_x(x,y)^2 + F_y(x,y)^2} \tag{6}$$

$$F_x(y,x) = \sum_{u=-w/2}^{w/2}\sum_{u=-w/2}^{w/2} g(x,y,\sigma)\left(\begin{array}{c}\bar{v}\cos(2\theta(y+u,x+v))-\\\bar{u}\sin(2\theta(y+u,x+v))\end{array}\right) \tag{7}$$

$$F_y(y,x) = \sum_{u=-w/2}^{w/2}\sum_{u=-w/2}^{w/2} g(x,y,\sigma)\left(\begin{array}{c}\bar{u}\cos(2\theta(y+u,x+v))-\\\bar{v}\sin(2\theta(y+u,x+v))\end{array}\right) \tag{8}$$

$$\bar{u} = \frac{u}{\sqrt{u^2+v^2}}, \bar{v} = \frac{v}{\sqrt{u^2+v^2}} \tag{9}$$

where M is the number of the pixels using in the filter, and w species the size of the filter $h1$. Fig. 3b and 3e present the results of the parabolic filter. In order to get stable feature with small size, we divide the complex image into no-overlapping block of size $s \times s$. Then, the filter is implemented by compute the 2-D scalar product for each center point of blocks. For example, if the size of the original image is $s \times s$, s = 16 and w = 16 , the size of resulted feature image will be $8 \times 8$. Fig. 3c and 3f are examples of such representations. A sequence of local symmetry energies of feature image of size of $8 \times 8$ is grouped in a one-dimensional (1-D) array, called symmetry energy vector (SEV), for coarse palmprint searching. Fig. 4 illustrate the process of SEV feature. Next subsection presents Palmprint search based on database clustering with SEV feature.

## 2.1 Palmprint Search Based on Database Clustering

The full palmprint search is time consuming in the on-line query process. Therefore, we employ the clustering technique to partition the database into a number of non-overlapping groups with more flexibility. The clustering based palmprint search consists of two phases: offline database clustering using SEV feature and online query processing.

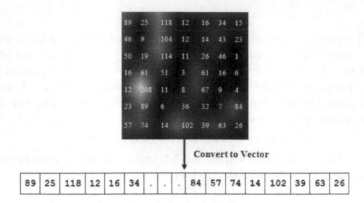

**Fig. 4** Process of SEV feature

## 2.2 The Offline Database Clustering

The clustering technique is applied to partition the orientation feature space into non-overlapping clusters for an efficient palmprint search. The K-means algorithm [14] is the most widely used clustering algorithm because of its high computational efficiency and low memory space requirement. This clustering algorithm represents each cluster with its mean vector and assign each pattern to the closest cluster iteratively. It terminates when the cluster labels do not change. Euclidean distance measure is used in the K-means clustering to assign each pattern to the closest cluster. Fig. 5 shows six cluster prototypes of K-means clustering method using SEV feature.

## 2.3 The Online Query Processing

The online query processing is to retrieve a subset of database templates close to the query palmprint. We propose a hierarchical query process that consists of two levels of search (see Fig. 6). In the first level, we search the clusters by comparing the query SEV vector with the cluster prototypes. Some ambiguous palmprints are located near the cluster boundary no matter how well the database is partitioned. To alleviate this problem, we retrieve multiple nearest clusters instead of only the nearest one. These coarse level searches can efficiently narrow down the search of database because the number of groups is much smaller than the number of templates. In the second level, the palmprint search is performed on the retrieved clusters using the SEV feature to further narrow down the search space. Therefore, the online query processing of palmprint search is accelerated by database clustering without compromising the effectiveness of palmprint search. For the cluster search, we compute the distances $d_c(sev_q, sev_l), (1 \leq l \leq K)$ between the query palmprint and cluster prototypes and retrieve the clusters with the distances smaller than a threshold. Since the clusters may be unevenly distributed in the SEV feature space,

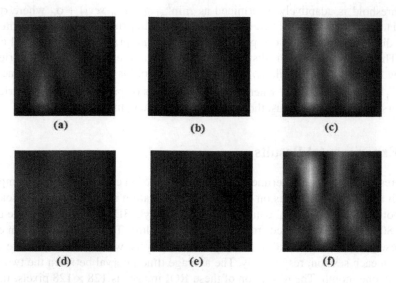

**Fig. 5** Six cluster prototypes of K-means clustering method using SEV feature

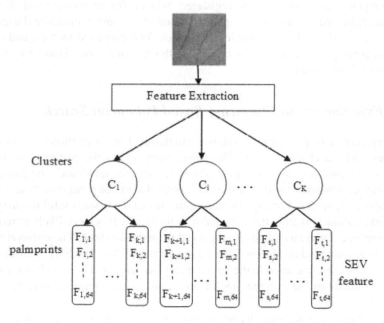

**Fig. 6** The hierarchical online query processing using SEV features

the threshold is adaptively determined as $\min_{l=1}^{K} d_C (sev_q, sev_l) + \sigma_c$ where $\sigma_c$ is tuned to adjust the retrieval neighborhood. In the palmprint search, we compute the SEV distances between query palmprint and all palmprints in the retrieved clusters. The palmprints with the distances smaller than a threshold are finally retrieved for the fine matching. Similarly, the retrieval threshold is adaptively determined as $\min_{j=1}^{N_p} d_C (sev_q, sev_j) + \sigma_p$ where $N_p$ is the number of palmprints in the retrieved clusters and is tuned to adjust the portion of retrieved palmprints.

## 3  Experimental Results

We present two sets of experiments: (1) Experiments on clustering based palmprint search and (2) Experiments on palmprint Identification using coarse-level search. For both, we use the PolyU database [20]. The PolyU 3D palmprint database contains 8000 samples collected from 400 different palms. Twenty samples from each of these palms are collected in two separated sessions, where 10 samples are captured in each session, respectively. The average time interval between the two sessions is one month. The resolution of these ROI images is $128 \times 128$ pixels. In the following tests, the registration database contains 3000 templates from 300 random different palms, where each palm has ten templates. The testing database contains 5000 templates from 300 different registered palms (3000 templates) and 100 different unregistered palms (2000 templates). None of palmprint images in the testing database is overlapped in registration databases. The proposed method and other methods are implemented using C# on a PC with CPU Intel Core2Duo 1.8 GHz and Windows 7 Professional.

### 3.1  Experiments on Clustering Based Palmprint Search

The performance of palmprint search is evaluated by the penetration rate, retrieval accuracy and search complexity. The penetration rate is the average portion of database retrieved over all query palmprints. It indicates how much the palmprint search can narrow down the database and is controlled by the parameter $\sigma_c$ and $\sigma_p$ in our proposed approach. For a query palmprint, the search is successful if one of the retrieved candidates is from the same palm as the query. It is more likely to retrieve the correct one if more templates are retrieved from the database. The retrieval accuracy is thus computed at various different penetration rates. The SEV comparisons cost most of the computation in the online query processing. The search complexity is thus evaluated by the average time of such comparisons required over all query palmprints.

In this experiment, the same feature set is used in the palmprint search procedures with and without clustering. In the query process of our clustering based palmprint search, the number of SEV comparisons is the number of clusters (60 in our experiments) plus the number of palmprints in the retrieved clusters. It varies at different penetration rates. Fig. 7 shows the results of the full palmprint search (without clustering) and our proposed palmprint search (with clustering). From Fig. 7a, we can see the search complexity is greatly reduced by using the clustering, especially at the low penetration rates. Moreover, the retrieval accuracy is slightly yet consistently improved by the K-means clustering (see Fig. 7b). This may be resulted by exploiting the similarities among the database templates through the clustering. The results demonstrate that the proposed clustering based approach not only speeds up the search process but also improves the retrieval accuracy.

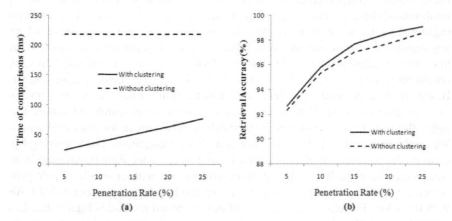

**Fig. 7** Results of full palmprint search (without clustering) and our palmprint search (with clustering): (a) search complexity and (b) retrieval accuracy

We also compare our palmprint search approach with Wu's exclusive classification [6] and Li's approach [3] according to the error rate at about the same penetration rate. Table 1 show the error rates of these methods and our method.

**Table 1** The error rates of the methods [3], [6] and our method

| Methods | Penetration rate | Retrieval accuracy |
|---|---|---|
| Wu's method [6] | 78% | 96.1% |
| Li's method [3] | 20% | 95.2% |
| Our method | 20% | 98.6% |

## 3.2 Experiments on Palmprint Identification Using Coarse-Level Search

In identification, we want to identify which class the query palmprint belongs to. Thus, identification is a process of comparing one query image against all training images and the label of the most similar images is obtained as the identification result. If a matching distance of two images from the same palm is smaller than the threshold, the match is a genuine acceptance. Similarly, if a matching distance of two images from different palms is smaller than the threshold, the match is a false acceptance. Each image in the testing database is matched with all images in the training databases to generate incorrect and correct identification distances. The minimum of the distances produced by the query and templates of the same registered palm is considered as correct identification distance. Similarly, we take the minimum of the distances produced by the query and all templates of the different registered palms as the incorrect identification distance. If the query does not have any registered images, we only obtain the incorrect identification distance. If we have 3000 queries of registered palms and 2000 queries of unregistered palms, we obtain 3000 correct identification scores and 5000 incorrect identification distance. Based on these distances, we obtain the identification results: the receiver operating characteristic curve (ROC curve). With the coarse-level search, we assume that only 20% of palm-print samples in the database will remain for fine-level search. We use DORIR feature [21] at fine-level search for recognition. The Receiving Operating Characteristic (ROC) curve of Genuine Acceptance Rate (GAR) and False Acceptance Rate (FAR) of identification without and with coarse-level palmprint search are presented in Fig. 8. Our proposed methods accuracy is about 97.8% GAR with 0% FAR. Therefore, the accuracy of our proposed method is higher than Li's method [3] (94.9% with the same penetration rate). The average testing time of the identification without and with clustering based palmprint search are com-pared in Table 2. Since the SEV feature at coarse-level search is much faster than DORIR feature of fine-level matching, the proposed layered search scheme speeds up the identification process.

**Table 2** Comparison of testing time

| Methods | Feature extraction (ms) | Identification time for one image (ms) |
|---|---|---|
| Identification without coarse-level search | 87 (DORIR[21]) | 735 (full search with DORIR[21]) |
| Identification with coarse-level search | 87 (DORIR[21]) + 45 (SEV) | 65 (coarse-level search) + 147(fine-level search) |

# 4   Conclusion

The problem of real-time identification in large databases is a great challenge. This paper proposes an efficient palmrint searching algorithm for personal identification based on database clustering, which explores the similarities among the database

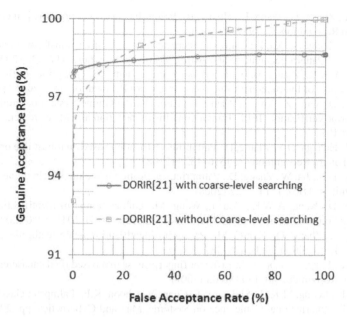

**Fig. 8** The ROC curves of identification without and with coarse-level search

templates and reduces the search space of fine matching. A symmetry energy vector, called SEV feature, computed by applying a complex filter to double orientation field is proposed for coarse-level search. A K-means clustering technique is applied to partition the symmetry feature space into clusters. Based on the offline database clustering, a hierarchical query processing is proposed to facilitate an efficient palmprint search. Experimental results on the public database of Hong Kong Polytechnic University show that our proposed method not only reduces the search complexity but also improves the retrieval and identification accuracy.

## References

[1] Jain, A.K., Ross, A., Prabhakar, S.: An introduction to biometric recognition. IEEE Trans. Circuits Syst. Video Technol. 14(1), 4–20 (2004)
[2] Kong, A., Zhang, D., Kamel, M.: A survey of palmprint recognition. Pattern Recognition 42, 1408–1418 (2009)
[3] Li, W., You, J., Zhang, D.: Texture-Based Palmprint Retrieval Using a Layered Search Scheme for Personal Identification. IEEE Transactions on Multimedia 7(5), 891–898 (2005)
[4] Zhang, L., Zhang, D.: Characterization of palmprints by wavelet signatures via directional context modeling. IEEE Transactions on Systems, Man and Cybernetics, Part B 34(3), 1335–1347 (2004)
[5] You, J., Kong, W.K., Zhang, D., Cheung, K.H.: On hierarchical palmprint coding with multiple features for personal identification in large databases. IEEE Transactions on Circuits and Systems for Video Technology 14(2), 234–243 (2004)

[6] Wu, X., Zhang, D., Wang, K., Huang, B.: Palmprint classification using principal lines. Pattern Recognition 37, 1987–1998 (2004)

[7] Li, W., Zhang, D., Xu, Z.: Palmprint identification by Fourier transform. International Journal of Pattern Recognition and Artificial Intelligence 16(4), 417–432 (2002)

[8] Kong, A.W.K., Zhang, D.: Competitive coding scheme for palmprint verification. In: Proceedings of International Conference on Pattern Recognition, pp. 520–523 (2004)

[9] Zhang, D., Guo, Z., Lu, G., Zhang, L., Zuo, W.: An Online System of Multispectral Palmprint Verification. IEEE Transactions Instrumentation and Measurement 59, 480–490 (2010)

[10] Jia, W., Huanga, D.S., Zhang, D.: Palmprint verification based on robust line orientation code. Pattern Recognition 41, 1316–1328 (2008)

[11] Huang, D.S., Jia, W., Zhang, D.: Palmprint verification based on principal lines. Pattern Recognition 41(5), 1514–1527 (2008)

[12] Zhang, D., Kong, A.W.K., You, J., Wong, M.: Online palmprint identification. IEEE Transactions on Pattern Analysis Machine Intelligence 25(9), 1041–1050 (2003)

[13] Kong, A., Zhang, D., Kamel, M.: Palmprint identification using feature-level fusion. Pattern Recognition 39(3), 478–487 (2006)

[14] Liu, M., Jiang, X., Kot, A.C.: Efficient fingerprint search based on database clustering. Pattern Recognition 40, 1793–1803 (2007)

[15] Fang, L., Leung, M.K.H., Shikhare, T., Chan, V., Choon, K.F.: Palmprint classification. In: IEEE International Conference on Systems, Man and Cybernetics, pp. 2965–2969 (2006)

[16] Bigun, J., Bigun, T., Nilsson, K.: Recognition by symmetry derivatives and the generalized structure tensor. IEEE Transactions on Pattern Analysis and Machine Intelligence 26(12), 1590–1605 (2004)

[17] Le, T.H., Van, H.T.: Fingerprint reference point detection for image retrieval based on symmetry and variation. Pattern Recognition 45, 3360–3372 (2012)

[18] Van, H.T., Tat, P.Q., Le, T.H.: Palmprint verification using GridPCA for Gabor features. In: Proceedings of the Second Symposium on Information and Communication Technology, SOICT 2011, pp. 217–225 (2011)

[19] Van, H.T., Le, T.H.: GridLDA of Gabor Wavelet Features for Palmprint Identification. In: SoICT 2012 Proceedings of the Third Symposium on Information and Communication Technology, pp. 125–134 (2012)

[20] Van, H.T., Le, T.H.: On Discriminant Orientation Extraction Using GridLDA of Line Orientation Maps for Palmprint Identification. In: Huynh, V.N., Denoeux, T., Tran, D.H., Le, A.C., Pham, B.S. (eds.) KSE 2013, Part I. AISC, vol. 244, pp. 237–248. Springer, Heidelberg (2014)

[21] PolyU 3D Palmprint Database, http://www.comp.polyu.edu.hk/~biometrics/2D_3D_Palmprint.htm

# Improving Table of Contents Recognition Using Layout-Based Features

Phuc Tri Nguyen and Dang Tuan Nguyen

**Abstract.** Table of content (TOC) recognition is an essential task in processing book contents for document retrieval applications. Existing methods focus on exploiting characteristic information of TOC page formats on specific types of books. However, we observe that many other normal layout based features of pages can also identify the nature of pages (TOC pages or not). In this paper we propose using some selected layout-based features for improving TOC pages recognition. To show the effectiveness of our proposed method, we conduct experiments on ICDAR Book Structure Extraction Datasets 2009, 2011 and 2013, on which it improves the state-of-the-art performance of current approach focusing on TOC pages based features only.

**Keywords:** table of content recognition, TOC pages detection, document structure extraction.

## 1 Introduction

Normally, a table of contents (TOC) of a book contains a list of headings and their corresponding page numbers of chapters, sections and sub-sections, which are presented in a hierarchical structure. Therefore, automatic TOC pages recognition is an important step towards TOC processing tasks. However, the challenges of TOC recognition are *i*) TOCs have many different layouts, and *ii*) the limitation of the optical character recognition (OCR) quality also affects the performance of TOC recognition.

Phuc Tri Nguyen · Dang Tuan Nguyen
Faculty of Computer Science,
University of Information Technology, VNU-HCM,
Ho Chi Minh City, Vietnam
e-mail: {phucnt,dangnt}@uit.edu.vn

© Springer International Publishing Switzerland 2015                                                405
V.-H. Nguyen et al. (eds.), *Knowledge and Systems Engineering*,
Advances in Intelligent Systems and Computing 326, DOI: 10.1007/978-3-319-11680-8_32

Previous methods only focus on exploiting specific information in formats of TOC pages [2], [9], [8]. Belong to approach using TOC pages based features, for example: TOC pages usually contain some characteristic words, TOC pages have multiple lines ending with numbers, or TOC pages can contain multiple-dotted lines, etc. Besides the TOC pages based features, we observe that many other normal layout features of TOC pages and non-TOC pages such as text spacing, text height, etc in a book also distinguish TOC pages from non-TOC pages. Therefore, we propose using some selected normal layout features for recognizing TOC. Furthermore, to increase performance of recognizing TOC pages, we use support vector machine classifier for recognizing TOC pages.

Our proposed method is experimentally validated on three public datasets in ICDAR Book Structure Extraction competition: 2009 [5], 2011 [6] and 2013 [4]. Books in these datasets are presented in different genres and domains. They are freely provided by Microsoft and the Internet Archive [7].

The remaining of this paper is organized as follows: Section 2 reviews related works on table of content recognition. Section 3 describes our proposed method in details. Experiments and results are given Section 4. Finally, Section 5 concludes the paper.

## 2 Related Works

There are various approaches for recognizing TOC pages. However, these approaches are primarily focus on exploiting TOC page based features.

Caihua Liu et al. [2] determine TOC recognition based on a rule-based approach. They observed that TOC pages are usually started with some specific words such as "Contents" or "Index", and they have also many lines ending with "numbers or Roman characters". In order to reduce processing time, they only consider pages in the first half of the document.

S.Mandal et al. [9] use information about the number of pages and the distribution of textual parts on pages to determine the indexed pages of the document based on machine learning decision tree algorithm.

Luo et al. [8] define that a TOC page consists of three main components: chapter and section/sub-sections headings, indexed page numbers, and "connectors". Their main idea is to identify the TOC page based on searching the predefined "connectors" such as "dot lines". Their method is applied for Japanese documents only.

## 3 Proposed Method

### 3.1 The Architecture of the TOC Pages Recognition System

In this section, we present the general framework of our TOC pages recognition system which is based on the fundamental machine learning (for classification) approach [1]. The framework is shown in Fig. 1.

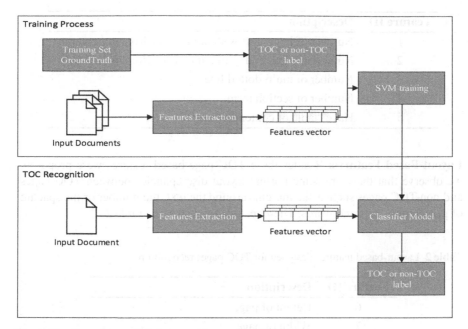

**Fig. 1** The flow chart of TOC recognition system (Cf. [1])

The input documents have been preprocessed with available information about the text and the bounding box of text, lines, paragraphs, and pages.

**Training process:** Extract features of each page in the documents from training set and build feature vectors. List of features is listed in features extraction section. SVM (Support Vector Machine) training is based on feature vectors to build the classifier model.

**TOC recognition:** Extract features of each page in the document from testing set, then use the classifier model to determine a page is a TOC page or a non-TOC page.

## 3.2  Features Extraction

The module "Features Extraction" extracts two kinds of features: TOC page based features and layout-based features.

**TOC Page Based Features:** We reuse TOC page based features designed by the previous works [2], [9], [8]. The detail of these features is shown in Table 1.

**Table 1** TOC page based features designed for TOC pages recognition (Source: [2], [9])

| Feature ID | Description |
|---|---|
| 1 | Number of lines start with numbers or Roman numerals |
| 2 | Number of lines end with numbers or Roman numerals |
| 3 | Number of multi dotted lines |
| 4 | Number of section terms |
| 5 | Have "table of content" term |

**Layout-Based Features:** Besides these TOC page based features were proposed, we observe that there are some normal layout discrepancies between TOC pages and non-TOC pages such as the margins around the text, the number or the spacing of paragraphs, lines and words. These features is showed in Table 2.

**Table 2** Layout-based features designed for TOC pages recognition

| Feature ID | Description |
|---|---|
| 6 | Height of page |
| 7 | Width of page |
| 8 | Top margin of textual region in page |
| 9 | Bottom margin of textual region in page |
| 10 | Left margin of textual region in page |
| 11 | Right margin of textual region in page |
| 12 | Page number of page |
| 13 | Number of paragraphs of page |
| 14 | Number of lines of page |
| 15 | Number of words of page |
| 16 | Average word spacing of page |
| 17 | Average line spacing of page |
| 18 | Average paragraph spacing of page |
| 19 | Average word height of page |

# 4   Experiments

## 4.1   Datasets

We use three datasets published by ICDAR Book Structure Extraction Competition from 2009 [5], 2011 [6] and 2013 [4] (see Table 3). These datasets contain books of different genres, and domains. Each book is provided in a searchable PDF and DjVu XML format. We conduct experiments on the DjVu XML books. The DjVu XML provides words, bounding boxes of words, lines, paragraphs, and page information.

**Table 3** ICDAR Book Structure Extraction Datasets 2009 [5], 2011 [6] and 2013 [4]

| Dataset | Books | TOC pages | Non-TOC pages |
|---------|-------|-----------|---------------|
| 2009 | 527 | 1.292 | 181.865 |
| 2011 | 513 | 1.283 | 180.285 |
| 2013 | 967 | 2.664 | 330.887 |

## 4.2 Evaluation Metrics

To analyze TOC recognition performance, we use three quality metrics: Precision, Recall and F-1 Score. Precision measures exactness of TOC recognition system: a higher Precision score means less incorrectly recognized TOC pages. Recall measures the coverage of TOC recognition system on testing datasets: a higher Recall score means less incorrectly recognized non-TOC pages. F-1 Score combines Precision and Recall scores. It is used for measuring TOC recognition performance.

## 4.3 Experiment Setup

The datasets are divided by the ratio 80% for training set and 20% used for testing set. We use LibSVM library [3], radial basis function (RBF) kernel with custom parameters selected by cross-validation strategies for optimal performance. With each dataset, we evaluate three approaches with different features used.

## 4.4 Result

Fig 2 shows our experiments for TOC recognition on three datasets of ICDAR Book Structure Extraction Competition: 2009 [5], 2011 [6] and 2013 [4]. TOC page based features was used in previous works [2], [9], [8]; and we propose using layout-based features.

We compare result of TOC recognition using the combination of two groups of features (TOC page based features and layout-based features) with each one individually. According to the results of experiments, the accuracy of TOC recognition would be better if we combine TOC page based features with layout-based features.

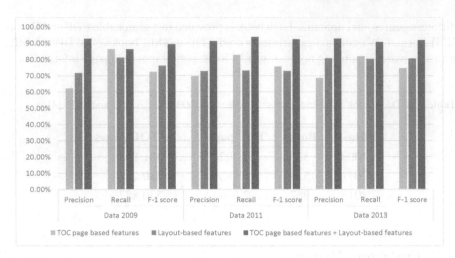

**Fig. 2** Result of TOC recognition on ICDAR Structure Extraction Datasets: 2009 [5], 2011 [6] and 2013 [4]

## 5 Conclusion

This paper proposed using some appropriate normal layout-based features for improving accuracy of TOC recognition. Experiments on three datasets of ICDAR Book Structure Extraction competition: 2009 [5], 2011 [6] and 2013 [4] show that the proposed method using both layout-based features and existing TOC page based features will improve the performance of TOC recognition.

## References

[1] Bird, S., Klein, E., Loper, E.: Natural Language Processing with Python, 1st edn. O'Reilly Media, Inc. (2009)
[2] Liu, C., Chen, J., Zhang, X., Liu, J., Huang, Y.: TOC structure extraction from OCR-ed books. In: Geva, S., Kamps, J., Schenkel, R. (eds.) INEX 2011. LNCS, vol. 7424, pp. 98–108. Springer, Heidelberg (2012)
[3] Chang, C.-C., Lin, C.-J.: LIBSVM: A library for support vector machines. ACM Transactions on Intelligent Systems and Technology 27:1–27:27, 1–4 (2011), Software available at http://www.csie.ntu.edu.tw/~cjlin/libsvm
[4] Doucet, A., Kazai, G., Colutto, S., Mühlberger, G.: Overview of the ICDAR 2013 Competition on Book Structure Extraction. In: Proceedings of the Twelfth International Conference on Document Analysis and Recognition (ICDAR 2013), Washington DC, USA, p. 6 (August 2013)
[5] Doucet, A., Kazai, G., Dresevic, B., Uzelac, A., Radakovic, B., Todic, N.: ICDAR 2009 Book Structure Extraction Competition. In: Proceedings of the Tenth International Conference on Document Analysis and Recognition, ICDAR 2009, Barcelona, Spain, pp. 1408–1412 (July 2009)

[6] Doucet, A., Kazai, G., Meunier, J.-L.: ICDAR 2011 Book Structure Extraction Competition. In: Proceedings of the Eleventh International Conference on Document Analysis and Recognition, ICDAR 2011, Beijing, China, pp. 1501–1505 (September 2011)

[7] Kazai, G., Doucet, A., Landoni, M.: Overview of the INEX 2008 book track. In: Geva, S., Kamps, J., Trotman, A. (eds.) INEX 2008. LNCS, vol. 5631, pp. 106–123. Springer, Heidelberg (2009)

[8] Luo, Q., Watanabe, T., Nakayama, T.: Identifying contents page of documents. In: Proceedings of the 13th International Conference on Pattern Recognition 1996, 3rd edn., pp. 696–700 (August 1996)

[9] Mandal, S., Das, A.K., Bhowmick, P., Chanda, B.: A unified algorithm for identification of various tabular structures from document images. Int. J. Digit. Library Syst. 2(2), 27–54 (2011)

# Predicting the Popularity of Social Curation

Binh Thanh Kieu, Ryutaro Ichise, and Son Bao Pham

**Abstract.** The amount and variety of social media content such as status, images, movies, and music are increasing rapidly. Accordingly, the social curation service is emerging as a new way to connect, select, and organize information on a massive scale. One noticeable feature of social curation services is that they are loosely supervised: the content that users create in the service is manually collected, selected, and maintained. A large proportion of these contents are arbitrarily created by inexperienced users. In this paper, we look into social curation, particularly, the Storify website[1]. This is the most popular social curation for creating stories included in various domains such as Twitter, Flicker, and YouTube. We propose a machine learning method with feature extraction to filter these contents and to predict the popularity of social curation data.

**Keywords:** curation, social curation, social network service, prediction, popularity.

## 1 Introduction

Along with the rapid growth of the Internet, social networks are increasingly attracting users, young people in particular. Therefore, the study of social networks is getting more and more attention. Social network services such as Facebook, Myspace, and Twitter have become viable sources of information for many online users. These websites are increasingly used for communicating breaking news, sharing eyewitness accounts, and organizing groups of people. At the most basic level,

Binh Thanh Kieu · Son Bao Pham
Faculty of Information Technology, University of Engineering and Technology,
Vietnam National University, Hanoi, Vietnam
e-mail: binhkt.vnu@gmail.com, sonpb@vnu.edu.vn

Ryutaro Ichise
National Institute of Informatics, Tokyo, Japan
e-mail: ichise@nii.ac.jp

[1] http://www.storify.com/

© Springer International Publishing Switzerland 2015
V.-H. Nguyen et al. (eds.), *Knowledge and Systems Engineering*,
Advances in Intelligent Systems and Computing 326, DOI: 10.1007/978-3-319-11680-8_33

a curation service offers the ability to manually collect, select, and maintain this social media information. This is very different from other social information sources,
and we can utilize this characteristic for efficient content mining.

The emergence of Web 2.0 and online social networking services, such as Digg,
YouTube, Facebook, and Twitter, has changed how users generate and consume online contents. For example, YouTube, well-known for its fast-growing user-generated
contents, reports 100 hours worth of video upload every minute [22]. Online social networking services, augmented with multimedia content support, sharing, and
commenting on other users' contents, constitute a significant part of the web experience by Internet users. The question is how do users find interesting contents? Or,
how do certain contents rise in popularity? If we can answer these questions, we can
predict the most likely contents to become popular and filter out others. Moreover,
when we can filter out unpopular contents that get little attention, good contents can
be used to build an automatic system for curating social content.

However, predicting the popularity of content is a difficult task for many reasons. Among these, the effects of external phenomena (e.g., media, natural, and
geo-political) are difficult to incorporate into models [16], and cascades of information are difficult to forecast [3]. Finally, the underlying contexts, such as locality,
relevance to users, resonance, and impact, are not easy to decipher [2].

The rest of the paper is organized as follows. In the second section, we explain
the social curation service, our target data source, and details of the dataset specifications. In the third section, we review related work. The fourth section is devoted
to the formulation of predicting view counts of a curation list. The fifth section describes experiments and the evaluation of our results. The last section concludes this
paper with a discussion about future work.

## 2 Social Curation

The word "curate" is defined as selecting, organizing, and looking after the items in
a collection or exhibition. The word is derived from the Latin root "curare" or "to
cure", which means "to care". Curation involves assembling, managing and presenting some types of collections. For example, curators of art galleries and museums
research, select, and acquire pieces for their institutions' collections and oversee interpretation, displays, and exhibits. Social curation is the collaborative oversight of
collections of web content organized around types of content such as Pinterest (a site
for sharing and organizing images) and Storify (a site for collecting and publishing
stories).

### 2.1 Social Curation Service

Social networks are spaces for dialog and conversation that have grown into ubiquitous information exchanges. Youth today refer to social networks, aggregators, and
mobile apps for most of their information instead of singling out specific media for
news, politics, personal communication, and leisure. In turn, social networks have

provided new functions that help users curate information in meaningful and productive ways. Social curation involves aggregating, organizing, and sharing the content created by others to add context, narrative, and meaning. Artists, changemakers, and organizations use social curation to showcase the full range of conversations around a topic, add more nuance to their own original content, and crowdsource content from their community members. The rise of social curation can be attributed to three broad trends.

- Firstly, people are creating a constant stream of social media content, including updates, location check-ins, blog posts, photos, and videos.
- Secondly, people are using their social networks to filter relevant content by following others who share similar interests.
- Thirdly, social media platforms are also curating content by giving curation tools to users (YouTube playlists, Flickr galleries, Amazon lists, Foodspotting guides), using editors and volunteers (YouTube Politics, Tumblr Tags), or using algorithms (YouTube Trends, Autogenerated YouTube channels, LinkedIn Today). As a result, a number of niche social curation platforms have emerged to enable people to curate different types of content, including links, photos, sounds, and videos.

We should emphasize that each curation list is a kind of loosely supervised but organized social dataset. This means that social media items in the same curation list are expected to share the same context to a certain degree: a curation list is manually generated to fully convey one idea to the consumer. This is a very distinct characteristic compared to other social media that are unorganized in many cases.

## 2.2 Storify

The website Storify is the most well-known site for people telling stories by curating social media. Storify was launched in September 2010 and accounts were invitation-only until April 2011. The site is now open to everyone and users only need a Twitter account. Storify provides a function to filter out poor content and unreliable sources. If social media changes or misinterprets context, Storify can help curators put it back together again [13]. Storify allows curators to embed dynamic images, text, tweets, and even Facebook status updates, and then knit these all together with background and context provided by the storyteller. It is an engaging way for us to learn how to work out what is true and what is speculation. We have also found that using Twitter has taught us how to look for sources and news and Storify has helped teach us how to think and write context and narrative. Each story is a curation list which shares some characteristics: manually collected (bundling a collection of content from diverse sources), manually selected (re-organizing them to give one's own perspective), manually maintained (publishing the resulting story for consumers).

The Storify data is in the form of lists of Twitter messages. An example of a list is shown in Figure 2. A list of tweets corresponds to what we called a story, which represents a manually filtered and organized bundle. Lists in Storify draw on Twitter as

**Fig. 1** Example of a Storify list

**Table 1** Statistics of curated domains

| Domain | Number of Elements | Proportion |
|--------|-------------------|------------|
| Twitter | 8,514,006 | 75.5% |
| Storify | 1,206,794 | 10.7% |
| YouTube | 190,611 | 1.7% |
| Facebook | 169,361 | 1.5% |
| Instagram | 155,762 | 1.4% |
| Flickr | 127192 | 1.2% |
| Others | 920,089 | 8% |

its source. The lists may be created individually in private or collaboratively in public as determined by the initial curator. In the Storify curation interface, the curator begins the list curation process by looking through his Twitter timeline (tweets from users that he or she follows), or directly searching tweets via relevant words/hashtags. The curator can drag-and-drop these tweets into a list, reorder them freely, and also add annotations such as a list header and in-place comments. We first provide some data statistics to get a feel for the curation data. We collected all the data from

**Table 2** Element types

| Types | Number | Proportion |
|-------|--------|------------|
| Quote | 7,715,616 | 68.4% |
| Text | 1,195,625 | 10.6% |
| Image | 1,436,673 | 12.7% |
| Video | 206,265 | 1.8% |
| Link | 732,096 | 6.5% |

**Table 3** Storify action statistics

| Action | Number | Average |
|--------|--------|---------|
| Views | 642,666,347 | 1823 per story |
| Comments | 21,306 | 0.06 per story |
| Element comments | 21,133 | 0.002 per element |
| Likes | 206,265 | 0.12 per story |

2010 to April 2013, which amounted to 63,419 users and 352,540 stories. This corresponds to a total of 11,283,815 elements from various domains. Table 1 describes the various domains used in the stories. Twitter is the largest domain source with more than 75% elements, and Flickr is the smallest specific source with 1.2% elements. The statistics of the element types is shown in Table 2. The five types of elements in stories are quote, text, image, video, and link. Because Storify users use a huge number of tweets, the number of quote contents accounts for a large percentage of nearly 70%. Media content as images and movies make up approximately 15%. Text contents are written by the curator to add more information, explain, or link elements. The Storify API provides the four main actions shown in Table 3. The Storify website allows users to comment on each element or on all parts of a story. However, the average numbers of comments, element comments and similar actions are quite small. Therefore, approaches utilizing user comments and actions are not suitable for this dataset.

# 3 Related Work

Several studies have investigated social curation as a new source of data mining. Pinterest[2] is the most popular website for sharing images and video, and the third most popular social network in the US behind Facebook and Twitter. The website is built around the activity of collecting digital images and videos and pinning them to a pin board. Each pin is essentially a visual bookmark and the pin boards are thematic collections of the bookmarks, where context is added to the collected information. Hall

---

[2] http://www.pinterest.com/

and Zarro described some of the user actions on Pinterest and created a dataset to find the pin content of Pinterest users across a wide variety of subject areas [8] [23]. Besides only curating images or video, other sites curate status, comments, news sources to write blogs, stories. Storyful[3] is a social media news agency established in 2010 with the aim of filtering news, or newsworthy content, from the vast quantities of noisy data on social networks such as Twitter and YouTube. Storyful invests considerable time into the manual curation of content on these networks. It sounds more or less like the same goal as Storify's but there is one important difference. Storyful aims to deliver content for news organizations, whereas Storify is more of a tool for journalists. It allows journalists to use its template to write stories that include relevant tweets and Facebook posts without losing the original formatting or links. Journalists can create interactive stories with clear links to original pictures or tweets. Greene et al. proposed a variety of criteria for generating user list recommendations based on content analysis, network analysis, and the "crowdsourcing" of existing user lists [7]. In addition, the Togetter website[4] is a rapidly growing social curation website in Japan. Togetter averaged more than 4 million user-views per month in 2011. The Togetter curation data mainly exist in the form of lists of Twitter messages. Ishiguro et al. used Togetter data for the automatic understanding and mining of images [11] and created a system [6] that suggests new tweets to increase the curator's productivity and breadth of perspective. Our research discovered another social curation website, Storify. The structure of a Storify list is quite similar to that of a Togetter list. The only difference is the language: the common language of Togetter is Japanese and Storify is English. However, we interested in another aspect which show the quality of curation list made by users.

The problem of predicting online content highlights how much attention it will ultimately receive. Research shows that user attention is allocated in a rather asymmetric way, with most content getting only some views and downloads, whereas a few receive a significant amount of user attention; thus, filtering these contents will help to save much time for viewers. There are different ways to formulate how much attention of contents. Many researchers interested in the number of views as the popularity of online content such as YouTube (the number of views [20]), Vimeo (the number of views [1]), Flickr (the number of views [24]). Otherwise, the popularity is presented by users' actions like Dig (the number of user votes [12]), Twitter (the number of retweets [10]). Moreover, others formulate the problem to a change of the number of views that contents receive over time.

Predicting the popularity of news articles is a complex and difficult task and different prediction methods and strategies have been proposed in several recent studies [20] [21]. The common point of all these methods is that they focus on predicting the exact attention that an article will generate in the near future. First, some researchers have studied features that describe the underlying social network of the users and contents that can be leveraged to predict popularity [9] [12] [18] [21]. The authors in [14] [16] [17] [21] studied features that take into account the

---

[3] http://www.storyful.com/
[4] http://www.togetter.com/

comments found in blogs to predict popularity. However, few other works forecast a value for the actual popularity of individual content. Lee et al. used survival analysis to evaluate the probability that a given content receives more than some $x$ number of hits [16] [17]. Hong et al. developed a coarse multi-class classifier-based approach to determine whether given Twitter hashtags are retweeted $x \leq (0; 100; 10000; \infty)$ times [10]. Similarly, Lakkaraju and Ajmera used support vector machines (SVMs) to predict whether a given content falls into a group that attracts $x \leq (10\%; 25\%; 50\%; 75\%; 100\%)$ of the attention in a system [15], while Jamali and Rangwala predicted the popularity of content by using an entropy measure [12]. Finally, Szabo and Huberman presented a linear regression model based on the number of views [20]; this method was applied to build predictive popularity by applying regression to different feature spaces [2] [9] [18] [21].

In this work, the popularity of Social Curation is shown by the number of views that the content will receive in the near future. We propose three groups for categorizing the popularity level of Social Curation. We build a predictor based on a machine learning method, SVM, with feature selection to classify into these groups.

## 4 Predicting the Popularity of Social Curation

### 4.1 Problem Formulation

Similar to normal content, the popularity of social curation is defined by the number of users' view. We predict how much view which stories will receive in the near future. However, it is difficult to predict exact amount of attention and people are almost interested in the popularity of content; thus, instead of predicting exactly the number, we cast the task as a multi-class classification problem that predicts the popularity that a curation list will receive after three months based on the number of views. Although our system cannot predict exactly the number of attention, but this system partly helps users to be able to identify popular contents and not popular contents. We divide the number of views into three different classes: class 1 – not popular, with the number of views less than 10, class 2 – less popular, with the number of views between 10 and 1000, class 3 – very popular, with the number of views more than 1000.

We used an SVM to classify these classes. LibSVM [4] with a radial basic function (RBF) kernel and default parameters, and the feature selection tool [5] were used to optimize the result. We extracted two types of features, namely curation features and curator features. Curator features are features of users who collect and organize elements from some domains and create curation lists. Curation features are features related to the content of the curation lists.

### 4.2 Features

Social curation lists contain many kinds of information that are useful for classifying. For example, if the curation list includes many Twitter contents, the view count of the contents is expected to increase; or, if elements match the context of the curation list, the content will attract much more attention.

In this study, as the social curation list included a large number of Twitter messages, we used applicable features for predicting the number of retweets and microblogging popularity. We divided the features into the two distinct sets mentioned above: curator features (which are related to the author of the story) and curation features (which encompass various statistics of the content in the story).

### 4.2.1  Curator Features

The following are the five curator features:

. The number of users who follow the curator of the content
. The number of users who the curator of the content follows
. The number of stories written by the curator
. The user's language (English or not)
. When the curator of the content started using Storify

These features were selected from the content creator features proposed by Ishiguro et al. [11]. We implemented these features as our baseline system. The number of followers and friends has been consistently shown to be a good indicator of retweetability, whereas the number of stories has not been found to have a significant impact [19]. Our prior analysis also showed that stories written in English are more likely to be viewed, so we used a binary feature indicating if the user's language is English. The date when a curator started using Storify shows their experience. Normally, longtime users have more experience producing more popular curator stories than do new users. We are not aware of any prior work that analyzes the effect of language or date on content popularity.

### 4.2.2  Curation Features

The following are the seven curation features:

. The number of hashtags
. The number of versions
. The number of embeds
. The story's language (English or not)
. The number of popular tweet elements/total elements (the number of retweets greater than 100)
. The number of popular image and video elements/total elements (the number of image views and video views greater than 1000)
. The total number of elements

As a large proportion of elements in the curation list is from the Twitter domain, hashtags therefore play an important feature for predicting the popularity. One paper showed that hashtags, URLs and mentions have a high correlation with predicting popular Twitter messages [19]. Although the Storify API provides hashtags, URLs and mentions of each story, URLs and mentions have an insignificant impact on the

**Table 4** Prediction accuracies for two feature types

| Type of feature | No. of features. | Classification (10-fold) |
|---|---|---|
| Curation features | 7 | 75.08% |
| Curator features | 5 | 80.20% |
| Combined features | 12 | 82.62% |

result. The version feature shows that users who modified their story can improve the story's quality and get more attention. The embed feature shows that more sharing is more popular. The English language is the most well-known language in the world, so stories written in English are read by more people than those in any other language. Although the feature is quite similar to the feature of the language of the curator, not all curators use their main language to write stories. According to our experiments, tests using this feature achieved higher results. Finally, the higher proportion of Twitter elements and media elements also increase the accuracy. Moreover, using many elements in a story draws more attention than stories with fewer elements.

To the best of our knowledge, no prior work analyzed the effect of these features on content popularity. Therefore, the features we proposed are based on the experiments and feature selection tool to acquire the highest result. The feature selection tool, combined with libSVM, uses the F-score for selecting features [5]. The F-score is a simple technique that measures the discrimination of two sets of real numbers. The larger the F-score is, the more likely this feature is more discriminative. Therefore, this score is used as a feature selection criterion. Moreover, libSVM also provides a feature scaling function in order to absorb the scale differences among feature values, then we re-scaled them between [0,1]. Finally, twelve of above features had the highest result for predicting the popularity of Storify data.

## 5 Experimental Results

### 5.1 The Experimental Dataset

We used Storify's streaming API to collect a random sample of public stories created from March 1, 2013 to March 31, 2013 with 34,810 curation lists. We suppose that these stories have the same published time. We crawled them in June 2013 so we predict how much attention of these contents in the three months later. Finally, we divided this dataset into 10 groups and ran 10 cross validations.

### 5.2 Results

The different popular levels are displayed followed by the three classes, as mentioned in Section 4.1. Statistically, nearly half of the stories are class 1, nearly 20% are class 2, and the remaining are class 3. The prediction accuracy for the two types of

**Table 5** Accuracy of 10 tests

| Test | Curator features | Combined features |
|------|------------------|-------------------|
| 1    | 83.61%           | 87.42%            |
| 2    | 82.26%           | 85.58%            |
| 3    | 79.88%           | 83.91%            |
| 4    | 78.55%           | 80.38%            |
| 5    | 82.23%           | 85.49%            |
| 6    | 73.56%           | 71.19%            |
| 7    | 76.31%           | 78.92%            |
| 8    | 87.60%           | 89.83%            |
| 9    | 68.44%           | 70.38%            |
| 10   | 88.78%           | 93.22%            |

features are shown in Table 4. The result of curation features (7 features) is the worst at 75.08%, curator features (5 features) at 80.02%, and the best result is combined features (combined between curator and curation features for a total of 12 features) at 82.62%. Therefore, both types of features are necessary for prediction with high accuracy.

Table 5 shows more detailed results for the 10 tests. The curator features (as baseline features) and combined features show different results. Most tests using combined features are more accurate than tests using only curator features except for test 6. We analyzed test 6 and realized that the percentage of class 2 is approximately 40%, which is double the normal percentage. It is shown that combined features cannot perform well for class 2. In addition, most tests using combined features attain roughly 83% accuracy except some tests such as tests 6, 9 (lower accuracy barely over 70%) and tests 8, 10 (high accuracy over 90%). Although the distribution ratio of classes in these tests is quite different from the others, the difference is irregular and not significant. This is an open problem in our research; finding the answer for this question would improve the result.

## 5.3  T-Test Evaluation

The (student's) t-test is a statistical examination of two population means. In simple terms, the t-test assesses whether the means of two groups are statistically different from each other. It is commonly used when the variances of two normal distributions are unknown and when an experiment uses a small sample size. In our case, we used the t-test to evaluate two group results of the above 10 tests (small sample size). The decision rule is a 95% confidence interval of the difference from $-3.8929$ to $-1.1271$ and we calculate our value $t = -4.1059$. Therefore, we conclude that this difference is considered to be very significant. It indicates that our proposal to use both features is effective to predict the popularity of social curation data.

## 6 Conclusion

In this paper, we presented a method to predict the popularity of social curation content as the first step for mining social curation. A key insight is that a curation list, which is unique compared to other social data, is the manual collection, selection, and maintenance by curators. We used a machine learning approach and selected key features. Analyzing the features, we found that social features (curator features) perform very well, but the system can be improved by combining the content features (curation features). A comparison by the t-test showed the significance.

On the other hand, the paper investigated only a specific curation dataset for a specific task. We are aware that there are many open problems. We have to investigate social features in a larger dataset or other domains. In addition, analyzing and explaining the effect of features for predicting the popularity of social curation could improve the result. Finally, our research is the first task for mining social curation data. Based on this research, we could consider future tasks such as an automatic system or a recommendation system for curating social data.

**Acknowledgment.** This work is partially supported by the Research Grant from Vietnam National University, Hanoi No. QG.14.04.

## References

[1] Ahmed, M., Spagna, S., Huici, F., Niccolini, S.: A peek into the future: Predicting the evolution of popularity in user generated content. In: Proc. WSDM 2013 (2013)

[2] Bandari, R., Asur, S., Huberman, B.A.: The pulse of news in social media: Forecasting popularity, CoRR, abs/1202.0332 (2012)

[3] Cha, M., Mislove, A., Gummadi, K.P.: A measurement-driven analysis of information propagation in the Flickr social network. In: Proc. WWW 2009(2009)

[4] Chang, C.C., Lin, C.J.: LIBSVM: A library for support vector machines. ACM Trans. Intelligent Systems and Technology 2 (2011)

[5] Chen, Y.-W., Lin, C.-J.: Combining SVMs with various feature selection strategies. In: Guyon, I., Nikravesh, M., Gunn, S., Zadeh, L.A. (eds.) Feature Extraction. STUDFUZZ, vol. 207, pp. 315–324. Springer, Heidelberg (2006)

[6] Duh, K., Hirao, T., Kimura, A., Ishiguro, K., Iwata, T., Au Yeung, C.-M.: Creating stories: Social curation on twitter messages. In: Proc. ICWSM (2012)

[7] Greene, D., Sheridan, G., Smyth, B., Cunningham, P.: Aggregating content and network information to curate twitter user lists. In: Proc. ACM RecSys workshop, RSWeb (2012)

[8] Hall, C., Zarro, M.: Social curation on the website Pinterest.com. In: Proc. ASIST 2012 (2012)

[9] Hogg, T., Lerman, K.: Social dynamics of Digg, CoRR, abs/1202.0031 (2012)

[10] Hong, L., Dan, O., Daviso, B.D.: Predicting popular messages in Twitter. In: Proc. WWW 2011 (2011)

[11] Ishiguro, K., Kimura, A., Takeuchi, K.: Towards automatic image understanding and mining via social curation. In: Proc. ICDM 2012 (2012)

[12] Jamali, S., Rangwala, H.: Digging Digg: Comment mining, popularity prediction, and social network analysis. In: Proc. WISM 2009 (2009)

[13] Fincham, K.: Storify. The National Association for Media Literacy Educations Journal of Media Literacy Education (2011)

[14] Kim, S.-D., Kim, S.-H., Cho, H.-G.: Predicting the virtual temperature of web-blog articles as a measurement tool for online popularity. In: Proc. CIT 2011(2011)

[15] Lakkaraju, H., Ajmera, J.: Attention prediction on social media brand pages. In: Proc. CIKM 2011(2011)

[16] Lee, J.G., Moon, S., Salamatian, K.: An approach to model and predict the popularity of online contents with explanatory factors. In: Proc. WI-IAT (2010)

[17] Lee, J.G., Moon, S., Salamatian, K.: Modeling and predicting the popularity of online contents with Cox proportional hazard regression model. Neurocomputing (2012)

[18] Lerman, K., Hogg, T.: Using a model of social dynamics to predict popularity of news. CoRR, abs/1004.5354 (2010)

[19] Suh, B., Hong, L., Pirolli, P., Chi, E.H.: Want to be retweeted? Large scale analytics on factors impacting retweet in Twitter network. In: Social Computing, SocialCom (2010)

[20] Szabo, G., Huberman, B.A.: Predicting the popularity of online content. Communications of the ACM 53, 80–88 (2010)

[21] Tsagkias, M., Weerkamp, W., de Rijke, M.: News comments:Exploring, modeling, and online prediction. In: Gurrin, C., He, Y., Kazai, G., Kruschwitz, U., Little, S., Roelleke, T., Rüger, S., van Rijsbergen, K. (eds.) ECIR 2010. LNCS, vol. 5993, pp. 191–203. Springer, Heidelberg (2010)

[22] YouTube Statistics,
      http://www.youtube.com/yt/press/statistics.html

[23] Zarro, M., Hall, C.: Exploring social curation. D-Lib Magazine 18(11/12) (November/December 2012)

[24] Van Zwol, R., Rae, A., Pueyo, L.G.: Prediction of favourite photos using social, visual, and textual signals. In: Proc. ACMMM (2010)

# Parameter Learning for Statistical Machine Translation Using CMA-ES

Viet-Hong Tran, Anh-Tuan Pham, Vinh-Van Nguyen,
Hoai-Xuan Nguyen, and Huy-Quang Nguyen

**Abstract.** Minimum error rate training (MERT) is probably still the most widely used parameter learning algorithm in statistical machine translation [1] (SMT). However, it does not support the use of large number of learning features (e.g. 30 features or more). Moreover, acting on parameter space, MERT is only a local optimization algorithm. In this paper, we investigate for the first time the use of meta-heuristics and global optimization techniques for the problem of learning parameters in SMT. In particular, We replace MERT with the well-known meta-heuristics for global optimization called Covariance Matrix Adaptation Evolution Strategy (CMA-ES) [2]. We test the effectiveness of CMA-ES by conducting SMT experiments on an English-Vietnamese corpus. The results show that the improved SMT system using CMA-ES achieved superior BLEU scores compared to the baseline SMT system using MERT both on the dev and test data sets.

**Keywords:** Statistical Machine Translation, Minimum Error Rate Training, Covariance Matrix Adaptation Evolution Strategy

Viet-Hong Tran
University of Economic and Technical Industries, Vietnam
e-mail: thviet@uneti.edu.vn

Viet-Hong Tran · Vinh-Van Nguyen
University of Engineering and Technology, Vietnam National University, Hanoi, Vietnam
e-mail: vinhnv@vnu.edu.vn

Anh-Tuan Pham
IT Center, Military Academy of Logistics, Vietnam
e-mail: anh.pt204@gmail.com

Anh-Tuan Pham · Hoai-Xuan Nguyen
IT R&D Center, Hanoi University, Vietnam
e-mail: nxhoai@gmail.com

Huy-Quang Nguyen
Faculty of IT, Vietnam - Germany Vocational College of Vinh Phuc, Vinh Phuc, Vietnam
e-mail: quanghuyinfor@gmail.com

© Springer International Publishing Switzerland 2015
V.-H. Nguyen et al. (eds.), *Knowledge and Systems Engineering,*
Advances in Intelligent Systems and Computing 326, DOI: 10.1007/978-3-319-11680-8_34

# 1 Introduction

One important phase in the process of statistical machine translation is to determine the weights for feature functions in the computation of the probability of a destination sentence (in destination language) given a source sentence (in source language) (see section 2 for more details). These weights are learnt based on training data and the problem could be casted as a global optimization task in the parameter space (usually the real vector space). To accomplish the task, Minimum Error Rate Training (MERT) for parameter learning in statistical machine translation (SMT) was introduced in Och [1].

Since its introduction, this line-search based method has probably been the most widely used parameter learning algorithm for SMT. MERT is the default option in Moses, the well-known statistical machine translation system that allows users to automatically train translation models on any language pairs [3]. However, MERT suffers from two weaknesses:

. It could not handle well large number of features (more than 30).
. It could only guarantee (if given enough time) to find the local optimum for the parameter optimization (learning) problem mentioned above.

Therefore, it is desirable to investigate alternatives to MERT that could handle larger number of features and seek for the global optimum in the parameter space.

In this paper, we present some first investigations on the use of a meta-heuristic and global optimization technique called Covariance Matrix Adaptation Evolution Strategy (CMA-ES) [2] as an alternative to MERT. CMA-ES is a stochastic, derivative-free, non-convex, and non-linear global optimization method in the field of evolutionary computation. CMA-ES has been proven to be an effective meta-heuristics for numerical global optimization tasks championing several optimization competitions for meta-heuristics techniques and evolutionary algorithms [2]. Our preliminary experimental results in this paper show that CMA-ES is a better choice compared to MERT on a statistical machine translation task with English-Vietnamese as the language pairs. This promising results could trigger further studies in the use of global meta-heuristic optimizers in solving the problem of parameter learning in statistical machine translation.

The rest of the paper is organized as follows. In the next section, we give some backgrounds on the problem of parameter learning in statistical machine translation and review some related works. Section 3 describes CMA-ES, the method we use in this paper for solving the problem of parameter learning in SMT. Section 4 contains our main experiments, results, and discussions. Section 5 concludes the paper and highlights some possible future extensions of this work.

# 2 Backgrounds

In statistical machine translation, the problem of translating from a source language to a destination language is solved by building a statistical model from training data for estimating the probability of a destination sentence given a source sentence. The statistical model is in the form of a log-linear model as follows [4].

$$P(e,d|f) = \frac{exp\sum_{k=1}^{K}\lambda_k h_k(e,d,f)}{\sum_{e',d'} exp\sum_{k=1}^{K}\lambda_k h_k(e',d',f')} \tag{1}$$

Where $P(e,d|f)$ is the conditional probability of a destination sentence $e$ and its derivation $d$ given a source sentence $f$. $h_k$ are feature functions and $\lambda_k$ are the feature weights. A feature function relates the source and the destination sentences in the manner defined by users and might be language dependent. For instance, we might want to use binary phrase feature function shown in equation (2) to measure how much a particular phrase pairs helps to discriminate between good and bad translations.

$$h_k(f_i,e_i) = \begin{cases} 1, & if f_i = "dieses\,Haus"and\,e_i = "this\,house" \\ 0, & otherwise \end{cases} \tag{2}$$

Given a set of feature functions (manually or automatically designed) the model training task is to estimate the weights $\lambda_k$ from the training data so as to maximize the likelihood of matched pairs of sentences in the training data. This could be casted as a parameter optimization problem. Generally, we could not assume any properties of the objective function of this optimization problem including: smoothness, convexity, or modality.

The most widely used numerical optimization algorithm in statistical machine translation system for solving the aforementioned optimization problem has been the Minimum Error Rate Training (MERT) algorithm Och [1]. However, as mentioned in the previous section, usually, MERT could not handle large feature sets (more than 30) and more seriously, it is just a local optimizer that could not guarantee to achieve the global optimum for the parameter learning task.

Alternative to MERT, there has been some other optimization approaches to solve the problem of parameter training in statistical machine translation models. The approach in [5] is claimed to be capable of handling large features sets, but not yet to receive a widespread adoption. Online methods [6] , [7] are recognized to be effective, but require substantial implementation efforts due to the difficulties in their parallelization. Margin Infused Relaxed Algorithm (MIRA) has been suggested for learning parameters in statistical machine translation systems with larger features sets. Watanabe et al. [7] and Chiang et al. [8] added thousands of features to their baseline systems and showed an improved translation quality after learning parameter of their enhanced models with MIRA. Arun and Koehn [9] explored training a phrase-based SMT system in a discriminative fashion with MIRA. McDonald et al. [10] were the first to apply MIRA to train a dependency parser.

While the main focus of the above approaches is to improve over MERT for handling larger feature sets, none of these is a global optimizer. In this paper, we investigate for the first time the use of a meta-heuristic and global optimizer, CMA-ES, as a substitute for MERT in statistical translation systems. CMA-ES is described in the next section.

## 3   Using CMA-ES for Training Parameter

The Covariance Matrix Adaptation Evolution Strategy [2] is an evolutionary algorithm for solving non-linear, non-separable, and non-convex optimization problem. The main idea of CMA-ES is to sample solutions from a multivariate Gaussian distribution. The solutions are then evaluated for their fitness (goodness in terms of objective values). Then, all solutions are sorted according to their fitness and the better solutions are used to update the mean vector and the covariance matrix of the Gaussian distribution. This process is repeated until the a certain number of iterations is reached or some conditions are satisfied. The pseudo-code for CMA-ES is as follows [2]:

---

**CMA-ES optimization Algorithm**

---

```
1. set λ // number of  samples per iteration, at least two, generally > 4
2. initialize m,  σ,  C = I,  p_σ = 0, p_c = 0 // initialize state variables
3. while not terminate //interate
4.    for I in {1…λ} // sample λ new solutions and evaluate them
5.        x_i = sample_multivariate_normal (mean = m, covariance_matrix = σ²C )
6.        f_i = fitness (x_i)
7.    x_1 … λ ← x_{s(1)…s(λ)} with s(i) = agrsort (f_{1…λ}, i) // sort solutions
8.    m' = m // we need later m' − m and x_i − m'
9.    m ← update_m(x_1, …, x_λ) // move mean to better solutions
10.   p_σ ← update_ps(p_σ, σ⁻¹C^{−½}(m' − m)) // update isotropic evolution path
11.   p_c ← update_ps(p_c, σ⁻¹(m' − m), ||p_σ||) // update anisotropic evolution path
12.   C ← update_c(C, p_c, (x_1 − m')/σ, …, (x_λ − m')/σ) // update covariance matrix
13.   σ ← update_sigma(σ, ||p_σ||) // update step_size using isotropic path length
14.return m or x_1
```

---

**Fig. 1**  Pseudo-code for CMA-ES Algorithm

CMA-ES is similar to (but not inspired by) Quasi-Newton methods [11] in that the estimation of covariance matrix is aimed to learn a second order model of the objective function, which is similar to the process of approximating the inverse Hessian in the Quasi-Newton method in classical optimization. However, contrary to Quasi-Newton methods, CMA-ES does not require much information about the objective function such as gradient information (which is not always easy to obtain in many cases including the problem of parameter learning in statistical machine translation systems). The adaptation of the covariance matrix in CMA-ES only requires the ranking information of the sampled solution. CMA-ES has shown to be capable global optimizer wining some optimization competitions in the field of evolutionary algorithms [2, 12]

**Table 1** The Summary statistical of data sets: English-Vietnamese

| Corpus | Sentence pairs | Training Set | Development Set | Test Set |
|--------|---------------|--------------|-----------------|----------|
| General | 56642 | 54642 | 200 | 499 |

| | | English | Vietnamese |
|--------------|----------------|---------|------------|
| Training | Sentences | 54620 | |
| | Average Length | 11.2 | 10.6 |
| | Word | 614578 | 580754 |
| | Vocabulary | 23804 | 24097 |
| Development | Sentences | 200 | |
| | Average Length | 11.1 | 10.7 |
| | Word | 2221 | 2141 |
| | Vocabulary | 825 | 831 |
| Test | Sentences | 499 | |
| | Average Length | 11.2 | 10.5 |
| | Word | 5620 | 6240 |
| | Vocabulary | 1844 | 1851 |

**Table 2** CMA-ES Settings

| Parameters | Value |
|------------|-------|
| Population size | 40 |
| Number of generations | 50 |
| Fitness | BLEU score |
| Number of runs | 1 |

# 4 Experiments and Results

In this section, we present our first experiments to test the effectiveness of CMA-ES when it is used as a substitution for MERT in a statistical machine translation system. Here, the language pairs chosen is English-Vietnamese.

## 4.1 Experimental Settings

To experimentally test the effectiveness of CMA-ES for the problem of parameter learning in statistical translation systems, We used an English-Vietnamese corpus (Nguyen et al., 2008) [13]. The corpus for experiments was collected from newspapers, books,... on the internet with topics such as social, sports, science. This data set is divided into three subsets: training set with 54642 sentence pairs, testing set with 499 sentence pairs, and development set with 200 sentence pairs. Table 1 summarizes the basic statistics of the corpus.

Some experiments are processed on the basis of Phrase-based Statistical Machine Translation model with MOSES [3] open-source decoder. For training data

**Table 3** Compare experiment results between running MERT and running CMA-ES with paramaters generated from MERT

| System | BLEU(%) |
|---|---|
| Baseline English-Vietnames SMT using MERT (Dev 200) | 36.97 |
| English-Vietnames SMT using CMA-ES (Dev 200) | 37.21 |
| Baseline English-Vietnames SMT using MERT (Test 499) | 37.30 |
| English-Vietnames SMT using CMA-ES (Test 499) | 37.69 |

**Table 4** Compare experiment results between running MERT and running CMA-ES with paramaters randomly generated in [0,1]

| System | BLEU(%) |
|---|---|
| Baseline English-Vietnames SMT using MERT (Dev 200) | 36.97 |
| English-Vietnames SMT using CMA-ES (Dev 200) | 38.08 |
| Baseline English-Vietnames SMT using MERT (Test 499) | 37.30 |
| English-Vietnames SMT using CMA-ES (Test 499) | 38.09 |

and training parameters, we used the available scripts in the Moses toolkit [14]. To build the language model, we used SRILM [15] toolkit with 4-gram. In this experiments, we evaluated the quality of the translation results by NIST and BLEU [16] score.

Our baseline system used MERT as the parameter training algorithm, while the improved system used CMA-ES. We used the CMA-ES system and code described in [17] and the experimental configuration setting for CMA-ES in our experiments is given in Table 2.

## 4.2 Results and Discussions

The performances of the statistical machine translation systems in our experiments are evaluated by the BLEU scores [4]. Table 3 displays the BLEU scores of the baseline SMT systems using CMA-ES. The results clearly show that the improved system (using CMA-ES as the parameter learning algorithm) is better than the baseline SMT system (using MERT as the parameter learning algorithm) by BLEU scores on both dev and test data sets.

In our experiment settings, the only difference between the two tested systems is the substitution of CMA-ES for MERT. Hence, this proves the effectiveness of using global optimizer techniques such as CMA-ES for learning/estimating parameters in our statistical machine translation system. This is important as local optimization methods such as MERT has widely been adopted in SMT systems for a long period of time having no comparable results from the methods that employ global optimization techniques. We believe that this is the first important step in trying global optimizers for learning parameters in SMT systems and that might lead to a wider adoption of them.

# 5 Conclusions

In this paper, we use a stochastic, meta-heuristic, and global optimization method called CMA-ES as an alternative to the tradition MERT algorithm for the task of learning parameter in statistical machine translation systems. Our experiment results, on a English-Vietnamese corpus, show that CMA-ES for tuning parameter in SMT is better than MERT for SMT by BLEU scores both on dev and test data sets. This warrants further studies in the applications of stochastic and global optimization techniques such as CMA-ES in SMT.

The Table 2 shows the settings for CMA-ES. With the first dataset, we conducted two experimentals:

. Individuals in CMA-ES were initialized from parameters that generated from MERT by adding each parameter with a noise. We used Gaussian noise with mean = 0 and standard deviation = 0.05.
. Individuals in CMA-ES were randomly initialized in [0, 1]

In the near future, we are planning to test the effectiveness of CMA-ES and other stochastic, global optimization techniques on more and larger copra of different language pairs.

**Acknowledgments.** This paper is partly funded by the Foundation for Science and Technology Development (NAFOSTED) under grant number 102.01-2011.08.

# References

[1] Och, F.J.: Minimum error rate training in statistical machine translation. In: Proceedings of the 41st Annual Meeting of the Association for Computational Linguistics, pp. 160–167. Association for Computational Linguistics, Sapporo (2003)

[2] Hansen, N.: The CMA evolution strategy: A comparing review. In: Lozano, J.A., Larrañaga, P., Inza, I., Bengoetxea, E. (eds.) Towards a New Evolutionary Computation. STUDFUZZ, vol. 192, pp. 75–102. Springer, Heidelberg (2006)

[3] Koehn, P., Hoang, H., Birch, A., Callison-Burch, C., Federico, M., Bertoldi, N., Cowan, B., Shen, W., Moran, C., Zens, R., Dyer, C., Bojar, O., Constantin, A., Herbst, E.: Moses: Open source toolkit for statistical machine translation. In: Proceedings of ACL, Demonstration Session (2007)

[4] Koehn, P.: Statistical Machine Translation. Cambridge University Press (2010)

[5] Smith, D.A., Eisner, J.: Minimum risk annealing for training log-linear models. In: Proceedings of the COLING/ACL 2006 Main Conference Poster Sessions, pp. 787–794. Association for Computational Linguistics, Sydney (2006)

[6] Huang, L., Mi, H.: Efficient incremental decoding for tree-to-string translation. In: Proceedings of the 2010 Conference on EmpiricalMethods in Natural Language Processing, pp. 273–283. Association for Computational Linguistics, Cambridge (2010)

[7] Suzuki, J., Tsukada, H., Watanabe, T., Isozaki, H.: Online large-margin training for statistical machine translation. In: Proceedings of EMNLP-CoNLL, Prague, pp. 764–773 (June 2007)

[8] Knight-and, K., Wang, W., Chiang, D.: 11,001 new features for statistical machine translation. In: Proceedings of Human Language Technologies: The 2009 Annual Conference of the NACL, Stroudsburg, PA, USA (June 2009)

[9] Arun, A., Koehn, P.: Online learningmethods for discriminative training of phrase based statistical machine translation. In: MT Summit XI, Copenhagen (September 2007)

[10] Crammer, K., McDonald, R., Pereira, F.: Online large-margin training of dependency parsers. In: Proceedings of ACL (2005)

[11] Teukolsky, S.A., Flannery, B.P., Press, W.H., Vetterling, W.T.: Numerical Recipes in C++: the art of scientific computing, 2nd edn. Cambridge University Press, New York (2002)

[12] Akimoto, Y., Nagata, Y., Ono, I., Kobayashi, S.: Bidirectional relation between CMA evolution strategies and natural evolution strategies. In: Schaefer, R., Cotta, C., Kołodziej, J., Rudolph, G. (eds.) PPSN XI. LNCS, vol. 6238, pp. 154–163. Springer, Heidelberg (2010)

[13] Ho, T.B., Nguyen, M.L., Nguyen, T.P., Shimazu, A., Van Nguyen, V.: A tree-to-string phrase-based model for statistical machine translation. In: Proceedings of the Twelfth Conference on Computational Natural Language Learning (CoNLL 2008), Manchester, England, pp. 143–150. Coling 2008 Organizing Committee (August 2008)

[14] Birch, A., CallisonBurch, C., Federico, M., Bertoldi, N., Cowan, B., Shen, W., Moran, C., Zens, R., Dyer, C., Bojar, O., Constantin, A., Koehn, P., Hoang, H., Herbst, E.: Moses: Open source toolkit for statistical machine translation. In: Proceedings of ACL, Demonstration Session (2007)

[15] Stolcke, A.: Srilm - an extensible language modeling toolkit. In: Proceedings of International Conference on Spoken Language Processing, Cambridge, MA, vol. 9, pp. 901–904 (2002)

[16] Roukos, S., Ward, T., Papineni, K., Zhu, W.J.: Bleu: A method for automatic evaluation of machine translation. In: ACL (2002)

[17] Fortin, F.A., De Rainville, F.M., Gardner, M.A., Parizeau, M., Gagné, C.: DEAP: Evolutionary algorithms made easy. Journal of Machine Learning Research 13, 2171–2175 (2012)

# Interacting with AutoMed-DM through Layers of Modelling Abstractions
## A Hierarchical, Event-Driven Design

Duc Minh Le

**Abstract.** We recently proposed AutoMed-DM – a transformation-based data modelling tool for heterogeneous data management system (HDMS). A key feature of the tool is the ability to not only support the activities needed to construct the schemas of the data sources connected to the HDMS, but also effectively and uniformly handle the heterogeneity inherent in those schemas and in the underlying data models. A primary design challenge in the development of AutoMed-DM is how to effectively enable the data model engineers to interact with the tool through its layers of modelling abstractions. In this paper, we address this challenge by proposing a novel hierarchical, event driven design model that formalises a compact and extensible UI hierarchy needed to support the user interaction through all the abstraction layers, and that uses interaction sequence to express the functional design logic. We present in our design model the specification of a key interactive function of AutoMed-DM named Create Schema.

## 1 Introduction

Heterogeneous data management systems (HDMS) are data management systems that capture and process data in a universe of discourse that are stored in various data sources, whose structures differ and/or are expressed in different data models. A common activity that is performed in the development of an HDMS is data modelling [11], which involves structuring the data requirements of the system to build a model of the data that the system needs to capture and process. Such a model, named schema [15] in this paper, helps the (data) model engineer/designer understand and verify (with the user) the structure of the data before implementing it.

Duc Minh Le
Faculty of Information Technology
Hanoi University, Hanoi, Vietnam
e-mail: duclm@hanu.edu.vn

© Springer International Publishing Switzerland 2015     433
V.-H. Nguyen et al. (eds.), *Knowledge and Systems Engineering,*
Advances in Intelligent Systems and Computing 326, DOI: 10.1007/978-3-319-11680-8_35

In [8], we proposed AutoMed-DM, a transformation-based data modelling tool to effectively support data modelling in an HDMS. We modelled AutoMed-DM as an extension of AutoMed [5, 14], an automatic mediator tool that uses a generic schema transformation approach to integrate heterogeneous data sources. In this paper, we will discuss the design of AutoMed-DM and demonstrate it using the following data modelling example that was also used in [8]. The three types of tool's users (namely designer, meta-model and model engineer [3, 8]) will uniformly be referred to as *model engineer*.

The example involves two Entity Relationship (ER) [7] schemas of an organisation named Pine Valley Furniture (PVF) [11], which model the data requirements about customers and the territories in which they operate. Fig. 1(a) is the basic ER schema $(S_t)$, while Fig. 1(b) is the enhanced ER schema $(S_T)$. The following business rules [11] underlie the relationships that are modelled in these schemas:

- $S_t$: each customer may do business in any number of these sales territories. A sales territory must have one or more customers
- $S_T$: only two types of customers exist: regular and national; only the former do business in sales territories, and a customer may be both a regular and a national

The two schemas in Fig.1 are generated by the AutoMed's schema visualiser function [14]. Both are graphs, whose elements are defined in terms of these four generic construct types: **nodal**, **link-nodal**, **link**, and **constraint**. The first three are termed **extensional constructs**, because they represent sets of data values (called **extents**) in some domain. More specifically, an entity is a nodal, whose graphical representation is a labelled rectangle. An attribute is a link-nodal that consists of an attribute node that must be attached via a link to an entity node. The graphical representation of the attribute node is a labelled circle, and that of the link is a line segment that connects the nodes. A relationship is a link that connects two or more entity nodes. A constraint restricts the sets of data values that may be assigned to the other three extensional constructs. For example, the many-to-many relationship "Does business in" has a cardinality constraint 1:N at the Customer end, which means that every instance of Territory that is in the extent of this relationship must be associated to at least one instance of Customer.

The above data modelling formalism of AutoMed has been proved suitable for most data models currently in use, including both structured models (*e.g.* ER, relational, UML [15]) and semi-structured ones (*e.g.* YATTA [14]). To effectively handle the transformation of schemas that are expressed in these so called **high-level data models**, AutoMed uses a generic low-level, **hypergraph-based data model(HDM)** [6] as a common data model for expressing them. HDM uses three basic constructs: node, edge and constraint, where node and edge are extensional constructs and edge can be nested. The four generic construct types mentioned above are mapped to the three HDM constructs as follows: nodal is mapped to node, link is mapped to edge, link-nodal is mapped to a combination of node and edge, and constraint is mapped to HDM constraint or a combination of HDM constraint and some suitable extensional HDM constructs.

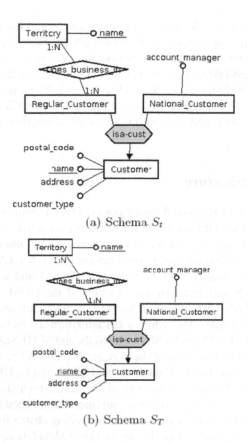

(a) Schema $S_t$

(b) Schema $S_T$

**Fig. 1** Two example ER schemas

The HDM, high-level models, and the schemas that are defined in these models form three modelling abstraction layers: meta-model, model and schema transformation. The first two layers are called meta-meta-model and meta-model in [9], respectively. The third layer is mapped to the model layer in [9] but differs in that it also includes transformation as a concept. The current AutoMed tool [14] provides model engineers with an application programming interface (API) and a graphical UI (GUI) to manipulate those as objects. However, as discussed in [8] the GUI is limited in that it only supports the schema transformation layer, while the API does not adequately address the essential requirements of AutoMed-DM. A primary design challenge of AutoMed-DM is how to effectively enable the model engineer to interact with the tool through its layers of modelling abstractions. In this paper, we address this challenge by proposing a novel **hierarchical, event driven design** that formalises a compact and extensible UI hierarchy needed to support the user interaction through all the abstraction layers, and that uses interaction sequence to express the functional design logic. We present in our design model the specification of a key interactive function of AutoMed-DM named Create Schema [8].

The rest of the paper is structured as follows. Section 2 reviews the modelling abstractions of AutoMed-DM (proposed in [8]) and describes a high-level architecture of the tool. Section 3 discusses the hierarchical, event-driven design model of the tool. Section 4 gives the design details of the function Create Schema. Section 5 presents the related work, and Section 6 gives some concluding remarks and plan for future work.

## 2   Layered Architecture

In [8], we proposed an ER, layered data model (LDM) to address the essential requirements of AutoMed-DM. In this LDM, we formalised four modelling abstractions of the tool as groups of related concept entities (*abbr.* **concepts**). Briefly, the **Meta-model** abstraction is consisted of one concept named META-MODEL, which defines the HDM meta-model and its constructs. The **Model** abstraction defines the high-level models and its constructs in terms of the HDM. It is consisted of one concept named MODEL, which is realised by three entities High-level Construct, Construct and Model. The **Schema transformation** abstraction is consisted of three concepts: SCHEMA, SCHEMA CHANGE, and SCHEMA MAPPING. The **Metadata** abstraction is consisted of the concept METADATA, which is added in AutoMed-DM to capture the metadata about SCHEMA and SCHEMA MAPPING.

Fig. 2 below is the layered architecture of the tool, which shows how a model engineer interacts with the tool through the user interface attached to the layers. Each layer is responsible for managing data for one modelling abstraction. The layers are stacked based on the dependency of the abstractions: Metadata depends on Schema Transformation, which depends on Model; Model, in turn, depends on Meta-model.

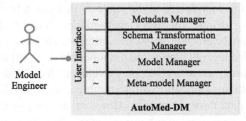

**Fig. 2** A layered architecture of AutoMed-DM (based on its LDM)

Let us now present a design for all interactive functions of the layers of the architecture. Function Create Schema, which we focus on in this paper, is a function of the second layer.

## 3  Hierarchical, Event-Driven Design

Two underlying design principles of the design model that we present in this section
are *event-driven design* [4] and *hierarchical GUI design* [12]. The former is a natural
fit for UI-based software systems. The latter specifies the use of a hierarchy of GUI
components (named UI hierarchy hereafter) for interactive software systems. As will
be explained below, UI hierarchy naturally corresponds to the concept layering of
the AutoMed-DM's LDM.

In addition, two design criteria that should be observed for this type of tool are:
compactness and extensibility [9]. Both are necessary because it is not possible to
engineer a tool that address all the possible modelling requirements. For *compactness*, we define the UI components needed to support the essential requirements
in [8]. For *extensibility*, we describe a design method which, together with the UI
components, is used for realising new requirements that emerge.

Fig. 3 is our proposed design model, expressed in the ER language. In the discussions concerning this model that follow, when it is necessary to make explicit the
entities in an ER schema and their instances, we will refer to an entity by its name
(*e.g.* Model) and an entity's instance(s) by a suitable article (*e.g.* a, the) followed by
the entity's name (the plural form of this name).

At the top of Fig. 3 are two abstract components: Event Handling and UI Hierarchy. The first component (the top-left, rounded rectangle) formalises the user's interaction in terms of the User Actions that are performed on the UI Components and
how the Events that result are handled by various Functions of the system. The actual
event handling logic performed by each Function is expressed using interaction sequence (explained below), which supports the specification of the UI Component(s)
needed to capture the data for the Layered Concept(s) of concern to the Function.
This relationship between UI Component and Layered Concept is modelled in the
second component (the top-right, rounded rectangle).

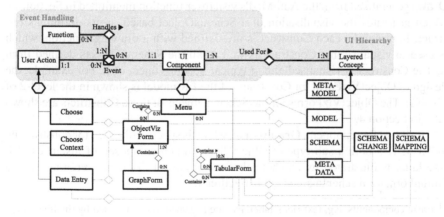

**Fig. 3**  The hierarchical, event-driven design model of AutoMed-DM

Displayed immediately below the two abstract components are the specialised entities and the relationships (the latter are unlabelled and depicted using thin grey lines.)

## 3.1 Layered Concept

Fig. 3 models six concepts of the AutoMed-DM's LDM as sub-types of the entity Layered Concept. To ease illustration of the containment relationships of the UI components (explained next), the sub-types are arranged in the figure in a somewhat different order from that of the architectural layers in Fig. 2.

## 3.2 UI Hierarchy and Components

Fig. 3 shows four structural UI components[1] that satisfy the LDM's requirements: Menu, ObjectVizForm, GraphForm, and TabularForm. The figure models the relationships inherent among the UI components as well as those between each UI component and the Layered Concept(s) for which it is used. In this paper, we will focus on the GUI-based designs of the UI components. First, **Menu** is used for capturing data at three layers. It is used, together with ObjectVizForm, at the top two layers for a pair of META-MODEL and MODEL[2]. The logical design of this is shown in Fig. 4(a) (more details will be given in Section 4.1). Menu is also used at the fourth layer for capturing the METADATA of a SCHEMA (explained shortly below).

Second, **ObjectVizForm** is used at the second and third layer for displaying visual representations of MODEL and SCHEMA. At the second layer, ObjectVizForm is used as part of a Menu as described above to visualise the Constructs of a Model. At the third layer, it is used as part of a GraphForm to visualise the SchemaObjects of a Schema. The visualisation formalism is captured by a function named Draw (generalised from the AutoMed's visualiser function mentioned in Section 1), which generates the visualisation of a SchemaObject based on its defining Construct. In principle, each Construct is pre-defined with a drawing function, which is used to visualise that Construct as well as all the SchemaObjects defined from it. The Construct is visualised after a typical SchemaObject of it. For example, the design of ObjectVizForm for Constructs of the ER model is shown in the level-2 of Fig 4(a). The ObjectVizForms for the SchemaObjects of these Constructs are shown later in Section 4.

The third UI component, **GraphForm**, which is specifically used for constructing SCHEMA, is both a sub-type of ObjectVizForm and composed of instances of it (the latter is illustrated by another relationship named Contains.) Fragments of a GraphForm for a schema are given in Section 4.

---

[1] Basic components, *e.g.* text field, label, *etc.* are assumed to be provided by the implementation platform for use as sub-components of these.

[2] Schemas may contain constructs from multiple models but we focus on single-model ones.

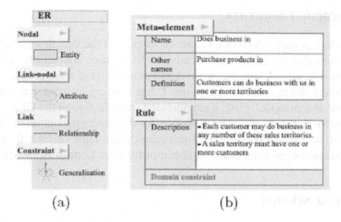

**Fig. 4** Logical GUI designs for (a) Menu and ObjectVizForm and (b) Menu and TabularForm

Unlike the other components, **TabularForm** is primarily used for the user to enter the data values of an entity's instance. This component, which can be nested (as illustrated by the reflexive relationship named Contains), is used at two layers. At the third layer, it is used as a stand-alone component for entering data for and displaying the instances of the SCHEMA CHANGE's entities. An example of this usage is described in [14]. It is also used at the third layer for the user to review and update the Semantic Mappings [5]. At the fourth layer, TabularForm is used as part of a Menu (as illustrated by yet another relationship named Contains) for entering data values of a METADATA. The logical design of this is shown in Fig. 4(b). In this figure, the level-1 items of the Menu are names of two entities Metadata Element and Rule [8], which model the metadata. The level-2 items are two TabularForms that capture data about an instance of each of these entities.

## 3.3 User Action

Logically, we identify three types of user actions (modelled in Fig. 3 as sub-types of User Action), that are performed when the user interacts with the UI Components. First, action **Choose** represents a mouse-click or key-press action on a Menu or ObjectVizForm. Second, action **Choose Context** captures a mouse or key-press action on an object that results in the display of its context menu. An example of this, which will be explained in Section 4, is the action that causes a context menu of a schema object being displayed. Finally, action **Data Entry**, which is related to both ObjectVizForm and TabularForm, represents the keyboard actions that are needed to enter data values on some sort of form field. Examples of this action will also be given in Section 4.

## 3.4   Interaction Sequence

Our design model uses interaction sequence to express the design logic of the system's functions that are performed in response to the user interaction. An **interaction sequence** is derived from the two-column description of concrete use case [13], but is written in the pseudocode language. It has two key features that are suitable for use in our design model. First, it supports the specification of the UI components that are used for capturing the input as well as for displaying the output of each function. Second, its pseudocode directly forms the design logic of each function. We present in Section 4 the interaction sequence of function Create Schema.

## 4   Create a Schema

In the development of HDMS, function Create Schema is the first design activity in which a schema of a given model is created from the user requirements [10]. The user would identify the concepts and rules from these requirements and create them as objects in the schema. In AutoMed-DM, the function intuitively proceeds as follows (the sub-function names are highlighted). First, the system **displays the high-level constructs** of a given model, so that the user can choose a construct to use to create each object. Then depending on the generic construct type of the chosen construct, the system performs one of the four sub-functions to **create a nodal, link-nodal, link** or **constraint object** in the schema. The process is stopped when there are no more objects to create.

## 4.1   Display High-Level Constructs

Fig. 4(a) showed a two-level Menu that contains a number of ObjectVizForms for displaying (and prompting the user to choose) the high-level constructs of a model. The first level contains names of the four generic construct types. This level is fixed for all high-level models supported by the tool. The second level includes the ObjectVizForms of the high-level constructs that are mapped to each generic one. The ObjectVizForm of a Construct is generated using the pre-defined drawing function of that Construct and a typical Schema Object whose name is the same as the Construct's. The example that we use here includes four ER model constructs that are used in the two PVF schemas (*c.f.* Fig. 1): Entity, Attribute, Relationship and Generalisation. The right arrow head attached to each level-1 construct means to allow the user to show/hide the high-level constructs at level-2.

## 4.2   Create Nodal, Link-Nodal, and Link Objects

Tab. 1 is an interaction sequence that defines the design of the three sub-functions that create nodal, link-nodal and link objects. The table is divided into two halves under the headings User and System. The User half is consisted of two sub-columns:

Actions (user actions) and UIs (the UI components on which the actions are performed.) These columns are left blank if there are no user actions required (*e.g.* in the row between steps ⑬ and ⑭). The System half is consisted of two sub-columns: Responses (system responses) and UIs (the UI components, if any, on which responses are displayed to the user). Two types of UI Component used in Tab. 1 are ObjectVizForm and GraphForm. The former is used for visualising SchemaObjects, while the later is used for visualising Schemas. The example UI components in Tab. 1 are drawn for schema $S_t$.

Steps ②, ⑦, and ⑫ in the Responses column of Tab. 1 list the pseudocodes of three sub-functions in question: CreateNodalObject, CreateLinkNodalObject, and CreateLinkObject. The steps in each pseudocode are broken down into segments, which are performed in response to events caused by user actions on UI components. These sub-functions invoke a function named CreateSchemaObject , which generalises the specification of the existing AutoMed's function [14] that creates a SchemaObject of a given Construct in a Schema.

**Table 1** An interaction sequence for creating nodal, link-nodal and link objects in a schema

| User | | System | |
|---|---|---|---|
| **Actions** | **UIs** | **Responses** | **UIs** |
| Choose a Nodal $c$ <br> ① (*e.g.* Entity := <br> Nodal⟨'entity'⟩) | Entity | ② **CreateNodalObject**(Schema <br> $s$, Nodal $c$): <br> SchemaObject⟨Nodal⟩ <br> $n$ := <br> CreateSchemaObject($s,c$,'_-<br> ') <br><br> Draw($n,c$.drawFunc) <br> $m_n$ := PromptForName($n$) | I |
| ③ Enter <br> SchemaObject:name | Customer| | set $n$.name = $m_n$ <br> return $n$ | Customer |
| Choose context of <br> a <br> ④ SchemaObject⟨Nodal⟩ <br> $o$ <br> (*e.g.* SchemaObject⟨Entity⟩) | Customer | ⑤ **ShowContextMenu**($o$) | Attribute <br> Relationship |

**Table 1** (*continued*)

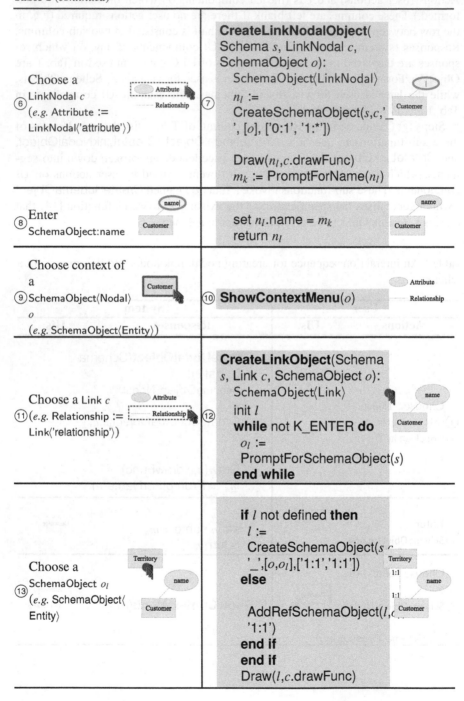

| | |
|---|---|
| ⑥ Choose a LinkNodal $c$ (*e.g.* Attribute := LinkNodal('attribute')) | **CreateLinkNodalObject(** Schema $s$, LinkNodal $c$, SchemaObject $o$): SchemaObject⟨LinkNodal⟩ $n_l$ := CreateSchemaObject($s$,$c$,'_', [$o$], ['0:1', '1:*']) ⑦ Draw($n_l$,$c$.drawFunc) $m_k$ := PromptForName($n_l$) |
| ⑧ Enter SchemaObject:name | set $n_l$.name = $m_k$ return $n_l$ |
| ⑨ Choose context of a SchemaObject⟨Nodal⟩ $o$ (*e.g.* SchemaObject⟨Entity⟩) | ⑩ **ShowContextMenu(**$o$**)** |
| ⑪ Choose a Link $c$ (*e.g.* Relationship := Link('relationship')) | **CreateLinkObject(**Schema $s$, Link $c$, SchemaObject $o$): SchemaObject⟨Link⟩ ⑫ init $l$ **while** not K_ENTER **do** $o_l$ := PromptForSchemaObject($s$) **end while** |
| ⑬ Choose a SchemaObject $o_l$ (*e.g.* SchemaObject⟨Entity⟩) | **if** $l$ not defined **then** $l$ := CreateSchemaObject($s$,$c$,'_',[$o$,$o_l$],['1:1','1:1']) **else** AddRefSchemaObject($l$,$c$,'1:1') **end if** **end if** Draw($l$,$c$.drawFunc) |

**Table 1** (*continued*)

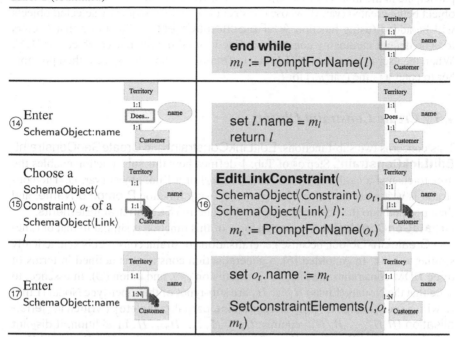

| | Event / Action | UI | | Code | UI |
|---|---|---|---|---|---|
| | | | | **end while** $m_l$ := PromptForName($l$) | Territory 1:1 name 1:1 Customer |
| (14) | Enter SchemaObject:name | Territory 1:1 Does... name 1:1 Customer | | set $l$.name = $m_l$ return $l$ | Territory 1:1 Does ... name 1:1 Customer |
| (15) | Choose a SchemaObject⟨Constraint⟩ $o_t$ of a SchemaObject⟨Link⟩ $l$ | Territory 1:1 name 1:1 Customer | (16) | **EditLinkConstraint(** SchemaObject⟨Constraint⟩ $o_t$, SchemaObject⟨Link⟩ $l$): $m_t$ := PromptForName($o_t$) | Territory 1:1 name 1:1 Customer |
| (17) | Enter SchemaObject:name | Territory 1:1 name 1:N Customer | | set $o_t$.name := $m_t$ SetConstraintElements($l$,$o_t$, $m_t$) | Territory 1:1 name 1:N Customer |

**CreateNodalObject** (step ②). Intuitively, this function first invokes function CreateSchemaObject to create an unnamed SchemaObject (var. $n$) of the specified Nodal construct (var. $c$). String '_' represents an unspecified name. It then draws this object by invoking the function Draw and using the custom drawing function (attrib. drawFunc) of $c$. After the object has been created, the system prompts the user to enter a name for $n$ (var. $m_n$). In the illustration of this step, $c$ is an Entity construct, whose visualisation is a rectangle shown in the UI column.

**CreateLinkNodalObject** (step ⑦). This function proceeds similarly to function CreateNodalObject, except for the use of two extra arguments in the invocation of the function CreateSchemaObject. The first argument (array [$o$] which holds var. $o$) specifies the object on which the newly-created LinkNodal object (var. $n_l$) depends (and thus refers to). In the illustration of this step, $o$ is an Entity-typed object and $n_l$ is an Attribute-typed object which depends on $o$. The second argument is the array ['0:1','1:*'], that defines the default cardinality constraints (in look-here semantics) associated to $n_l$. Each constraint has the form '$l;u$', which specifies the lower ($l$) and upper ($u$) bounds. In this example, the two constraints mean that $n_l$ defines a non-mandatory attribute (*i.e.* the value of the attribute node can be null).

**CreateLinkObject** (step ⑫). This function creates an n-ary link of a specified type (var. $c$), between a given object (var. $o$) and one or more other schema objects in the schema. In the illustration of this step, $c$ is Relationship and $o$ is an Entity-typed object. To achieves this, it defines a while loop which repeatedly asks the user, using function PromptForSchemaObject, to choose a new schema object (var. $o_l$) to participate in the link. The first object chosen will result in a new unnamed, schema object being created (var. $l$), whose construct is $c$. All subsequently selected objects will be added (using function AddRefSchemaObject) to $l$ as those that $l$ refers to. The default cardinality constraint defined for all the referenced objects is '1:1'. When the user presses the ENTER key to finalise the link, the system then prompts her to enter a name (var. $m_l$) for $l$.

## 4.3 Create Constraint Objects

This concerns two sub-functions: EditLinkConstraint and CreateGenConstraint.
**EditLinkConstraint.** Step ⑯ of Tab. 1 defines how this sub-function enables the user to change a cardinality constraint (var. $o_t$) of a Link object (var. $l$). The constraint is written in the name of $o_t$ (var. $m_t$), using function PromptForName, and then updated into the reference of $o_t$ in $l$ by function SetConstraintElements.
**CreateGenConstraint.** The design logic of this function is similar to that of function CreateLinkObject, because generalisation constraint construct is syntactically similar to Link. In AutoMed [6], a generalisation constraint is defined in terms of three HDM constraints: inclusion ($\subseteq$), exclusion ($\emptyset$), and union ($\cup$). In essence, to state that the Nodals (Links) $H_1,\ldots,H_n$ are sub-types of the super-type Nodal (Link) $H$ with a specification $D \subseteq \{\underline{o}verlap, \underline{d}isjoint, partial, \underline{c}omplete\}$ (written as generalisation $(H,H_1,\ldots,H_n,D)$, requires: $H_1 \subseteq H, \ldots, H_n \subseteq H$. In addition, if disjoint $\in D$ then requires $H_1 \emptyset \ldots \emptyset H_n$; and if total $\in D$ then requires $H = H_1 \cup \ldots \cup H_n$.

In the ER model $H$ and the $H_i$s are entity type; while in the UML model they are class or association. For example, the constraint in schema $S_T$ of Fig. 1(b) is: generalisation (Customer,Regular_Customer,National_Customer, {overlap, complete}).

Thus, CreateGenConstraint proceeds similar to the steps ⑪ – ⑭ of Tab. 1 to create a constraint object with a default specification $D$; then this specification is changed, similar to the steps ⑮– ⑰, by the user entering a subset of the specification elements in the object's name.

## 5 Related Work

The related work of this paper fall into the category of **designing a data modelling tool for heterogeneous data management systems**. In particular, we will focus on two tools, DB-MAIN [9, 10] and MDM [1, 2, 3], which were highlighted in our previous work [8] as being comparable to AutoMed-DM in three key aspects: (*i*) use of a generic common data model, (*ii*) data modelling formalism and (*iii*) schema transformation formalism. Let us review the design models of these tools and compare and contrast them with AutoMed-DM's.
**Support for the abstraction layers.** Both tools support the top three abstraction layers (meta-model, model, and schema), although DB-MAIN uses a different (but compatible) naming for the layers (meta-meta-model, meta-model, and model). Unlike AutoMed-DM, MDM considers metadata in the third layer. A more recent work

of this tool [2] supports an additional layer for data instances. First, we argue that separating metadata from the schema transformation layer as in our model helps make the design easier to understand without altering its meaning; because, as discussed earlier in this paper, the layers are logical groupings of concepts. Second, our model can easily be extended to support the data layer, as the design logic of this layer is similar to that of the metadata layer. We are considering this extension in our future work.

**Design model.** Our event-driven design model is novel compared to the designs of MDM and DB-MAIN in that these do not follow the event-driven principle [4] and are not based on a formalisation of the abstraction layers. Further, these designs do not make a clear distinction between logical and physical design. A component-based architecture of MDM is first described in [3]. Later in [2], a multilevel dictionary to support the abstraction layers is presented using a relational database. In [3], the logic of the core schema and data translation functions of MDM is discussed using a pseudocode language; while in [1] Datalog is used to define the basic translation rules. As for DB-MAIN, a layered architecture of the tool and details about its meta-meta-model are presented in [9]. The architecture refers to the abstraction layers along two dimensions: basic and meta. Also discussed in [9] is a functional language named Voyager2, which is used for writing the functions logic.

**UI design.** Our logical UI design is similar in spirit to MDM and DB-MAIN in the use of menus for organising the tools features and the use of diagrams for visualising the schemas. However, MDM [1] and DB-MAIN [9] do not formalise the UI hierarchy nor do they identify the essential set of UI components needed to support the requirements of the abstraction layers.

## 6 Conclusions

In this paper, we proposed a novel hierarchical, event-driven design for AutoMed-DM to effectively enable the data model engineers to interact with the tool through its layers of modelling abstractions. We showed how our design formalises a compact and extensible UI hierarchy needed to support the user interaction through all the abstraction layers, and how the functions' design logic is expressed using interaction sequence. We presented in our design model the specification of a key interactive function of AutoMed-DM named Create Schema.

Our plan for future work is to apply the design model to the remaining essential functions of AutoMed-DM, and to improve the UI design to support mixed-model schemas. We would also consider supporting the data layer. Last but not least, we plan to use the design to develop a prototype of the tool in Java (possibly re-using the existing core components of the AutoMed tool).

## References

[1] Atzeni, P., et al.: Model-independent schema translation. VLDB Journal 17(6), 1347–1370 (2008)

[2] Atzeni, P., Cappellari, P., Bernstein, P.A.: Model-independent schema and data translation. In: Ioannidis, Y., Scholl, M.H., Schmidt, J.W., Matthes, F., Hatzopoulos, M., Böhm, K., Kemper, A., Grust, T., Böhm, C. (eds.) EDBT 2006. LNCS, vol. 3896, pp. 368–385. Springer, Heidelberg (2006)

[3] Atzeni, P., Torlone, R.: Management of multiple models in an extensible database design tool. In: EDBT 1996, pp. 79–95. Springer (1996)

[4] Bastide, R., Palanque, P.: A Petri Net based environment for the design of event-driven interfaces. In: DeMichelis, G., Díaz, M. (eds.) ICATPN 1995. LNCS, vol. 935, pp. 66–83. Springer, Heidelberg (1995)

[5] Boyd, M., Kittivoravitkul, S., Lazanitis, C., Mçbrien, P., Rizopoulos, N.: AutoMed: A BAV data integration system for heterogeneous data sources. In: Persson, A., Stirna, J. (eds.) CAiSE 2004. LNCS, vol. 3084, pp. 82–97. Springer, Heidelberg (2004)

[6] Boyd, M., McBrien, P.: Comparing and transforming between data models via an intermediate hypergraph data model. Data Semantics IV, 69–109 (2005)

[7] Chen, P.C.: The Entity Relationship Model - Toward a unified view of data. ACM Transactions on Database Systems 1(1), 9–36 (1976)

[8] Duc, M.L.: AutoMed-DM: A transformation-based data modelling tool for heterogeneous data management systems. Technical report, Hanoi University (2014)

[9] Englebert, V., Hainaut, J.-L.: DB-MAIN: a next generation meta-case. Information Systems 24(2), 99–112 (1999)

[10] Hainaut, J.L.: Transformation of Knowledge. In: Information and Data: Theory and Applications, Ch. 1, pp. 1–28. IGI Global (2005)

[11] Hoffer, J.A., et al.: Modern Database Management, 10th edn. Prentice Hall (2011)

[12] Krasner, G.E., Pope, S.T.: A description of the Model-View-Controller user interface paradigm in the Smalltalk-80 system. Journal of Object Oriented Programming 1(3), 26–49 (1988)

[13] Larman, C.: Applying UML And Patterns: An Introduction to Object-Oriented Analysis and Design and the Unified Process, 2nd edn. Prentice Hall (2001)

[14] McBrien, P.: AutoMed in a Nutshell, Release 1.0. Technical report, Imperial College London (2006)

[15] Brien, P.M., Poulovassilis, A.: A uniform approach to inter-model transformations. In: Jarke, M., Oberweis, A. (eds.) CAiSE 1999. LNCS, vol. 1626, p. 333. Springer, Heidelberg (1999)

# Human Action Recognition Using 2DPCA-DMM Representation and GA-SVM in Depth Sequences

Vo Hoai Viet, Ly Quoc Ngoc, and Tran Thai Son

**Abstract.** Automatic human action recognition is an interesting and challenging problem in computer vision. Furthermore, it has wide range of real-world applications such as human-machine interaction, surveillance system, data-driven automation, smart home and robotics. In the recent years, the availability of 3D sensors has recently made it possible to capture depth maps in real time, which simplifies variety of visual recognition tasks, including action classification, 3D reconstruction, etc? We address here the problem of human action recognition in depth sequences. On one hand, we present a novel for human action recognition based on 2DPCA applied to DMM. Then, we project feature matrixes into difference spaces to create robust action representation. Finally, we use GA to create the coefficients for multi-SVM classifiers. Our approach is systematically examined on benchmark datasets such as MSR-Gesture3D and MSR-Action3D which results with an overall accuracy of 91.32% on the MSR-Action3D and 94.89% on the MSR-Gesture3D. The experimental results also indicated that the extraction of features and representation are effective and shows the effective of our proposal.

**Keywords:** Human Action Recognition, Depth Motion Map, Depth Sequences, 2D PCA, Support Vector Machine, Genetic Algorithm.

## 1 Introduction

Recognizing action is currently of the most challenging tasks and has been active researches through many years with wide range of real-world applications such as human-machine interaction, surveillance system, data-driven automation, smart home and robotics. In the past, there have been many approaches dealing with this problem in many ways. However, there still remain many different layers of

Vo Hoai Viet · Ly Quoc Ngoc · Tran Thai Son
Computer Vision and Robotics Department, University of Science - VNU - HCMC,
Ho Chi Minh, Vietnam
e-mail: {vhviet,lqngoc,ttson}@fit.hcmus.edu.vn

© Springer International Publishing Switzerland 2015    447
V.-H. Nguyen et al. (eds.), *Knowledge and Systems Engineering,*
Advances in Intelligent Systems and Computing 326, DOI: 10.1007/978-3-319-11680-8_36

complexity for studying due to the extensive range of possible human motions, variations in scene, within-class variation, occlusion, background clutter, pose and lighting changes. In the past decades, the methods focus on action recognition using 2D visual information captured by single or multiple cameras [1, 3, 5, 6, 7, 8, 9]. They focus on using RGB data for feature extraction and representation. In order to derive effective actions representation from image sequences, these approaches focus on discovering two properties are shape and motion of objects. The shape information of the object, features are statistics about the overall distribution of visual information such as texture, shape, edge... And motion information of object, features are statistics about optical flow. However, the problem, especially robust and viewpoint independent recognition of diverse human actions in a real environment is far from being solved.

Depth sensors have been available in many years. Though, they are used limitation in many applications because of high cost and complexity of operations. However, the recent proliferation of affordable low-cost 3D sensors such as Kinect with low cost has alleviated the hardness of the traditional action recognition problem, by allowing an extreme enrichment of visual color-depth data. With its advanced sensing techniques, this technology opens up an opportunity to significantly increase the capabilities of many automated vision-based recognition tasks [12]. And, it excited interest within the vision and robotics community for its broad applications [12]. In fact, this is a significant motivation for computer scientists to get deep in this research field to find out effective ways to utilize benefits from both the available depth and color information. Compared with conventional color data, depth maps provide several advantages, such as the ability of reflecting pure geometry and shape cues, or insensitive to changes in lighting conditions. Moreover, the range sensor provides 3D structural information of the scene, which offers more discerning information to recover postures and recognize actions. These properties help depth data provide more natural and discriminative vision cues than color or texture. Furthermore, the depth images provide natural surfaces which can be extracted to capture the geometrical structure of the observed scene in a rich descriptor. For instance, it was recently shown in [19] that the shape of an object can be better characterized using the field of normal vectors in depth images, instead of the gradients in color images.

In 2004, Yang et al. [10] has proposed an extension of PCA technique for face recognition using gray-level images. 2D-PCA treats image as a matrix and computes directly on the so-called image covariance matrix without image-to-vector transformation. The eigenvector estimates more accurate and computes the corresponding eigenvectors more efficiently than PCA. In a previous contribution, Abdelwahab and Mikhael [15] introduced the 2D-PCA for human action recognition in videos using the 2D silhouettes extracted for humans in videos. This method reduces computational complexity by two order of magnitude, while maintaining high recognition accuracy and minimum storage requirements compared to existing methods.

In this paper, we focus on action recognition by using depth data from Kinect. We present robust action representation using 2DPCA-DMM in depth data and effective classifier using GA and SVM. In specific, the contributions of this paper are three-folds. Firstly, we propose use 2D-PCA apply to DMM create feature matrixes. Secondly, we use different space projection to create robust action representation. Thirdly, we propose GA algorithm to optimize the weighting of multi-classifiers using SVM. This paper is organized as follows: in section 2, we review related works. In section 3, we introduce our approach for action recognition. In section 4, we show some results from our experiments and discussion. We conclude in section 5.

## 2 Related Works

Comprehensive reviews of the previous researches can be found in [9]. Our discussion in this section is restricted to a few influential and relevant parts of literature, with a focus on hand-crafted feature extraction and representation.

Most previous researches on action recognition focus on using 2D video [9]. The first idea of temporal templates is introduced by Bobick and Davis [3]. The authors presented a new approach for action representation. A binary motion energy image (MEI) which represents where motion has occurred in images sequence is generated. A motion history image (MHI) which is a scalar-valued image that its intensity is a function of the temporal history of motion. They use the two components (MEI and MHI) for representation and recognition of human movement. State-of-the-art approaches [1, 5, 6] have reported good results on human action datasets. Among most methods, local spatio-temporal features and bag-of-features (BoF) representations achieve remarkable performance for action recognition. Laptev et al. [6] are the first to introduce space-time interest point by extending 2D Harris-Laplace detector. To produce denser space-time feature points, Dollar et al.[18] use a pair of 1D Gabor-filter to convolve with a spatial Gaussian to select local maximal cuboids.

Although most previous works on feature extraction focus on using 2D videos [1, 3, 5, 6, 7, 8, 9], several approaches extract features from 3D videos have been proposed in the past few years. One of the earliest works on action recognition using a depth sensor is presented in [13]. Li et al. propose a bag-of-3D-points feature representation for action recognition from depth map sequences, where the 3D points are sampled from the silhouettes of the depth maps [13]. Since the color-based spatio-temporal interest points detectors, such as Harris3D [8] or Hessian3D [5], etc. are not certainly reliable in depth sequences. To describe action in 4D space, J. Wang al [11] proposed Random Occupancy Pattern (ROP) features which were extracted from randomly sampled 4d sub-volumes with different sizes and at different locations. In [17], Omar et al. present a 4-dimensional extension of the surface oriented normal vectors [19], and a quantization method using 4D 600-cell polychoron (i.e. 120 uniform projectors). This quantization method is inspired by the successful work of [1] which uses a 3D polyhedron to quantize 3D oriented gradient vectors. The authors in [17] further carry out a process of quantization step refinement, using the classification score in training phases to generate some perturbations to the original uniform projectors, since they prove that non-uniform quantization performs much better. As a consequence, though bypass the use of a skeleton tracker, their method still result with state-of-the-art performances on several standard benchmarks (e.g. MSRAc-tion3D, MSRGesture3D) at the present of this paper. However, this work already includes a heavy process of projectors refinement for surface normals quantization stage. Similar to Bobick's idea, Yang et al. [21] propose a method to recognize human actions from sequences of depth maps. They project the depth maps onto three orthogonal planes and accumulate the whole sequence generating a depth motion map (DMM). Histograms of oriented gradients (HOG) are obtained for each DMM. The concatenation of the three HOG serves as input feature to a linear SVM classifier. This method captures motion and shape information to represent for action. There are two important properties for action representation.

From above researches, our proposal methods fall in global features and data analysis. In this work, we use DMM to extract motion and shape of objects in performing action. In order to obtain robust action representation, we apply 2DPCA to matrix features which are projected on different spaces to create discriminative representation. Moreover, we use GA to build the coefficients for multi classifiers using SVM to achieve good performance for action classification.

## 3 Our Approach

The proposed action recognition system consisted of a structure is shown in Fig 1. Firstly, we use bilateral filter to remove noise is called pre-processing. Secondly, DMM feature on XY, XZ and YZ views are parallel extracted for depth sequences. Thirdly, reducing data dimension decreases computational cost and create compact matrix features with 2D-PCA. The next, combining the matrix features that are projected in the same space into a vector features. Finally, we use GA to build the coefficients for multi- SVM classifiers are used to identify the most likely class for input sequence.

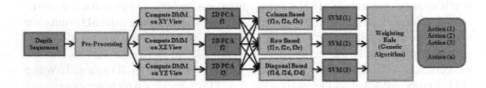

**Fig. 1** Our proposal approach for human action recognition on depth sequences

### 3.1 Pre-Processing

The 3D sensors such as Kinect use structured light to estimate depth information, it is prone to be affected by noises due to reflection issues. These affects of noise could significantly decrease the overall performance of depth-based action recognition framework. Therefore, we firstly apply a smooth to relieve the effect of noise from the depth channel is necessary step. As a result at [16], we adopted bilateral filter for smooth depth channel. A bilateral filter [16] is a combination of a domain kernel, which gives priority to pixels that are close to the target pixel in the image plane, with a range kernel, which gives priority to the pixels which have similar labels as the target pixel. This filter is often useful when one wants to preserve edge information because of the range kernel advantages.

## 3.2 Depth Motion Map

Yang et al. [21] developed the so-called Depth Motion Maps (DMM) to capture the aggregated temporal motion energies. More specifically, the depth map is projected onto three pre-defined orthogonal Cartesian planes and then normalized. For each projected map, a binary map is generated by computing and thresholding the difference of two consecutive frames. The binary maps are then summed up to obtain the DMM for each projective view. So, DMM have two properties that are useful for action representation such as motion and shape of object. Moreover, DMM are projected on three views to create more discriminative for distinguishing between actions. DMM are computed as follows:

- Step 1: Apply bilateral filter to remove noise.
- Step 2: Project depth map into XY, XZ and YZ view.
- Step 3: Compute DMM for each view.

$$DMM_v = \sum_{i=1}^{N-1} \left( |map_v^{i+1} - map_v^i| > \varepsilon \right) \tag{1}$$

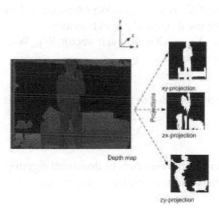

**Fig. 2** Illustration of depth map projection

## 3.3 Two Dimension Principle Component Analysis (2D PCA)

PCA [14] is a classical technique widely used in pattern recognition. The goal of this method is a method that uses to reduce the dimension of data that keep important information for feature vector. PCA-based techniques usually operate in vectors.

That is, before applying PCA, the 2D pattern matrices should be mapped into pattern vectors by concatenating their columns or rows. The pattern vectors generally lead to a high-dimensional space. In such a high-dimensional vector space, computing the eigenvectors of the covariance matrix is very time-consuming. 2D-PCA [10] is extended version of PCA and is a more efficient technique for dealing with 2D pattern, as 2D-PCA works on matrices rather than on vectors. Therefore, 2D-PCA does not transform a pattern into a vector, but rather it constructs a pattern covariance matrix directly from the original pattern matrices. In contrast to the covariance matrix of PCA, the size of the pattern covariance matrix of 2D-PCA is much small.

As a result, 2D-PCA has two important advantages over PCA. First, it is easier to evaluate the covariance matrix accurately. Second, less time is required to determine the corresponding eigenvectors. Moreover, 2D-PCA works on matrixes in computing process that keeps spatial information is useful for action representation. Let given training data set consists of N pattern with size of $m \times n$. $X_1, X_2 ..., X_N$ are the matrixes of samples. The 2D-PCA proposed by Yang et al. [10] is as follows:

- Step 1: Compute the average pattern M of all training samples

$$M = \frac{1}{N} \sum_{i=1}^{N} X_i \tag{2}$$

- Step 2: Compute the pattern covariance matrix C

$$C = \frac{1}{N} \sum_{i=1}^{N} (X_i - M)^T \times (X_i - M) \tag{3}$$

- Step 3: Compute d orthonormal vectors $W_1, W_2, ..., W_d$ corresponding to the d largest eigenvalues of C. $W_1, W_2, ..., W_d$ construct a d-dimensional projection subspace, which are the d optimal projection axes.
- Step 4: Project $X_1, X_2, ..., X_N$ on each vector $W_1, W_2, ..., W_d$ to obtain the principal component vectors:

$$P_i^j = X_j W_i \tag{4}$$

- Step 5: The reconstructed pattern of a sample pattern $A_j$ is defined as:

$$A_j = \sum_{i=1}^{N} P_i^j W_i^T \tag{5}$$

In this paper, we adopted the numbers of dominant eigenvectors were selected to maintain 95% of the energy of the dominant eigenvalues.

## 3.4 Action Representation

Action representation is important step for action recognition system. A robust representation will help to discriminate between actions and improve the performance of action classification. After a transformation by 2D-PCA, a feature matrix is obtained for each video. In order to get action representation, feature matrix variants to project the pattern into different spaces with different grouping strategies. An action representation will be projected to 3 presentation spaces by 2D-PCA (column-based, row-based and diagonal-based). Each of above presentation spaces yields to the feature vectors. Therefore, a feature matrix will be presented by F1, F2 and F3. In particular, F1 is the feature vector of column-based matrix, F2 is the feature vector of row-based matrix, and F3 is the feature vector of diagonal-based matrix. So, we create 3 vector features for action representation. Each vector feature is extract by combining that same projected space on three views (XY, XZ and YZ).

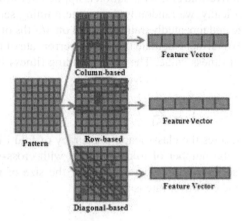

**Fig. 3** Action representation using 2D-PCA

## 3.5 Multiple Classifiers

Multiple features that are projected on three views and three 2D-PCA spaces are used to improve the performance of action recognition. So, we use multiple SVM classifiers are called late fusion strategy. Each classifier will classify the pattern based on the responsive feature. To compose the classified result into final label, we can use the selection method, average combination method or create the reliability coefficients. In this paper, we use GA [4] algorithm to build the coefficient of each classifier for optimal classification.

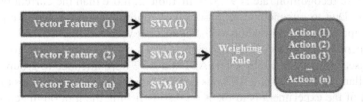

**Fig. 4** Multi-Classifiers for Human Action Recognition

As a significant part in GA algorithm, fitness function evaluates how well a program is able to solve the problem. Since the basic principle of GA is to maximize the performance of individual solutions, we design our fitness function by using the classification error rate computed by multi support-vector-machine (SVM) classifiers on the training set. We use non-linear SVM [20] with a RBF kernel which have shown good performance in many researches and the one-against-rest approach for multi-classification. To obtain a more reliable and fair fitness evaluation, for each candidate classifier, a five-fold cross-validation is applied to estimate the classification error rate. Specifically, we randomly divide the training set into five sub-sets with the identical size and repeatedly train the SVM on 4/5-ths of the set and test on the remaining fifth. We further calculate the average error rate of the five-fold cross-validation as the final fitness value. The corresponding fitness function is defined as follows:

$$E_r = \left( 1 - \sum_{i=1}^{n} \frac{classifier[i]}{n} \right) \times 100\% \tag{6}$$

where *classifier[i]* denotes the classification accuracy of fold i by the multi-SVM and n indicates the total number of folds executed with cross-validation. Here n is equal to 5. We also adopted GA parameters are the size of population is 100, mutation rate is 0.2 and crossover rate is 0.8.

## 4 Experiences

### 4.1 MSR-Action3D Dataset

The MSR-Action3D dataset, which is introduced in [13], provides both skeleton and depth information. It contains twenty actions (e.g. high arm wave, horizontal arm wave, side kick, pick up, throw, etc.), ten subjects, and a total of 567 action samples. The dataset is challenging due to small interclass variations among actions, while the skeleton tracker fails often, and contains significant noises. In order to have fair comparison with the other works, we use two experiments to compare with others.

The first setting, we use five cross-validation on whole dataset to evalutate the general ability of our method compare with the same experiment settings. It should be noted that different action representation have different experimental results and the fusion strategy achieve peak performance at Table 1. In the experiment, our method have recognition rate is 91.32% in Table 2, more than the current best rate by 2.43%. Due to our method propose robust representation for action on multi-view, it capture more information to distinguish between actions.

The second setting, we use experiments as [13, 21] to split 20 categories into three subsets as listed in Table 3. The AS1 and AS2 were intended to group actions with similar movement, while AS3 was intended to group complex actions together. The goal of the experiment is to focus on distinguishing the motion in action .The experiment results at Table 4 show that our method have a robust representation to distinguish between actions that the similar movement. The average accuracy of

**Table 1** Results on MSR-Action3D dataset with five cross-validation

| Action Representation | Accuracy (%) |
|---|---|
| Colum-based | 88.36 |
| Row-based | 85.71 |
| Diagonal-based | 90.47 |
| **Our approach** | **91.32** |

**Table 2** Compare with State of Art Methods on MSR-Action3D dataset with five cross-validation

| Methods | Accuracy (%) |
|---|---|
| **Our method** | **91.32** |
| Omar et al. [21] | 88.89 |
| J. Wang et al. [11] | 86.5 |

**Table 3** Action subsets for experiments

| Action Set 1 (AS1) | Action Set 2 (AS2) | Action Set 3 (AS3) |
|---|---|---|
| Horizontal Wave | High Wave | High Throw |
| Hammer | Hand Catch | Forward Kick |
| Forward Punch | Draw X | Side Kick |
| High Throw | Draw Tick | Jogging |
| Hand Clap | Draw Circle | Tennis Swing |
| Bend | Hands Wave | Tennis Serve |
| Tennis Serve | Forward Kick | Golf Swing |
| Pickup Throw | Side Boxing | Pickup Throw |

our method increases more than 1% on all of the datasets in this experiment. Our approach focus on using motion and shape information of performing action from multi-view that increases distinguishing between actions.

From two experiment results, this show our approach have a good performace in recognition rate. In specific, we use advantages of 2D-PCA to create matrix features instead of the traditional methods focus on a vector feature extraction in 1D. Action represenation from 2D-PCA keep significant information to seperate between actions. Moreover, we use GA to optimize the weighting of multiple SVM classifiers

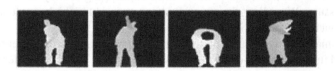

**Fig. 5** Examples from the MSR-Action3D dataset

**Table 4** Compare with State of Art Methods on MSR-Action3D dataset with three subsets

| Dataset | Bag-of-3D-Points [13] | DMM-HOG[21] | Our Methods |
|---------|-----------------------|-------------|-------------|
| AS1-One | 89.5 | 97.3 | 97.56 |
| AS2-One | 89.0 | 92.2 | 94.15 |
| AS3-One | 96.3 | 98.0 | 98.0 |
| AS1-Two | 93.4 | 98.7 | 98.7 |
| AS2-Two | 92.9 | 94.7 | 96.1 |
| AS3-Two | 96.3 | 98.7 | 98.7 |
| AS1CrSub | 72.9 | 96.2 | 97.35 |
| AS2CrSub | 71.9 | 84.1 | 87.44 |
| AS3CrSub | 79.2 | 94.6 | 95.57 |

**Table 5** Results on MSR-Gesture3D dataset

| Action Representation | Accuracy (%) |
|-----------------------|--------------|
| Colum-based | 91.89 |
| Row-based | 90.69 |
| Diagonal-based | 93.39 |
| **Our approach** | **94.89** |

**Table 6** Compare with State of Art Methods on MSR-Gesture3D dataset

| Methods | Accuracy (%) |
|---------|--------------|
| **Our method** | **94.89** |
| Omar et al. [21] | 92.45 |
| J. Wang et al. [11] | 88.2 |

in increasing recognition rate. These results manifest that our approaches give a robust representation and classifier to distinguish well actions. Moreover, this proves that our method is generalized well and does not depend on specific training data.

## 4.2  MSR-Gesture3D Dataset

The Gesture3D dataset [11] is a depth-only dynamic hand gesture dataset, containing 12 American Sign Language (ASL) gestures (e.g. "bathroom", "pig", "store", "where", etc.). As total, the MSR-Gesture3D dataset contains 333 depth sequences performed by 10 subjects, and is considered challenging mainly because of self-occlusion issues. We follow the experiment setup in [11] which conducts subject leave-one-out cross-validation.

Table 5 manifest the results of the action representations on diffence feature speaces and fusion strategry using GA-SVM. It should be noted that different action representation have different experimental results and the fusion strategy achieve

**Fig. 6** Examples from the MSR-Gesture3D dataset

peak performance. The results prove that is to improve the performance that should have a good strategy when combining the different models.

Table 6 compare our method result with the state-of-art result on MSR-Gesture3D dataset. Our recognition rate is 94.89%, more than the current best rate by 2.44%. The above results are to get, beause we use advantages of 2D-PCA to create matrix features instead of the traditional methods focus on a vector feature extraction in 1D. We also have used the features are projected on different spaces of 2D-PCA to increase discriminant between actions and multi classifier with optimal cofficients of final classifier using GA to improve action classification.

## 5 Conclusion

In this paper, we propose a novel approach for human action recognition based on 2D-PCA applied to DMM. We create robust action representation by project on different space of matrix features to obtain multi representation for each action. In order to improve the performance of system, we use GA to build the coefficients for multi-SVM classifier. The experiments show that our proposed approach achieves superior performance to the state-of-the-art algorithm on MSR-Action3D and MSR -Gesture3D datasets. The recognition rates are 91.32% on MSR-Action3D and 94.89% on MSR-Gesture3D. Meanwhile, it also demonstrates the potential to easily extend for future works in combination with other color-based representations for the task of multi-modal color-depth action recognition. Moreover, we can focus on combining body joints to create robust features for action representation. Besides, we will consider learning features to use advantage of learning methods in feature extraction from raw data.

**Acknowledgments.** This research is funded by Vietnam National University, Ho Chi Minh City (VNU-HCM) under grant number B2014-18-02.

## References

[1] Klaser, A., Marszalek, M., Schmid, C.: A spatio-temporal descriptor based on 3d-gradients. In: BMVC (2008)

[2] Vieira, A.W., Nascimento, E.R., Oliveira, G.L., Liu, Z., Campos, M.F.M.: STOP: Space-time occupancy patterns for 3D action recognition from depth map sequences. In: Alvarez, L., Mejail, M., Gomez, L., Jacobo, J. (eds.) CIARP 2012. LNCS, vol. 7441, pp. 252–259. Springer, Heidelberg (2012)

[3] Bobick, A., Davis, J.: The Recognition of Human Movement Using Temporal Templates. IEEE Trans. on Pattern Analysis and Machine Intelligence (2001)

[4] Reeves, C.R., Rowe, J.E.: Genetic Algorithm: Principles and Perspectives. Kluwer (2002)

[5] Willems, G., Tuytelaars, T., Van Gool, L.: An efficient dense and scale-invariant spatio-temporal interest point detector. In: Forsyth, D., Torr, P., Zisserman, A. (eds.) ECCV 2008, Part II. LNCS, vol. 5303, pp. 650–663. Springer, Heidelberg (2008)

[6] Wang, H., Klaser, A., Schmid, C., Liu, C.-L.: Dense trajectories and motion boundary descriptors for action recognition. International Journal of Computer Vision (March 2013)

[7] Laptev, I.: On space-time interest points. Int. J. Comput. Vision 64(2-3), 107–123 (2005)

[8] Laptev, I., Marszałek, M., Schmid, C., Rozenfeld, B.: Learning realistic human actions from movies. In: CVPR, IEEE (2008)

[9] Aggarwal, J., Ryoo, M.: Human activity analysis: A review. ACM Comput. Surv. 43(3), 16:1–16:43 (2011)

[10] Yang, J., Zhang, D., Frangi, A.F., Yang, J.-Y.: Two dimensional PCA: a new approach to appearance-based face representation and recognition. IEEE Transactions on Pattern Analysis and Machine Intelligence 26, 131–137 (2004)

[11] Wang, J., Liu, Z., Chorowski, J., Chen, Z., Wu, Y.: Robust 3D action recognition with random occupancy patterns. In: Fitzgibbon, A., Lazebnik, S., Perona, P., Sato, Y., Schmid, C. (eds.) ECCV 2012, Part II. LNCS, vol. 7573, pp. 872–885. Springer, Heidelberg (2012)

[12] Cruz, L., Lucio, D., Velho, L.: Kinect and RGBD Images: Challenges and Ap-plications. In: SIBGRAPI Tutorials, pp. 36–49 (2012)

[13] Li, W., Zhang, Z., Liu, Z.: Action Recognition based on A Bag of 3D Points. In: IEEE Workshop on CVPR for Human Communicative Behavior Analysis (2010)

[14] Turk, M., Pentland, A.: Eigenfaces for Recognition. Journal of Cognitive Neuroscience 3(1) (1991)

[15] Naiel, M.A., Abdelwahab, M.M., Mikhael, W.B.: Human action recognition employing 2DPCA and VQ in the spatio-temporal domain. In: PPSN 2000, pp. 381–384 (June 2010)

[16] Camplani, M., Salgado, L.: Efficient spatio-temporal hole filling strategy for kinect depth maps. In: Baskurt, A.M., Sitnik, R. (eds.) SPIE, vol. 8290(1), p. 82900E (2012)

[17] Oreifej, O., Liu, Z.: Hon4d: Histogram of oriented 4d normals for activity recognition from depth sequences. In: CVPR (2013)

[18] Dollar, P., Rabaud, V., Cottrell, G., Belongie, S.: Behavior recognition via sparse spatio-temporal features. In: VS-PETS (October 2005)

[19] Tang, S., Wang, X., Lv, X., Han, T.X., Keller, J., He, Z., Skubic, M., Lao, S.: Histogram of oriented normal vectors for object recognition with a depth sensor. In: Lee, K.M., Matsushita, Y., Rehg, J.M., Hu, Z. (eds.) ACCV 2012, Part II. LNCS, vol. 7725, pp. 525–538. Springer, Heidelberg (2013)

[20] Vapnik, V.: Statistical learning theory. John Wiley and Sons, New York (1998)

[21] Yang, X., Zhang, C., Tian, Y.: Recognizing actions using depth motion maps based histo-grams of oriented gradients. In: ACM International Conference on Multimedia 2012, pp. 1057–1060 (2012)

# A Study of Feature Combination
# in Gesture Recognition with Kinect

Ngoc-Quan Pham, Hai-Son Le,
Duc-Dung Nguyen, and Truong-Giang Ngo

**Abstract.** Human gesture recognition is an interdisciplinary problem, with many important applications. In the structure of a gesture recognition system, feature extraction, without doubt, is one of the most important factor affecting the performance. In this paper, we desired to improve the covariance feature, which is the current state-of-the-art feature extraction method, by integrating other frame-level features extracted in the data captured by Microsoft Kinect, and experimenting the features with various classification methods such as Random Forest (RF), Multi Layer Perceptron (MLP), Support Vector Machines (SVM). The leave-person-out experiments showed that feature combination is beneficial, especially with Random Forest, to achieve the highest score in recognition, which is improved by 2%, from 90.9% to 93.0%. However, the dimensional increase sometimes exacerbated the performance, indicating the side effect of feature combination.

**Keywords:** Kinect, gesture recognition, feature extraction, Support Vector Machine, Random Forest, Multi Layer Perceptron.

## 1 Introduction

Gesture recognition is one of the important topics in human-machine interaction. The target of gesture recognition is to interpret the human gestures, which are captured in the form of video sequences or captured motions, into meaningful actions. Applications that utilize gesture recognition vary from sign language recognition, ambient intelligent system, remote control, or alternative computer interfaces. One

Ngoc-Quan Pham · Hai-Son Le · Duc-Dung Nguyen
Institute of Information Technology
Vietnam Academy of Science and Technology, Hanoi, Vietnam

Truong-Giang Ngo
Hai Phong Private University, Hai Phong, Vietnam
e-mail: {quanpn,lehaison,nddung}@ioit.ac.vn, giangnt@hpu.edu.vn

© Springer International Publishing Switzerland 2015                                        459
V.-H. Nguyen et al. (eds.), *Knowledge and Systems Engineering,*
Advances in Intelligent Systems and Computing 326, DOI: 10.1007/978-3-319-11680-8_37

of the obstacle that hinders the research is the exorbitant costs of data acquisition, which relies on expensive sensors and depth cameras. Besides, it is difficult to gather a large amount of data from such devices.

The recent appearance of Microsoft Kinect has made touchless body movement interfaces available to a much larger audience [1], by providing a depth sensor to capture the skeleton-based motions in 3D. The affordable cost and the ability to widespread the technology lead to the abundant data resources, with the acceptable loss in accuracy. The real-time data recorded by Kinect is easier to be analyzed, and insensitive to changes in lighting conditions [2], which facilitates the gesture recognition task. The feature extraction step, which analyzes the motion data to acquire the feature vectors or matrices containing the relevant information of the actions, plays an important part to decide the performance of the system.

Among many researches towards feature extraction in gesture recognition, the approach based on constructing a covariance matrix over the action is proved to be the state-of-the-art approach with the best results produced [3]. Since this feature was only experimented with only one type of frame-level feature (the joint coordinates) and one classification method (SVM with Linear kernel), two questions are reasonably raised:

. By providing more information with other frame-level features can the performance be enhanced? Joint angles were used as features in [1] and outperformed the coordinates. Besides, the joint velocities and angular velocities contain temporal dependencies of the consecutive frames, which can be useful as features.
. Excluding SVM, how do other learning methods work with the covariance features?

In order to answer these questions, in this paper we would like to conduct experiments, with the objective to achieve the highest recognition rate, by further investigating the covariance matrix. Specifically, multiple types of frame-level features extracted in the data provided by Microsoft Kinect are used to create the covariance features, and produce comparative results. Moreover, other classification methods namely SVM with different kernels, Random Forest and Multi Layer Perceptron (MLP) are also experimented with the covariance features as well.

The organization of our paper is as follows: Section 2 notes several related approaches on feature extraction in gesture recognition, especially with Kinect data. Section 3 will cover the fundamentals of the covariance feature, including the formulas to create the covariance matrices, the temporal covariance matrices, and the way we utilized to combine the frame-level features. Section 4 is dedicated to explain the learning process by the learning methods. The description of the experiments lies in Section 5, which provides the simulation performance to compare different feature combinations with different learning methods, and analysis the learning results. Finally, we conclude the paper in Section 6.

## 2 Related Work

Concerning the related works in feature extraction, there are two main challenges to be addressed:

- . Extracting the frame-level feature - the individual feature descriptor, which contains the most information of the action
- . Finding a discriminative feature descriptions for action sequences, which not only is the aggregation of the individual feature descriptor, but also is able to model the temporal properties of the actions.

The first challenge is the task to extract the frame-level features from the data. The motion samples from different people do not share the same properties, such as positions, trajectories, or speed. Even when a person performs a gesture repetitively and consecutively, the difference between gestures is inevitable. The features, therefore, are required to be invariant with rotation, position and scale. For the image-based data, the frames were often segmented into identical religions, which are then processed to acquire individual features [4]. Color histogram was also commonly used for feature extraction [5], [6], [7]. The later steps involve using locating the recognition focus, which are the most active human body parts such as hands and heads.

The skeleton-based data acquisition has resulted in the increase of the features extracted from mocap and skeletal data. The most commonly used features include 3D coordinates and the angular information of the joints. The feature vector can be purely created from the 3D coordinates [8], or the angles [9], or by stacking the individual features [1]. Instead of absolute coordinates, relative positions between pairs of skeleton joints can also be used as features [10].

The second challenge is to represent the whole action, from the frame-level features. The most considerable hindrance in this problem is the temporal dependency of the frames, which is hard to be represented. The common approach for feature aggregation is to stack all of the individual features of all frames in the action [1, 7]. In [10], temporal modeling is represented using Fourier Temporal Pyramids. To better represent the temporal properties, state-space-based models, notably, Hidden Markov Model [6], [11] and Conditional Random Field (CRF) [12, 13, 14, 15] are also utilized.

## 3 The Covariance Feature

As said, the goal of the frame-level feature extraction in GR is to create a fixed-length vector representation for each instance of gestures. In this section, we will briefly present how to use covariance matrix to extract a feature. The detailed description can be found in [3]. The covariance matrix for a set of $N$ random variables is a $N \times N$ matrix, the elements of which are the covariance between every pair of the variables. Assuming that an action $\mathbf{S}$ is performed over $T$ frames, and $S_t$ is the individual feature vector at frame $t$, which is the feature extracted from a single

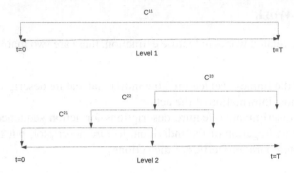

**Fig. 1** The temporal hierarchical structure of the covariance feature

video/motion frame. The sample covariance over the frame sequence is computed
with the equation:

$$C(\mathbf{S}) = \frac{1}{T-1} \sum_{t=1}^{T} (S_t - \bar{S})(S_t - \bar{S})^T, \qquad (1)$$

where

$$\bar{S} = \frac{1}{T} \sum_{t=1}^{T} S_t \qquad (2)$$

Equation 1 shows how to compute the covariance matrix for an individual feature.
It is notable that the covariance matrices needed to be reshaped into one-dimension
matrices before being used as features, or combined. After being computed, the
covariance matrix is collapsed into a vector, which is then considered as a feature.

Temporal Covariance Feature

The disadvantage of the covariance matrix feature is that, when the order of the
frames in the sequence is changed, the matrix still maintains the same. In order to
represent the temporal dependency of the frames, a hierarchy of covariance matrix
was employed. The fundamental idea of hierarchical covariance matrices is to create
levels to the covariance matrices computed over a sequence. Each level of covariance matrix is the combination of the matrices which are computed over the smaller
windows of that sequence. As it is shown in Fig. 1, a covariance matrix at level $l$ is
computed over $T/2^l$ frames. The overlapping length that we used is half of the window length, which provides more detail in the overall feature. The final temporal
covariance matrix is created by the concatenation of all of the member covariance
matrices, including concatenating the matrices within a level. It is notable that, the
higher the level is, the more temporal information is contained in the feature.

Feature Combination for Covariance Feature

Since the target of covariance features is to capture the dependence of a particular feature during the action, it is reasonable to compute the covariance matrix of each feature separately. After that, the final feature is created by concatenating those matrices. For example, in Fig. 2, covariance matrices of Coordinates and Angles are computed, then concatenated and collapsed into a vector to form a final feature.

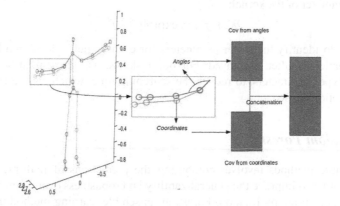

**Fig. 2** Feature combination graph

# 4 The Classification Methods

In this section, we will provide the description of how to train classification methods considered in this article, namely, SVM, Random Forest and MLP. Basically, their task is by taking as input any type of fixed-length feature extracted to represent an instance of gesture $f$, to assign the most likely gesture class to that instance.

## 4.1 Support Vector Machine

The practice of Support Vector Machine (SVM) in almost every problem of Machine Learning and Computer Vision is popular, thanks to its effectiveness regardless to the high dimensional spaces, or even if the dimensional size is larger than the number of samples (the Kinect problem) [16].

As mentioned above, the feature extraction step involves computing the covariance matrices over the action sequences and then, combining them. In order to combine different types of features, e.g., Coordinates and Angles, normalization on the feature space is often a crucial step for SVM [17], [18]. As shown later with the experimental results, for this task considered here, normalization is actually the important factor that makes feature combination feasible.

There are two factors that affect the performance of SVM. The first one is the kernel selection, i.e., choose the best one among the most common used: Linear,

Polynomial, Radial Basis Function (RBF). The second one concerns how to tune hyper-parameters for each type of kernel. For two types of kernel studied in this paper, we have:

- The Linear Kernel's performance depends on the sole hyper-parameter $C$, which controls the trade-off between a low error on the training data and minimising the norms of the weights.
- For the RBF Kernel, another hyper-parameter needed to be tuned is $\gamma$, which is the parameter of the kernel:

$$\mathcal{K}(f, f') = \exp(\gamma \|f - f'\|_2^2) \tag{3}$$

In order to identify the hyper-parameters for each kernel, grid-search based on cross-validation is often employed. For our task, we use the validation data for choosing hyper-parameters, to reduce the searching time, which can be massive if cross-validation is used.

## 4.2  Random Forest

The ensemble methods involves combining the predictions of multiple learning models in order to improve the generalizability and robustness over a single model. Here, we concentrate on Random Forest, an ensemble learning method using Decision Tree as a single model. The advantages of Random Forest is its ability to handle with large databases with a good learning speed [19]. Each tree from the forest is built from a subset of the training data (a bootstrap sample). For the feature extraction step, contrary to SVM, data normalization does not have any impact on the performance of Random Forest.

The parameters that need to be tuned in Random Forests:

- The number of trees in the forest (M)
- The maximum number of features accounted when splitting
- The maximum depth of the trees

In our experiment, we fixed the maximum number of features as the square root of the number of features, and allowed the trees to split until the leaf node. Therefore, the only parameter to find in grid-search is the number of trees. Similar to the SVM, a grid-search on $M$ is conducted by considering the performance on the validation data, At the end, we have the optimized number of trees.

## 4.3  Multi Layer Perceptron

Multi Layer Perceptron (MLP) is a feed-forward neural network with one or more hidden layers. Here, a neural network takes as input a vector $f$, a feature used to represent an instance and output a probability for each class (gesture). Its structure is inspired by the architecture of the human brain which processes information in a hierarchical fashion with multiple levels of data representation and transformation. However, due to some practical training issues, the so-called *deep* architectures were often outperformed by *shallow* ones with a single hidden layer [20], which is very similar to SVM. The introduction of the *Deep Learning* based on Deep Belief Networks with a more sophisticated learning algorithm [21] can be considered as a breakthrough. The main idea is to pre-train the network layer by layer in an unsupervised way using Restricted Bolzmann Machines (RBM) in order to increase the generalization capacity. Deep learning has been recently applied with success in many fields, especially in the domain of image processing [21] or recently in acoustic modeling [22]. However, it seems that due to the small number of training examples, MLP trained with a pre-training stage does not bring any significant improvement. Nevertheless, it remains interesting to carry out experiments with MLP in order to make a comparison with other methods. The structure of MLP plays a crucial role. In our experiment, the performance of the system reaches a peak with MLP with two hidden layers of 1000.

MLP is usually trained by backpropagating derivatives of a lost function measuring the mismatch between the desired outputs and the actual outputs of the model. For large training set, rather than computing derivatives on the whole training set, it is typically more efficient to use a Stochastic Gradient Descent method. MLP is trained by passing through the training data for some epochs until the convergence or the performance on the validation data decreases (early-stopping strategy).

# 5 Experiment

## 5.1 Setup

### 5.1.1 Dataset

The dataset used in our paper is the Microsoft Research Cambridge 12 (MSRC-12), provided by Microsoft. This dataset contains twelve actions which are performed by 30 people, with a total of 6204[1] samples. Each sequence in the dataset is a sequence of an action only, performed consecutively by an actor.

Concerning the annotation, this dataset is annotated with action points. According to [23], an action point is a single time instance at which the present of the action is clear and can be uniquely determined for all instances of the action. Action spotting using action points are employed in [1], when the skeletal data from 35 consecutive frames ending with action points. However, this approach has several disadvantages:

. The location of the action point is very sensitive, depending on the individuals and the actions, as well as the opinion of the annotator.
. The action length varies between 13 and 492 frames. Fixing the length at 35 frames are not reasonable.

---

[1] After eliminating 3 files, which contain errors.

For that reason, it is better to have action boundaries marked to guarantee the trustworthiness of the evaluations. Thanks to [3], the manual annotation of MSRC-12 dataset is available under the format of begin-end markers.

### 5.1.2   Setup Description

We were able to extract the following frame-level features from the data channels of Kinect:

- 60 coordinates for 20 skeleton joints, in the space defined in Kinect, where $(0, 0, 0)$ is the location of the camera.
- 60 velocities at the skeleton joints: the differences between the coordinates of corresponding joints between each pair of adjacent frames
- 35 joint angles: the angle between the segments on either side of the joints
- 35 joint angular velocities: the difference of angles between the adjacent frames

Besides, we utilized the action boundaries included in the annotations by [3] which enabled us to define the temporal regions of the motion sequences with labels.

We employed the leave-person-out experiments as the main approach to assess the efficiency of the features and methods. As the name suggests, leave-person-out involves using a set of observations belonging to one person as the evaluation data, and the remaining observations as the training data. We also used data from another person as validation data, which serves for the parameter selection steps in the training phase. Specifically, when the data from person $i$ is evaluated, the person $i + 1$ is used for validation, with the exception that person 1 would serve as the validation data for person 30.

It should be noted that, we used Torch7 [24] for MLP and Scikit-learn package [25] for other classification methods to carry out experiments.

## 5.2   Results

### 5.2.1   Feature Normalization

As said, for some classification methods, feature normalization may be the important factor in order to combine the covariance matrices computed from each individual feature. To verify it, we carried out the preliminary experiment with a simpler setup. We found that it is actually the case for SVM and MLP. For instance, F1-score of SVM increases from 85.0% to 91.2%, which is quite significant. On the contrary, for Random Forest, it has no effect on the system performance. Due to the important affect of feature normalization in general, we decide to apply normalization for all methods in the later experiments.

### 5.2.2   Feature Comparisons

In order to compare the ability to contain information of the four individual features, we performed the leave-person-out experiments with each feature. Random forest classification is used in this comparison, thanks to its speed and low power consumption. As illustrated from table 1, the coordinates contain the most information, followed by angles, joint velocities and angular velocities. Therefore, we would mostly use coordinates and angles to compare the efficiency of the learning methods.

**Table 1** F1-score(%) of Random Forest with each feature

| Method | Coordinates | Angles | Joint velocities | Angular velocities |
|---|---|---|---|---|
| Random Forest | **92.4** | 87.8 | 85.8 | 84.0 |

### 5.2.3 Method Comparisons

We managed to implement the classification modules with the most popular learning methods: SVMs, Random Forest and MLP. For each method, the final result is the average result after experimenting with 30 people. As mentioned before, the parameter selection steps are performed on the validation set, to choose the final model to evaluate the test set. At first, only the coordinates were used as features for all of the methods. After that, the angular feature was added to observe the progressive results of the methods.

Level 1 Features

First, the level 1 features were experimented. As it was shown in Table 2, most of the classification methods did not receive any benefits from feature combination. While the scores of Random Forest and SVM Linear slightly rose, the performance of SVM with RBF Kernel and MLPs were decreased after angles were integrated.

**Table 2** F1-score(%) with features level 1

| Method | Coordinates | Coordinates + Angles |
|---|---|---|
| SVM (RBF Kernel) | **91.4** | **91.1** |
| SVM (Linear Kernel) | 89.0 | 90.1 |
| Random Forest | 90.4 | 90.9 |
| Multi-layer Perceptrons | 90.1 | 89.8 |

Level 2 Features

Feature combination was further investigated with level 2 features. Table 3 shows that feature combination was effective with all of the methods. It is notable that our experiments produced different results with [3] due to the differences in experimental setups and toolkits. Their best setup is corresponding to the setup with the feature produced from Coordinate level 2 and learned with SVM Linear, whose score is 90.9%.

**Table 3** F1-score(%) with features level 2

| Method | Coordinates | Coordinates + Angles |
|---|---|---|
| SVM (RBF Kernel) | 91.0 | 91.9 |
| SVM (Linear Kernel) | 90.9 | **92.7** |
| Random Forest | **92.4** | **92.7** |
| Multi-layer Perceptrons | 89.8 | 91.9 |

Random Forest and SVM with Linear Kernel tend to be the most effective methods for gesture recognition with the covariance features. While Random Forest was the best method in both feature types, SVM Linear and MLP are likely to benefit the most from feature combination. Adding another covariance matrix created by joint angles increased the scores of SVM and MLP by approximately 1.8%. Besides, the increase in Random Forest performance is the same with the features level 1, showing the consistency in feature combination of this method.

All Feature Combined

To further investigate the capability of feature combination, all of the frame-level features are combined in the following experiments. Surprisingly, Random Forest managed to achieve the highest performance, with the score of 93.0%, while SVM Linear witnessed a decrease by 1.2%. Again, the robustness of Random Forest is proved, and SVM is likely to suffer from dimensional increasing.

**Table 4** F1-score(%) with all features

| Method | Coordinates + Angles | All features |
|---|---|---|
| SVM (Linear Kernel) | **92.7** | 91.5 |
| Random Forest | **92.7** | **93.0** |

Fig. 3 indicates that the performance rate per subject can be perfect (in the cases of person 14, 16, 17), while the most difficult cases to be recognized are the cases of person 2 and 9. Besides, with the same features, the difference between the results of SVM (Linear) and Random Forest for each subject is negligible, excluding the cases of person 4, 24 and 29.

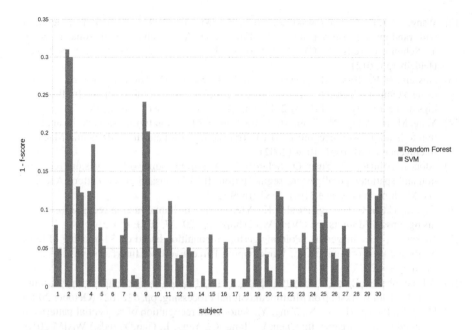

**Fig. 3** Leave-Person-Out results of SVM Linear and Random Forest

# 6   Conclusion

Feature combination with covariance matrices was presented in our paper, along with the experiment results with various classification methods. The leave-person-out experiments illustrated that feature combination did not show its effectiveness until level 2 features were applied. More importantly, Random Forest seems to be the most appropriate method, which not only produced the highest score in performance, but also is robust to feature combination, which tremendously increases the input dimension. Both SVM and MLP are inconsistent to feature combination and sometimes suffered from the dimension increase. In brief, we managed to increase the recognition performance from 90.9% to 93.0%, thanks to feature combination and Random Forest.

**Acknowledgments.** This work was supported in part by the National Key Lab for Networking, Technology and Multimedia, Institute of Information Technology, Vietnamese Academy of Science and Technology.

# References

[1] Fothergill, S., Mentis, H., Kohli, P., Nowozin, S.: Instructing people for training gestural interactive systems. In: Proceedings of the SIGCHI Conference on Human Factors in Computing Systems, ser. CHI 2012, pp. 1737–1746. ACM, USA (2012)

[2] Wang, J., Liu, Z., Chorowski, J., Chen, Z., Wu, Y.: Robust 3D action recognition with random occupancy patterns. In: Fitzgibbon, A., Lazebnik, S., Perona, P., Sato, Y., Schmid, C. (eds.) ECCV 2012, Part II. LNCS, vol. 7573, pp. 872–885. Springer, Heidelberg (2012)

[3] Hussein, M.E., Torki, M., Gowayyed, M.A., El-Saban, M.: Human action recognition using a temporal hierarchy of covariance descriptors on 3d joint locations. In: Proceedings of the Twenty-Third IJCAI 2013, pp. 2466–2472. AAAI Press (2013)

[4] Yang, M.-H., Ahuja, N., Tabb, M.: Extraction of 2d motion trajectories and its application to hand gesture recognition. IEEE Transactions on Pattern Analysis and Machine Intelligence 24(8), 1061–1074 (2002)

[5] Alon, J., Athitsos, V., Yuan, Q., Sclaroff, S.: A unified framework for gesture recognition and spatiotemporal gesture segmentation. IEEE Transactions on Pattern Analysis and Machine Intelligence 31(9), 1685–1699 (2009)

[6] Xia, L., Chen, C.-C., Aggarwal, J.K.: View invariant human action recognition using histograms of 3d joints. In: CVPR Workshops, pp. 20–27. IEEE (2012)

[7] Nickel, K., Stiefelhagen, R.: Pointing gesture recognition based on 3d-tracking of face, hands and head orientation. In: Workshop on Perceptive User Interfaces, pp. 140–146. ACM Press (2003)

[8] Adistambha, K., Ritz, C., Burnett, I.: Motion classification using dynamic time warping. In: IEEE 10th Workshop on Multimedia Signal Processing, pp. 622–627 (October 2008)

[9] Deng, L., Leung, H., Gu, N., Yang, Y.: Automated recognition of sequential patterns in captured motion streams. In: Chen, L., Tang, C., Yang, J., Gao, Y. (eds.) WAIM 2010. LNCS, vol. 6184, pp. 250–261. Springer, Heidelberg (2010)

[10] Wang, J., Liu, Z., Wu, Y., Yuan, J.: Mining actionlet ensemble for action recognition with depth cameras. In: CVPR, pp. 1290–1297 (June 2012)

[11] Kim, D., Nguyen-Duc-Thanh, N., Lee, S.: Two-stage hidden markov model in gesture recognition for human robot interaction. Int. J. Adv. Robot. Syst., 9–39 (2012)

[12] Sminchisescu, C., Kanaujia, A., Li, Z., Metaxas, D.: Conditional models for contextual human motion recognition. In: Tenth IEEE International Conference on Computer Vision, ICCV 2005, vol. 2, pp. 1808–1815 (2005)

[13] Wang, S.B., Quattoni, A., Morency, L., Demirdjian, D., Darrell, T.: Hidden conditional random fields for gesture recognition. In: IEEE Computer Society Conference on Computer Vision and Pattern Recognition, vol. 2, pp. 1521–1527 (2006)

[14] Elmezain, M., Al-Hamadi, A., Sadek, S., Michaelis, B.: Robust methods for hand gesture spotting and recognition using hidden markov models and conditional random fields. In: ISSPIT, pp. 131–136 (2010)

[15] Vinh, L., Lee, S., Le, H., Ngo, H., Kim, H., Han, M., Lee, Y.-K.: Semi-markov conditional random fields for accelerometer-based activity recognition. Applied Intelligence 35(2), 226–241 (2011)

[16] Oommen, T., Misra, D., Twarakavi, N., Prakash, A., Sahoo, B., Bandopadhyay, S.: An objective analysis of support vector machine based classification for remote sensing. In: Mathematical Geosciences, vol. 40(4), pp. 409–424 (2008)

[17] Graf, A.B.A., Borer, S.: Normalization in support vector machines. In: Radig, B., Florczyk, S. (eds.) DAGM 2001. LNCS, vol. 2191, p. 277. Springer, Heidelberg (2001)

[18] Ali, S., Smith-Miles, K.A.: Improved support vector machine generalization using normalized input space. In: Sattar, A., Kang, B.-H. (eds.) AI 2006. LNCS (LNAI), vol. 4304, pp. 362–371. Springer, Heidelberg (2006)

[19] Breiman, L.: Random forests. Machine Learning 45(1), 5–32 (2001)

[20] Bengio, Y.: Learning deep architectures for ai. Found. Trends Mach. Learn. 2(1), 1–127 (2009)

[21] Hinton, G.E., Salakhutdinov, R.R.: Reducing the dimensionality of data with neural networks. Science 313(5786), 504–507 (2006)

[22] Seide, F., Li, G., Yu, D.: Conversational speech transcription using context-dependent deep neural networks. In: Proceedings of the 12th INTERSPEECH, pp. 437–440 (2011)

[23] Nowozin, S., Shotton, J.: Action points: A representation for low-latency online human action recognition. In: TechReport MSR-TR-2012-68. 7 J J Thomson Ave, CB30FB Cambridge, UK: Microsoft Research Cambridge (July 2012)

[24] Collobert, R., Kavukcuoglu, K., Farabet, C.: Torch7: A Matlab-like Environment for Machine Learning. In: BigLearn NIPS Workshop (2011)

[25] Pedregosa, F., Varoquaux, G., Gramfort, A., Michel, V., Thirion, B., Grisel, O., Blondel, M., Prettenhofer, P., Weiss, R., Dubourg, V., Vanderplas, J., Passos, A., Cournapeau, D., Brucher, M., Perrot, M., Duchesnay, E.: Scikit-learn: Machine learning in Python. Journal of Machine Learning Research 12, 2825–2830 (2011)

# Formalising Concurrent UML State Machines Using Coloured Petri Nets

Étienne André, Mohamed Mahdi Benmoussa, and Christine Choppy

**Abstract.** While UML state machines are widely used to specify dynamic systems behaviours, their semantics is described informally, which prevents the complex systems verification. In this paper, we propose a formalisation of concurrent UML state machines using coloured Petri nets. We consider in particular concurrent aspects (orthogonal regions, forks, joins, shared variables), the hierarchy induced by composite states and their associated activities, and entry/exit/do behaviours.

## 1 Introduction

UML [10] became the *de facto* standard for modelling systems, and features a very rich syntax with different diagrams to model the different aspects of a system. UML behavioural state machine diagrams (SMDs) are transition systems used to express the behaviour of dynamic systems in response to external interactions. Although UML is widely used in the industry, its semantics is not formally expressed, which prevents formal verification to be performed.

Related Work

Verification of SMDs has been often tackled. Some approaches directly give UML a semantics. The closest to the set of syntactic constructs we consider here is [6], where entry/exit behaviours, activities, synchronisation and history states are considered; however, global variables are discarded, and no model checking is performed. In [9], an operational semantics is proposed for UML SMDs with synchronisation. Most syntactic aspects are taken into account, except real-time aspects and object-oriented issues. A formalisation has also been proposed using UML-B [11]. However, these approaches require the implementation of a standalone, dedicated tool.

Étienne André · Mohamed Mahdi Benmoussa · Christine Choppy
Université Paris 13, Sorbonne Paris Cité, LIPN, CNRS, Villetaneuse, France

© Springer International Publishing Switzerland 2015
V.-H. Nguyen et al. (eds.), *Knowledge and Systems Engineering,*
Advances in Intelligent Systems and Computing 326, DOI: 10.1007/978-3-319-11680-8_38

Other approaches translate UML specification into an intermediate model of some model checker. Most of these approaches consider quite restrictive subsets of the UML syntax as defined by the OMG [10]. Out of many approaches, for lack of space, we cite only the most recent and related to our work. A translation of SMDs to SPIN is considered in [7] with both hierarchical and non-hierarchical cases; history, fork, join pseudo-states, entry and exit activities and complex data structures are not supported. A semantics is defined in [4] for a syntax subset, including deferring of messages, concurrent composite states and choice pseudo-state; verification is performed using NuSMV. An automated translation from SMDs into CSP♯ (an extension of CSP) is proposed in [13]; modelling techniques such as use of data structures, join/fork, history pseudo-states, entry/exit behaviours (but with no variable) are considered, and properties are checked. Note that these formalisms do not provide a graphical representation. Other approaches (e.g. [8]) use an intermediary model (e.g., flat state machines) to formalise SMDs to coloured Petri nets (CPNs) [5]. In [8], concurrency (fork/join, concurrent composite states) is taken into account. The main difference with our approach is that [8] requires to flatten the SMD, hence losing the hierarchy. A translation from SMDs to Petri nets is proposed in [3] were SMDs include synchronisation, limited aspects of hierarchy, join and fork (with no inter-level transitions) but history pseudo-states and variables are not considered.

Contribution

We introduce here a translation of concurrent SMDs into CPNs. CPNs offer a detailed view of the process with a graphical representation, and benefit from powerful tools (such as CPN Tools [12]) to test and check the model. Our SMDs can communicate on synchronised events; we also take into account the most common syntactic features, i.e. state hierarchy with entry/exit/do behaviours, history pseudo-states, synchronised events, fork/join, shared variables and local/external/internal transitions. Our approach partially relies on the work presented in [2] where we proposed a first attempt of formalising non-concurrent SMDs using CPNs. Here, we extend this previous approach to the concurrency (forks, joins, nested composite states). We also add internal transitions.

Outline

We present in Section 2 the formalisms we use (viz. SMDs and CPNs). In Section 3, we describe our translation: we first describe our general translation scheme, and recall the assumptions we make (Section 3.1); then, we define functions used in our algorithms (Section 3.2), and give in details our algorithms translating states and behaviours (Section 3.3), and transitions (Section 3.4). We conclude and give some perspectives in Section 4.

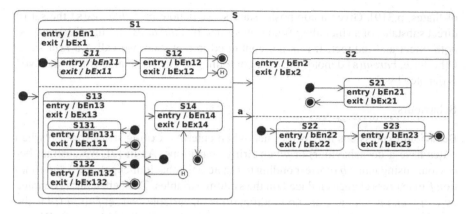

**Fig. 1** Example of a state machine diagram

## 2 Basic Concepts Used in This Work

### 2.1 Our Assumptions on UML State Machine Diagrams (SMDs)

The underlying paradigm of UML SMDs [10] is that of a finite automaton, that is each entity (or subentity) is in one state at any time and can move to another state through a well-defined conditional transition. For lack of space, we assume the reader's knowledge on SMDs, and we only present here technical issues required for our work. The UML provides an extended range of constructs for SMDs, and we take into account the following syntactical elements: simple/composite states, entry/exit/do behaviours, concurrency (regions in composite states, fork/join transitions), shared variables, shallow history pseudostates (not detailed for lack of space) and hierarchy (of states and behaviours). In the following, we recall these elements together with some assumptions we need to set for our work. We use the SMD in Fig. 1 as a running example.

States

We consider two kinds of states: simple and composite. A composite state is a state that contains at least one region and can be a simple composite state or an orthogonal state. A simple composite state has exactly one region, that can contain other states, allowing to construct hierarchical SMDs. An orthogonal state (e.g. S1, S2 in Fig. 1) has multiple regions (regions can contain other states), allowing to represent concurrency. *regions*(s) denotes the set of regions in composite state s, and *NbRegions*(s) the number of regions in s. We assume that a composite state must not be empty ("A composite State contains at least one region" [10, p.322]), and that each region contains at least one state.

"Any state enclosed within a region of a composite state is called a *substate* of that composite state. It is called a *direct substate* when it is not contained by any other state; otherwise, it is referred to as an *indirect substate*." [10, Section Kinds of States, p.319]. Given a composite state **s**, we denote by *SubStates*(**s**) the set of direct substates of **s** (including final states), and by *SubStates*$^*$(**s**) all substates of **s**, both direct and indirect. If **s** is not contained in any state, we call it a *root* state. Otherwise, *parent*(**s**) denotes the state containing **s**. The set of all states of an SMD is denoted by $\mathcal{S}$.

## Behaviours

Behaviour expressions may be defined when entering, exiting states (or also when states have a do behaviour) or when firing transitions. As in [2], we abstract behaviours using name $b$ (corresponding to the actual behaviour expression) and function $f$ to express changes induced on the system variables. The behaviour is denoted by $(b,f)$. and when there are no modifications on variables we set $f$ to *id* (*id* is the identity function). When a transition has no behaviour $(b,f) = (none, id)$. We assume that a do behaviour is an atomic behaviour that can be executed as many times as wished. This is a rather strong assumption in our setting. Furthermore, to simplify our algorithms (and avoid complicated subcases), we make two further assumptions, (i) only simple states can have a do behaviour, and (ii) each state (be simple or composite) always has an entry and an exit behaviour. These two assumptions could be lifted with no difficulty.

## Initial Pseudostates

We require that each region contains one and only one (direct) initial pseudostate, which has one and only one outgoing transition. We also consider that the active state of the system cannot be an initial pseudostate. Note that this is a modelling choice only: if one wants to model an SMD where the system can *stay* in an initial pseudostate, it suffices to add another state between the initial pseudostate and its immediate successor.

## Final States

In each region **r** of a composite state, we allow exactly one final state (denoted by $\mathbf{r}^F$), whereas the specification allows zero or one ("Each region [...] may have its own initial Pseudostate as well as its own FinalState." [10, p.321]) Note that final states are simple states, and thus are included in relation *SubStates*. We define function *ToFinal*(**s**) that returns the set of simple states with transitions to final states of each region in composite state **s**. Similarly, *ToFinal*$^*$(**s**) returns the set of composite states (including **s**) that have transitions with the final states of each region in composite state **s** or its substates.

## Variables

We allow any kind of variables in any behaviour and transition guard. Such variables (integers, lists, etc.) are often met in practice [10].

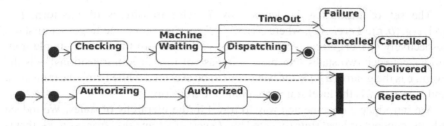

**Fig. 2** Example of transitions taken into account

Transitions

A transition can have a guard, a synchronization event, and a behaviour; transitions can have as source and destination any (composite or simple) state, with some restrictions (e.g. a transition from an initial state cannot have a behaviour, etc.). Due to concurrency, various kinds of transitions exist: completion transitions (e.g. the transition from Machine to Delivered in Fig. 2) have no event and exit a composite state when all its regions are in their final state. Exiting a composite state machine through an event (e.g. the transition from Machine to Cancelled in Fig. 2) results "in the exiting of all substates of the composite State, executing any defined exit Behaviors starting with the innermost States in the active state configuration" [10, Section Transitions, p.325]. This is an exception-like transition. A join transition (e.g. the transition from Checking and Authorizing to Rejected in Fig. 2) exits a given state in each composite state region. All these transitions are taken into account in our work. The "implicit join" (e.g. the transition from Waiting to Failure in Fig. 2) exits some of the regions from a given state; the other regions are exited whatever is their current active state. For readability sake, we do not take them into account in our algorithms, but our general scheme can perfectly adapt to this case.

Concurrency can also appear when entering a composite state (cf. left-most transition in Fig. 2). "If the Transition terminates on the edge of the composite State (i.e. without entering the State), then all the Regions are entered using the default entry rule." [10, Section Entering a State, p.323]. We take into account forks (a transition with a given state in each region as destination), but not "implicit forks" (where only some of the regions have an explicit destination), although our algorithms could be trivially extended, at the cost of more complicated subcases. We also take into account internal/external/local transitions.

Finally, we make the following assumption: the execution of different transitions in different regions of the same state is done in parallel. For instance, the transition from Checking to Waiting in the upper region of Machine in Fig. 2 (and that could involve exit, do and entry behaviours) is performed in an interleaving manner with, e.g. the transition from authorizing to authorized in the lower region. Although this not entirely clear in [10], this assumption might not conform to the notion of run-to-completion step. We believe that interleaving is indeed a natural mechanism between two subsystems executed in parallel.

The set of transitions is denoted by $T$ with transitions of the form $t = (S1, e, g, (b,f), sLevel, S2)$, where $S1, S2 \subseteq S$ are the source and target set of states respectively ($S1$ or $S2$ can contain at least one state – in the case of simple transitions between two states – or more – in case of fork or join transitions), $e$ is the event, $g$ is the guard, $(b,f)$ is the behaviour to be executed while firing the transition, and $sLevel \in S$ is the level state containing the transition.

We introduce a documented way to travel in the hierarchy of states. We indeed add the concept of *level state* (denoted by $sLevel$) of a transition from $s_1$ to $s_2$, that is the innermost state in the hierarchical SMD structure that contains the transition (the level state can be the SMD if $s_1$ and $s_2$ are root states). For example, let us consider that S1 and S2 in Fig. 1 are root states and the transition labelled by "a" between those states is encompassed by the state machine S. Then S is the level state for this transition because it is the innermost common ancestor of S1 and S2 that contains transition "a". Similarly, for the transition (say t) between S13 and S14, we have $sLevel(t) = S1$.

## 2.2 Coloured Petri Nets with Global Variables

CPNs [5] are also a kind of automaton represented by a bipartite graph with two kinds of nodes, places (e.g. $p_1$ in Fig. 3 (left)) and transitions (e.g. $t$ in Fig. 3 (left)). Places hold tokens, possibly of a complex value, and that should be of the place type (e.g. type $\mathbb{N} \times \mathbb{B}$ in Fig. 3 (left)).

**Fig. 3** Global variable notation (left) and corresponding semantics (right)

We use here the concept of *global variables*, a notation that does not add expressive power to CPNs, but renders them more compact. Global variables can be read in guards and updated in transitions, and are supported by some tools (such as CPN Tools). Otherwise, one can simulate a global variable using a "global" place, in which a single token (of the variable type) encodes the current value of the variable. An example is given in Fig. 3 (left). Variable $v$ (of type $\mathbb{N}$) is a global variable updated to the expression $v + i$.

This construction is equivalent to the one in Fig. 3 (right). When a global variable is read in a guard, the token with value $v + i$ is put back in place $p_v$.

The CPN current state (or marking) is the information on which tokens are present in which places. The state evolves when a transition is fired, and tokens are consumed from its source places (according to the input arc expressions) and generated to its target places (according to the output arcs).

# 3   Translation of Concurrent State Machines

We first introduce our general translation scheme (Section 3.1). Then, we define functions used in our algorithms (Section 3.2). We then introduce the algorithms translating states and behaviours (Section 3.3), and transitions (Section 3.4).

## 3.1   General Translation Scheme

In this section, we present the general view of our translation. We define a translation scheme where simple states, final and history pseudostates are translated into places, whereas behaviour expressions (entry, exit and do) and events are translated into CPN transitions. Further places and transitions will also be defined to connect these places and transitions together. The firing of transitions in SMDs is represented by the firing of transitions with tokens in CPNs. Note that composite states will not be translated as such (no CPN place will correspond to a composite state), but their behaviours and all their simple substates will be considered. This is inline with the fact that the active state of an SMD is a (set of) simple state(s). Similarly, initial pseudostates will not be translated as such.

Encoding and Factoring Transitions

A main issue in our translation is to encode transitions between composite states with different regions, in particular the variants of forks and joins. Each such transition (in particular for joins) may in fact correspond to a large number of transitions. For example, in Fig. 1, firing the transition labelled with event "a" corresponds to 18 different ways to leave S1 (3 possible active states in the upper region of S1, multiplied by 4 in S13 and 2 others in the lower region), and hence in 18 CPN transitions, together with all the entry/exit behaviours. Given a transition $t$, let us denote by *combinations*($t$) the function that computes all possible combinations of outgoing substates. We could translate each transition separately; but this would quickly result in an explosion of the number of CPN transitions corresponding to the *same* entry/exit behaviours and actions.

For sake of readability, maintainability, and size of the translated CPN, this should be factored. Hence, as in [2], we shall propose a scheme such that each behaviour is encoded into only one CPN transition. Since many SMD transitions go through the same behaviours, we need a "memory" mechanism to remember from which place we originate so as to find the correct target. Unfortunately, the factoring scheme of [2] does not easily extend to the concurrent case, and further studies showed us that it could lead to a state space explosion when exploring branches that would lead to no end. We reuse the idea of having one unique transition for each behaviour in the SMD, and we define a structure encoding the hierarchy of all entry (resp. exit) behaviours together, in a sort of tree (resp. reversed tree). An example of two such trees will be given in Fig. 5, with the blue part enclosed in a dashed box on the left (resp. turquoise part enclosed in a dotted box on the right) corresponding to the entry (resp. exit) behaviours. Now, in contrast from [2], we will navigate in

these "trees" of behaviours using a memory mechanism, based on guards. For each
possible way to leave a composite state when firing a transition, we will create a
"path" in these trees: this can be easily obtained using typed tokens (e.g. using a
unique integer identifier for each transition), and guards checking the type of the
token along all the CPN transitions on the considered path. Checking a guard is
very cheap when doing model checking, and this mechanism avoids state space explosion, while keeping the CPN small. For example, in Fig. 5, when exiting from
pF11 and pF12, the token will reach tbEn2, and will be duplicated in two tokens
in pbEn2; then, although there are three possible outputs, i.e. tbEn21, tbEn22 and
tbEn23, we expect from our translation that one token fires tbEn21 and the other
one tbEn22, since these transitions correspond to the entry behaviours of the initial
states of the two regions of S2. (For sake of conciseness, we do not depict explicitly
in the figures this mechanism based on guards and tokens.)

Note that, in the case of the 18 combinations corresponding to the transition
labelled with event "a", we still have 18 CPN transitions; but all exit and entry
behaviours are factored, which considerably reduces the resulting CPN size.

Finally, we add a synchronisation CPN transition (e.g. TEnS2_t1 in Fig. 5) just
before entering the places corresponding to the destination simple states, which ensures that only one transition fires at a time in a given region (different transitions
can still fire in an interleaving manner in different regions).

Encoding Shared Variables

In [2], the non-concurrency allowed us to store the value of all shared variables in
a single token that encoded also the current active simple state of the SMD. In the
concurrent case, this approach is no longer possible since several simple states can
be active at the same time; encoding the values of the same shared variables in different tokens would lead to consistency problems. Hence, we use global variables
in CPNs (see Section 2.2) to encode the value of shared variables. Note that concurrent reading/writing of such variables can lead to "strange" executions (when a
region reads a variable and then immediately changes its value, but in between another region concurrently changed its values), but these executions are also possible
in the UML semantics.

## 3.2 Functions

We now describe the functions used in our algorithms. First, we consider two special places $p_{in}$ (resp. $p_{out}$), corresponding to the root of the tree of entry (resp.
exit) behaviours, that will be used by our algorithms. We assume function $p$ associates place $p(s)$ with each state $s$, and place $p(b)$ to each behaviour $b$. We also
assume function $t$ associates CPN transition $t(b)$ with each behaviour $b$. Functions
$SupEN(s)$ and $SupEX(s)$ return the places representing entry behaviour $(p(b^{EN}))$
and exit behaviour $(p(b^{EX}))$ of a given state, respectively. Function $SubEN(s)$ returns the transition encoding the entry behaviour in case of a simple state, or the
entry behaviour of each substate in case of a composite state.

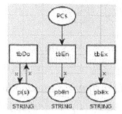

**Fig. 4** A simple state (left) and its translation (right)

In the hierarchy of entry or exit behaviours, we need to know from which behaviour we enter and from which behaviour we leave the state. This depends on the level of the transition. We thus define two functions $OutTo(sLevel, s_1)$ and $InFrom(sLevel, s_2)$ that return the entry and exit behaviour to enter/leave the hierarchy, respectively. Note that $s_1$ of $OutTo(sLevel, s_1)$ (respectively $s_2$ of $InFrom(sLevel, s_2)$) can be a set of states if the transition is a fork transition (respectively a join transition).

$$OutTo(sLevel, s_1) = \begin{cases} p_{out} & \text{When } sLevel \text{ is the overall SMD} \\ SupEX(sLevel)) & \text{otherwise} \end{cases}$$

$$InFrom(sLevel, s_2) = \begin{cases} p_{in} & \text{When } sLevel \text{ is the overall SMD} \\ SupEN(sLevel) & \text{otherwise} \end{cases}$$

We finally assume a function $init^*(s)$ that returns the initial state of $s$, or the (recursive) set of initial states of the initial state if $s$ is composite.

## 3.3 Translating States and Behaviours

We describe here the translation of the states and their behaviours (do, entry and exit). One the one hand, we translate each simple, final and history (pseudo)state into a place. On the other hand, we translate the purely hierarchical structure of the SMD, so that to get a tree of entry and exit behaviours, that will be used later when connecting transitions. Each behaviour expression is represented by a transition and a place connected by an arc. We also connect the places corresponding to simple states with their "do" behaviour, if any. Connecting the entry/exit behaviours with the places corresponding to simple states depends on the transitions and will be done by Algorithm 6.

First, as an example, Fig. 4 shows a state $s$ and its translation into a CPN. State $s$ is represented by place $p(s)$, its entry behaviour $b^{EN}$ is represented by place pbEn and transition tbEn, and its exit behaviour $b^{EX}$ is represented by place pbEx and transition tbEx. If the state has a do behaviour $b^{DO}$ then it will be represented by transition tbDo.

We use in our algorithms the following graphical notation. Places and transitions generated at a given point are denoted by a *solid* line, whereas places and transitions already generated are denoted by a *dotted* line.

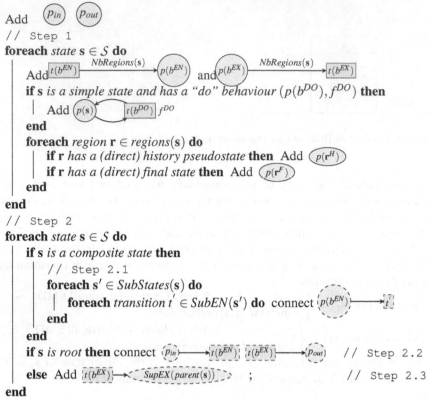

**Algorithm 5.** Encoding the states and behaviours

Algorithm 5 translates states and behaviours. Two places $p_{in}$ and $p_{out}$ are created (line 5), then there are two main steps: Step 1 generates the CPN part corresponding to do/entry/exit behaviours as well as history pseudostates and final states. Step 2 is composed of 3 substeps. Step 2.1 adds an arc between the entry behaviour place of a composite state and the entry behaviour transitions of its substates. This represents the fact that, after executing an entry behaviour, we will execute the entry behaviours of the substates. Step 2.2 adds an arc from place $p_{in}$ to the entry behaviour transitions of the root states, and from the exit behaviours of the root states to place $p_{out}$. Step 2.3 adds an arc from the exit behaviour transition to the exit behaviour place of its parent.

After executing Algorithm 5, we apply the following initialisations to the result of the translation: All guards are initialised to false (and may be later modified during the translation).

Fig. 5 shows the application of Algorithm 5 to a part of the example in Fig. 1. The blue part enclosed in a dashed box on the left of Fig. 5 represents the entry behaviours of S2 in Fig. 1. The dotted/turquoise part on the right of Fig. 5 represents

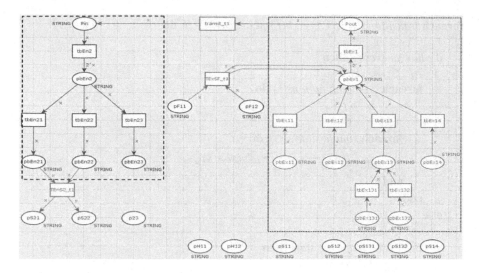

**Fig. 5** Application of Algorithms 5 and 6

the exit behaviours of S1 in Fig. 1. We also added places to represent final states (e.g. pF11), history pseudostates (pH11) and states (e.g. pS11, pS22).

## 3.4 Translating Transitions

Algorithm 6 describes the translation of UML transitions, using three steps followed by an initialisation function. Step 1 deals with exit behaviours: Step 1.1 deals with event triggered transitions, while other transitions are processed in Step 1.2 (exit from a composite state) and 1.3 (exit from a simple state). For a given transition, Step 2 establishes the connection between its source state exit behaviour and its destination state entry behaviour. Step 3 expresses the destination state entry.

Line 6 updates all relevant guards, so as to guide the token within the hierarchy of behaviour places and transitions. Finally, we use a function *Initialisation*() for token initialisation (line 6) to model the initial state in the global SMD.

Fig. 5 shows the application of Algorithms 5 and 6 to S1 and S2 and also the transition without an event in Fig. 1. In this translation, we consider only the entry behaviours of S2, the exit behaviour of S1, the final states of S1, to get a clear picture to illustrate the algorithm. Step 1 adds transition TExSF_t1 between places pF11 and pF12 (that correspond to the region final states) and place pbEx1 that correspond to the exit behaviour of S1. Step 2 adds transition transit_t1 to connect exit place Pout to entry place Pin. Step 3 adds the transition TEnS2_t1 to link places of S2 substates and their entry behaviour.

**foreach** *transition* $t = (S1, e, g, (b,f), sLevel, S2) \in T$ **do**
   // Step 1
   **if** *t has an event e* **then**
      // Step 1.1
      **foreach** $c \in combinations(t)$ **do**
         $V := f(V)$
        Add $\boxed{e}$
        **foreach** *simple state* $\mathbf{s} \in c$ **do**
           Add $(p(\mathbf{s})) \longrightarrow \boxed{e} \longrightarrow (SupEx(\mathbf{s}))$
        **end**
      **end**
   **else**
      // Step 1.2
      **if** $S1$ *is a composite state* **then**
        Add $\boxed{TExS1\_t}$
        **foreach** *state* $\mathbf{s} \in ToFinal(S1)$ **do**
           Add $(p(\mathbf{s})) \longrightarrow \boxed{TExS1F\_t} \longrightarrow (SupEx(S1))$
        **end**
      **else**
        // Step 1.3
        Add $(p(S1)) \longrightarrow \boxed{TExS1\_t} \longrightarrow (SupEx(S1))$
      **end**
   **end**
   // Step 2
   Add $(OutTo(sLevel, S1)) \longrightarrow \boxed{transit\_t} \longrightarrow (InFrom(sLevel, S2))$
   // Step 3
   Add $\boxed{TEnS2\_t}$
   **foreach** *state* $\mathbf{s_2} \in S2$ **do**
      **foreach** *simple* $\mathbf{s} \in init^*(\mathbf{s_2})$ **do**
        Add $(SupEn(\mathbf{s})) \longrightarrow \boxed{TEnS2\_t} \longrightarrow (p(\mathbf{s}))$
      **end**
   **end**
   Update all relevant guards
**end**
*Initialisation*()

**Algorithm 6.** Encoding the transitions

## 4    Conclusion and Future Works

We present here a formalization of UML concurrent SMDs by translating them into CPNs. We take into account different syntactic elements in our translation, specially the concurrency, including simple and composite states, most kinds of transitions (including forks and joins), behaviours (entry, exit, do), history pseudo-states, etc. Once our implementation is completed (see below), this will allow for automated model checking of SMDs.

Our main future work is to develop a tool automating the translation so as to be able to perform formal verification of SMDs. We implemented a first prototype using Acceleo, but this technology turned out to be slightly inaccurate for our framework [1]. Although we could translate toy examples, we will likely build a home-made tool. Of course, this implementation must extend to the general case the assumptions made in order to simplify the description of our algorithms (e.g. each state always has both an entry and an exit behaviours).

Most syntactic aspects not taken into account in our work could be added in a rather straightforward manner – except for timing aspects. Adding them to our translation is an interesting future work. Recall that some tools (including CPN Tools) allow one to defined time(d) coloured Petri nets.

Finally, although it goes beyond of the scope of this paper, a challenging future work would be to formally prove the equivalence between the original SMD and the resulting CPN. Of course, the problem is that the OMG does not formally define a formal semantics for SMDs. However, we could reuse the operational semantics that we recently proposed for SMDs [9], and define a trace equivalence taking into account active states, behaviours and events.

# References

[1] André, E., Benmoussa, M.M., Choppy, C.: Translating UML state machines to coloured Petri nets using Acceleo: A report. In: ESSS. EPTCS (2014)

[2] André, E., Choppy, C., Klai, K.: Formalizing non-concurrent UML state machines using colored Petri nets. ACM SIGSOFT Soft. Eng. Notes 37(4), 1–8 (2012)

[3] Choppy, C., Klai, K., Zidani, H.: Formal verification of UML state diagrams: a Petri net based approach. ACM SIGSOFT Soft. Eng. Notes 36(1), 1–8 (2011)

[4] Dubrovin, J., Junttila, T.A.: Symbolic model checking of hierarchical UML state machines. Technical Report B23, Helsinki University of Technology (2007)

[5] Jensen, K., Kristensen, L.M.: Coloured Petri Nets – Modelling and Validation of Concurrent Systems. Springer (2009)

[6] Jin, Y., Esser, R., Janneck, J.W.: A method for describing the syntax and semantics of UML statecharts. Software and System Modeling 3(2), 150–163 (2004)

[7] Jussila, T., Dubrovin, J., Junttila, T., Latvala, T., Porres, I.: Model checking dynamic and hierarchical UML state machines. In: MDV (2006)

[8] Kerkouche, E., Chaoui, A., Bourennane, E.B., Labbani, O.: A UML and colored Petri nets integrated modeling and analysis approach using graph transformation. Journal of Object Technology 9, 25–43 (2010)

[9] Liu, S., Liu, Y., André, É., Choppy, C., Sun, J., Wadhwa, B., Dong, J.S.: A formal semantics for complete UML state machines with communications. In: Johnsen, E.B., Petre, L. (eds.) IFM 2013. LNCS, vol. 7940, pp. 331–346. Springer, Heidelberg (2013)

[10] OMG. Unified Modeling Language Superstructure, Version 2.5, beta 1(October 2012), http://www.omg.org/spec/UML/2.5/Beta1/PDF/

[11] Snook, C.F., Butler, M.J.: UML-B: Formal modeling and design aided by UML. ACM TSEM 15(1), 92–122 (2006)
[12] Westergaard, M.: CPN tools 4: Multi-formalism and extensibility. In: Colom, J.-M., Desel, J. (eds.) PETRI NETS 2013. LNCS, vol. 7927, pp. 400–409. Springer, Heidelberg (2013)
[13] Zhang, S., Liu, Y.: An automatic approach to model checking UML state machines. In: SSIRI-C, pp. 1–6. IEEE (2010)

# Emotional Facial Expression Analysis in the Time Domain

Thi Duyen Ngo, Thi Chau Ma, and The Duy Bui

**Abstract.** Emotions have been studied for a long time and results show that they play an important role in human cognitive functions. In fact, emotions play an extremely important role during the communication between people. And the human face is the most communicative part of the body for expressing emotions; it is recognized that a link exists between facial activity and emotional states. In order to make computer applications more believable and friendly, giving them the ability to recognize and/or express emotions are research fields which have been much focused on. Being able to perform these tasks, firstly, we need to have knowledge about the relationship between emotion and facial activity. Up to now, there have been proposed researches on this relationship. However, almost all these researches focused on analyzing the relationship without taking into account time factors. They analyzed the relationship but did not examined it in the time domain. In this paper, we propose a work on analyzing the relationship between emotions and facial activity in the time domain. Our goal is finding the temporal patterns of facial activity of six basic emotions (happy, sad, angry, fear, surprise, disgust). To perform this task, we analyzed a spontaneous video database in order to consider how facial activities which are related to the six basic emotions happen temporally. From there, we bring out the general temporal patterns for facial expressions of each of the six emotions.

**Keywords:** emotional facial expression, analysis, time domain, temporal pattern, FACS.

## 1 Introduction

One particularity of humans is to have emotions, this makes people different from all other animals. Emotions have been studied for a long time and results show that

Thi Duyen Ngo · Thi Chau Ma · The Duy Bui
University of Engineering and Technology, Vietnam National University, Hanoi,
Hanoi, Vietnam
e-mail: {duyennt,chaumt,duybt}@vnu.edu.vn

© Springer International Publishing Switzerland 2015                                     487
V.-H. Nguyen et al. (eds.), *Knowledge and Systems Engineering,*
Advances in Intelligent Systems and Computing 326, DOI: 10.1007/978-3-319-11680-8_39

they play an important role in human cognitive functions. Picard has summarized this in her "Affective Computing" [19]. In fact, emotions play an extremely important role during the communication between people. People usually assesses others' emotional states, probably because of their good indication of how the person feels, what the person could do next, and how he is about to act. For this assessment, the human face is the most communicative part of the body for expressing emotions [7]. It is recognized that a link exists between facial activity and emotional states. This is asserted in Darwin's pioneer publication "The expression of the emotions in man and animals" [4].

Recognizing the importance of emotions to human cognitive functions, Picard [19] concluded that if we want computers to be genuinely intelligent, to adapt to us, and to interact naturally with us, then they will need the ability to recognize and express emotions, to model emotions, and to show what has come to be called "emotional intelligence". In order to make computer applications more believable and friendly, giving them the ability to recognize and/or express emotions are research fields which have been much focused on. As mentioned above, the human face is the most communicative part of the body for expressing emotions. Therefore, being able to provide computer applications with the ability to recognize and/or express emotions, firstly, we need to have knowledge about the relationship between emotion and facial activity. Up to now, there have been proposed researches on this relationship(e.g., [8, 12, 9, 5, 8, 22, 17]). However, almost all these researches focused on analyzing the relationship without taking into account time factors. They analyzed the relationship but did not examined it in the time domain.

In this paper, we propose a work on analyzing the relationship between emotions and facial activities in the time domain. We focus on an analysis of how emotional facial activities vary over time. Our goal is finding the temporal patterns of facial activities of six basic emotions (happy, sad, angry, fear, surprise, disgust). To perform this task, we analyze a spontaneous video database using facial expression recognition techniques. Our hypothesis is that the facial expressions happen in series with decreasing intensity when a corresponding emotion is triggered. For example, when an event happens, which triggers the happiness of a person, he/she would not smile in full intensity during the time the happiness lasts. Instead, he/she would express a series of smiles in decreasing intensity. In order to verify our hypothesis, movements of features on the face are detected and matched automatically with predefined patterns.

The rest of the paper is organized as follows. Section 2 presents a summary on related work. Then, in Section 3, we describe our facial expression analysis process. After that, Section 4 shows analysis results and evaluation. Finally, Section 5 shows conclusion.

## 2 Related Works

We first review psychological works on the relationship between emotion and facial activity. Most of the research on this relationship follows one of three main views:

the basic emotions view, the cognitive view and the dimensions view. The basic emotions view (e.g., [23, 7, 6, 13] is the most popular one which assumes that there is a small set of emotions that can be distinguished discretely from one another by facial expressions. In supporting the basic emotions view, there is considerable evidence (see [6]) showing that distinct prototypical facial signals corresponding to the six basic emotions of happiness, sadness, surprise, disgust, anger and fear can be reliably recognized across a variety of cultures. The cognitive view on emotional facial expressions was proposed by researchers working on emotions from the cognitive perspective such as Arnold [2] and Scherer [21]. The cognitive perspective assumes that emotions are triggered by a cognitive evaluation/appraisal process of an individual's situation. Different from the basic emotions view's assumption that facial actions are produced in patterns because a specific emotion has been triggered, promoters of appraisal theory assume that the outcomes of the appraisals are associated with changes in activity in many systems of the body, including the face. The dimensions view was proposed by researchers who believe that emotional states are fundamentally differentiated on a small number of dimensions, and that facial activity is linked to these dimensions (e.g., [20]). It assumes that the basic dimensions of an underlying emotional state are reflected in facial behavior.

There are everlasting controversies among three views on emotional facial expressions. However, although each view on emotional facial expressions has its own predictions, it is not impossible that there are common ideas among them. According to [16], the relationship of the appraisal components to basic dimensions and to discrete emotions seems to be complicated but highly plausible. For example, Arnold [2] proposed the valence dimension in the cognitive view, which is clearly related to the basic valence dimension found in all dimensional approaches. It seems that no researchers from any view can provide evidence to fully defend their hypothesis. Nevertheless, the psychological studies from these views have a significant effect on our understanding of the link between emotion state and facial activity. These studies also play a very important role in the task of simulating and recognizing emotional facial expressions on computers. According to Kappas [14], the basic emotions view is most useful in the context of diagnosing emotions from facial actions. Compared to research within the cognitive view and the dimensions view, research within the basic emotions view provides more empirical evidence on the relationship between emotion and facial activity. Moreover, the predictions of the basic emotions view are usually so clear to confirm or reject. In comparison, many predictions of the appraisal view and the dimensions view are not specific enough. In our opinion, the results from research within the basic emotions view are most useful in simulating the relationship between emotion and facial activity.

In order to objectively capture the richness and complexity of facial expressions, behavioral scientists found it necessary to develop objective coding standards. The Facial Action Coding System (FACS) [8] is one of the most widely used expression coding system in the behavioral sciences. FACS was developed by Ekman and Friesen to identify all possible visually distinguishable facial movements; it makes for a clear, compact representation of the muscle activation of a facial expression. FACS involves identifying the various facial muscles that individually or in groups cause changes in facial behaviors. These changes in the face and the underlying (one or more) muscles that caused these changes are called Action Units (AU). AUs are elementary movements and can be regarded as the "phonemes" of facial expressions. The FACS is made up of several such action units. Related to the relationship between emotions and facial activity, each AU codes the fundamental actions of individual or groups of muscles typically seen while producing facial expressions of emotion. For example, AU 4 defines the contraction of two muscles resulting in the lowering of the eyebrows. This AU is typically observed in expressions of sadness, fear, and anger. FACS provides an objective and comprehensive language for describing facial expressions and relating them back to what is known about their meaning from the behavioral science literature.

Up to now, there have been quite many proposed researches followed the basic emotion view to simulate the relationship between emotion and facial activity. There have been also proposed works on analyzing and recognizing facial expressions. Extensive research in [8] showed that certain combinations of action units are linked to the six "universal" facial patterns of the emotions anger, disgust, fear, sadness, surprise, and happiness. EMFACS (Emotional Facial Action Coding System) [12] was proposed by Friesen and Ekman, which is similar to FACS but considers only emotion-related facial actions. Ekman and Hager [9] also presented a database called facial action coding system affect interpretation database (FACSAID), which allows to translate emotion related FACS scores into affective meanings. Emotion interpretations were provided by several experts, but only agreed affects were included in the database. In [5], all images in a database which consisted of pictures in neutral, six basic, and fifteen compound emotions were FACS coded in order to analysis the relationship between emotions and facial activity. The AU analysis results of the six basic emotion were consistent with those given in ref. [8]. Besides, the research also listed the AUs for each of the fifteen compound emotion categories. Tian et al. have developed the Automatic Face Analysis (AFA) system which can automatically recognize six upper face AUs and ten lower face AUs [22]. This helped in determining the expression and in turn the emotional state of the person.

In simulating the emotion-facial activity relationship and analyzing, recognizing facial expressions, most of the proposed works attempt at dealing with basic emotions and some attempts at dealing with non- basic emotions. However there have been very few attempts at considering the temporal dynamics of the face. Temporal dynamics refers to the timing and duration of facial activities. The important terms that are used in connection with temporal dynamics are: onset, apex and offset [10]. These are known as the temporal segments of an expression. Onset is the instance when the facial expression starts to show up, apex is the point when the expression is at its peak and offset is when the expression fades away (start-of-offset is when the expression starts to fade and the end-of-offset is when the expression completely fades out). Similarly onset-time is defined as the time taken from start to the peak, apex-time is defined as the total time at the peak and offset-time is defined as the total time from peak to the stop. Pantic and Patras have reported successful recognition of facial AUs and their temporal segments [17]. By doing so, they have been able to recognize a much larger range of expressions. However, this research only analyzing single AUs, it did not mention to facial expression patterns for emotions in the time domain.

## 3  Emotional Facial Activity Analysis

### 3.1  Database

It is the fact that expressions can be posed or spontaneous. Posed expressions are the artificial expressions that a subject will produce when he or she is asked to do so. This is usually the case when the subject is under normal test condition or under observation in a laboratory. In contrast, spontaneous expressions are the ones that people give out spontaneously. These are the expressions that we see on a day to day basis, while having conversations, watching movies etc. Many psychologists have proved that posed expressions are different from spontaneous expressions, in terms of their appearance, timing and temporal characteristics [10]. Furthermore, it also turns out that many of the posed expressions that researchers have used in their works are highly exaggerated.

In our work, we use a spontaneous facial expression database which consist of video sequence selected from three databases namely MMI [1, 18], FEED-TUM [25], DISFA [15]. MMI Facial Expression Database [1, 18] contains posed and spontaneous expressions but only spontaneous ones being used in our research. These expressions belong to part IV and part V of the MMI database. These two parts consist of video sequences that express six basic emotions. These video sequences were collected through experiments in which researchers showed the participants images, videos, short clips of cartoons and comedy shows, or sound of the stimuli to induce emotions. The FEEDTUM database [25] consisting of elicited spontaneous emotions of 18 subjects. Besides the neutral state the content of the database covers the emotions anger, disgust, fear, happiness, sadness and surprise for each contained subject, which have been recorded three times. To elicit the emotions as natural as possible it was decided to play several carefully selected stimuli videos and record the participants' reactions. In the DISFA database [15], twenty-seven young adults were video recorded by a stereo camera while they viewed video clips intended to elicit spontaneous emotion expression.

To form the database used in our work, from the three above databases, video sequences in which the human face begins with a neutral expression, proceeds to a peak expression which is fully FACS coded, and then gets back to the neutral state were selected. Because our goal is finding the temporal "pattern" of facial activity of emotion, such selected video sequences will be suitable to be used in the analyzing process. These videos are arranged into six categories according to the six emotions they belong to.

## 3.2    Facial Activity Analysis Process

Our emotional facial activity analysis process is illustrated in Figure 1. The input of the system is a video sequence which is processed by the system frame by frame. For each frame, the Face Detector detects the face and returns its location. Then the ASM Fitting perform fitting task and returns ASM shape of the face. From this shape, Face Normalization module carries out the normalizing task in order to change the shape to the common size. Finally, the AUs Intensity Extractor module extract AUs intensity related to each of six basic emotions using feature points obtained from ASM Fitting and then Face Normalization modules. The construction and detail operation of the four modules are presented in the following.

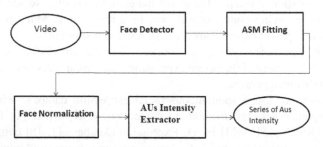

**Fig. 1** Block diagram of the emotional facial activity analyzing system

### A. Face Detector

We start by detecting the face inside the scene. For each frame of the input video, the Face Detector module check whether or not is there a human face, and provide the approximate location and size of the detected face. In this work, we have selected Viola Jones algorithm [24] for face detection because it is known to work robustly for large inter-subject variations and illumination changes. The result of face detection algorithm is illustrated in Figure 2(a).

### B. ASM Fitting

This module extracts feature points from the detected face using ASM fitting algorithm. Within the face region passed from the Face Detector, we search the exact location of facial landmarks using Active Shape Model [3]. Among the various types of deformable face models, we use Active Shape Model (ASM) for several reasons. ASM is arguably the simplest and fastest method among deformable models, which fit our need to track a large number of frames in multiple videos. Furthermore, ASM is also known to generalize well to new subjects due to its simplicity. We trained

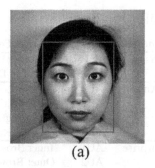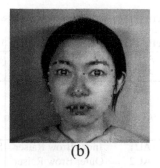

(a) (b)

**Fig. 2** (a)Face Detection; (b) Facial landmarks

ASM with manually collected 68 landmark locations from a set of still images. The ASM method detects facial landmarks through a local-based search constrained by a global shape model which statistically learned from training data. The output of ASM Fitting module is location of 68 feature points (facial landmarks)(we call this as ASM shape) as illustrated in Figure 2(b).

C. Face Normalization

Due to variations in head pose and/or camera positions, the face size in frames of the same video sequence maybe not same. Since then the ASM shapes of these frame are also not at the same size. This may lead to the less accuracy in analysis results. So it needs performing normalization task in order to set all the shapes into a common size. In our work, we use the distance between the centers of eyes for normalization. All the shapes will be normalized so that in their normalized reproductions the distance between the centers of eyes equal to that in the other ASM shapes.

D. AUs Intensity Extractor

This component extracts facial features related to each of six basic emotions using feature points (facial landmarks) obtained from ASM Fitting and then Face Normalization modules. It uses normalized landmark locations to calculate the intensity of Action Units (AUs) which are related to the emotion style of the input video. We follow the basic emotion view and base on research of Ekman [8], Shichuan [5] about the link between combinations of action units and the six universal facial patterns of the emotions anger, disgust, fear, sadness, surprise, and happiness. For each emotion, there is a set of related AUs to classify it from the others, as showed in the table 1.

**Table 1** Description of six basic emotions ( [8, 5])

| Emotion | Action Unit | Facial Feature | Emotion | Action Unit | Facial Feature |
|---------|-------------|----------------|---------|-------------|----------------|
| Happiness | AU12 | Lip Corner Puller | Disgust | AU15 | Lip Corner Depressor |
|  | AU25 | Lips Part |  | AU16 | Lower Lip Depressor |
| Sadness | AU1 | Inner Brow Raiser | Anger | AU4 | Brow Lowerer |
|  | AU4 | Brow Lowerer |  | AU5 | Upper Lid Raiser |
|  | AU15 | Lip Corner Depressor |  | AU7 | Lid Tightener |
|  |  |  |  | AU17 | Chin Raiser |
| Fear | AU1 | Inner Brow Raiser | Surprise | AU1 | Inner Brow Raiser |
|  | AU2 | Outer Brow Raiser |  | AU2 | Outer Brow Raiser |
|  | AU4 | Brow Lowerer |  | AU5 | Upper Lid Raiser |
|  | AU5 | Upper Lid Raiser |  | AU25 | Lips Part |
|  | AU20 | Lip Stretcher |  | AU26 | Jaw Drop |
|  | AU25 | Lips Part |  |  |  |
|  | AU26 | Jaw Drop |  |  |  |

## 4   Results of Analysis

For a video of each emotion, from the landmark locations, the intensity of each related AU is calculated frame by frame. As a result, we have a temporal series of intensity values for each AU. This series is extracted and graphing. Finally, series and graphic results of all AUs from all videos of one emotional style are used to generalize the temporal pattern for facial expressions of that emotion. By observing these graphics, we bring out a hypothesis that the facial expressions happen in series with decreasing intensity when a corresponding emotion is triggered. Thence, we proposed pre-defined temporal patterns for facial expressions of six basic emotions. The temporal pattern for facial expressions of the happiness and the sadness is depicted in Figure 3(a) and the temporal pattern for facial expressions of the disgust, angry, fear, and surprise emotions is depicted in Figure 3(b).

In this pattern, we see that there are solid line part and dash line part. The difference between these two parts is that the solid line part is always present while the dash line part may be absent. This can be explained as follows. The internal

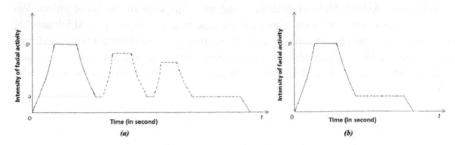

**Fig. 3** (a): Temporal pattern for facial expressions of happiness and sadness. (b): Temporal pattern for facial expressions of fear, angry, disgust, and surprise emotions.

emotional states are the cause of the appearance of external facial activities which occur in order to produce facial expressions expressing that emotion. When the internal emotional states with sufficient intensity take place in a duration, it will lead to emergence of the external facial activities and then facial expressions in this duration. If this duration is short, the facial expressions also appear in short time; then only the solid line part in the pattern appears. Conversely, if this duration is long, the facial expressions also appear in long time. Then in addition to the solid line part, the dash line part in the pattern also appears. As shown in the pattern, although the internal emotional state may have constant sufficient intensity in a long time, the corresponding facial expressions are not always at the same intensity in this long duration. On the other hand, the facial expressions appear with the intensity corresponding to the intensity of the emotion, then stay in this state for a while, and then fall near the initial state. We call this process is a cycle. With happiness and sadness, this cycle repeats several times with decreasing intensity, then the facial expressions are kept at a constant low intensity until the end of the long duration. With four remaining emotions, the facial expressions often occur in only one cycle and then the facial expressions are kept at a constant low intensity.

We define a cycle of facial expressions as:

$$E = (P, Ts, Te, Do, Dr)$$

where $P$ defines the target intensity of the expressions; $Ts$ and $Te$ are the starting time and the ending time of the cycle; $Do, Dr$ are onset duration (determines how long the facial movement takes to appear) and offset duration (determines how long the facial movement takes to disappear) of the expressions, respectively. The process in which the expressions occur in a cycle is described as a function of time:

$$F_e(t) = \begin{cases} P.\phi_+(t - Ts, Do) & \text{if } (Ts < t < Ts + Do) \\ P & \text{if } (Ts + Do \le t \le Te - Dr) \\ P.\phi_-(t - Te + Dr, Dr) & \text{if } Te - Dr < t < Te \end{cases}$$

where $\phi_+$ and $\phi_-$ are the functions that describe the onset and offset phase of expressions. We follow Essa's work [11] to use exponential curves to fit the onset and offset portions of expressions. A function of the form $(e^{bx} - 1)$ is suggested for the onset portion, while a function of the form $(e^{c-dx} - 1)$ is suggested for the offset portion. Basing on the suggested functions, we derive two functions for the onset and offset portions. For the onset portion, we want to choose b so that:

$$\phi_+(0, Do) = e^{b.0} - 1 = 0$$
$$\phi_+(Do, Do) = e^{b.Do} - 1 = 1$$

From the second equation, we obtain:

$$e^{b.Do} = 2$$

and so

$$b = \frac{ln2}{Do}$$

Replacing b with the obtained value, the derived function to describe the onset portion is defined as:

$$\phi_+(x, Do) = exp(\tfrac{ln2}{Do}x) - 1$$

For the offset portion, we want to choose c and d so that:

$$\phi_-(0, Dr) = e^{c-d.0} - 1 = 1$$
$$\phi_-(Dr, Dr) = e^{c-d.Dr} - 1 = \tfrac{a}{P}$$

From the first equation, we obtain:

$$e^c = 2, \; c = ln2$$

Replacing c with the obtained value to the second equation, we can infer:

$$e^{ln2-d.Dr} - 1 = \tfrac{a}{P}$$

and so

$$d = \frac{ln2 - ln(\tfrac{a}{P} + 1)}{Dr}$$

Replacing $c$ and $d$ with the obtained value, the derived function to describe the offset portion of a parameter activity is defined as

$$\phi_-(x, Dr) = exp(ln2 - \frac{ln2 - ln(\tfrac{a}{P} + 1)}{Dr}x) - 1$$

In order to verify the reasonableness of the pre-defined temporal patterns, we have performed the fitting task for all temporal AU profiles. Figure 4 shows an example of the temporal AU25 profiles of a representative subject in the surprise

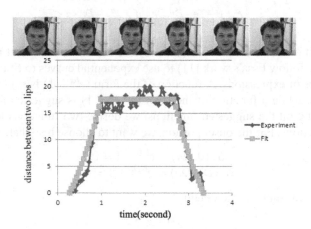

**Fig. 4** Experiment and fitting temporal AU25 profiles of a subject in surprise emotion. TOP: Captured video frames at several time points from the whole video. BOTTOM: The temporal profiles show the distance between two lips which characterizes the intensity of AU25.

emotion. The subject displayed gradual increase of Lips Part (AU25) which is typical of a surprise expression. In the figure, the top part depicts captured video frames at several time points from the whole video. In the bottom part of the figure, the temporal profiles show the distance between two lips which characterizes the intensity of AU25. The darker points and darker line represent data obtained from experiment analysis. The paler points and paler line show fitting data using the pattern and function described above. If the distance between the centers of two eyes is normalized to 1, the sum of squares due to error (SSE) of the fit is 0.0207. Performing the fitting task for all temporal AU profiles, we found that the average of the sum of squares due to error (SSE) was 0.055 with the standard deviation was 0.078. These values show that the above temporal patterns and the fitting function are reasonable.

Our analysis results showed that in the happiness, the average duration of a cycle is about 3.5 seconds. It is usually not less than 1.5 seconds and not more than 6 seconds. In the sadness, the average duration of a cycle is about 5.3 seconds, and it is usually not less than 2 seconds and not more than 7 seconds. The analysis results also found that the average duration of a cycle for disgust emotion is about 3.6 seconds; the average duration of a cycle for angry and fear emotions is about 3 seconds; the average duration of a cycle for surprise emotion is about 2.7 seconds.

## 5 Conclusion

This paper has presented the work on analyzing the relationship between emotions and facial activity in the time domain. We analyzed a spontaneous video database in order to consider how facial activities which are related to the six basic emotions happen temporally. For each video frame, the Face Detector detects the face and returns it's location. Then the ASM Fitting perform fitting task and returns ASM shape of the face. From this shape, Face Normalization module carries out the normalizing task in order to change the shape to the common size. Finally, the AUs Intensity Extractor module extracts AUs intensity related the video's emotional style using feature points obtained from ASM Fitting and then Face Normalization modules. For each video of each emotional style, there is a temporal series of intensity values for each AU. This series was extracted and graphing. By observing these graphics, we proposed pre-defined temporal patterns for facial expressions of six basic emotions. We then verified the reasonableness of these temporal patterns by performing the fitting task for all temporal AU profiles. The results show that the pre-defined temporal patterns and the fitting function are reasonable.

## References

[1] http://mmifacedb.eu/
[2] Arnold, M.B.: Emotion and personality. Psychological aspects, vol. 1(2). Columbia University Press, New York (1960)
[3] Cootes, T.F., Taylor, C.J., Cooper, D.H., Graham, J.: Active shape models-their training and application. Computer Vision and Image Understanding 61(1), 38–59 (1995)

[4] Darwin, C.: The expression of the emotions in man and animals. Univerity of Chicago Press, Chicago (1872/1965)

[5] Du, S., Tao, Y., Martinez, A.M.: Compound facial expressions of emotion. In: Heeger, D.J. (ed.) Proceedings of the National Academy of Sciences, New York University, New York (2014)

[6] Ekman, P.: Emotion in the human face. Cambridge University Press, Cambridge (1982)

[7] Ekman, P., Friesen, W.V.: Unmasking the face: A guide to recognizing emotions from facial clues. Prentice-Hall, Englewood Cliffs (1975)

[8] Ekman, P., Friesen, W.V.: The facial action coding system: A technique for the measurement of facial movement. Consulting Psychologists Press, Inc, San Francisco (1978)

[9] Ekman, P., Hager, J.: Facial action coding system affect interpretation database (facsaid), http://face-and-emotion.com/ dataface/facsaid/description.jsp

[10] Ekman, P., Rosenberg, E.L.: What the face reveals: basic and applied studies of spontaneous expression using the facial action coding system (FACS), Illustrated Edition. Oxford University Press (1997)

[11] Essa, I.A.: Analysis, Interpretation, and Synthesis of Facial Expressions. PhD thesis, Massachusetts Institute of Technology, Cambridge, MA (1994)

[12] Friesen, W., Ekman, P.: EMFACS-7: Emotional Facial Action Coding System. Unpublished manual, University of California, California (1983)

[13] Izard, C.: Emotions and facial expressions: A perspective from differential emotions theory. In: Russell, J., Fernandez-Dols, J. (eds.) The Psychology of Facial Expression, Maison des Sciences de l'Homme and Cambridge University Press (1997)

[14] Kappas, A.: What facial Activity can and cannot tell us about emotions. In: Katsikitis, M. (ed.) The Human Face: Measurement and Meaning, pp. 215–234. Kluwer Academic Publishers, Dordrecht (2003)

[15] Mavadati, S.M., Mohammad, H.M., Bartlett, K., Trinh, P., Cohn, J.F.: Disfa: A spontaneous facial action intensity database. IEEE Transactions on Affective Computing 4(2), 151–160 (2013)

[16] Ortony, A., Clore, G.L., Collins, A.: The Cognitive Structure of Emotions. Cambridge University Press, Cambridge (1988)

[17] Pantic, M., Patras, I.: Detecting facial actions and their temporal segments in nearly frontal-view face image sequences. In: Proc. IEEE conf. Systems, Man and Cybernetics, vol. 4, pp. 3358–3363 (2005)

[18] Pantic, M., Valstar, M.F., Rademaker, R., Maat, L.: Web-based database for facial expression analysis. In: Proc. 13th ACM Intl Conf. Multimedia, pp. 317–321 (2005)

[19] Picard, R.: Affective Computing. MIT Press, Cambridge (1997)

[20] Russell, J.A.: Reading emotions from and into faces: Resurrecting a dimensional-contextual perspective. In: Russell, J.A., Fernndez-Dols, J.M. (eds.) The Psychology of Facial Expression, Cambridge University Press, New York (1997)

[21] Scherer, K.R.: What does facial expression express? In: Strongman, K. (ed.) International Review of Studies on Emotion, vol. 2, Wiley, Chichester (1992)

[22] Tian, Y., Kanade, T., Cohn, J.: Recognizing action units for facial expression analysis. IEEE Trans. Pattern Analysis and Machine Intelligence 23(2), 97–115 (2001)

[23] Tomkins, S.S.: Affect, Imagery, Consciousness. The Positive Affects, vol. 1. Springer, New York (1962)

[24] Viola, P., Jones, M.: Robust real-time object detection, tech. rep.,Cambridge Research Laboratory Technical report series (2) (2001)

[25] Wallhoff, F.: The facial expressions and emotions database homepage (feedtum), www.mmk.ei.tum.de/~waf/fgnet/feedtum.html

# An Efficient Method for Automated Generating Models of Component-Based Software

Hoang-Viet Tran, Chi-Luan Le, Quang-Trung Nguyen, and Pham Ngoc Hung

**Abstract.** This paper proposes an efficient method for automated generating models of component-based software. This method accepts regular expressions that describe behaviors of software components. The proposed method uses the CNNFA algorithm to parse the regular expressions in order to generate corresponding models. This method can generate minimal accurate models of the software components. A tool is implemented and applied for some typical systems to show the efficiency of this method. The obtained experimental results show that this method is faster than existing methods. The generated models play an important role in making such model-based approaches as model checking and model-based testing more feasible in practice.

## 1 Introduction

Component-based software (CBS) development plays an important role in software industry thanks to their solid advantages. However, it is very hard for a single organization to fully develop large scale software. In fact, organizations usually focus only on developing the main business parts of systems. For other parts like user interfaces, difficult algorithms or common libraries, large system interactive libraries, etc. can be purchased from third parties. Moreover, one of the key issues for developing such systems is how to ensure that all components from different providers can cooperate correctly in order to obtain the goal of systems. Furthermore, software components always evolve along software life cycles. Consequently, software companies must pay high costs for rechecking of the evolved systems in the context of software evolution.

Hoang-Viet Tran · Chi-Luan Le · Pham Ngoc Hung
VNU University of Engineering and Technology, Hanoi, Vietnam

Chi-Luan Le
University of Transport Technology, Hanoi, Vietnam

Quang-Trung Nguyen
Vietnam University of Commerce, Hanoi, Vietnam

© Springer International Publishing Switzerland 2015       499
V.-H. Nguyen et al. (eds.), *Knowledge and Systems Engineering,*
Advances in Intelligent Systems and Computing 326, DOI: 10.1007/978-3-319-11680-8_40

There are many model-based approaches such as model checking and model-based testing have been known to solve the above issues. In order to apply these approaches, we must have models that exactly describe the behaviors of systems under checking. However, these approaches generally assume that the mechanism for obtaining models of the components are available and accurate. It means that the component models are available and accurate. In fact, these assumptions may not always hold in practice due to the frequent lack of information. Therefore, we need a method for generating models of component-based systems.

Currently, there are several methods for generating software models. The method presented in [11] generates the component models from sets of traces resulting from doing experiment on components bases on the Thompson algorithm [1]. The method in [10] generates Extended Finite State Machines (EFSMs) from interactive traces. The work proposed in [8] presents a method that generate finite state models from source code of software program. The method [9] generates models for software environment bases on an assumption that user will provide the information of that environment. The method mentioned in the [12] learns the components specification which are regular expressions and generate corresponding models. This method uses the VC algorithm [3] to answer the membership queries of $L^*$ and check the conformation of the candidate proposed by $L^*$ algorithm. The above methods have several disadvantages so that they are difficult to be applied in practice.

This paper proposes an efficient method to generate component-based software models. The key idea of this method is to apply an algorithm named CNNFA [5, 6] in order to analyze specifications of components which are in form of regular expressions. This is the same context as in [12]. After CNNFA returns non-deterministic finite automata (NFA) models, this method applies some optimization algorithms in order to obtain minimal models represented by deterministic finite automata (DFA) that exactly describe behaviors of the corresponding components. CNNFA has been known as a more efficient algorithm than both of $L^*$ and Thompson algorithms in both terms of time and space theoretically. The generated models are used as inputs of model-based approaches in improving software quality.

The rest of this paper is organized as follow. At first, we present some backgrounds in the section 2. After that, we propose an efficient method for generating models of CBS in section 3. Section 4 shows techniques for optimizing the generated models. The experimental results are shown in section 5. Finally, we conclude the paper in section 6.

## 2 Background

In this section, we present some basic concepts used in this paper as follow.

**Component Specification:** In this paper, we use a regular expression to describe the behaviors of a software component. This regular expression can be produced by design activities as described in [12] or can be associated with a component provided by a third party as the component specification. Whenever we have such regular expressions, we can use the CNNFA algorithm described in [5] and [6] to generate the needed software models.

**Definition 1 (Regular expression).** *Regular expressions are used to represent regular languages. Let $\Sigma$ be an alphabet, the regular expression over $\Sigma$ is defined inductively as follow.*

- *$\lambda, a \in \Sigma$ are regular expressions corresponding to regular languages $\emptyset, \{a\}$.*
- *If $r$ and $s$ are the regular expressions corresponding to regular languages $L_r$ and $L_s$, then $(r|s), (r.s), (r^*)$ are regular expressions corresponding to regular languages $L_r \cup L_s$, $L_r.L_s$, $L_r^*$.*
- *Nothing is a regular expression except the above two clauses.*

**Definition 2 (Length of a regular expression).** *The length of a regular expression $R$, denoted $|R|$, is the number of symbol occurrences in $R$ including $\lambda$, alphabet symbols, parentheses, star, and alternation operator.*

**Note 1.** *The product operator is sometimes implicit. In that case, it is not counted to the length of a regular expression.*

**Definition 3 (Basic blocks and non-basic blocks).** *A Basic block is a valid subexpression of $R$ that contains at least one alphabet symbol occurrence. Non-basic blocks are regions of expression $R$ separated out by basic blocks.*

**Component Model:** In this paper, the models of software components describing the components behaviors are represented by using finite automata. Let $\mathcal{A}ct$ be the set of observable actions and let $\varepsilon$ denote a local action unobservable to a component's environment.

**Definition 4 (Finite automata).** *A finite automaton $M$ representing a component model is a five-tuple $\langle Q, \Sigma, \delta, q_0, F \rangle$, where:*

- $Q$ is a non-empty set of states,
- $\Sigma \subseteq \mathcal{A}ct$ is a finite set of observable actions called the alphabet of M,
- $\delta \subseteq Q \times \Sigma \cup \{\varepsilon\} \times Q$ is a transition relation,
- $q_0 \in Q$ is the initial state, and
- $F \in Q$ is a set of accepting states.

**Definition 5 (Deterministic and Non-Deterministic Finite automata).** *A finite automaton $M = \langle Q, \Sigma, \delta, q_0, F \rangle$ is a Non-Deterministic Finite automaton (NFA) if it contains $\varepsilon - transition$ or if $\exists (q, a, q'), (q, a, q'') \in \delta$ such that $q' \neq q''$. Otherwise, it is called a Deterministic Finite automaton (DFA). In case of NFA, if it contains $\varepsilon - transition$ then it is called $\varepsilon - NFA$.*

**Definition 6 (Size of finite automata).** *Given a finite automaton $M$, size of $M$, denoted $|M|$, is the number of states in $Q$ (i.e., $|M| = |Q|$).*

**Traces.** A trace $\sigma$ of a finite automaton M is a sequence of observable actions that will put the automaton in the accepting state at the end of the sequence starting from the initial state.

**Definition 7 (Trace).** *A trace* $\sigma$ *of an automaton* $M = \langle Q, \Sigma, \delta, q_0, F \rangle$ *is a finite sequence of actions* $a_1 a_2 ... a_n$, *such that there exists a sequence of states starting at the initial state (i.e.,* $q_0, q_1, ..., q_n$*) such that for* $1 \leq i \leq n$, $(q_{i-1}, a_i, q_i) \in \delta$ *and* $q_n \in F$.

**Note 2.** *The length of a trace* $\sigma$ *is the number of actions of* $\sigma$ *denoted* $|\sigma|$. *The set of all traces of* $M$ *is called the language of* $M$, *denoted* $L(M)$.

**Definition 8 (Equivalence between DFAs).** *Two DFAs* $M$ *and* $M'$ *are equivalent if and only if* $L(M) = L(M')$.

**Definition 9 (Minimal DFA).** *A DFA* $M = \langle Q, \Sigma, \delta, q_0, F \rangle$ *is called minimal if and only if there is no equivalent DFA* $M' = \langle Q', \Sigma', \delta', q_0', F' \rangle$ *such that* $|Q'| < |Q|$.

## 3   Model Generation

In this section, we first review some concepts and data structure related to the generation process. After that, we present equations that allow us to calculate CNNFA representations for subexpressions: $\lambda, a \in \Sigma, J|K, J.K$, and $J^*$. Then, we present the regular expression parsing algorithm that uses the previous equations to generate the CNNFA that is equivalent to R. At last, we generate the needed models from its CNNFA.

### 3.1   Related Concepts

**Definition 10 (NNFA).** *A normal NFA (NNFA) is an NFA in which all edges leading into the same state have the same label. We can represent an NNFA as a 6-tuple* $\langle \Sigma, Q, \delta, I, F, A \rangle$, *where:*

- $\Sigma$ *is an alphabet,*
- $Q$ *is a set of states,*
- $\delta \subseteq Q \times Q$ *is a set of (unlabeled) edges,*
- $I \subseteq Q$ *is a set of initial states,*
- $F \subseteq Q$ *is a set of final states, and*
- $A : Q \to \Sigma$ *maps states* $q \in Q$ *into labels* $A(q)$ *belonging to alphabet* $\Sigma$.

**Note 3.** *A McNaughton/Yamada NNFA (MYNNFA) is an NNFA with only one initial state with zero in-degree.*

**Definition 11 (The tail of an MYNNFA).** *The tail of an MYNNFA* $M = \langle \Sigma, Q, \delta, I = \{q_0\}, F, A \rangle$ *is an NNFA* $M^T = \langle \Sigma^T, Q^T, \delta^T, I^T, F^T, A^T \rangle$, *where* $\Sigma^T = \Sigma$, $Q^T = Q - \{q_0\}$, $\delta^T = \{[x, y] \in \delta | x \neq q_0\}$, $I^T = \{y : [q_0, y] \in \delta\}$, $F^T = F - \{q_0\}$, $A^T = A$.

**Definition 12 (A couple of sets).** *Given two sets* $G$ *and* $H$, $pair(G, H) = \{[G, H]\}$ *if both of* $G$ *and* $H$ *are not empty, otherwise* $pair(G, H) = \emptyset$.

**Note 4.** *Let* $nred_M = F_M^T I_M^T - \delta_M^T$, *we have its lazy evaluation called lazynred. We also use lazy evaluation of* $\delta$ *called lazy* $\delta$. *Each of the lazynred and lazy* $\delta$ *is a list of couple of sets.*

**Definition 13 (The CNNFA).** *The compressed NNFA (CNNFA) of the tail of an MYNNFA* $M^T = \langle \Sigma_M^T, Q_M^T, \delta_M^T, I_M^T, F_M^T, A_M^T \rangle$ *is the compressed representation which is defined as a 7-tuple* $P^T = \langle \Sigma_P^T, Q_P^T, lazynred_P^T, lazy\delta_P^T, I_P^T, F_P^T, A_P^T \rangle$, *where:*

- $\Sigma_P^T = \Sigma_M^T$ *is an alphabet,*
- $Q_P^T = Q_M^T$ *is a set of states,*
- $lazynred_P^T$ *is a lazy evaluation representation of* $nred_M = F_M^T I_M^T - \delta_M^T$,
- $lazy\delta_P^T$ *is a lazy evaluation representation of* $\delta_M^T$,
- $I_P^T = I_M^T$ *is a set of initial states,*
- $F_P^T = F_M^T$ *is a set of final states, and*
- $A_P^T = A_M^T$.

From the tail of an **MYNNFA** $M^T = \langle \Sigma^T, Q^T, \delta^T, I^T, F^T, A^T \rangle$, we can construct the original **MYNNFA** $M = \langle \Sigma, Q, \delta, I, F, A \rangle$ as follows:

$$\Sigma = \Sigma^T, Q = Q \cup \{q_0\}, \delta = \delta^T \cup \{[q_0, y] | y \in I^T\}, I = \{q_0\},$$

$$F = F^T \cup \{q_0\} null_M, A = A^T, \quad (1)$$

*where* $\{q_0\} null_M = \{q_0\}$ *if* $null_M = \{\lambda\}$, *otherwise* $\{q_0\} null_M = \emptyset$.

**Note 5.** *We sometimes omit* $\Sigma$ *in NNFA, MYNNFA, tail MYNNFA and CNNFA specifications when it is obvious. Sometimes, we allow the label map A to be undefined on states with zero in-degree.*

In CNNFA, we represent the set of $F$ and $I$ by using the data structures called *F-Forest* and *I-Forest* described below:

- One set is represented as a binary tree.
- Leaves of a tree are linked together by doubly linked list.
- The root of a tree contains a pointer to the list of its leaves.
- The union of two sets is a new tree with a new root, those two sets become the left child and right child of the new tree. This union operation can be done in a unit time when the two sets are disjoint.

## 3.2 Generating CNNFAs of Tail Machines for Subexpressions

Now we can construct the CNNFA representations of tail machines for subexpressions: $\lambda, a \in \Sigma, J|K, J.K,$ and $J^*$ as follows:

$$M_\lambda^T = \langle Q_\lambda^T = \emptyset, lazynred_\lambda^T = \emptyset, lazy\delta_\lambda^T = \emptyset, I_\lambda^T = \emptyset,$$

$$F_\lambda^T = \emptyset, A_\lambda^T = \emptyset \rangle, null_\lambda = \{\lambda\}. \quad (2)$$

$$M_a^T = \langle Q_a^T = \{q\}, lazynred_a^T = pair(F_a^T, I_a^T), lazy\delta_a^T = \emptyset,$$

$$I_a^T = \{q\}, F_a^T = \{q\}, A_a^T = \{[q, a]\} \rangle, null_a = \emptyset, \quad (3)$$

*where* $a \in \Sigma$, *and* $q$ *is a distinct state.*

In order to use induction, we assume that $J$ and $K$ are two arbitrary regular expressions that equivalent respectively to CNNFA $M_J^T$ and $M_K^T$ with

$$M_J^T = \langle Q_J^T, lazynred_J^T, lazy\delta_J^T, I_J^T, F_J^T, A_J^T \rangle \text{ and } M_K^T = \langle Q_K^T, lazynred_K^T, lazy\delta_K^T, I_K^T,$$
$$F_K^T, A_K^T \rangle, \text{ where } Q_J^T \text{ and } Q_K^T \text{ are disjoint. We have the following formulas.}$$

$$M_{J|K}^T = \langle Q_{J|K}^T = Q_J^T \cup Q_K^T, lazynred_{J|K}^T = lazynred_J^T \cup lazynred_K^T \cup$$
$$pair(F_J^T, I_K^T) \cup pair(F_K^T, I_J^T), lazy\delta_{J|K}^T = lazy\delta_J^T \cup lazy\delta_K^T,$$
$$I_{J|K}^T = I_J^T \cup I_K^T, F_{J|K}^T = F_J^T \cup F_K^T, A_{J|K}^T = A_J^T \cup A_K^T \rangle, \tag{4}$$
$$null_{J|K} = null_J \cup null_K.$$

$$M_{JK}^T = \langle Q_{JK}^T = Q_J^T \cup Q_K^T, lazynred_{JK}^T = null_K lazynred_J^T \cup$$
$$null_J lazynred_K^T \cup pair(F_K^T, I_J^T), lazy\delta_{JK}^T = pair(F_J^T, I_K^T) \cup lazy\delta_J^T \cup$$
$$lazy\delta_K^T, I_{JK}^T = I_J^T \cup null_J I_K^T, F_{JK}^T = F_K^T \cup null_K F_J^T, A_{JK}^T = A_J^T \cup A_K^T \rangle, \tag{5}$$
$$null_{JK} = null_J null_K.$$

$$M_{J*}^T = \langle Q_{J*}^T = Q_J^T, lazynred_{J*}^T = \emptyset, lazy\delta_{J*}^T = lazy\delta_J^T \cup lazynred_J^T,$$
$$I_{J*}^T = I_J^T, F_{J*}^T = F_J^T, A_{J*}^T = A_J^T \rangle, null_{J*} = \{\lambda\}. \tag{6}$$

## 3.3 Parsing Regular Expressions

The following parsing algorithm [6] is used to parse a regular expression into basic blocks and non-basic blocks. In the meantime, it constructs equivalent CNNFAs of tail machines for each of the created basic blocks using equations (2-6). When the algorithm halts, if there is only one basic block remains, then the given regular expression is valid and we have the CNNFA of the result tail machine. Otherwise, it is invalid. The algorithm depends on a stack of elements, each of which is either a symbol from $R$, or a record $N_P$ that stores $Q_P^T, lazynred_P^T, lazy\delta_P^T, I_P^T, F_P^T, A_P^T$ and $null_P$ for some subexpression occurrence $P$ of $R$.

. Initialize the stack to empty.
. For each input symbol $c$ in a left-to-right scan through $R$, do the following:
   a. Push $c$ onto the stack.
   b. Repeat the following steps until it can no longer be applied:
     *(Reduction)* If the topmost elements of the stack match one of following cases, replace them in the prescribed way. Otherwise, terminate the *Reduction* steps.
      i. **(case $\lambda$)** Replace by $N_\lambda$ according to the equation (2).
      ii. **(case a, an alphabet symbol)** Replace by $N_a$ according to the equation (3).
      iii. **(case $N_J | N_K$)** Replace by $N_{J|K}$ using the equation (4).
      iv. **(case $N_J N_K$)** Replace by $N_{JK}$ using the equation (5).
      v. **(case $N_J$ \*)** Replace by $N_{J*}$ using the equation (6).
      vi. **(case $(N_J)$)** Replace by $N_J$.

. If the stack contains only one entry, which is a record, then that entry is $N_R$, which stores components of the CNNFA representation of $M_R^T$. From that CN-NFA representation, we can compute the software component as shown in the next section.

## 3.4 Generating Models of Components

Now we have a full CNNFA representation of the tail machine equivalent to $R$. To construct the needed NFA that is the model of the given component from the CNNFA, we must be able to calculate the set of $\delta_M$ from $lazy\delta_M$. The following equations allow us to do that. Let $V \subseteq Q$, we need to calculate $\delta(V,a)$, $\forall a \in \Sigma$.

$$domain\, E = \{x : \exists y \mid [x,y] \in E\} \tag{7}$$

$$E[S] = \{y : \exists x \in S \mid [x,y] \in E\}.\, E(S)\ is\ called\ image\ of\ S. \tag{8}$$

Let $frontier(X)$ be the set of leaves of a tree with the root is $X$. We have the following equations:

$$F\_domain(V) = \{x \in domain\, lazy\delta | V \cap frontier(x) \neq \emptyset\} \tag{9}$$

$$I\_image(V) = lazy\delta_R[F\_domain(V)] \tag{10}$$

$$\delta_R(V,\Sigma) = \{z : \exists y \in I\_image(V) | z \in frontier(y)\} \tag{11}$$

For each $a \in \Sigma$, we have:

$$\delta_R(V,a) - \{q \subset \delta_R(V,\Sigma) \mid A(q) = a\} \tag{12}$$

We now have the full representation of the tail machine equivalent to $R$. To generate the needed software model, we apply the equation (1) to construct the MYNNFA and then create the needed software model.

**The Complexity:** With the regular expression of traces of actions of length r and s occurrences of alphabet symbols, we can calculate CNNFA in $O(r)$ time and use $O(s)$ auxiliary space [5].

## 4 Model Optimization

The result of the model generation process presented in section 3 is an NFA, in general, representing the behaviors of the given component. That NFA is DFA only if each of the alphabet symbol occurs in the regular expression only once. In almost all of other cases, the results of CNNFA are NFAs. Therefore, we must perform some extra tasks to optimize the model we have. Those tasks are converting NFA to DFA and then minimize the result DFA to have the optimal software model.

## 4.1 Converting from NFA to DFA

In this section, we use the algorithm presented in [11] to convert from NFA to an equivalent DFA. Given an NFA $M = \langle Q, \Sigma, \delta, q_0, F \rangle$, the algorithm produces an equivalent DFA $M' = \langle Q', \Sigma', \delta', q_0', F' \rangle$. The DFA $M'$ is constructed as follow:

- $Q' \subseteq 2^Q, q_0' = \{q_0\}, F' = \{q' \in Q' | q' \cap F \neq \emptyset\}$, and
- $\delta'$ is defined as follows: $\delta'(q', a) = \bigcup_{q \in q'} \delta(q, a)$, where $q' \in Q'$.

---

**Algorithm 1** Converting from NFA to equivalent DFA

---

**Input:** $M = \langle Q, \Sigma, \delta, q_0, F \rangle$ is an NFA.
**Output:** $M' = \langle Q', \Sigma', \delta', q_0', F' \rangle$ is an equivalent DFA.

1:    $q' := \{q_0\}, Q' := \{q_0\}$.
2:    **repeat**
3:        with $q' \in Q'$
4:        **if** $q'$ *is not marked* **then**
5:           **begin**
6:           **for** $i := 1$ **to** *length of* $\Sigma$ **do**
7:              **begin**
8:              $a :=$ *the $i^{th}$ element of* $\Sigma$.
9:              $p' := \delta'(q', a) := \bigcup_{q \in q'} \delta(q, a)$.
10:            **if** $p' \notin Q'$ **then**
11:                add $p'$ to $Q'$.
12:            **end**
13:        **end**
14:      Set $q'$ *is marked*.
15: **until** *No new element is added to* $Q'$.
16: $F' = \{q' \in Q' | q' \cap F \neq \emptyset\}$.
17: return $M'$.

---

The detail of this algorithm is shown in the algorithm 1. Initially, the algorithm sets $q'$ to $\{q_0\}$, $Q'$ to $\{q_0\}$ (line 1). Let $q'$ be an element of $Q'$ (line 3). If it is not marked, the algorithm find states which are transformed from the state $q'$ by symbols in the alphabet $\Sigma$ and add these states to $Q'$ if they are not in $Q'$ (from line 4 to line 12). After that, the algorithm sets the state $q'$ is marked (line 14). The process from line 3 to line 14 is repeated until no new element is added to $Q'$. Then, the set $F'$ of accepting states is constructed from elements in $Q'$ that contain accepting states of the original NFA (line 16).

**The Complexity:** Let n be the number of states in the NFA $M$. The time complexity of the algorithm 1 is $O(n^2 2^n)$.

## 4.2 Minimizing DFA

In most of the cases, the DFA software model received from the algorithm presented in the section 4.1 is not the minimal one. Therefore, we must minimize it in order to have the last and optimal model. In this paper, we use the Hopcroft algorithm [2, 11]. Given a DFA $M = \langle Q, \Sigma, \delta, q_0, F \rangle$, the algorithm constructs an equivalent minimal DFA $M' = \langle Q', \Sigma', \delta', q_0', F' \rangle$. A string $x \in \Sigma^*$ is called distinguishes two states $p, q \in Q'$ if either $\delta(p, x) \in F$ and $\delta(q, x) \notin F$ or $\delta(p, x) \notin F$ and $\delta(q, x) \in F$. In this case, the two states p and q are called distinguishable; otherwise, they are called indistinguishable or equivalent. In order to obtain $M'$, we find all of the equivalence classes of indistinguishable states and join all of those states in one class into one state of $M'$. Let [p] be an equivalence class. To determine transitions, we pick up any state p in [p] and define $\delta'([p], a) = [q]$ where p in [p], $a \in \Sigma$, [q] is an equivalent class that contains $q = \delta(p, a)$.

---

**Algorithm 2** Minimizing DFA

---

**Input:** $M = \langle Q, \Sigma, \delta, q_0, F \rangle$: a given DFA.
**Output:** $M' = \langle Q', \Sigma', \delta', q_0', F' \rangle$: an equivalent minimal DFA.
1: Initially, mark (p,q) where $p \in F$ and $q \notin F$.
2: **repeat**
3:     **for** $i := 1$ **to** *length of non − marked pair list* **do**
4:         **begin**
5:         Let $(p, q)$ be the $i^{th}$ element of the non-marked pair list.
6:         **for** $j := 1$ **to** *length of $\Sigma$* **do**
7:             **begin**
8:             Let $a$ be the $j^{th}$ element of $\Sigma$.
9:             **if** *The pair* $(\delta(p, a), \delta(q, a))$ *is marked* **then**
10:                mark the pair (p,q).
11:             **end**
12:         **end**
13: **until** *No new pair is marked.*
14: **for** $i := 1$ **to** *length of Q* **do**
15:     **begin**
16:     Let $q$ is the $i^{th}$ of Q.
17:     Create equivalence class [q] of q consists of equivalent states of q.
18:     Add [q] to $Q'$.
19:     **if** [q] *contains a accepting state of M* **then**
20:         Add [q] to F'.
21:     **end**
22: Set the initial state $q_0' = [q_0]$.
23: **for** $j := 1$ **to** *length of $\delta$* **do**
24:     **begin**
25:     Let $(q_1, a, q_2)$ be the $j^{th}$ element of $\delta$.
26:     **if** $([q_1], a, [q_2])$ *is not in $\delta'$* **then**
27:         Add $([q_1], a, [q_2])$ to $\delta'$.
28:     **end**
29: Return $M'$.

---

The detail of the algorithm is shown in algorithm 2. Initially, the algorithm marks all distinguishable pair of states (line 1). Obviously, if $p \in F$ and $q \notin F$ then p and q are distinguishable. Therefore, the algorithm marks those pairs distinguishable. After that, for each of $(p,q)$ which is not marked, the algorithm finds $a \in \Sigma$ that $\delta(p,a)$ and $\delta(q,a)$ are distinguishable, then we mark the pair (p,q) distinguishable as well (from line 5 to line 11). This process is repeated until no more pair is marked. We construct the minimal DFA $M'$ as follow. The set of states contains all equivalence classes determined in line 17 and 18. The set of accepting states are equivalence classes that contains accepting states of M (line 19 and 20). The initial state is $[q_0]$ (line 22). The transition rules are defined from line 23 to line 28. The generated DFA is minimal DFA [2, 4].

**The Complexity:** Let n be the number of states in the DFA $M$. The time complexity of the algorithm 2 is $O(n \log n)$ [2].

## 5  Experiments

In order to show the advantages of the presented software model generation method over other methods in [12, 11], we have implemented the model generation tool called MG[1] to compare these three model generation methods. When doing experiments, we used the same set of experiment data as the method shown in [11]. The data is for several typical systems shown in [7] (i.e., the Sender component, the Controller component of the Cruise Control system, and the Jitter component). The experiments are carried out on a normal computer with Processor Intel(R) Core(TM) i3-2120 CPU @ 3.30GHz, 3300 MHz, 2 Core(s), 4 Logical Processor(s), Total Physical Memory 3.40 GB, and Microsoft Windows 7 Ultimate OS. The number of symbols in the alphabet $|\Sigma|$ and the number of traces in the language $L$ in column $|L|$ are shown in the table 1.

**Table 1**  Experimental Data

| No. | Id | Applied Systems | $|\Sigma|$ | $|L|$ |
|---|---|---|---|---|
| 1 | S1 | jiter-8-21 | 3 | 21 |
| 2 | S2 | cruiseControl | 11 | 19 |
| 3 | S3 | jiter-16-101 | 3 | 101 |
| 4 | S4 | sender-len18 | 3 | 18 |
| 5 | S5 | sender-len9 | 3 | 9 |

To have the experimental results shown in the tables 2 and 3, we carried out experiment 20 times for each of the applied systems and each of the software model generation methods. The obtained experimental results shown in the tables 2 and 3

---

[1] http://www.uet.vnu.edu.vn/~hungpn/MG

**Table 2** Experimental results for comparing the proposed method with the Thompson method

| Id | Thompson method | | | | | | | CNNFA method | | | | | |
|---|---|---|---|---|---|---|---|---|---|---|---|---|---|
| | $|\sigma|$ | $|M|$ | $|\delta|$ | $|M_R|$ | $|\delta_R|$ | Time | Memory | $|M|$ | $|\delta|$ | $|M_R|$ | $|\delta_R|$ | Time | Memory |
| S1 | 8 | 136 | 155 | 11 | 12 | 00.089 | 771,630 | 114 | 113 | 11 | 12 | 00.049 | 568,926 |
| S2 | 9 | 133 | 150 | 11 | 17 | 0.079 | 761,366 | 113 | 112 | 11 | 17 | 0.066 | 735,675 |
| S3 | 16 | 1290 | 1389 | 42 | 49 | 73.415 | 2,025,680 | 1188 | 1187 | 42 | 49 | 42.168 | 4,226,692 |
| S4 | 18 | 191 | 207 | 19 | 18 | 0.235 | 1,323,565 | 172 | 171 | 19 | 18 | 0.074 | 925,113 |
| S5 | 9 | 56 | 63 | 10 | 9 | 00.008 | Too fast to count | 46 | 45 | 10 | 9 | 00.005 | Too fast to count |

**Table 3** Experimental results for comparing the proposed method with the L*-based method

| Id | L*-based method | | | | | | | CNNFA method | | | | | |
|---|---|---|---|---|---|---|---|---|---|---|---|---|---|
| | $|\sigma|$ | $|M|$ | $|\delta|$ | $|M_R|$ | $|\delta_R|$ | Time | Memory | $|M|$ | $|\delta|$ | $|M_R|$ | $|\delta_R|$ | Time | Memory |
| S1 | 8 | 14 | 48 | 10 | 11 | 00.135 | 5,963,545 | 114 | 113 | 11 | 12 | 00.049 | 568,926 |
| S2 | 9 | - | - | - | - | - | Out | 113 | 112 | 11 | 17 | 0.066 | 735,675 |
| S3 | 16 | - | - | - | - | - | Out | 1188 | 1187 | 42 | 49 | 42.168 | 4,226,692 |
| S4 | 18 | - | - | - | - | - | Out | 172 | 171 | 19 | 18 | 0.074 | 925,113 |
| S5 | 9 | 5 | 15 | 4 | 4 | 0.065 | 4,477,565 | 46 | 45 | 10 | 9 | 00.005 | Too fast to count |

are the average of the above 20 times of experiments. The tables 2 and 3 contain of the following information: The maximum length of traces for the two methods using $L^*$ and Thompson algorithms in column called $|\sigma|$, the size of the generated models right after using $L^*$, Thompson and CNNFA algorithms in column $|M|$, the number of transitions in those models in column $|\delta|$, the size of the final optimal models in column $|M_R|$, the number of transition in those models in column $|\delta_R|$, the time in seconds a method used to generate final optimal models in column Time (seconds) and the memory in bytes allocated to each of the model generation methods during processing in column Memory (bytes). In the obtained results, "Out" means "Out of memory". This fact is clear to show the efficiencies of the proposed method.

From the experiment result data, we have the following comments:

- The proposed method is much faster than the other two methods.
- The proposed method is not limited by the length of the traces of actions of the software component like methods proposed in [12, 11].
- The proposed method uses less memory than the method using L* algorithm. It uses less memory than the method using Thompson algorithm with small test data. With big test data, it uses more memory than the method using Thompson algorithm.
- The size of the generated models using $L^*$ algorithm is more compact than the generated models using Thompson and CNNFA algorithms.

# 6   Conclusion

This paper has presented an efficient method for generating models of component-based software by using the CNNFA algorithm. The proposed method is not only independent on the maximum length of traces of actions of the component but also has smaller time and space complexity in comparison with methods proposed in [12, 11]. The generated models are minimal and exactly describe behaviors of the corresponding components of the system under checking. The experimental results clearly show the efficiencies of the proposed method. Therefore, this method makes the model-based approaches higher applicable in practice.

We are investigating to apply this method for some practical systems in order to show its efficiency. We are also studying a method for automated generating regular expressions from design documents. This is for providing inputs for the proposed method. The most challenging problem is how to integrate this method with a model checker in order to provide a full framework for model checking of systems from design documents.

**Acknowledgments.** This work is supported by the project no. QG.12.50 granted by Vietnam National University, Hanoi (VNU).

# References

[1] Thompson, K.: Regular expression search algorithm. Communications of the ACM 11(6), 419–422 (1968)
[2] Hopcroft, J.E.: An nlogn algorithm for minimizing states in a finite automaton, Tech. Report, Stanford University, Stanford, CA, USA (1971)
[3] Chow, T.S.: Testing software design modeled by finite-state machine. IEEE Trans. on Software Engineering 4(3), 17–187 (1978)
[4] Hopcroft, J.E., Ullman, J.D.: Introduction to Automata Theory, Languages, and Computation, 1st edn. Addison-Wesley Longman Publishing Co., Inc, Boston (1990)
[5] Chang, C.: From regular expressions to DFA's using compressed NFA's, Ph.D. Thesis, New York University, New York (1992)
[6] Chang, C., Paige, R.: From regular expressions to DFA's using compressed NFA's. Theoretical Computer Science 178, 1–36 (1997)
[7] Magee, J., Kramer, J.: Concurrency: State Models & Java Programs. John Wiley & Sons (1999)
[8] Corbett, J.C., Dwyer, M.B., Hatcliff, J., Laubach, S., Pasareanu, C.S., Robby, Zheng, H.: Bandera: extracting finite-state models from Java source code. In: Proceedings of the 2000 International Conference on Software Engineering, pp. 439–448 (2000)
[9] Tkachuk, O., Dwyer, M.B., Pasareanu, C.S.: Automated environment generation for software model checking. In: Proceedings. 18th IEEE International Conf. on Automated Software Engineering, pp. 116–127 (2003)
[10] Lorenzoli, D., Mariani, L., Pezzè, M.: Automatic generation of software behavioral models. In: ACM, Proceedings of the 30th International Conference on Software Engineering, pp. 501–510 (2008)

[11] Cuong, L.B., Hung, P.N.: A Method for Generating Models of Black-box Components. In: 4th International Conference on Knowledge and Systems Engineering, pp. 177–222. IEEE Computer Society Press (2012)

[12] Duong, H.M., Trinh, L.K., Hung, P.N.: An Assume-Guarantee Model Checker for Component-Based Systems. In: The 10th IEEE-RIVF International Conference on Computing and Communication Technologies, pp. 22–26. IEEE Computer Society Press (2013)

# A Lightweight Formal Approach for Component Reuse

Khai T. Huynh*, Thang H. Bui, and Tho T. Quan

**Abstract.** Component reuse is playing a crucial role in today software design. However, the current approaches used in industry are quite effort-consuming due to the lack of an effective mechanism to describe and capture semantics in software components. In this paper, we propose a formal approach to overcome this problem, which is based on First-Order Logic (FOL). In one hand, FOL is sufficiently expressive to describe semantics in various software domains, from generic to specific ones. In the other hand, FOL also supports automatic searching, matching and inferring mechanism by computer-based tools and provers. Thus, our approach both effectively supports expert human to describe system components and computer programs to reuse those components, and even to compose new components from existing ones for further usage. We realize our approach as a framework which can be applied in various situations of software designs, as illustrated in some case studies.

**Keywords:** Component-based development, rapid software development, formal specification, component reuse.

## 1 Introduction

Component reuse is one of the most promising field of software engineering [1]. Enhanced productivity (as less code needs to be written), increased quality (since assets proven in one project can be carried through to the next) and improved business performance (lower costs, shorter time-to-market) are the main benefits of developing software from a stock of reusable components [2][3].

Khai T. Huynh · Thang H. Bui · Tho T. Quan
Faculty of Computer Science and Engineering
Ho Chi Minh City University of Technology
Ho Chi Minh City, VietNam
e-mail: 551220035@stu.hcmut.edu.vn

* Corresponding author.

© Springer International Publishing Switzerland 2015      513
V.-H. Nguyen et al. (eds.), *Knowledge and Systems Engineering,*
Advances in Intelligent Systems and Computing 326, DOI: 10.1007/978-3-319-11680-8_41

However, there are problems associated with reuse. There is a significant cost associated with understanding whether or not a component is suitable for reuse in a particular situation, and in testing that component to ensure its dependability. These additional costs will increase the overall development costs. There are some problems associates with reuse software components [3] as follows: (i) current software tools do not support development with reuse; (ii) the cost for creating, maintaining, and using a component repository is typically quite expensive; and (iii) there is currently lacking a practical and formal mechanism for finding, understanding, and adapting reusable components. The common reason for all of above problems is the lack of an effective mechanism to encode semantics to the descriptions of software components. There are several attempts reported to formally describe software design with embedded semantic information [4][5]. However, those approaches typically suffered from high computational cost thus causing some difficulties when applied in real life applications.

This observation urges us to consider an approach which should be at the same time conveniently lightweight for the human users and formally expressive for machine understanding and processing. This idea is realized as a framework proposed in this paper, which formally facilitates reusing existing components previously developed in a lightweight manner. It is achieved by the following technical aspect:

- Using first-order logic-based (FOL) [6] to describe components.
- Using machine learning-based approach for similarity evaluation between FOL formulas.
- Employing and enhancing the Artificial Intelligence (AI) planning technique to compose a design fulfilling an input request. During doing so, additional components can be composed and added to the current component library.

The rest of the paper is organized as follows. Section 2 discusses some background knowledge. Section 3 presents our proposed framework for rapid software development based on reusing formal specification components. Section 4 illustrates the capability of framework through two case studies. We present the related works in Section 5 and conclude the paper in Section 6.

## 2   Background

### 2.1   *FOL-Based Component Specification*

First-order logic (FOL) [7] is a well-known formal representation which extends the classical proposition logic with predicate, quantifier and variable. Thus, FOL allows more flexible and compact representation of knowledge [8]. In our framework, FOL is used to specify software components. The component is considered as a black box and stored in repository as a structure in Listing 1 with *expression* is a FOL-based expression.

*Listing 1: Structure of a component stored in repository.*

```
<component name=comp_name>
    <input>variable: type_name</input>
    <output>variable: type_name</output>
    <pre>expression</pre>
    <post>expression</post>
</component>
```

Listing 2 shows an example specification of *binarySearch* function considered as a component stored in the repository. The components are organized in a hierarchical organization to provide a faster means for browsing and searching.

*Listing 2: Specification of binarySearch function.*

```
<component name=java.util.Arrays.binarySearch>
    <input>a: Object[], x: Object</input>
    <output>i: Integer</output>
    <pre><![CDATA[
        a != null && (\forall i:int; 0<i && i<a.length; a[i-1]<=a[i])
    ]]></pre>
    <post><![CDATA[
        ((\forall j:int; 0<=j && j<a.length; a[j]!=x) && i==-1) ||
        (\exists j:int; 0<=j && j<a.length; a[j]==x) && i==j))
    ]]></post>
</component>
```

## 2.2 FOL-Based Similarity Computation

Once software components are represented as FOL formulas stored in a repository, it is needed to develop a similarity measure between those formulas to support searching the required component over the repository. To achieve this, we rely on the following definitions [9]:

Let $\Gamma$ be the set of FOL formulas, $T = \{t_1, ..., t_m\}$ is set of all symbols and terms that appear in $\Gamma$ and c, p are two FOL formulas in $\Gamma$.

**Definition 1. Feature matrix**

The feature matrix $\Phi = \Gamma \times \{1, ..., m\} \to \{0, 1\}$

$$\Phi(c, i) = \begin{cases} 1 \text{ if } t_i \text{ appear in } c \\ 0 \text{ otherwise} \end{cases}$$

**Definition 2. Feature function**

The feature matrix gives rise to the *feature function* $\varphi$.

Define $\varphi: \Gamma \to \{0, 1\}^m$ which for $c \in \Gamma$ is the vector $\varphi^c$ with entries in $\{0, 1\}$ satisfying:

$$\varphi_i^c = 1 \leftrightarrow \Phi(c, i) = 1$$

**Definition 3. Classifier function**

For each $p$ in $\Gamma$, the classifier function is defined:

$$C_p(.) : \Gamma \rightarrow R$$

Given a conjecture $c$, $C_p(c)$ estimates how useful $p$ is for proving $c$. The classifier function is useful when we need to decide if a premise $p$ would then be useful to prove a required goal $c$. It would be the case if $C_p(c)$ is above certain threshold. A common approach to ranking is to use classification, and to combine the real-valued classifiers [10]. The premises for a conjecture $c$ are ranked by the values of $C_p(c)$, and we choose a certain number of the best ones.

With vector $\varphi$ has been calculated, we can develop methods of calculating the similarity $C_p(c)$. These methods can be as simple as the angle between the vector calculus, or others methods such as Naive Bayes [11], depending on the complexity of the problem.

*Listing 3: Specification of three components: f1, f2 and f3.*

```
<component name=f1>
  <input>x: double</input>
  <output>y: double</output>
  <post>y== 2 * x</post>
</component>
<component name=f2>
  <input>x: double</input>
  <output>y: double</output>
  <post>y==-x</post>
</component>
<component name=f3>
  <input>x: double</input>
  <output>y: double</output>
  <post><![CDATA[(x>0)=>(y==x+1) && (x<=0)=>(y==log(x))]]></post>
</component>
```

For example, with the components $f1$, $f2$ and $f3$ are described in Listing 3, if we denote the terms $p_1$, $p_2$, $p_3$ and $p_4$ corresponding to the set of operators: $\{+, -\}$, $\{*\}$, $\{>, <=\}$ and $\{log\}$, we will have the feature matrix as follow:

$$
\begin{array}{c|cccc}
\varphi & p_1 & p_2 & p_3 & p_4 \\
\hline
f1 & 0 & 1 & 0 & 0 \\
f2 & 1 & 0 & 0 & 0 \\
f3 & 1 & 0 & 1 & 1 \\
\end{array}
$$

Then, we will calculate the feature vectors for each component based on the matrix $\varphi$. For example, we have the vector representing the component $f2$ is $\varphi^{f2} = [1,0,0,0]$ (value 1 in $i$th column mean that $f2$ has the feature $p_i$). With the calculated vectors $\varphi^{fi}$, we can develop the techniques for calculating the similarity between two vectors. This technique can be as simple as calculation the angle between two vectors, or maybe a machine learning method such as Naive Bayes [11], depending on the complexity of each domain.

## 2.3 FOL-Based AI Planning for Component Composition

AI planning is the area of study concerned with the automatic generation of a plan to solve a problem within a particular domain. At its simplest, a plan is a sequence of actions. Given an initial state, the planner tries to find the actions required to achieve some goal conditions [12].

According to the STRIPS representation [13], the system consists of a set of *states*, *goals* and *operations*. Each operation has *operation name, preconditions* - a sentence describing the conditions that must occur so that the operator can be executed and *effect* - a sentence describing how the world has change as a result of executing the operator. The algorithm of classical AI planning with STRIPS representation is presented in [13].

With classical AI planning, the matching mechanism between the two FOL formulas is exact matches. In fact, the matching is not so simple. It is easy to see that in the mathematical formulas representation, two different formulas may have the same computation result but there are many different representation ways, example $(2*x)$ and $(x+x)$. Hence, to make the matching is more precise, we enhance the classical AI planning that allows to prove the truth of two FOL formulas have different representation with a theorem prover, such as Z3 [14].

## 3 The Framework

Fig.1 describes the framework for automatic search and reusage components based on formal specification. The framework has three modules as follows.

### 3.1 Component Selection Module

This module acts as a pre-processing step of framework. It computes the similarity between the logic formulas representing the components, as discussed in Section 2.2. Thus, we can narrow the component search space. The narrowing is obtained by similarity computation between formal specifications of the components in the repository.

### 3.2 Component Composition Module

The second module of framework is the component composition module. This is the most important part of framework. The searching and synthetic components module based on AI-planning works on the following principle. Starting from initial requirement $(c)$ and the set of components $(P)$ which have a degree of similarity in a predefined threshold for $(c)$, the planner will choose a component $p$ from $P$ (with the priority of the degree of similarity). As discussed in Section 2.3, we rely on a mathematical prover to judge if the precondition of $p$ is equivalent to that of $c$. If it is the case, the search process stops and returns results (return $p$). In contrast, the

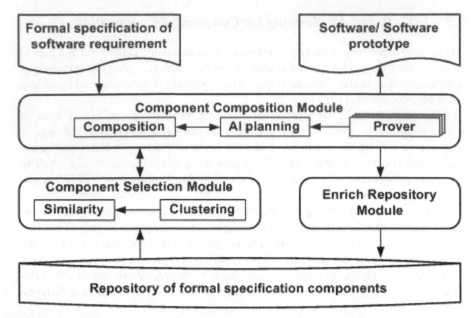

**Fig. 1** Structure of framework

algorithm will consider the precondition of $p$ as new postcondition and the search process will repeated. In case of deadlock, the algorithm will backtrack with the function $p'$ in $P$ ($p' \neq p$).

## 3.3  Enrich Repository Module

This module of framework performs enriching the component specification repository for future use. So, this module helps us to optimize the system performance when encountering a search request that is similar to those previously performed.

## 4  Case Studies

### 4.1  Apply the Framework to Find Function(s) Satisfy the Requirement

In this section, we present an illustrative example of the automatic API functions searching which satisfy the requirement specifications.

We want to generate a program segment which returns the list of students who to be received scholarship from a list of students. List of students received scholarship is defined as the list which has a maximum of $n$ items (usually, $n$ is 10% of total students) taken in order from top to bottom by GPA; The student in this list must have a GPA greater than or equal 7.0/10.0 and every subject must has the result is greater than or equal 5.0/10.0. With that requirement, the specification is given as follows:

```
//@in s: StudentList, n: int
//@out x: StudentList
//@require listnotnull(s)
//@ensure x.len<=n && orderlist(x) &&
//@   (forall i: int; i>=0 && i<x.len; x[i].GPA>=7.0) &&
//@   (forall i: int; i>=0 && i<x.len;
//@     (forall j:int; j>=0&&j<x[i].subjs.len; x[i].subjs[j]>=5.0))
```

In our system repository, we have a list of API functions:

The function returns the list of students who have the GPA greater than or equal the value *n*:

```
<component name=get_StudentList_By_GPA>
  <input>s:StudentList, n:float</input>
  <output>x: StudentList</output>
  <pre>listnotnul(s)</pre>
  <post><![CDATA[ (forall i: int; i>=0 && i<x.len; x[i].GPA>=n)
    ]]></post>
</component>
```

The function returns the list of students who have the result of every subject is greater than or equal the value *n*:

```
<component name=get_StudentList_By_SubjectScore>
  <input>s:StudentList, n:float</input>
  <output>x: StudentList</output>
  <pre>listnotnul(s)</pre>
  <post><![CDATA[ (forall i: int; i>=0 && i<x.len;
      (forall j:int; j>=0&&j<x[i].subjs.len; x[i].subjs[j]>=n))
    ]]></post>
</component>
```

The function returns the first *n* students from the list of Student. If the number of items in student list is less than *n*, all item of student list will be returned.

```
<component name=get_n_Elements>
  <input>s:StudentList</input>
  <output>x: StudentList</output>
  <pre>listnotnul(s)</pre>
  <post><![CDATA[ x.len<=n ]]></post>
</component>
```

The function sorts the list in descending order of GPA:

```
<component name=sort_DescList_By_GPA>
  <input>s:StudentList</input>
  <output>x: StudentList</output>
  <pre>listnotnul(s)</pre>
  <post>orderlist(x)</post>
</component>
```

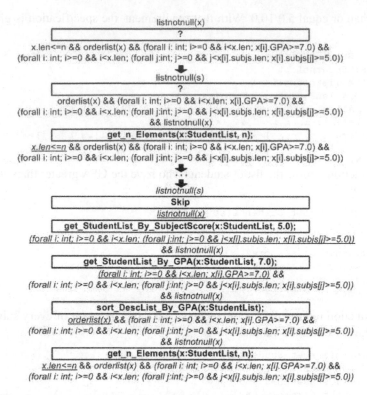

**Fig. 2** Planning process to generate list of Student received scholarship

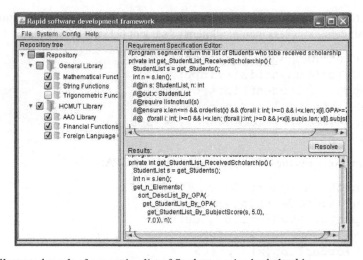

**Fig. 3** Illustrated result of generating list of Student received scholarship

With the above requirement and API function repository, the planning steps perform as in Fig.2 and Fig.3 illustrates the result produced from framework. Note that the composed result can be further added to the repository for future reuse if needed.

## 4.2 Automatically Searching for Web Template

In practice, when developing a website, we want to find the templates which can apply to our website. Searching the template matching with the requirement is a boring and time consuming work. With our framework, we can perform this searching process automatically. Suppose that we have a repository of website templates which contains three specified templates as follows:

*Template_*1 has three parts: *header*, *body*, and *footer*. The *header* contains *banner*, *logo* and *menu*. *Body* of the template is divided into three columns: *left_col*, *center_col* and *right_col*. The *menu* contains three links: *Home*, *Intro* and *Contact*. This template is specified:

```
<component name=Template_1>
   <post><![CDATA[ page(header, body, footer) &&
      header(banner, logo, menu) &&
      body(left_col, center_col, right_col) &&
      menu(Home, Intro, Contact) ]]></post>
</component>
```

*Template_*2 has three parts: *header*, *body*, and *footer*. The *header* contains *banner* and *logo*. *Body* of the template is divided into three columns: *left_col*, *center_col* and *right_col*. The *left_col* contains a *vertical_menu* which has three links: *Home*, *Intro* and *Contact*.

```
<component name=Template_2>
   <post><![CDATA[ page(header, body, footer) &&
      header(banner, logo) &&
      body(left_col, center_col, right_col) &&
      left_col(vertical_menu)) &&
      vertical_menu(Home,Intro,Contact) ]]></post>
</component>
```

*Template_*3 has three parts: *header*, *body*, and *footer*. The *header* contains *banner* and *logo*. *Body* of the template is divided into two columns: *left_col* and *main_col*. The *left_col* contains a *vertical_menu* and the *vertical_menu* has three links: *Home*, *Intro* and *Contact*. This template is specified:

```
<component name=Template_3>
   <post><![CDATA[ page(header, body, footer) &&
      header(banner, logo) &&
      body(left_col, main_col) &&
      left_col(vertical_menu)) &&
      vertical_menu(Home,Intro,Contact) ]]></post>
</component>
```

User wants to find a website template which has three parts: *header*, *body* and *footer*. The *header* contains the *banner* and *logo*. Body of the site is divided into three columns: *left_col*, *center_col* and *right_col*. The *left_col* contains a vertical menu which has links: *Home*, *News*, *Products* and *Contact*. The specification of user requirements is as follows:

```
//@  ensure page(header, body, footer) && header(banner, logo) &&
//@       body(left_col, center_col, right_col) &&
//@       left_col(vertical_menu)) &&
//@       vertical_menu(Home,News,Products,Contact)
```

With above requirement, the similarity computation between the specification of requirement and templates in repository will return the *Template_2* is the most suitable template with a similarity degree of 80% (matching 4 of 5 terms). The *Template_1* has a similarity degree of 40% (matching 2 of 5 terms) and the *Template_3* has similarity degree of 60% (matching 3 of 5 terms).

Thus, the *Template_2* is the needed template. Of course, this template does not completely satisfy to the requirement, but only the one that the best fits the requirement. Then, user may need to further modify before deploying.

## 5 Related Work

Intelligent Software Architecture Reuse Environment (ISARE) [15] is a framework on the aspects of software Architecture on quality attributes. It is mainly directed on reuse of available software architecture evaluation methods. ISARE strengthen the Software Architecture repository and automate the software architecture selection and evaluation methods for efficient and reliable systems. The main input of ISARE is the set of architectural styles and their comprehensive analysis against the set of quality attribute.

Trusted National Software ResoUrce Sharing and Co-operaTeng EnvironmEnt (TRUSTIE) [16] is a tool that is used to construct a large-scale software production environment for trust worthy resource sharing and development cooperation. TRUSTIE is used as a command-line tool for component development. A new tool has to be developed to provide advantages of meta-model to model-editor generators. A research issue presents the close connection with Service Oriented Architecture [16].

N Md Jubair Bash et al. [17] has proposed a framework studio for effective component reusability which provides the selection of components from framework studio and generates source code based on the stakeholders needs. This framework studio is implemented in using swings which are integrated onto the Netbeans IDE which helps in faster generation of the source code.

# 6   Conclusion

In this paper, we propose a framework which allows automatic reuse and composition of software components. In our framework, we first suggest to use First-order Logic (FOL) for describing software components. By doing so, our frameworks can serve various situations of software design, ranging from function APIs to GUI templates searching. One of the key properties of our framework is that it can be applied in a lightweight manner, i.e. users are not restricted to a certain convention when using FOL to describe the components. Then, due to the solid mathematical foundation of FOL, the predefined FOL-based components can be processed effectively for searching, matching or even incremental composition by computer-based tools and provers. We have demonstrated the effectiveness of our framework in two case studies reflecting two common demands of component reuse in practice, which are Java API and Web interface searching.

# References

[1] Basili, V., Rombach, H.D.: Support for comprehensive reuse. IEEE Software Engineering Journal, vol 6(5), 303–316 (1991)
[2] Sametinger, J.: Software engineering with reusable components. Springer, New York (1997)
[3] Sommerville, I.: Software Engineering, 9th edn., ch. 16. Addison-Wesley (2010) ISBN-10: 0137035152
[4] Liu, S., et al.: SOFL: A formal engineering methodology for industrial applications. IEEE Transactions on Software Engineering 24(1), 24–45 (1998)
[5] Zschaler, S.: Formal specification of non-functional properties of component-based software systems. Software & Systems Modeling 9(2), 161–201 (2010)
[6] Barwise, J.: An introduction to first-order logic. Studies in Logic and the Foundations of Mathematics 90, 5–46 (1977)
[7] Srivastava, S.M.: Syntax of First-Order Logic. A Course on Mathematical Logic, pp. 1–13. Springer, New York (2013)
[8] Ayorinde, I.T., Akinkunmi, B.O.: Application of First-Order Logic in Knowledge Based Systems. African Journal of Computing & ICT 6(3) (2013)
[9] Jesse, A., et al.: Premise selection for mathematics by corpus analysis and kernel methods. Journal of Automated Reasoning, 1–23 (2011)
[10] Richard, M.D., Richard, P.L.: Neural network classifiers estimate Bayesian a posteriori probabilities. Neural Computation 3(4), 461–483 (1991)
[11] Irina, R.: An empirical study of the naive Bayes classifier. In: IJCAI 2001 Workshop on Empirical Methods in Artificial Intelligence, vol. 3, pp. 41–46 (2001)
[12] Russell, S.: Artificial intelligence: A modern approach, 2nd edn., vol. ch. 11. Pearson Education India (2003)
[13] Richard, E., Nilsson, J.: Strips: A new approach to the application of theorem proving to problem solving. Artificial Intelligence 2, 189–208 (1971)
[14] de Moura, L., Bjørner, N.S.: Z3: An efficient SMT solver. In: Ramakrishnan, C.R., Rehof, J. (eds.) TACAS 2008. LNCS, vol. 4963, pp. 337–340. Springer, Heidelberg (2008)
[15] Rizwan, A., Saif, R.K., Aamer, N., Tai-hoo, K.: IASRE: An Integrated Software Architecture Reuse and Evaluation Framework. In: ASEA 2010, pp. 174–187 (2010)

[16] Xinyu, Z., Li, Z., Cheng, S.: The Research of the Component-based Software Engineering. In: Sixth International Conference on Information Technology: New Generations, pp. 1590–1591. IEEE (2009)

[17] Jubair, B.N.M., Salman, A.M.: A Framework Studio for Component Reusability. CS & IT-CSCP-2012, 325–335 (2012)

# Part III
# KSE 2014 Special Sessions

# Dynamic Behavior of Uterine Contractions: An Approach Based on Source Localization and Multiscale Modeling

Catherine Marque, Ahmad Diab, Jérémy Laforêt,
Mahmoud Hassan, and Brynjar Karlsson

**Abstract.** Among the complex systems of systems of the human body, the uterus is one of the least understood. The sequence of contraction of the myometrium results from numerous interconnected control systems (electric, hormonal, mechanical). In this paper we try to understand what the electrical activity of the uterus can tell us about uterine synchronization. This work is the first attempt to locate uterine EMG sources by means of source localization tools. These tools are tested on signals simulated by a multiscale model of the uterine EMG. As a preliminary result, we present the source localization for one pregnancy and one labor contraction recorded on the same woman. The localized sources do not demonstrate any "propagation like" behavior. This supports the hypothesis that the global organ-level communication is probably not done via action potential propagation, but may be via a coupling between the electrical and the mechanical systems controlling the uterus.

## 1 Introduction

"Every object that biology studies is a system of systems" [16]. Among the complex system of system (SoS) of the human body, many questions remain open concerning the human uterus. One of these questions concerns how the uterus exactly operates as an organ that remains quiescent during most of pregnancy, and then contracts in a very organized way during labor. The contraction of the uterus starts from a chemical phenomenon: ionic exchanges trough specific membrane channels of the uterine cell, from a "pacemaker" cell area. These exchanges are controlled by different

Catherine Marque · Ahmad Diab · Jérémy Laforêt
Université de Technologie de Compiègne - CNRS UMR 7338 BMBI, Compiègne, France
e-mail: catherine.marque@utc.fr

Ahmad Diab · Brynjar Karlsson
Reykjavik University, Reykjavik, Iceland

Mahmoud Hassan
Université de Rennes1, LTSI, Rennes, France

© Springer International Publishing Switzerland 2015                                527
V.-H. Nguyen et al. (eds.), *Knowledge and Systems Engineering*,
Advances in Intelligent Systems and Computing 326, DOI: 10.1007/978-3-319-11680-8_42

factors: electrical and hormonal environments, mechanical stretching [6]... They induce a cell depolarization, producing thus two effects: i) activation of the contractile proteins, producing the mechanical contraction; ii) electrical activity propagating to neighboring cells. During labor, the uterine synchronization occurs via a mechano-electrical coupling phenomenon: the contracting cells produce an increase in the intra-uterine pressure (IUP), inducing a stretching of the uterine tissue, that produces then an increase in cells excitability, leading to a new "pacemaker" area... This control involves different multi-physic systems (electric, hormonal, mechanical) interconnected at different scales (intracellular, cellular, tissue, organ). The uterine electrical activity, measured noninvasively on the woman's abdomen by means of surface electrodes is referred to as the electrohysterogram (EHG= uterine electromyogram) [3]. In order to better understand the uterine activity generation and control, a physiology-based model has been developed to represent the main ionic dynamic responsible for the genesis of the uterine electrical activity [35]. Then, in order to understand the links existing between the physiological phenomena determining the efficiency of uterine contraction (excitability, synchronization) and the EHG characteristics, we developed a biophysics based multiscale model of the EHG, going from the cell to the electrical signal measured on the abdomen, in an integrative physiology approach [20, 25]. This model represents both phenomena involved in the complex behavior of the uterine organ: cell excitability and propagation of the electrical activity, as well as the woman's abdomen conducting volume present between the uterine muscle and the abdominal electrodes. A lot of work has been done in order to investigate the uterine synchronization by means of the propagation of EHG recorded at the abdominal surface. This work was based on various classical methods such as correlation [24, 4, 12] and propagation velocity [32, 28, 34, 23]. Recently we have shown that the nonlinear correlation between EHG signals increases from pregnancy to labor [13], associated with a coupling direction concentration toward cervix. But none of this parameter has been able to represent in a clinical efficient way, the complex phenomenon involved in the uterine synchronization during labor. The aim of our research is to be able to understand what the electrical activity of the uterus can tell us about this complex uterine synchronization, in order to find tools that can be used for labor detection and prediction of preterm labor. We tackle in this paper the problem of uterine source localization (origin of the activity at the uterine muscle level) rather than the more classical approach of studying the coupling and the uterine activity propagation at the abdominal level. Source localization has been widely applied to EEG [1, 22, 29, 14], MEG [26, 2, 15] field, but to our knowledge it was never applied to EHG signal. Source estimation is a general tool for analyzing spatiotemporal dynamics of organ or system. Uterine EMG would permit us to observe the dynamics of uterine contractile activity on a time-scale of milliseconds, reflecting synchronous activity of cells within the uterus muscle. It could therefore provide a dynamic characterization of uterus activity that is not possible with other direct and indirect modalities [7]. The work presented in this paper is the first attempt to locate uterine EMG sources. We will develop a source localization tool suitable for EHG source localization. Then, this method is tested on signals simulated by our biophysics multiscale model of the EHG, with known source location. And then, as a preliminary result, we present the result of source localization for one pregnancy and one labor contraction recorded on the same woman at different pregnancy terms.

## 2 Methods

### 2.1 Multiscale EHG Model

Cellular Level

We base our model on a cellular model previously developed and described in [35]. This model uses ten variables. We reduced this model complexity by replacing the expression of calcium current $I_{Ca}$, which is the most computational intensive, by the one proposed by Parthimos et al. [30]. We thus obtain a less complex three-variable reduced model described by the following equations:

$$\frac{dV_m}{dt} = \frac{1}{C_m} \left( I_{stim} - I_{Ca} - I_K - I_{KCa} - I_{leak} \right),$$

$$\frac{dn_K}{dt} = \frac{h_{K_\infty} - n_K}{\tau_{n_K}}, \tag{1}$$

$$\frac{d[Ca^{2+}]}{dt} = f_c \left( -\alpha I_{Ca} - K_{Ca}[Ca^{2+}] \right),$$

with $V_m$ the trans-membrane potential, $n_K$ the potassium activation variable, $K_{Ca}$ the Calcium extraction factor and $[Ca^{2+}]$ the intracellular calcium concentration. The ionic currents are defined as follows:

$I_{Ca}$ Voltage dependent calcium channel current: $J_{back} - G_{Ca}(V_m - E_{Ca}) \frac{1}{1+\exp\left(\frac{V_{Ca}-V_m}{R_{Ca}}\right)}$

$I_K$ Voltage dependent potassium channel current: $G_k n_k(V_m - E_k)$

$I_{KCa}$ Calcium dependent potassium channel current: $G_{kCa} \frac{[Ca^{2+}]^2}{[Ca^{2+}]^2+k_d^2} (V_m - E_k)$

$I_L$ Leak current: $G_L(V_m - E_L)$

with $J_{back}$ the background current, $E_i$ and $G_i$ the respective Nernst potential and the conductance for each ion $i$, $R_{Ca}$ the maximum slope of the voltage operated calcium channel (VOCC) activation, $k_d$ the half-point potassium concentration. For more information, see [21].

Tissue Level

To represent a sample of tissue, we simulate a 2D rectangular grid of cells. Each individual cell is described with the reduced model and is electrically coupled with its neighbors. For this study, we use a rectangular grid, with an equal distance between cells in all directions. The electrical coupling between the cells is computed by using a gap junction model, as proposed by Koenigsberger et al. [17]. It adds a coupling term to the $\sum I_x/C_m$ reaction term initially described by Red3:

$$\frac{dV_{m_i}}{dt} = \frac{\sum I_x}{C_m} - \nabla \cdot D\nabla(V_i), \tag{2}$$

where $I_x$ represent the different currents, $C_m$ is the membrane capacitance and $D$ the diffusion tensor.

Organ Level

In order to achieve a first approximation of the uterine anatomy, we first define a balloon like 3D structure. This structure is far from being representative of the human uterus but it permits to simulate the propagation of the activity to the whole organ

Abdominal Level

To model the propagation of signals towards the skin surface, we adapt the model proposed in [33]. It models the surface EHG, in the spatial frequency domain, as the product of an electrical source, the myometrial activity coming from the tissue level, and of an analytical expression representing the transfer function of the volume conductor. For this transfer function definition, the volume conductor is considered as made of parallel interfaces separating the 4 different abdominal tissues, namely: the myometrium, where the source is located at a depth y=y0, the abdominal muscle, fat, and skin. The description of the volume conductor is derived by solving the Poisson equation at each interface. This description shows that the volume conductor effect depends on the tissue thicknesses, their conductivities, and the source depth, y0. All the tissues are assumed to be isotropic with the exception of the abdominal muscle. (see [20] for more details). The original volume conductor model was developed for striated muscle, whose model is mainly 1D. We then adapt it for 2D modeling (tissue level source). We also add to the model the filtering effect of the recording electrodes and the presence of measurement noise [19].

## 2.2   EHG Source Localization

Methods for localization are termed inverse solutions. Indeed, the procedure for the electrical source localization deals with two problems: 1) a forward problem to find the skin potentials for the current source at the origin of the activity (definition of the lead-field), 2) an inverse problem to estimate the source(s) that fit with the given potential distribution at the skin electrodes [36]. The existing methods developed to solve the inverse problem are divided into two main families, the non parametric (over-determined) and parametric (under-determined) methods [9]. In our application on EHG, we do not have a priori information on source number, strength and location. Therefore we focused our interest in the non-parametric methods. In our case, uterine activity can be estimated from EHG by solving an ill-conditioned inverse problem that is regularized using neuroanatomical, computational, and dynamic constraints [31]. 1) The forward problem is to define the rules of the signal propagation from the source to the recorded site (here the abdominal skin). This problem involves calculating the electric potentials generated by known current sources for a given anatomical situation. Forward modeling is done in our case based on a volume-conduction model that describes the geometrical and electrical properties of the tissue in the abdomen above the uterus. This model requires the geometrical description of tissue boundaries in the abdomen. We assume here that the abdomen above the uterus consists of a set of meshes, triangulated surfaces in 3D-space, representing the uterine muscle, abdomen muscle, fat, and skin. If the conductivities within each of these regions are isotropic and constant, the electric potentials can be expressed in terms of surface integrals. The forward EHG problems can then be solved numerically using a boundary-element method (BEM) [18, 8, 10]. We used in this step the same geometrical and electrical definitions as used for the physiological multiscale model (figure 1). The volume conductor is considered to be made of parallel interfaces separating the four different abdominal tissues, namely: the myometrium (where the source is located) with conductivity = 0.2 S/m and depth = 0; the abdominal muscle with conductivity 0.2 S/m and 0.4 S/m respectively in the x and y directions, and thickness = 0.936 cm; fat with conductivity = 0.04 S/m and thickness = 1.132 cm; and skin with conductivity = 0.5 S/m and thickness = 0.2 cm. To limit computation time we only simulate a square of 10x10 cm of the uterus muscle and the above abdomen conductor volume is simulated. This area is supposed to be near the median axis of the uterus, where the surfaces are supposed to be more or less flat and parallel. The mesh is done with a precision of 51x51 elements. The dimension of one cell in the mesh is 0.049 cm.

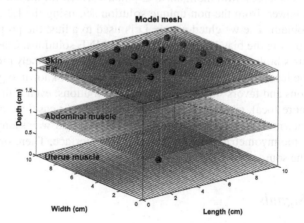

**Fig. 1** The uterus EHG model representation of the volume conductor for the physiological multiscale model, with the four layers (uterine muscle, abdominal muscle, fat and skin). The 16 black disks above the skin layer represent the electrodes. The black disk in the uterus muscle layer represents an example for a source located at the center of the uterine muscle.

BEM calculates the potentials/fields of the non-intersecting homogeneous regions bounded by the uterine muscle, abdominal muscle, fat, and skin surface boundaries, each one having a conductivity values of 0.2, 0.2 and 0.4, 0.04 and 0.33 S/m respectively. The resulting forward model (lead-field) is then used to solve the inverse problem. 2) The inverse problem: An increasing interest in current-density reconstruction algorithms has occurred during the past few years. All these algorithms have in common that elementary dipoles are distributed on regular grids [5]. The calculation of the strengths and position of these dipoles usually leads to a highly under-determined system of equations - the number of unknown dipole components is greater than the number of electrodes and sensors. Thus, it requires additional mathematical constraints (e.g., minimum norm and variance-weighted minimum-norm) to yield to a unique solution. A generalized formulation for the weighted minimum-norm solution of the inverse problem with a squared deviation, $\Delta^2$, and the dipole component vector, $J$, can be written as follows [27]:

$$\Delta^2 = |D(M - L*J)|^2 + \lambda^2 |C*J|^2 \tag{3}$$

The data term, $|D(M - L*J)|^2$, (measuring the closeness of the solution obtained to the data) and the constraining model term, $|C*J|^2$, (measuring the closeness to a given source model) are optimized simultaneously. Both are linked using a regularization parameter, $\lambda$. The choice of this parameter is crucial; in this preliminary work $\lambda$ is equal to 1e-1 based on the analysis of the signal to noise ratio of the surface signals. M is the spatiotemporal measured data matrix (m x n), lead-field matrix L coming from the forward problem (m x c current dipole components), D is an (m x m) weighting matrix of the sensors, and C is a (c x c) weighting matrix of the current dipole components; where m is the channels number, n is signal point number, and c is the number of sources. Minimum norm estimates search for the solution that has the minimum power, from the non-unique solution set, using the L2 norm to regularize the problem. The weighted matrix D is used to adjust the properties of the solution by reducing the bias inherent to classical MNE solutions. Classically, D is a diagonal matrix built from matrix L with non-zero terms inversely proportional to the norm of the lead field vectors. This standard solution is known to generate very smooth solutions and favors superficial source distributions, even if the true source is a deeper, more focal, current generator. However in our case, we consider here that the myometrium layer contains only one cell layer and we assume the source to be close to the myometrium–abdominal muscle interface. Therefore we are not interested in the source depth.

## 2.3   Real Signals

After validation on simulated signals, we apply the source localization algorithm to real uterine EHG signals in order to study the uterine dynamic of one pregnancy contraction (recorded 3 weeks before labor, 3WBL) and one labor contraction recorded on the same woman, in order to get free from anatomical changes. We used a 4x4 electrode recording matrix (16 monopolar channels) located on the woman's abdomen (see [12] for details). The signals were recorded with a sampling frequency of 200 Hz during pregnancy and labor at the Landspitali University Hospital in Iceland, following a protocol approved by the relevant ethical committee (VSN 02-0006-V2). The bipolar and monopolar EHG signals were segmented manually to extract the activity bursts related to uterine contractions. Then due to their very low signal to noise ratio, the monopolar bursts are filtered using a CCA-EMD method developed by our team [11]. The analysis below was applied to these segmented and filtered monopolar uterine bursts.

# 3   Results

## 3.1   Simulated Signals

We have simulated two cases of source generation: i) two asynchronous sources with circle propagation waveform; ii) a longitudinal bar of sources that propagates longitudinally. This "plane wave" type propagation corresponds to a source similar

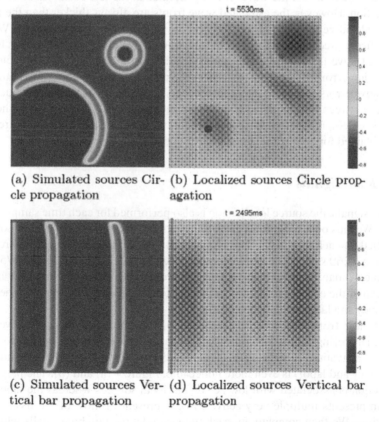

(a) Simulated sources Circle propagation

(b) Localized sources Circle propagation

(c) Simulated sources Vertical bar propagation

(d) Localized sources Vertical bar propagation

**Fig. 2** Simulation of two asynchronous (a) and vertical bars (c) sources, and their estimated localization and propagation (b) and (d) respectively, for a single instant. The colored disks in (b) and the red bar in (d) represent the position of the original sources.

to the ones in case i), but viewed at a point distant from the source. The results of sources localization are presented Figure 2 (a) and (b) for the first case (asynchronous sources which have a delay between their onset), Figure 2 (c) and (d) for the second case (bar of sources that propagate longitudinally). Each picture of Figure 2 represents a single instant of the signal generated in each case. Subfigures (a) and (c) represent the original sources generated by the multiscale physiological model; subfigures (b) and (d) represent the corresponding sources estimated by our algorithm. For figures 2-4, the color scales indicate the source intensity: red = maximal source intensity, blue = no identified source. It is clear from Figure 2 that the implemented forward/inverse algorithm is able to localize the given sources in all cases. We can see in Figure 2 (a) and (b) approximately the same wave front shapes, after the collision of the two circle waves generated by the two asynchronous sources. In figure 2 (b), a yellow circle shape is first generated by the source propagating from the bottom left corner, represented by the blue disk. The second source generated, at the top right corner Figure 2 (a), represented by the magenta disk, is also well localized with some error in its position. In Figure 2 (c) and (d), the method was also able to detect the waves fronts generated at the left of the tissue and their propagation. However, the estimated wave fronts are always thicker than the actual simulated source ones introduced into the physiological multiscale model. We can also notice a precision error in the localization, as well as a smaller power for the estimated wave front than for the initial simulated sources. We thus computed the localization error for two time instants of the asynchronous sources. We computed the relative errors on both, x and y, dimensions of the mesh, as well as the Euclidian distance between the simulated and the estimated sources. The position of the estimated source(s) is the one(s) of the maximum value(s) for each instant. The relative error obtained for both sources and both directions ranges from 7% to 30%.

## 3.2   Real Signals

For real signals, the source localization is also performed for each time sample of the signal. We thus obtain the temporal evolution of the dynamic of the uterine sources, identified for each time instant, for one contraction recorded at 3 weeks before labor (duration = 260 s), and one contraction recorded during labor (duration = 175 s) for the same woman. To be able to visualize the results we plot in Figure 3 (a) and (b) the mean of the estimated localization on the total samples of the real uterine EHG pregnancy and labor bursts respectively.

It is clear from Figure 3 that the power of the estimated source is lower at 3 WBL than during Labor. Furthermore, if we plot one instant representative of the maximal activation for each contraction, this difference between the two situations (pregnancy and labor) is even more noticeable, Figure 4 (a) and (b). The pregnancy contraction is associated with low active and local sources whereas the labor contraction presents multiple very active sources present in most of the tissue at the same time. We then compute, for each time sample, the number of cells which activity is greater than 50% of the maximal source intensity, identified for each whole

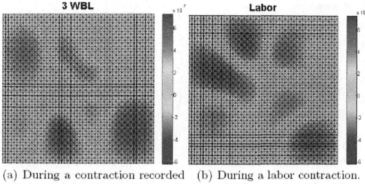

(a) During a contraction recorded    (b) During a labor contraction.
at 3 weeks before labor.

**Fig. 3** Mean of source localization on all the samples in two contractions recorded on the same woman

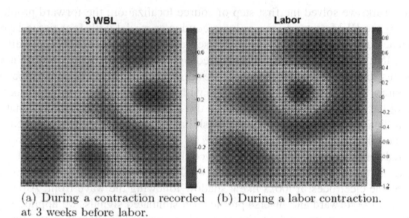

(a) During a contraction recorded    (b) During a labor contraction.
at 3 weeks before labor.

**Fig. 4** Source localization for one sample selected as representative of the maximum activity for two contractions recorded on the same woman

contraction. The results (Figure 5) present an indicator of the contraction dynamic. We can see that the number of active cells is much greater for the labor contraction than for the pregnancy one. The percentage of time when this activity is present is also much greater for the labor (18 %) than for the pregnancy (3 %) contraction. Both results indicate a more synchronous activity of the whole uterine tissue investigated by our electrode grid during labor.

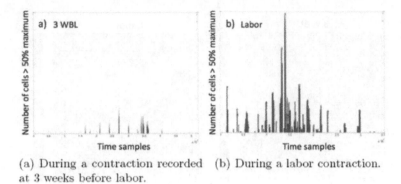

(a) During a contraction recorded    (b) During a labor contraction.
at 3 weeks before labor.

**Fig. 5** Evolution of the number of active cells along the contraction duration, both recorded on the same woman

## 4 Discussion

In this work, we solved the first step of source localization, the forward problem, by using the BEM method for a volume conductor model with four tissue layers. Simulated and real signals where then used as inputs of the inverse problem by using MNE method. The study on simulated signals presented two cases: two asynchronous sources, and a bar of source. Results show that MNE is able to localize the sources and to detect their propagation with some precision error. We think that most of the localization error comes from the down sampling of the mesh, that we used to limit memory usage, since the mesh has been decimated by 4 to speed and ease the calculation. Also for all the cases of simulated signals, the wave front is detected thicker than its actually size. This could be due to the high velocity of wave propagation used in the model and to the well known 'cone shape' of the volume seen by the electrodes. It looks like a shadow following the wave front. After validation of the forward and inverse problem on simulated signals, we applied these methods to a pregnancy and a labor contraction, in order to attempt to localize the origin of the contractions and to detect its propagation, in order to better understand the functioning of uterus during pregnancy and labor. We found that the sources intensity of labor contraction is higher than for pregnancy one. But in all cases, the intensity of the identified sources remained very small. This problem could come from the characteristic of the wMNE method itself (minimum norm least-squares) that searches for the solution that presents the minimum power. We could solve this drawback by using another method, such as sLOTERA or MUSIC-based algorithms that gives also information on the depth of the source. We clearly evidenced a multiple source activity during labor of greater extend than during pregnancy. This is in agreement with that was expected: labor contractions are supposed to be more global (propagating fast to the whole uterus) whereas pregnancy contractions are supposed to remain more or less localized (smaller propagation). But the localized sources did not demonstrate any "propagation like" behavior. Indeed, the dynamic of the sources evidenced for each time sample of each contraction appeared more like a blowing or an erupting one than a propagated one. This could lend support to the hypothesis proposed by Young [37] who stated that the mechanism of global organ-level communication is probably not via action potential propagation, but may be via a hydrodynamic-stretch activation mechanism, including a communication and a coupling between the electrical and the mechanical systems controlling the uterus. This complex behavior should be studied by improving the current existing EHG model and introducing a mechanical component, making this model multi scale and multiphysics.

## 5   Conclusion and Perspectives

A first preliminary study on EHG source localization problem was presented in this work. A validation of forward and inverse methods was done on simulated signals in different cases simulated by a multiscale EHG model. Then we applied these methods on real uterine EHG bursts, corresponding to contractions recorded during pregnancy and labor. The methods were able to identify the sources injected in the model for the simulated signals, with some error, related to the mesh decimation. On real EHG signals, the methods gave logical results that fit with the hypothesis posed for long time in the literature. In a future work, we will improve the used methods: varying the velocity and of the regularization parameter $\lambda$, use the whole mesh, compare different inverse methods to see which one is more suited to our EHG source localization problem. We will also compare the results obtained on more real EHG signals, in different situations (normal and risk pregnancies, labor). An evolution in the model is also needed to introduce a mechanical component, making this model multi scale and multiphysics, one step closer to the real complexity of the uterine organ.

**Acknowledgments.** The authors wish to thanks the ERASysBio+ European program for funding, and the Labex MS2T, funded by the French Government, through the program "Investments for the future" managed by the National Agency for Research (Reference ANR-11-IDEX-0004-02).

## References

[1] Becker, H., Albera, L., Comon, P., Haardt, M., Birot, G., Wendling, F., Gavaret, M., Bénar, C.G., Merlet, I.: EEG extended source localization: Tensor-based vs. conventional methods. NeuroImage 96, 143–157 (2014)

[2] Dale, A.M., Sereno, M.I.: Improved localizadon of cortical activity by combining EEG and MEG with MRI cortical surface reconstruction: A linear approach. Journal of Cognitive Neuroscience 5(2), 162–176 (1993)

[3] Devedeux, D., Marque, C., Mansour, S., Germain, G., Duchene, J.: Uterine electromyography: a critical review. American Journal of Obstetrics and Gynecology 169(6), 1636–1653 (1993), PMID: 8267082

[4] Euliano, T.Y., Marossero, D., Nguyen, M.T., Euliano, N.R., Principe, J., Edwards, R.K.: Spatiotemporal electrohysterography patterns in normal and arrested labor. American Journal of Obstetrics and Gynecology 200(1), 54.e1–54.e7 (2009)

[5] Fuchs, M., Wagner, M., Köhler, T., Wischmann, H.A.: Linear and nonlinear current density reconstructions. Journal of Clinical Neurophysiology: Official Publication of the American Electroencephalographic Society 16(3), 267–295 (1999), PMID: 10426408

[6] Garfield, R.: Control and assessment of the uterus and cervix during pregnancy and labour. Human Reproduction Update 4(5), 673–695 (1998)

[7] Garfield, R.E., Maul, H., Shi, L., Maner, W., Fittkow, C., Olsen, G., Saade, G.R.: Methods and devices for the management of term and preterm labor. Annals of the New York Academy of Sciences 943(1), 203–224 (2001)

[8] Gramfort, A., Papadopoulo, T., Olivi, E., Clerc, M.: OpenMEEG: opensource software for quasistatic bioelectromagnetics. BioMedical Engineering OnLine 9(1), 45 (2010), PMID: 20819204

[9] Grech, R., Cassar, T., Muscat, J., Camilleri, K.P., Fabri, S.G., Zervakis, M., Xanthopoulos, P., Sakkalis, V., Vanrumste, B.: Review on solving the inverse problem in EEG source analysis. Journal of NeuroEngineering and Rehabilitation 5(1), 25 (2008), PMID: 18990257

[10] Hall, W.S.: The Boundary Element Method. Kluwer (1994)

[11] Hassan, M., Boudaoud, S., Terrien, J., Karlsson, B., Marque, C.: Combination of canonical correlation analysis and empirical mode decomposition applied to denoising the labor electrohysterogram. IEEE Transactions on Biomedical Engineering 58(9), 2441–2447 (2011)

[12] Hassan, M., Terrien, J., Karlsson, B., Marque, C.: Interactions between uterine EMG at different sites investigated using wavelet analysis: Comparison of pregnancy and labor contractions. EURASIP J. Adv. Signal Process., 17:1–17:10 (March 2010)

[13] Hassan, M.M., Terrien, J., Muszynski, C.: A Alexandersson, C. Marque, and B. Karlsson. Better pregnancy monitoring using nonlinear correlation analysis of external uterine electromyography. IEEE Transactions on Biomedical Engineering 60(4), 1160–1166 (2013)

[14] Hauk, O.: Keep it simple: a case for using classical minimum norm estimation in the analysis of EEG and MEG data. NeuroImage 21(4), 1612–1621 (2004)

[15] Hauk, O., Wakeman, D.G., Henson, R.: Comparison of noise-normalized minimum norm estimates for MEG analysis using multiple resolution metrics. NeuroImage 54(3), 1966–1974 (2011)

[16] Jacob, F.: The logic of life: a history of heredity. Vintage books. Vintage Books (1976)

[17] Koenigsberger, M., Sauser, R., Lamboley, M., Bény, J.-L., Meister, J.-J.: Ca2+ dynamics in a population of smooth muscle cells: modeling the recruitment and synchronization. Biophys. J. 87(1), 92–104 (2004)

[18] Kybic, J., Clerc, M., Abboud, T., Faugeras, O., Keriven, R., Papadopoulo, T.: A common formalism for the integral formulations of the forward EEG problem. IEEE Transactions on Medical Imaging 24(1), 12–28 (2005)

[19] Laforet, J., Marque, C.: uemg: Uterine emg simulator. In: 7th International Workshop on Biosignal Interpretation, BSI 2012, Como, Italy (2012)

[20] Laforet, J., Rabotti, C., Mischi, M., Marque, C.: Improved multi-scale modeling of uterine electrical activity. IRBM 34(1), 38–42 (2013)

[21] Laforet, J., Rabotti, C., Terrien, J., Mischi, M., Marque, C.: Toward a multiscale model of the uterine electrical activity. IEEE Transactions on Biomedical Engineering 58(12), 3487–3490 (2011)

[22] López, J.D., Litvak, V., Espinosa, J.J., Friston, K., Barnes, G.R.: Algorithmic procedures for bayesian MEG/EEG source reconstruction in SPM. NeuroImage 84, 476–487 (2014)

[23] Lucovnik, M., Maner, W.L., Chambliss, L.R., Blumrick, R., Balducci, J., Novak-Antolic, Z., Garfield, R.E.: Noninvasive uterine electromyography for prediction of preterm delivery. Am. J. Obstet. Gynecol. 204(3), 228.e1–228.10 (2011)

[24] Marque, C., Duchene, J.: Human abdominal EHG processing for uterine contraction monitoring. Biotechnology 11, 187–226 (1989)

[25] Marque, C., Laforet, J., Rabotti, C., Alexandersson, A., Germain, G., Gondry, J., Karlsson, B., Leskosek, B., Mischi, M., Muszinski, C., Oei, G., Peuscher, J., Rudel, D.: A multiscale model of the electrohysterogram the BioModUE_PTL project. In: 2013 35th Annual International Conference of the IEEE Engineering in Medicine and Biology Society (EMBC), pp. 7448–7451 (July 2013)

[26] Mattout, J., Phillips, C., Penny, W.D., Rugg, M.D., Friston, K.J.: MEG source localization under multiple constraints: An extended bayesian framework. NeuroImage 30(3), 753–767 (2006)

[27] Mideksa, K.G., Hellriegel, H., Hoogenboom, N., Krause, H., Schnitzler, A., Deuschl, G., Raethjen, J., Heute, U., Muthuraman, M.: Source analysis of median nerve stimulated somatosensory evoked potentials and fields using simultaneously measured EEG and MEG signals. In: Conference Proceedings:... Annual International Conference of the IEEE Engineering in Medicine and Biology Society. IEEE Engineering in Medicine and Biology Society. Conference, pp. 4903–4906 (2012), PMID: 23367027

[28] Miyoshi, H., Boyle, M.B., MacKay, L.B., Garfield, R.E.: Gap junction currents in cultured muscle cells from human myometrium. American Journal of Obstetrics and Gynecology 178(3), 588–593 (1998)

[29] Montes-Restrepo, V., Mierlo, P.V., Strobbe, G., Staelens, S., Vandenberghe, S., Hallez, H.: Influence of skull modeling approaches on EEG source localization. Brain Topography 27(1), 95–111 (2014)

[30] Parthimos, D., Edwards, D.H., Griffith, T.M.: Minimal model of arterial chaos generated by coupled intracellular and membrane ca2+ oscillators. Am. J. Physiol. 277(3 Pt. 2), H1119–H1144 (1999)

[31] Pirondini, E., Babadi, B., Lamus, C., Brown, E.N., Purdon, P.L.: A spatially-regularized dynamic source localization algorithm for EEG. In: Conference Proceedings:... Annual International Conference of the IEEE Engineering in Medicine and Biology Society. IEEE Engineering in Medicine and Biology Society. Conference, pp. 6752–6755 (2012), PMID: 23367479

[32] Planes, J.G., Morucci, J.P., Grandjean, H., Favretto, R.: External recording and processing of fast electrical activity of the uterus in human parturition. Medical and Biological Engineering and Computing 22(6), 585–591 (1984)

[33] Rabotti, C., Mischi, M., Beulen, L., Oei, S.G., Bergmans, J.W.M.: Modeling and identification of the electrohysterographic volume conductor by high-density electrodes. IEEE Trans. Biomed. Eng. 57, 519–527 (2010)

[34] Rabotti, C., Mischi, M., Oei, S.G., Bergmans, J.W.M.: Noninvasive estimation of the electrohysterographic Action-Potential conduction velocity. IEEE Transactions on Biomedical Engineering 57(9), 2178–2187 (2010)

[35] Rihana, S., Terrien, J., Germain, G., Marque, C.: Mathematical modeling of electrical activity of uterine muscle cells. Medical & Biological Engineering & Computing 47(6), 665–675 (2009)

[36] Shirvany, Y., Edelvik, F., Persson, M.: Multi-dipole EEG source localization using particle swarm optimization. In: Conference Proceedings:... Annual International Conference of the IEEE Engineering in Medicine and Biology Society. IEEE Engineering in Medicine and Biology Society. Conference, pp. 6357–6360 (2013), PMID: 24111195

[37] Young, R.C.: Myocytes, myometrium, and uterine contractions. Ann. N. Y. Acad. Sci. 1101, 72–84 (2007)

# Improving the Operability of Personal Health Record System by Dynamic Dictionary Configuration for OCR

Atsuo Yoshitaka, Shinobu Chujyou, and Hiroshi Kato

**Abstract.** Information is managed by computers, which enabled us to handle, maintain, and distribute huge amount of data. However, data transfer from the source to the destination of information is not always achieved in ideal way. One good example in Japan is personal health related information, such as the report of health check-up led by government. Paper-printed report is considered to be a reasonable way for distributing the report especially before the era when the internet becomes one of the infrastructures of the society. However, the Japanese government recognized the need of individual-based health maintenance for extending healthy life expectancy. In accordance with this context, PHR or personal health record systems are developed and operated for experimental proof. One of the difficulties revealed is low operability of data migration from printed data. One of the promising solutions for that issue is to apply OCR for printed document. In this paper, we propose a method of dictionary optimization to improve the accuracy of OCR by adaptable switching of dictionary parameter. Experimental results showed that the recognition accuracy improved for both Japanese letters and figures.

**Keywords:** Personal health record, data migration, OCR, dictionary optimization.

Atsuo Yoshitaka
School of Information Science, Japan Advanced Institute of Science and Technology
1-1 Asahidai, Nomi, Ishikawa, 923-1292, Japan

Shinobu Chujyou
goowa inc.
3-4 Unetanaka, Kanazawa, Ishikawa 920-0343, Japan

Hiroshi Kato
Life Care on Demand
3-13-5 Asahimachi, Kanazawa, Ishikawa 920-0941, Japan

© Springer International Publishing Switzerland 2015                                  541
V.-H. Nguyen et al. (eds.), *Knowledge and Systems Engineering,*
Advances in Intelligent Systems and Computing 326, DOI: 10.1007/978-3-319-11680-8_43

# 1 Introduction

Japan is one of the countries where the life expectancy is high in the world. The ratio of aged person is increasing year by year, and Japanese government is recognizing that policies for improving healthy life expectancy are more important than those for improving life expectancy. This idea includes the idea of improving the quality of life especially for aged persons.

Japanese system for medical care for the elderly is well consolidated. For the next step for matured aged society, Japanese government has put more focus on developing infrastructure for the people to maintain health by personal effort. In accordance with this movement, various aspects of EHR[1] or PHR systems have been studied. Computers have enabled to manage information such as medical record, medical prescription, images of X-ray picture, CT-scanner, MRI images within a hospital or a medical inspection institute. However, data sharing among hospitals or clinics is not well established yet, since widely accepted standardization of medical data is still in discussion as in HL7 CDA, for example [5]. Current situation of medical data sharing among hospitals is limited. EHR or PHR systems are considered as breakthrough of this issue. EHR infrastructure includes hospitals, clinics, medical inspection institutes, local governments, and citizen. Several practical experiments for EHR system are going under the cooperation with local government. When we mention PHR system, more emphasis is on how the citizen maintains his/her own health record based on his/her own effort, using personal computers with PHR service maintained by PHR service provider. In this paper, we focus on PHR systems rather than EHR systems, since we are more motivated on how we can offer more effective platform for PHR data management.

As mentioned, PHR systems[2] [3] [4] are designed to consolidate health information such as height, weight, BMI (Body Mass Index), blood type, blood pressure, allergic food, medication, history of disease, and so on. Final goal of the development of PHR infrastructure is to offer the environment to maintain such information based on continuous individual effort, and to be able to share such information with his/her family and doctors even if he/she has his/her illness diagnosed regardless of his/her location. If the infrastructure of PHR is established, it means that it is possible to refer to personal health information anytime and anywhere needed without unwilled duplication of basic medical check.

There are several studies and demonstration experiments on PHR system. In Uchinada town in Ishikawa prefecture, gLife care on dementh project has been carried out, which is led by the local government of Uchinada town. The gLife care on demandh project has offered a system for PHR service called gLicoh. The Lico is a web based system for maintaining personal health record by individual effort. When a user enters personal health record such as the report of annual health checkup, he/she has to enter all the figures of items manually one by one into fields of the web page with a keyboard, since distribution of such data by electric data is not common yet. This task of data migration is often time-consuming, stressful, and error-prone. The experience of the Life care on demand project revealed that most users of the Lico wished to avoid the above-mentioned data migration task.

One possible solution for this issue is to exploit an image scanner connected with a personal computer to capture a printed form of health check-up to apply OCR (Optical Character Recognition). This solution requires a user to prepare an image scanner, which will be an additional investment for maintaining PHR data. In addition to this, an OCR engine in general image scanner also reads out characters in each row, without proper recognition of the logical structure such as a table. Moreover, it does not read out printed information with acceptable accuracy of character recognition. Because of these reasons, simple installation of a general image scanner will not be a promising solution.

Recently, more people are using so called a tablet PC, which equips CCD camera and a LCD with a touch panel. From the point of view of minimizing the investment for equipping devices which may be specialized for PHR data migration, assuming to utilize such a tablet PC can be considered as one of the solution for this issue. In this study, we assume for a user of PHR system to use a tablet PC for PHR service, especially for data migration. Our scenario is as follows. The user will take photos of a printed report of health checkup with a built-in camera of the tablet PC. Then the images are referred to by PHR front end for OCR. After that, PHR data is transferred to PHR server, and he/she can display PHR data on the display of the tablet PC. Compared with the case of maintaining PHR data with laptop (or desktop) PC with a keyboard and a mouse, that hardware configuration is advantageous since LCD with touch panel is able to offer intuitive user interface which will be more familiar with general users.

As mentioned, one of the major issues in data migration from printed form with OCR is how we assure acceptable rate of accuracy of character recognition. This issue is more serious than in the case of assuming image scanner, since images shot with a hand-held tablet may be blurred, which results to deteriorate the accuracy of character recognition. The primary objective for considering OCR for data migration is to reduce the burden of user. This objective will not be achieved is the accuracy of character recognition is not satisfactory; unacceptable amount of manual error correction for PHR data recognized by OCR process is equal to or worth than entering all the data items manually.

In this paper, we focus on the issue of improving the accuracy of character recognition by OCR engine for PHR data migration. Since there are still obstacles for realizing the distribution of PHR data to end user from medical institution, it is inevitable to assume migrating PHR data from a printed form. Though this technology is considered to be a transient technique, however, we believe this is significant in order to offer better PHR platform as soon as possible.

The reminder of this paper is as follows. Section 2 introduces the current implementation of a PHR system Lico. In section 3, we will present the design of our PHR platform which migrates PHR data into PHR server by taking photos of printed forms. We describe the detail of dynamic dictionary configuration for OCR in Section 4. The experimental result of the proposed method is presented in Section 5. Concluding remarks and future work is mentioned in Section 6.

## 2 An Example of PHR System, Lico

### 2.1 Overview

The population ratio of senior is becoming higher in Japan, therefore, it is an urgent issue to develop and offer comprehensive medical/health support system. There are several trials of demonstrational experiments. One of them is led by a local government of Uchinada town in Ishikawa prefecture, Japan. The project there is called gLife care on demandh [6], and the system for PHR data management is named gLicoh. Lico system offers web based PHR service which enables a user to enter PHR data such as a report of annual health checkup, and to view stored PHR data in the form list of graph. The gLife care on demandh project was supported by the subsidy for promoting utilization and exploitation regional information communication technology that is led by the Ministry of Internal Affairs and Communications.

Figure 1 shows a screen snapshot of the interface for entering personal health record in the current implementation of Lico system. As you see in the figure, the user interface for entering PHR data is designed using ordinary interaction widget of text boxes that are commonly used for data input form. In this example, a user is expected to enter the date of health checkup, note, height, weight, abdominal circumstance, and body fat percentage.

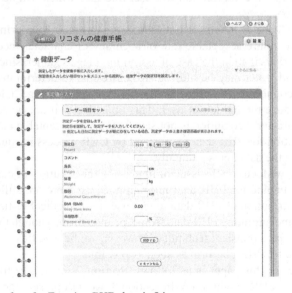

**Fig. 1** User Interface for Entering PHR data in Lico

Approximately, from 30 to 40 items are measured in an annual health checkup, it is easy to image that entering all of such data into the form as shown in Figure 1 is time-consuming, stressful, and error-prone for the user. Figure 2 and 3 are examples of displaying PHR data stored in PHR server. Figure 2 shows the case of displaying the transition of measurement in the form of table, and Figure 3 is an alternative way of displaying one of the data item in the form of graph. Note that the design of data migration interface as shown in Figure 1 is based on the current limitation that end users cannot obtain PHR data in electric form from medical institutes directly.

**Fig. 2** Viewing PHR data by table format

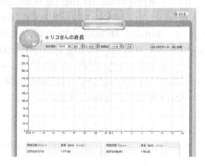

**Fig. 3** Displaying the transition of height by graph

## 2.2 PHR Data Migration

Currently, over 600 accounts are registered at Lico system. Form the experience and feedback from the user, it is revealed that the insufficient operability of data migration, i.e., migrating printed data into PHR server manually, discourages the user to continue to use the system. One reason is that the task of typing all the data manually is stressful and time-consuming. Another reason for difficulty relates to the unfamiliarity of terms such as MCV, MCH, GOT, or GPT, all of which are the indices on blood quality or soundness of liver. This causes another aspect of discouraging the motivation of using the system, since information which is difficult to understand is stress for a user to face, and therefore, human error in the process of migration is more likely to occur.

Since electric data distribution from the medical institutes to end users is not established because of the immaturity of environments for that, one of the most reasonable solutions is to transform printed PHR data into electric data by OCR. There are mainly two methods to obtain electric data by OCR; one is to exploit an image scanner, and another is to take photos for applying OCR. The former requires a user to prepare an image scanner, which is more special and less versatile device that the device to take photo such as a camera module in tablet PC or smart phone. In addition to this, general users are more familiar to use a camera than an image scanner. Since most tablet PCs and smart phones equip camera module to take photos, we chose a tablet PC as a device for PHR data migration. Another advantage of assuming tablet PC is that we can provide a user with intuitive user interface which offers direct manipulation on LCD display with touch panel.

The accuracy of character recognition by OCR process is not always perfect, in spite of the maturity of character recognition algorithm. One reason of imperfect accuracy is in the insufficient resolution of source image. Another is the case where an image is captured with blur, especially in the case where it is captured by holding the camera device with hands. There is also a problem where a character is misclassified into different character but visual appearance is quite similar. An example is that a single glh (the lower case of the alphabet gLh) is misrecognized as a single g1h (the letter which corresponds to the number goneh). This problem is inevitable if the OCR engine does not hold a dictionary of the letters specific to the printed fonts that are to be recognized. In addition to above-mentioned problems to be solved, there is also a problem in recognizing spatially structured data such as a table in a document. If an image including tables with items with null values, output of recognition will be a sequence of characters in accordance with the order of scan, the logical structure or correspondence between the item name and item value will be lost.

In this paper, we focus on the issue of improving the accuracy of character recognition for PHR data migration, assuming to capture a printed form with a camera mounted in a tablet PC.

# 3 PHR Data Management Front-End

## 3.1 System Organization

The system organization of the PHR management platform is illustrated in Figure 4. The software installed in a tablet PC consists handles both data migration by capturing printed form of **PHR** data and data presentation by accessing **PHR** server. The form capturing module guides a user to capture proper region of the printed **PHR** data such as the report of annual health checkup, and the captured image data is transferred to OCR processor. After applying OCR processing to obtain electric data of **PHR**, the resultant data is stored into **PHR** server via **PHR** manager. OCR processor is placed on the server side in this figure, however, it may also be placed in the tablet PC as one of the components of **PHR** front-end, if the tablet PC has processing performance enough to offer acceptable response time due to OCR processing.

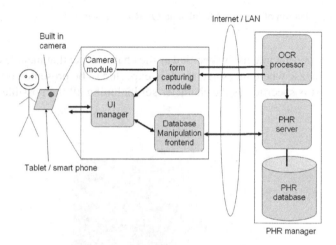

**Fig. 4** The Architecture of the PHR Front-end

## 3.2 Form Capture Assistance and OCR Processor

Figure 5 illustrates the detail of form capture module and OCR processor. In accordance with a form being captured, form skeleton to indicate regions to capture is selected and superimposed onto the image that is going to be captured by the built-in camera. When an image is captured by touching a eshutterf button, it is then segmented into sub-images in accordance with the spatial structure of the form. After that, each sub-image with the spatial information in the form is sent to OCR processor. Since spatial information which indicates the position in the form is associated with meta-data of data type (e.g., enumerical dataf, eevaluation represented by Japanese wordsf), dictionary for character recognition is configured with regard to the meta-data.

Dictionary configuration is carried out by switching ewhite listf in accordance with the meta-data of information type. The white list configures the set of candidate characters to be recognized, which specifies the set of characters supposed to appear in the sub-image to be processed. If a region where the figures of measured value is

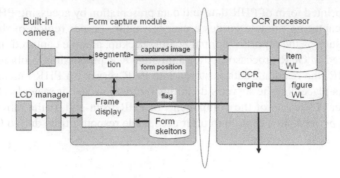

**Fig. 5** The Organization of Capture Module and OCR Processor

expected to appear, a while list is specified as the character set that includes numbers and decimal. Incase where measured result is expected to be presented by Japanese word, white list is specified as the character set including all the possible Japanese characters.

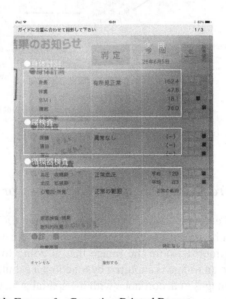

**Fig. 6** Displaying Guide Frames for Capturing Printed Report

Figure 6 is an example of guidance frame display which is shown when a user is going to capture a part of printed form of annual health checkup. A user is guided to adjust the position of a tablet PC so that the guide frames will be overlapped onto the corresponding areas of the printed form.

After touching a eshutterf button, the image is captured and then it is segmented for OCR processing with adapted dictionary configuration. The result of character recognition is displayed as shown in Figure 7. The captured image is displayed on the left and the values recognized by OCR process are displayed on the right. Scrolling on the table is carried out simply by dragging a finer on the table. Captured image and displayed form are synchronized to each other in scrolling so that a user can compare to check recognition error easily.

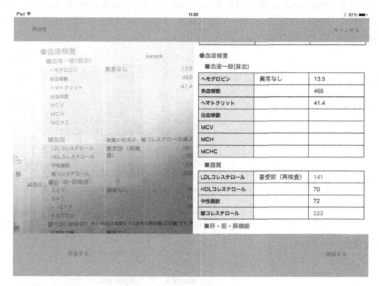

**Fig. 7** An Example in Confirmation of Captured Data

## 4 Dynamic Dictionary Configuration for OCR

Dynamic Dictionary Configuration is carried out in the process of character recognition. In the process of character recognition for general documents, it is not proper to expect the appearance of a word or a figure in a document converges to a predefined subset of words or letters. However, we can expect a word or figure is one of the elements in a set consisting of limited number of instances. For example, the possible words appeared in a report of annual health check-up corresponds to item names such as eheightf, eweightf, eBMIf, eblood pressuref, and so on. The number of possible word is quite limited compared with the case of general documents.

The idea of dynamic dictionary configuration is based on this observation. Since the number of kanji (i.e., Chinese letters used in Japan) designated for everyday use is more than 2000, excluding other Chinese letters that are not used for annual health check-up will eventually contribute to improve the accuracy of OCR, since the accuracy of character recognition will be improved by excluding Chinese letters that will never appear from the dictionary for OCR.

Figure 8 illustrates how adaptive OCR configuration is performed in data migration process. We assume that a user will place a tablet so that the form skeleton is overlapped onto the corresponding part of the printed report of health check-up. Possible set of character, i.e., Chinese letter, Japanese character, and/or figures is defined for each sub-region in the form skeleton as shown in Figure 8. In this Figure, left side part for the area of anthropometric is associated with OCR region 1, and word/value dictionary as well as item WL (white list)/figure WL which define possible words/range of characters and values, respectively. In OCR process, that region is segmented and sent to OCR engine by applying above mentioned white list. After that, word/value based consistency is checked by referring to word/value dictionary so as to correct outlier.

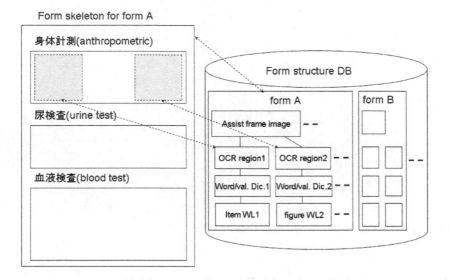

**Fig. 8** Organization of Capture Assist Frame and OCR Configuration

## 5   Experimental Result

The experimental result to evaluate the effectiveness of adaptive configuration of recognition dictionary is shown in Table 1. We used Tesseract-OCR [7] for OCR engine. Standard configuration of dictionary is specified by applying the option as e-l jpnf in case for recognizing Japanese characters, alphabet, and figure. Another option of e-l engf is preferable in case for recognizing figures, simply because there are less characters which may be misrecognized. We configured dictionary so that the OCR engine refers to only the characters that are possible to be appeared in item name in case of item name recognition, and configured it so that it refers to only the figure (i.e., number from 0 to 9) and decimal.

In order for obtaining best performance with standard configuration of dictionary, we applied e-l jpnf for item name recognition, and applied e-l engf for figures. The accuracy of recognizing item name with standard configuration was 52%, and that of figure was 82%. Fractions beside the percentage denote the number of correctly recognized characters out of characters being recognized.

The accuracy obtained by applying adaptive dictionary configuration is shown in the rightmost column in Table 1. The result shows that approximately 27% and 9% improvement are achieved in item name recognition and figures, respectively. Evaluating the number correctly recognized characters by adaptive configuration out of misrecognized characters by standard configuration, 56% and 48% of misrecognized characters of item name and figures are correctly recognized by adaptive configuration, respectively.

**Table 1** Comparison of Accuracy

| type | Standard Config. | Adaptive Config. |
|------|------------------|------------------|
| Item name | 52%(456/876) | 79%(692/876) |
| numeric value | 82%(570/695) | 91%(630/695) |

## 6  Conclusion

In this paper, we proposed the method for improving the accuracy of character recognition for the PHR front-end system. Guide frame display in capturing printed form let a user know where and how to take printed form with desirable resolution, direction, and region. Captured image is segmented in accordance with the guide frame so as to make sub-images based on the spatial structure of the form. Each sub-image is then applied OCR processing for transferring printed information to electric data. Adaptive dictionary configuration is carried out with regard to the position on the form, which excludes characters that will not appear in the position. This improves the accuracy of character recognition for both Japanese letters and figures.

This result contributes to alleviate burden for correcting misrecognized characters of figures on screen, however, user test for evaluating whether the current accuracy of recognition is acceptable or not is part of our future work.

**Acknowledgement.** The authors would like to express appreciation to the members of this project and Ms. Y. Nakada at the Industrial Collaboration Promotion Center, JAIST, for their valuable discussion. We would also like to acknowledge Mr. Y. Nishikawa, Mr. S. Urabe, and Mr. H. Yamashina for their effort of developing the system. This work is supported by the Strategic Information and Communications R&D Promotion Programme (SCOPE) that is supervised by the Ministry of Internal Affairs and Communications, Japan.

## References

[1] Tsiknakis, M., Katehakisa, D., Orphanoudakis, S.C.: A health information infrastructure enabling secure access to the life-long multimedia electronic health record. In: CARS 2004 - Computer Assisted Radiology and Surgery, Proceedings of the 18th International Congress and Exhibition, vol. 1268, pp. 289–294 (2004)

[2]  Dean, F.: Sittig: Personal health records on the internet: a snapshot of the pioneers at the end of the 20th Century. International Journal of Medical Informatics 65(1), 1–6 (2002)

[3]  Kahn, J.S., Aulakh, V., Bosworth, A.: What It Takes: Characteristics Of The Ideal Personal Health Record. Health Affairs 28(2), 369–376 (2009)

[4]  Reti, S.R., Feldman, H.J., Ross, S.E., et al.: Improving personal health records for patient-centered care. Journal American Medical Informatics Association 17, 192–195 (2010)

[5]  Dolin, R.H., Alschuler, L., et al.: The HL7 Clinical Document Architecture. Journal American Medical Informatics Association 8, 552–569 (2001)

[6]  Lico PHR service, https://www.1-cod.com/

[7]  tesseract-ocr, https://code.google.com/p/tesseract-ocr/

# Real-Time Rehabilitation System of Systems for Monitoring the Biomechanical Feedbacks of the Musculoskeletal System

Tien Tuan Dao, Philippe Pouletaut, Didier Gamet, and Marie Christine Ho Ba Tho

**Abstract.** Functional rehabilitation aims at recovering the locomotion dysfunction of the human body by the physical therapy. The objective of this present study was to develop a system of systems (SoS) for monitoring the biomechanical feedbacks of the musculoskeletal system during the rehabilitation exercises. Wireless sensors (Kinect camera sensor, Shimmer kinematics and Electromyography (EMG) sensors) coupled with a biomechanical model was used to supervise the joint kinematics, muscle forces and joint contact area of the musculoskeletal system of the human body. A calculation time reduction strategy for joint contact area analysis was established to reach the constraints of real time simulation. A pilot case study was applied on a patient with post-polio residual paralysis. Joint kinematics and contact area at the knee level were supervised during flexion motion using video-based non-contact Kinect sensor. For the first time, an enhanced rigid multi-bodies model integrating joint contact area behavior was used for a rehabilitation purpose. Our system would be of great interest in the supervision of physical therapy exercises in non-clinical environments (e.g. rehabilitation at home).

**Keywords:** Functional Rehabilitation, System of Systems, Musculoskeletal System, Markerless Motion Capture, Real-time Simulation.

## 1 Introduction

Functional rehabilitation aims at recovering the locomotion dysfunction of the human body by the physical therapy exercises [1], [2], [3], [4]. This involves performing controlled physical and occupational therapy interventions with or without the assistance of physiotherapist to improve musculoskeletal strength and flexibility as well as the range of motion. At the moment, functional rehabilitation has been com-

Tien Tuan Dao · Philippe Pouletaut · Didier Gamet · Marie Christine Ho Ba Tho
UTC CNRS UMR 7338, Biomechanics and Bioengineering (BMBI),
University of Technology of Compiègne, BP 20529, 60205 Compiègne Cedex, France

© Springer International Publishing Switzerland 2015                                553
V.-H. Nguyen et al. (eds.), *Knowledge and Systems Engineering*,
Advances in Intelligent Systems and Computing 326, DOI: 10.1007/978-3-319-11680-8_44

monly realized in clinical environment under supervision of physiotherapists [5]. However, the supervision and the evaluation of a rehabilitation motion pattern remain a medical and engineering challenge due to the lack of biofeedback information about the effect of the rehabilitation motion on the human biological tissues and structures. Recently, rehabilitation systems using immersive virtual reality technologies have been developed to provide useful reinforced biofeedbacks (e.g. functional measurements) during rehabilitation exercises such as motion velocity (speed), duration of motion pattern (time), ergonometric measurement, video data or joint patterns [6], [7], [8], [9], [10], [11], [12]. These quantitative biofeedbacks may be used to identify the musculoskeletal impairments and assess the quality of the rehabilitation motion as well as to assist the patient or the physiotherapist to correct the motion patterns. Virtual simplified avatar has been usually created to represent the patient body [7]. However, these systems provided only external information (e.g. kinematics) of the musculoskeletal system during rehabilitation motion. In fact, the acquisition of internal information (muscle forces, joint contact behavior) inside the musculoskeletal system is still a challenging problem for such useful rehabilitation systems.

The kinematics of the musculoskeletal system is commonly acquired using accelerometer [13], [14] or 3D motion capture system (e.g. VICON, OptoTrack, Motion Analysis) [9], [15], [16]. These systems provide kinematical data with a good precision but the set up protocol is complex (e.g. definition of skin-based marker clusters or acquisition requirement in well-controlled environment) and time-consuming. Recently, Kinect sensor has been widely used to provide kinematics data with a quasi-similar precision as 3D motion capture systems for planar motions [9]. Moreover, this technology has been used to develop rehabilitation system for monitoring locomotion [7], [8], [9], [10] and cognitive rehabilitation exercises [17]. The main advantages of this video-based non-contact technique are the portability, the ease-to-use capacity, the cheap price and especially the possibility to develop a home-based rehabilitation system, which may lead to reduce significantly medical cost, infrastructures and human resources. However, the use of only Kinect sensor leads to limited biofeedback information as only planar kinematics was acquired and supervised. Consequently, the first objective of this present study was to couple this video-based non-contact system with Shimmer kinematics sensors to obtain 3D joint kinematics. Moreover, an enhanced musculoskeletal model integrating joint contact behavior was developed to provide EMG-driven muscle forces as well as joint contact area during the rehabilitation motions in an immersive virtual reality environment. Joint contact behavior could be accurately acquired using finite element modeling [18]. However, this modeling approach is very time-consuming and this does still not satisfy the requirement of a real time biomechanical simulation with rapid reinforced biofeedbacks. Consequently, we proposed an enhanced rigid multi-body model allowing the musculoskeletal animation and joint contact area tracking during rehabilitation motions.

Rehabilitation system is a complex system with dynamic behavior of physical therapy exercises in space and time. Moreover, the motion pattern is uncontrolled and unpredictable due to the human variations. Furthermore, the integration of a biomechanical model into a rehabilitation system leads to a computational and biological complexity [19], especially in real time simulations. Emergent events and chaos may be appeared for bad motion patterns. From system engineering point of view, rehabilitation system may be classified in the most complex systems [20], [21], [22]. The development of such a complex rehabilitation system integrating a biomechanical model needs an innovative and flexible engineering methodology. Recently, we introduced a biomechanical system of systems (SoS) to deal with the complexity of biomechanical data and models [23]. In fact, the second objective of this present study was to develop software architecture of our rehabilitation system using system of systems approach for monitoring the biomechanical feedbacks (muscle forces, joint kinematics and contact area) of the musculoskeletal system in real time conditions.

## 2  System Description and Software Specifications

The flow chart of our system of systems for monitoring the kinematics and joint patterns of the musculoskeletal system is shown in Fig. 1. It consists of a data acquisition and management system, a multi-physical modeling system and a graphical user interface (GUI) system. The workflow of our system for a patient/subject under investigation consists of the following steps: 1) geometrical data acquisition using medical imaging at clinical center (hospital, clinics); 2) medical imaging processing to reconstruct the 3D geometrical model; 3) development of the biomechanical model derived from image-based 3D geometries; and 4) real-time monitoring of the functional rehabilitation motions.

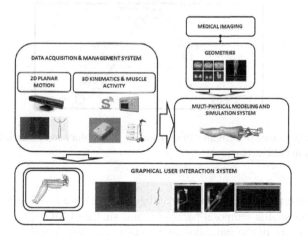

**Fig. 1** Flow chart of our system of systems for the monitoring biomechanical feedbacks of the musculoskeletal system during functional rehabilitation motions

## 2.1 Data Acquisition and Management System

A Kinect camera was used to acquire the kinematic data of the musculoskeletal system in real-time conditions. This marker-less motion capture system has a RGB (red, green, and blue) camera and a pair of depth sensors including an infrared laser projector and a monochrome CMOS (complementary metal-oxide-semiconductor) sensor. The Kinect system may capture the 3D geometrical data and 2D planar kinematics in ambient light conditions. The non-commercial Kinect software development kit (SDK) v1.7.0 for Windows and Visual C# (Microsoft, USA) were used as programming languages to access into Kinect capabilities (e.g. raw sensor streams, skeletal tracking) to develop our system. The 2D planar kinematics was acquired using the available skeletal tracking algorithm. The complete skeleton model has 20 joints (1 head, 3 shoulders (center, left, right), 2 elbows (left, right), 2 wrists (left, right), 2 hands (left, right), 1 spine, 3 hips (center, left, right), 2 knees (left, right), 2 ankles (left, right), 2 foots (left, right)) and 19 related segments. Then, 3D coordinates of joints of interest are stored for further processing. For this present study, only lower limb region was tracked and then its kinematics was extracted for the modeling and simulation purpose (Fig. 2).

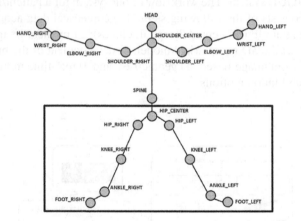

**Fig. 2** Full skeleton model including 19 segments and 20 joints and focused lower limb region of interest in bold rectangle

Shimmer kinematics (accelerometer and gyroscope) and EMG sensors were used to acquire joint spatial-temporal parameter and surface EMG signals representing the muscle activity of the musculoskeletal system during functional rehabilitation motions. Based on the acceleration information, joint angles were estimated. Raw EMG signals were preprocessed (filtering, rectification) and used as muscle activation profile for forward dynamics simulation to estimate the muscle forces according to a specific rehabilitation motion. Specific device, in which sensors are embedded, is developed to facilitate the data acquisition process.

## 2.2 Graphical User Interaction (GUI) System

An enhanced virtual reality interaction system was developed using Visual C# as programming language. Three user interfaces were created (Fig. 3). The first one is the main graphical interface with a menu system and a Kinect-based 3D skeletal tracking viewer. This interface allows the user to view the result of 3D skeletal tracking process from Kinect sensors. The second one is a 3D biomechanical model viewer to display and animate the real-time simulation. This interface allows the user to view and interact with the model during real time rehabilitation motions. The third interface is the joint kinematics and muscle forces viewer to plot the biofeedbacks of the musculoskeletal system in real-time.

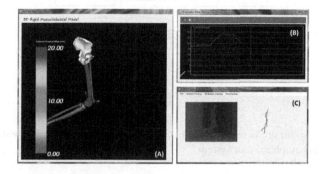

**Fig. 3** Graphical user interfaces of our system of systems: (A) 3D biomechanical model viewer, (B) joint kinematics and muscle forces plot viewer and (C) Kinect-based skeletal tracking viewer within the main interface

## 2.3 Multi-physical Modeling and Simulation System

To obtain the information inside the human musculoskeletal system during functional rehabilitation motions, an advanced 3D rigid multi-bodies model was developed. Medical imaging such as computed tomography (CT) scan or magnetic resonnance imaging (MRI) were used to acquire the anatomical data of the patient/subject under investigation. Then, medical imaging processing (2D segmentation, 3D reconstruction) was applied to develop 3D geometrical model. Kinematics data at specific joints of interest were extracted from Kinect sensorsand then fused with joint kinematics derived from Shimmer kinematics sensors. The recommendations of the International Society of Biomechanics were used to define the joint coordinate frame [24].

The block diagram of the modeling and simulation system in interaction with data acquisition system and graphical user system is shown in Fig. 4. First, Kinect and Shimmer sensors are activated and the biomechanical model developed from medical imaging data is generated. Second, a calibration process was set up to align the image-based geometries and the joint kinematics data. Third, musculoskeletal tracking is started to animate the musculoskelletal model as well as to provide joint kinematics and muscle forces data for plotting and contact area retrieval from a look-up table. Then the threshold-driven distance map between the two surfaces of the joint is displayed in an interactive manner. Finally, the system is looped until the user decides to quit the system. This modeling and simulation system was implemented using Visual C#.

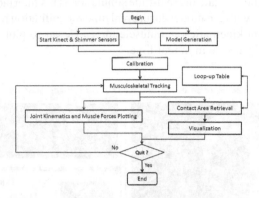

**Fig. 4** Block diagram of the modeling and simulation system in interaction with data acquisition system and graphical user system

To track the contact area information at the joint level during the functional rehabilitation motions, a look-up table including all values of contact areas within the feasible range of motion was pre-computed and stored. Then, during real time rehabilitation motion, at each time step, one joint angle value was used to look up the corresponding contact value; this information was displayed in the graphical user interface. A point-to-point distance principle with a geometrical threshold was applied to compute the contact area according to a specific joint position [25].

At each time step, the 3D joint (point) coordinates are generated from Kinect-based skeletal tracking algorithm. Then, the law of cosines was applied to compute the angle between three joint (point) coordinates. For example, right knee angle was computed using the coordinates of the **HIP_RIGHT, KNEE_RIGHT**, and **AN-KLE_RIGHT** points (Fig. 2). The skeletal tracking algorithm relates to a body recognition problem using per-pixel classification algorithm. This two-step process includes a depth map computing using structured light derived from speckle pattern of infrared sensor and a body recognition using randomized decision forest approach [26]. Then, these information were fused with joint kinematics data derived from Shimmer kinematics sensors to provide accurate 3D joint angles for musculoskeletal simulation. Based on the pre-processed EMG signals, forward dynamics was used to estimate the muscle forces [25], [27].

## 3 A Pilot Case Study

To illustrate the workflow of our system, a pilot case study was designed on a patient with post-polio residual paralysis (male, 26 years old, 1m70 height, and 66kg body mass). All necessary steps are addressed in the following subsections. It is important to note that we focused only on the tracking of joint kinematics derived from Kinect sensor and joint contact area at the knee level.

### 3.1 Geometrical Data Acquisition Using Medical Imaging at Clinical Center

Computed tomography (CT) scanner images were acquired using a spiral-imaging scanner (GE Light Speed VCT 64) at the Polyclinique St Côme of Compiègne (France). The subject signed an informed consent agreement before participating into this present study. The CT scan protocol included 3mm thickness and a matrix of $512 \times 512$ pixels leading to have 384 joint slices (Fig. 5A).

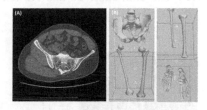

**Fig. 5** (A) Raw CT image and (B) 3D reconstruction of bone tissues of the lower limbs structures

### 3.2 Geometrical Data Acquisition Using Medical Imaging at Clinical Center

Semi-automatic segmentation was applied using 3D Slicer software to extract the bone tissues of the lower limbs (Fig. 5B).

### 3.3 Biomechanical Model Development and Simulation

The enhanced 3D musculoskeletal model interating joint contact behavior of the patient was developed (Fig. 6A). We considered that the center of knee joint is located at the middle point of the epicondylar axis connected between the epicondylar peaks (Fig. 6B).

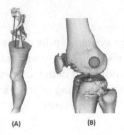

(A)                          (B)

**Fig. 6** (A) Enhanced 3D musculoskeletal model and (B) definition of knee joint center based on the epicondylar peaks (spheres in green color)

## 3.4   Real-Time Monitoring of the Functional Rehabilitation Motions

The calibration is needed to align the biomechanical model and the Kinect-based kinematic data (Fig. 7A). The animation of this flexion motion is shown in Fig. 7B.

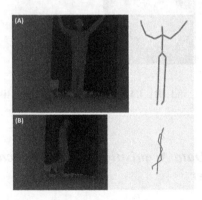

**Fig. 7** (A) Calibration process and (B) a flexion motion

Contact areas at the knee joint during a flexion motion with plausible range from 0 degree (full extension) to 45 degree were computed and presented in Fig. 8 and Fig. 9. We observed that the evolution of knee contact area decreases over the range of flexion motion with higher values on the lateral side. Thus, threshold-driven color map of the distance between femur and tibia revealed such evolution pattern (Fig. 8). Moreover, the contact area is more extended on the lateral side. Furthermore, the choice of the geometrical threshold has an important impact on the range of values of the knee contact area. In fact, the increase of the threshold leads to the increase of the knee contact area (as shown in Fig. 9).

**Fig. 8** Evolution of distance maps in mm for 0 to 45 degrees of flexion (left to right) of the knee joint with the threshold of 10 mm

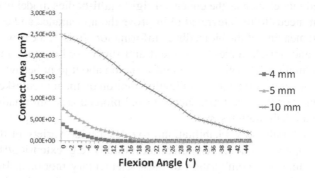

**Fig. 9** Evolution of contact areas from 0 to 45 degrees of flexion with three different thresholds

As concerns the runtime of the process, we can note that the computing of one contact value at one specific joint angle costs about 1.9 seconds on a 64-bits Intel Xeon 3.5 GHz computer while the retrieval time for one contact value at one specific angle is about 6 milliseconds.

## 4 Discussion

Musculoskeletal rehabilitation is commonly prescribed for patients with musculoskeletal impairments or disabilities to optimize their functional capacities and performances as well as to reduce disability symptoms. Moreover, these physical exercises also contribute to improve the well-being nature of the patient [1]. The efficiency of a rehabilitation program is evaluated using functional measurements (e.g. joint kinematics parameters) about the effect of a rehabilitation motion on the musculoskeletal system. Recently, enhanced virtual reality technologies have been used to develop immersive rehabilitation systems to provide such useful information [7], [8]. Current systems allow only external visual biofeedback information to be supervised during rehabilitation motion. The only use of external information is not sufficient to have a reliable judgment on the efficiency of a rehabilitation motion, especially for patients who suffered commonly bone deformation and muscle disabilities. In our study, a novel rehabilitation system of systems was developed to provide both external and internal visual biofeedback information. Thus, Kinect-based non-contact, Shimmer kinematics and EMG sensors were used to provide external and internal biofeedbacks data. An enhanced rigid multi-bodies model integrating joint contact behavior was incorporated into our system of systems to provide muscle forces as well as internal evolution of joint contact area during rehabilitation motion. In the literature, existing rehabilitation systems provide also some kinds of virtual avatar to represent the human body in immersive virtual reality environment [17], [28], [29], [30]. However, these simplified avatar models are so far to be realistic according to the biological complexity of the musculoskeletal system. Thus, physics-based model such as the enhanced rigid multi-bodies model in our rehabilitation system needs to be integrated to improve the appearance and especially the physiological meaning of the biofeedback information. In fact, the evolution of joint contact area and muscle forces are tracked and supervised over the range of each rehabilitation motion. This gives useful information about joint behaviors and muscle activities under the effect of rehabilitation motion on the musculoskeletal tissues and structures. For the first time, such internal biofeedback information could be elucidated in a rehabilitation system.

Conventional/traditional rehabilitation provides a wide variety of therapy exercises with complex functional rehabilitation motions (e.g. extension/flexion, axial rotations, bending or a combination of these elementary motions). However, this deals with the limited time and non-controlled nature of the rehabilitation training for a patient due to high medical treatment cost and human resources (e.g. experimented clinicians and therapists) as well as the lack of objective and visual biofeedback of the rehabilitation effect. It is well known that the intensive and well-controlled use of rehabilitation program/training leads to significant improvement of musculoskeletal dysfunctions. Thus, a rehabilitation system providing an immersive virtual reality environment in which visual objective and visual biofeedback about the effect of rehabilitation motion on the musculoskeletal tissues and structures could be of great clinical interest. In particular, this assistive technology could be a valuable assistant to the patient to perform more precisely and accurately the exercises/motions of interest. Moreover, the system could allow the patient to perform his rehabilitation exercises at home in an intensive manner leading to maximize the benefit of the rehabilitation program.

Internal musculoskeletal information is commonly acquired using rigid multibodies or continuum finite element modeling approaches [16], [31]. Joint contact behavior may be accurately provided from finite element model [32], [33], [34]. However, the finite element simulation is very time-consuming and this does still not satisfy the computational requirement of a real time simulations. In fact, we proposed an enhanced rigid multi-bodies model integrating geometrical joint contact behavior to provide contact area information at the joint level. Moreover, a calculation time reduction strategy was established to reach the constraints of real time simulation. In fact, feasible range of motion was simulated and then related joint contact areas were pre-computed and stored in a look-up table. Finally, during real time simulation, the joint contact area is retrieved and displayed from this table. This strategy reduces significantly the processing time to satisfy the computational constraints of a real time simulation.

Research topics related to system of systems have been intensively investigated in the last decade. The notion of system of systems has been recently introduced in the engineering system field. From an engineering point of view, a system is defined as a group of functionally, physically and/or behaviorally interactive, independent, material or non-material components. A system of systems (SoS) is a set of useful systems integrated into a larger system to achieve a unique set of tasks. In our present study, we developed a comprehensive rehabilitation system of systems, which is a combination of software and hardware systems for data acquisition, model development and user interaction purposes. The benefit from the coordination of these complex systems is to achieve a common healthcare goal with higher clinical significance and relevance for functional rehabilitation purpose. In fact, each constituent system such as data acquisition and management system or multi-physical modeling system or graphical user interface system may work independently without a common goal but only their integration into a system of systems provides an innovative solution for functional rehabilitation. Moreover, the use of low-cost and portable Kinect sensor provides a potential solution of enhanced virtual reality games tailored for a specific rehabilitation program in clinical or home-based settings leading to reduce significantly the medical cost and infrastructures [35]. In particular, the Kinect sensor provides a safely interaction with impaired patients leading to improve the clinical benefit of such useful rehabilitation system of systems.

## 5 Conclusions

In conclusion, we developed a novel rehabilitation system of system to provide reinforced biofeedback information of external and internal behavior of the musculoskeletal system during rehabilitation motions.

**Acknowledgments.** This work was carried out in the framework of the Labex MS2T, which was funded by the French Government, through the program **Investments for the future** managed by the National Agency for Research (Reference ANR-11-IDEX-0004-02).

## References

[1] Orlin, M.N., Cicirello, N.A., O'Donnell, A.E., Doty, A.K.: The Continuum of Care for Individuals With Lifelong Disabilities: Role of the Physical Therapist. Phys. Ther. (in Press, 2014), doi:http://dx.doi.org/10.2522/ptj.20130168

[2] Twist, D.J., Ma, D.M.: Physical Therapy Management of the Patient with Post-Polio Syndrome: A Case Report. Phys. Ther. 66, 1403–1406 (1986)

[3] Tiffreau, V., Rapin, A., Serafi, R., Percebois-Macadré, L., Supper, C., Jolly, D., Boyer, F.C.: Post-polio syndrome and rehabilitation. Ann. Phys. Rehabil. Med. 53 (1), 42–50 (2010)

[4] Keawutan, P., Bell, K., Davies, P.S., Boyd, R.N.: Systematic review of the relationship between habitual physical activity and motor capacity in children with cerebral palsy. Res. Dev. Disabil. 35 (6), 1301–1309 (2014)

[5] Andreoni, G., Mazzola, M., Perego, P., Standoli, C.E., Manzoni, S., Piccini, L., Molteni, F.: Wearable monitoring devices for assistive technology: case studies in post-polio syndrome. Sensors 14 (2), 2012–2027 (2014)

[6] Kiper, P., Agostini, M., Luque-Moreno, C., Tonin, P., Turolla, A.: Reinforced in virtual environment for rehabilitation of upper extremity dysfunction after stroke: preliminary data from a randomized controlled trial. Biomed. Res. Int. (2014), doi: http://dx.doi.org/10.1155/2014/752128; 752128

[7] Ibarra, J.M.Z., Tamayo, A.J., Sánchez, A.D., Delgado, J.E., Cheu, L.E., Arévalo, W.A.: Development of a system based on 3D vision, interactive virtual environments, ergonometric signals and a humanoid for stroke rehabilitation. Comput. Methods Programs Biomed. 112 (2), 239–249 (2013)

[8] Chang, Y.J., Han, W.Y., Tsai, Y.C.: A Kinect-based upper limb rehabilitation system to assist people with cerebral palsy. Res. Dev. Disabil. 34 (11), 3654–3659 (2013)

[9] Bonnechère, B., Jansen, B., Salvia, P., Bouzahouene, H., Omelina, L., Moiseev, F., Sholukha, V., Cornelis, J., Rooze, M., Van Sint Jan, S.: Validity and reliability of the Kinect within functional assessment activities: comparison with standard stereophotogrammetry. Gait Posture 39 (1), 593–598 (2014)

[10] Cipresso, P., Serino, S., Pedroli, E., Gaggioli, A., Riva, G.: A virtual reality platform for assessment and rehabilitation of neglect using a kinect. Stud. Health. Technol. Inform. 196, 66–68 (2014)

[11] Avola, D., Spezialetti, M., Placidi, G.: Design of an efficient framework for fast prototyping of customized human-computer interfaces and virtual environments for rehabilitation. Computer Methods and Programs in Biomedicine 110 (3), 490–502 (2013)

[12] Giansanti, D., Morelli, S., Maccioni, G., Brocco, M.: Design, construction and validation of a portable care system for the daily telerehabiliatation of gait. Computer Methods and Programs in Biomedicine 112 (1), 146–155 (2013)

[13] Liu, K., Inoue, Y., Shibata, K., Cao, E.: Ambulatory Estimation of Knee-Joint Kinematics in Anatomical Coordinate System Using Accelerometers and Magnetometers. IEEE Transactions on Biomedical Engineering 58(2), 435–442 (2011)

[14] Bugané, F., Benedetti, M.G., Casadio, G., Attala, S., Biagi, F., Manca, M., Leardini, A.: Estimation of spatial-temporal gait parameters in level walking based on a single accelerometer: Validation on normal subjects by standard gait analysis. Computer Methods and Programs in Biomedicine 108 (1), 129–137 (2012)

[15] Dao, T.T.: Modeling of Musculoskeletal System of the Lower Limbs: Biomechanical Model vs. Meta Model (Knowledge-based Model). PhD Thesis, University of Technology of Compigne, France (2009)

[16] Dao, T.T., Marin, F., Pouletaut, P., Aufaure, P., Charleux, F., Ho Ba Tho, M.C.: Estimation of Accuracy of Patient Specific Musculoskeletal Modeling: Case Study on a Post Polio Residual Paralysis Subject. Computer Method in Biomechanics and Biomedical Engineering 15 (7), 745–751 (2012)

[17] González-Ortega, D., Díaz-Pernas, F.J., Martinez-Zarzuela, M., Antón-Rodríguez, M.: A Kinect-based system for cognitive rehabilitation exercises monitoring. Computer Methods and Programs in Biomedicine 113 (2), 620–631 (2014)

[18] Courtecuisse, H., Allard, J., Kerfriden, P., Bordas, S.P.A., Cotin, S., Duriez, C.: Real-time simulation of contact and cutting of heterogeneous soft-tissues. Medical Image Analysis 18, 394–410 (2014)

[19] Csete, M.E., Doyle, J.C.: Reverse Engineering of Biological Complexity. Science 295 (5560), 1664–1669 (2002)

[20] Foote, R.: Mathematics and Complex Systems. Science 318 (5849), 410–412 (2007)

[21] Soodak, H., Iberall, A.: Homeokinetics: A Physical Science for Complex Systems. Science 201(4356), 579–582 (1978)

[22] Ottino, J.M.: Engineering complex systems. Nature 427, 399–399 (2004)

[23] Dao, T.T.: Ho Ba Tho, M.C. Uncertainty Modeling of Input Data for a Biomechanical System of Systems. In: Conf. Proc. IEEE Eng. Med. Biol. Soc., pp. 4581–4584 (2013)

[24] Wu, G., Siegler, S., Allard, P., Kirtley, C., Leardini, A., Rosenbaum, D., Whittle, M., D'Lima, D.D., Cristonfolini, L., Witte, H., Schmid, O., Stokes, I.I.: recommendation on definitions of joint coordinate system of various joints for the reporting of human joint motion–part I: ankle, hip, and spine. Journal of Biomechanics 35(4), 543–548 (2002)

[25] Dao, T.T., Pouletaut, P., Goebel, J.C., Pinzano, A., Gillet, P., Ho Ba Tho, M.C.: In vivo Characterization of Morphological Properties and Contact Areas of the Rat Cartilage derived from High Resolution MRI. Biomedical Engineering and Research (IRBM) 32(3), 204–213 (2011)

[26] Shotton, J., Fitzgibbon, A., Cook, M., Sharp, T., Finocchio, M., Moore, R., Kipman, A., Blake, A.: Real-Time Human Pose Recognition in Parts from Single Depth Images. Machine Learning for Computer Vision. Studies in Computational Intelligence 411, 119–135 (2013)

[27] Erdemir, A., McLean, S., Herzog, W., van den Bogert, A.J.: Model-based estimation of muscle forces exerted during movements. Clin. Biomech. 22, 131–154 (2007)

[28] Kifayat, K., Fergus, P., Cooper, S., Merabti, M.: Body Area Networks for Movement Analysis in Physiotherapy Treatments. In: IEEE 24th International Conference on Advanced Information Networking and Applications Workshops, pp. 866–872 (2010)

[29] Kurillo, G., Koritnik, T., Bajd, T., Bajcsy, R.: Real-time 3D avatars for tele-rehabilitation in virtual reality. Stud. Health. Technol. Inform. 163, 290–296 (2011)

[30] Cameirao, M.S., Bermúdez, I.B.S., Duarte, O.E., Verschure, P.F.: The rehabilitation gaming system: a review. Stud. Health Technol. Inform. 145, 65–83 (2009)

[31] Dao, T.T., Rassineux, A., Charleux, F., Ho Ba Tho, M.C. A.: Robust Protocol for the Creation of Patient Specific Finite Element Models of the Musculoskeletal System from Medical Imaging Data Computer Methods in Biomechanics and Biomedical Engineering: Imaging & Visualization (in press, 2014)

[32] Bei, Y., Fregly, B.J.: Multibody dynamic simulation of knee contact mechanics. Med. Eng. Phys. 26 (9), 777–789 (2004)

[33] Fukubayashi, T., Kurosaw, H.: The contact area and pressure distribution pattern of the knee. A Study of Normal and Osteoarthrotic Knee Joints. Acta Orthop. Scand. 51, 871–879 (1980)

[34] Shiramizu, K., Vizesi, F., Bruce, W., Herrmann, S., Walsh, W.R.: Tibiofemoral contact areas and pressures in six high flexion knees. Int. Orthop. 33(2), 403–406 (2009)

[35] Chan, M., Estève, D., Escriba, C., Campo, E.: A review of smart homes-Present state and future challenges. Computer Methods and Programs in Biomedicine 91(1), 55–81 (2008)

# Uncertainty Modeling and Propagation in Musculoskeletal Modeling

Tien Tuan Dao and Marie Christine Ho Ba Tho

**Abstract.** Biomechanical input data of in silico models are subject to uncertainties due to subject variability, experimental protocol and computing technique. Traditional perturbation analysis showed important drawbacks such as unobvious definition of the true range of value for the sensitivity analysis. In this present study, we used a novel framework to model the uncertainties of thigh mass property as well as to quantify their impact on the thigh muscle force estimation. A simplified patient specific musculoskeletal model (3 segments, 2 joints and 8 hip flexor muscles) of a post-polio residual paralysis subject was developed. Knowledge-based fusion p-box was used to model the uncertainties of the thigh mass property. Then, a Monte Carlo simulation was performed to quantify their impact on the thigh muscle force estimation through forward dynamics simulation. The global range of value of the rectus femoris force is from $2327.59 \pm 39.32$ N to $3353.16 \pm 383.8$ N at the peak level. The global range of value of the gracilis force is from $143.53 \pm 2.35$ N to $159.27 \pm 8$ N at the peak level. Cumulative probability functions of these ranges were presented and discussed. Our study suggested that under input data uncertainties, the musculoskeletal simulation results needs to be determined within a global range of values. Consequently, the clinical use of such global range will make the decision making more reliable. Thus, our study could be used as a guideline for such a purpose.

**Keywords:** Uncertainty Modeling, Uncertainty Propagation, Thigh Mass Property, Muscle Force Estimation, Patient Specific Musculoskeletal Model, OpenSIM, Monte Carlo Simulation.

Tien Tuan Dao · Marie Christine Ho Ba Tho
UTC CNRS UMR 7338, Biomechanics and Bioengineering (BMBI),
University of Technology of Compiègne, BP 20529, 60205 Compiègne Cedex, France

© Springer International Publishing Switzerland 2015                567
V.-H. Nguyen et al. (eds.), *Knowledge and Systems Engineering*,
Advances in Intelligent Systems and Computing 326, DOI: 10.1007/978-3-319-11680-8_45

## 1 Introduction

*In silico* medicine is one of the most challenging research topics in the field of biomechanics for the last decades [1]. Computer-aided numerical models like rigid or deformable musculoskeletal models have been used for better diagnosis of musculoskeletal disorders [2], [3], [4], [5], [6], [7]. However, the development of these models, especially patient specific models deals with engineering challenges such as input data uncertainties due to subject variability, measuring protocol or computing method [8], [9], [10]. In fact, the clinical use of any numerical model should take these input data uncertainties into consideration leading to have more reliable decision makings.

Input data of musculoskeletal models deals with random and epistemic uncertainties. The random uncertainty relates to the repeatability, the reproducibility errors and the variability due to the fact that data obtained from different protocols/population (races, origins)/experimental techniques. The epistemic uncertainty concerns the accuracy level of the measuring protocols and computing methods [10].

In the musculoskeletal modeling literature, the sensitivity of input data has been performed using traditional approaches such as variation and perturbation analysis [11], [12], [13], [14], [15], [16], [17] or Monte Carlo analysis [18]. However, there are some drawbacks dealing with these approaches even if they are easy to be implemented and to be used. The first one is the uncertain determination of the true perturbation thresholds which have been selected in a subjective manner according to each study. The second problem is the difficult consideration of data uncertainties derived from multiple acquisition sources (e.g. experimental measurement, literature-based extraction) with their proper accuracy levels. Consequently, there is a lack of a generic mathematical framework to model both random and epistemic uncertainties of input data derived from multiple acquisition sources. Recently, we have initiated a novel framework to model the biomechanical data uncertainties using knowledge-based fusion probability box and their propagation using Monte Carlo simulation [10]. The first example of this framework was performed on the uncertainty modeling of muscle morphological properties (physical cross-sectional area (pCSA) and muscle tension coefficient). Moreover, the impact of these uncertainties on the computing of peak muscle force was quantified. Even if this example is simple, the obtained results showed the potential use of such useful framework for more complex musculoskeletal simulation. The aims of this present study was to apply this novel framework to model the uncertainties of the thigh mass as property well as to quantify their impact on the estimation of thigh muscle forces during swing-like motion of the right leg.

## 2 Materials and Methods

A 3-steps methodology was developed to model the uncertainty of input data as well as to quantify its impact on the musculoskeletal simulation results (Fig. 1). The first step deals with the development of the musculoskeletal model derived from medical imaging. The second step includes the data collection, uncertainty representation and computing components. The third step consists of the forward dynamics simulation, Monte Carlo simulation, and the results analysis.

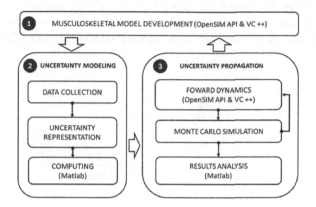

**Fig. 1** Workflow used for the model development, uncertainty modeling and propagation

## 2.1 Development of a Musculoskeletal Model Derived from Medical Imaging

A musculoskeletal model was developed for a post-polio residual paralysis (PPRP) subject (male, 26 years old, 1m70 height, and 66kg body mass). Computed tomography (CT) scanner images were acquired using a spiral-imaging scanner (GE Light Speed VCT 64) at the Polyclinique St Cme of Compiegne (France). The subject signed an informed consent agreement before participating into this present study. The CT scan protocol included 3mm thickness and a matrix of 512512 pixels leading to have 384 joint slices (Fig. 2A). Semi-automatic segmentation was applied using ScanIP software (Simpleware, UK) to extract the bone and soft tissues of the lower limbs (Fig. 2B).

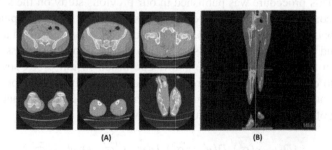

**Fig. 2** Draw CT images (A) and (B) segmented image

A simplified 3-segment (pelvis, right femur and thigh envelop, right tibia and leg envelop) musculoskeletal model with 6 degrees of freedom was developed using the OpenSIM 3.1 API (Application Programming Interface) [19] (Fig. 3A). The model includes 2 joints (i.e. right hip and knee) modelled as ball-and-socket and hinge joints respectively. The model includes 8 right hip flexor muscles: rectus femoris, sartorius, psoas, tensor fasciae latae, gluteus minimus, gluteus medius, pectineus, gracilis. Each muscle was modelled as a Thelen2003Muscle model [20].

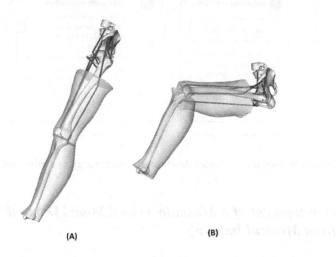

(A)                    (B)

**Fig. 3** Simplified musculoskeletal model: initial position (A) and final position of swing-like motion (B)

## 2.2 Uncertainty Modeling of the Thigh Mass Property

The thigh mass property of the developed musculoskeletal model was collected from 3 acquisition sources. The first one relates to the in vivo geometrical computing using segmented images, mass density of the soft tissue and accumulated pixels principle. This procedure was published in our previous study on the same patient [5]. The second data source is the use of a regression table [21] to compute the thigh mass in proportion to the body mass. The third data source is the use of a regression equation and geometrical characteristics [22] to compute the thigh mass.

Based on the collected data, the knowledge-based fusion p-box structure [10] was used to create an uncertainty representation of the thigh mass property. The knowledge-based fusion p-box is a fused structure from multiple parametric p-boxes enveloping of four normal distributions as follows:

$$(D(\mu_i^l, \sqrt{\sigma_i^l}), D(\mu_i^l, \sqrt{\sigma_i^u}), D(\mu_i^u, \sqrt{\sigma_i^l}), D(\mu_i^u, \sqrt{\sigma_i^u}))$$

where mean $[\mu_i^l, \mu_i^u]$ and standard deviation $[\sigma_i^l, \sigma_i^u]$ intervals (l means lower bound and u means upper bound) were computed from the used three data sources.

The computing of the knowledge-based fusion p-box was performed using Matlab R2010.b (Mathworks, USA).

## 2.3 Uncertainty Propagation of the Thigh Mass Property through the Forward Dynamics

To estimate the distribution of the muscle force estimation results, an interval-based Monte Carlo simulation with 20 samples was performed using the lower and upper probability non-decreasing functions of the knowledge-based fusion p-box of the thigh mass property [10]. At each iteration with a thigh mass value selected randomly, a forward simulation was performed using computed muscle excitation to estimate the different thigh muscle forces during a swing-like (85-degrees hip flexion and 55-degrees knee flexion) motion [20], [23] (Fig. 3B). Note that the forward dynamics and Monte Carlo simulations were realized using OpenSIM 3.1 API [19] and Visual C++ (Microsoft, USA) on a Dell computer (Precision T3500, 2.8Ghz, 3GB RAM). The results were analyzed using Matlab R2010.b (Mathworks, USA).

## 3 Computational Results

Computed thigh mass from 3 different acquisition sources is from 4.462 to 7.419 kg. Note that the computing methods are very different from one to other source. The regression table was established from the measurements on 7 cadavers ($69 \pm 17$ yo, $61.1 \pm 10.9$ kg) with standard measuring protocol and devices (dissection, balance). The thigh ratio is about $10.008 \pm 1.197$ % of the body mass. The regression equation ($m_{thigh} = 0.074 * M + 0.138 P_{25} - 4.641$) (M is the whole body mass and $P_{25}$ is the circumference of upper thigh (= 52 cm for this present patient)) was created from the measurements on 13 cadavers. The patient specific procedure used in vivo medical images, image processing and geometrical computing. The thigh ratio is about $7.145 \pm 0.386$ % of the body mass.

Using the collected data, the knowledge-based fusion p-box of the right thigh mass property was established. Each range of value from each acquisition source allows a knowledge-based p-box to be established (Fig. 4). Then a conservative rule was applied to create a fused probability box (Fig. 5).

The estimated forces of rectus femoris and gracilis muscles at the peak level corresponding to 85-degrees hip flexion and 55-degrees knee flexion considering the uncertainties of thigh mass into account are shown in Fig. 6 and Fig. 7. The muscle forces were expressed by cumulative probability functions established from the results of Monte Carlo simulations. The range of value of the rectus femoris force is from $2327.59 \pm 39.32$ N to $3353.16 \pm 383.8$ N at the peak level. The range of value of the gracilis force is from $143.53 \pm 2.35$ N to $159.27 \pm 8$ N at the peak level. The total run time of all Monte Carlo simulations (20 samples) is around 200 minutes.

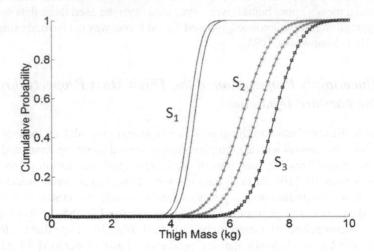

**Fig. 4** Knowledge-based fusion p-box of the right thigh mass property: 3 separate sources

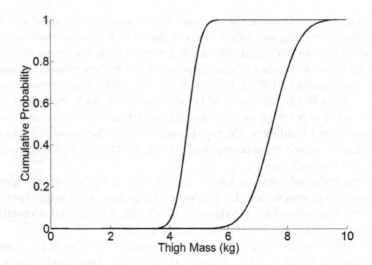

**Fig. 5** Knowledge-based fusion p-box of the right thigh mass property: fused result

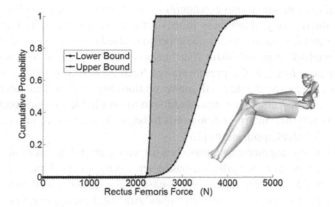

**Fig. 6** Cumulative probability of the estimated rectus femoris forces at the peak level corresponding to the 85-degrees hip flexion and 55-degrees knee flexion

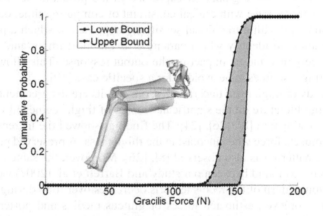

**Fig. 7** Cumulative probability of the estimated gracilis forces at the peak level corresponding to the 85-degrees hip flexion and 55-degrees knee flexion

## 4 Discussion

The reliability of the input and output data of the musculoskeletal simulations plays an essential role in the promotion of musculoskeletal model in clinical routine practices [5]. Traditional approaches commonly used perturbation analysis to quantify the sensitivity of input data on the simulation results. In addition to the unobvious choice of the true range of value for the variation process, these approaches couldn't provide enough output information to establish a probabilistic distribution of the output simulation results [24]. In this present study, the use of a more generic mathematical structure as knowledge-based fusion probability box [10] allows a global range of value to be taken into consideration in the uncertainty modeling of a parameter of interest (e.g. thigh mass in this present study). Then, its impact on the musculoskeletal simulation results could be quantified using a Monte Carlo method leading to establish a probabilistic distribution of the simulation results (e.g. thigh estimated muscle forces in this present study). Such useful information is of great interest to provide more reliable result interpretations for clinical decision making. In fact, the use of one-value input data could lead to unreliable musculoskeletal simulation results and their interpretation needs to be performed with caution, especially in the case of clinical applications [25].

There are many engineering approaches for accounting data uncertainty into numerical models such as perturbation-based sensitivity, interval analysis, statistical inference method as bootstrapping, worst-case analysis, fuzzy arithmetic, possibility theory, belief theory, and probability boxes. Among these approaches, probability boxes (p-boxes) are flexible mathematical structures for the modeling of both random and epistemic uncertainties. The advantage of this approach deals with the use of an objective integration of multiple data uncertainties from different data sources. Data set (mean ± standard deviation) derived from each data acquisition source could be combined into a global set expressed in the probabilistic space. In particular, for a specific case with critical constraint of computing time, our approach could be combined with a traditional sensitivity approach in which a perturbation study could allow to identify which parameter is the most important, and then to model its uncertainty and its impact on the output response. This allows unnecessary computing resources to be avoided for a specific case [10].

For the study of thigh mass uncertainties, our results are in agreement with those reported in the literature on the significant impact of thigh segmental mass on the muscle force estimation [8], [26], [24]. The findings showed the increasing behavior of thigh muscle force when increasing the thigh mass. A preserved profile of the muscle force pattern was also observed [8], [26]. Moreover, the same muscle activation pattern was found between our study and Barrett et al. (2007) work [27]. In fact, activation pattern of the rectus femoris decreases over time during the swing-like motion. Moreover, estimated forces of gluteus medius and posterior gluteus minimus muscle decrease from the initial position to the final position (85-degrees hip flexion and 55-degrees knee flexion) of a swing-like motion. This profile is concordant with the finding reported by Redl et al. (2007) [28]. However, our approach estimated thigh muscle forces within a global range through a probabilistic distribution function.

## 5   Conclusions

In conclusions, our present study used a novel mathematical framework to model the uncertainties of musculoskeletal modeling input data as well as to quantify their impact on the musculoskeletal simulation results. Our study suggested that under input data uncertainties, the musculoskeletal simulation results needs to be determined within a global range of values expressed by a more powerful mathematical structure (e.g. probability box in our present study). Consequently, the clinical use of such global range will make the decision making more reliable. Thus, our study could be used as a guideline for such a purpose.

**Acknowledgments.** This work was carried out in the framework of the Labex MS2T, which was funded by the French Government, through the program **Investments for the future** managed by the National Agency for Research (Reference ANR-11-IDEX-0004-02).

# References

[1] Viceconti, M., Clapworthy, G., Van Sint Jan, S.: The Virtual Physiological Human - a European initiative for in silico human modelling. Journal of Physiological Sciences 58(7), 441–446 (2008)

[2] Périé, D., Sales De Gauzy, J., Ho Ba Tho, M.C.: Biomechanical evaluation of Cheneau-Toulouse-Munster brace in the treatment of scoliosis using optimisation approach and finite element method. Med. Biol. Eng. Comput. 40(3), 296–301 (2002)

[3] Chabanas, M., Luboz, V., Payan, Y.: Patient specific Finite Element model of the face soft tissue for computer-assisted maxillofacial surgery. Medical Image Analysis 7(2), 131–151 (2003)

[4] Arnold, A.S., Delp, S.L.: Computer modeling of gait abnormalities in cerebral palsy: application to treatment planning. Theoretical Issues in Ergonomics Science 6, 305–312 (2005)

[5] Dao, T.T., Marin, F., Pouletaut, P., Aufaure, P., Charleux, F.: Ho Ba Tho, M.C. Estimation of Accuracy of Patient Specific Musculoskeletal Modeling: Case Study on a Post Polio Residual Paralysis Subject. Computer Method in Biomechanics and Biomedical Engineering 15(7), 745–751 (2012)

[6] Fregly, B.J., Boninger, M.L., Reinkensmeyer, D.J.: Personalized neuromusculoskeletal modeling to improve treatment of mobility impairments: a perspective from European research sites. Journal of NeuroEngineering and Rehabilitation 9(18), 1–11 (2012)

[7] Dao, T.T., Pouletaut, P., Charleux, F., Lazáry, Á., Eltes, P., Varga, P.P., Ho Ba Tho, M.C.: Estimation of Patient Specific Lumbar Spine Muscle Forces Using Multi-Physical Musculoskeletal Model and Dynamic MRI. In: Huynh, V.N., Denoeux, T., Tran, D.H., Le, A.C., Pham, B.S. (eds.) KSE 2013, Part II. AISC, vol. 245, pp. 425–438. Springer, Heidelberg (2014)

[8] Dao, T.T., Marin, F., Ho Ba Tho, M.C.: Sensitivity of the anthropometrical and geometrical parameters of the bones and muscles on a musculoskeletal model of the lower limbs. In: Proceedings of the 31th Annual International Conference of the IEEE Engineering in Medicine and Biology Society, pp. 5251–5254 (2009)

[9] Pàmies-Vilà, R., Font-Llagunes, J.M., Cuadrado, J., Alonso, F.J.: Analysis of different uncertainties in the inverse dynamic analysis of human gait. Mechanism and Machine Theory 58, 153–164 (2012)

[10] Dao, T.T.: Ho Ba Tho, M.C. Uncertainty Modeling of Input Data for a Biomechanical System of Systems. In: Proceedings of 35th Annual International Conference of the IEEE Engineering in Medicine and Biology Society, pp. 4581–4584 (2013)

[11] Nagano, A., Gerritsen, K.G.M., Fukashiro, S.: A sensitivity analysis of the calculation of mechanical output through inverse dynamics: a computer simulation study. Journal of Biomechanics 33(10), 1313–1318 (2000)

[12] Silva, M.P.T., Ambrósio, J.A.C.: Sensitivity of the results produced by the inverse dynamic analysis of a human stride to perturbed input data. Gait & Posture 19(1), 35–49 (2004)

[13] Rao, G., Amarantini, D., Berton, E., Favier, D.: Influence of body segments' parameters estimation models on inverse dynamics solutions during gait. Journal of Biomechanics 39(8), 1531–1536 (2006)

[14] Scovil, C.Y., Ronsky, J.L.: Sensitivity of a Hill-based muscle model to perturbations in model parameters. Journal of Biomechanics 39(11), 2055–2063 (2006)

[15] Lawson, S.E.M., Chteau, H., Pourcelot, P., Denoix, J.M., Crevier-Denoix, N.: Sensitivity of an equine distal limb model to perturbations in tendon paths, origins and insertions. Journal of Biomechanics 40(11), 2510–2516 (2007)

[16] Chen, L., Ren, L.: The Influence of Intrinsic Muscle Properties on Musculoskeletal System Stability: A Modelling Study. Journal of Bionic Engineering 7, S158–S165 (2010)

[17] Carbone, V., van der Krogt, M.M., Koopman, H.F.J.M., Verdonschot, N.: Sensitivity of subject-specific models to errors in musculo-skeletal geometry. Journal of Biomechanics 45(14), 2476–2480 (2012)

[18] Ackland, D.C., Lin, Y.C., Pandy, M.G.: Sensitivity of model predictions of muscle function to changes in moment arms and muscle-tendon properties: A Monte-Carlo analysis. Journal of Biomechanics 45(8), 1463–1471 (2012)

[19] Delp, S.L., Anderson, F.C., Arnold, A.S., Loan, P., Habib, A., John, C.T., Guendelman, E., Thelen, D.G.: OpenSim: Open-source Software to Create and Analyze Dynamic Simulations of Movement. IEEE Transactions on Biomedical Engineering 54(11), 1940–1950 (2007)

[20] Thelen, D.G., Anderson, F.C., Delp, S.L.: Generating dynamic simulations of movement using computed muscle control. J. Biomech. 36, 321–328

[21] Dempster, W.T., Gaughran, G.R.L.: Properties of body segments based on size and weight. American Journal of Anatomy 120, 33–54 (1967)

[22] Clauser, C.E., McConville, J.T., Young, J.W.: Weight, volume and center of mass of segments of the human body. AMRL-TR-69-70. Aerospace Medical Research Laboratory, Wright-Patterson Air Force Base, Ohio (1969)

[23] Erdemir, A., McLean, S., Herzog, W., van den Bogert, A.J.: Model-based estimation of muscle forces exerted during movements. Clin. Biomech. 22, 131–154 (2007)

[24] Wesseling, M., de Groote, F., Jonkers, I.: The effect of perturbing body segment parameters on calculated joint moments and muscle forces during gait. J. Biomech. 47(2), 596–601 (2014)

[25] May, B., Saha, S., Saltzman, M.: A three-dimensional mathematical model of temporomandibular joint loading. Clinical Biomechanics 16, 489–495 (2001)

[26] van Den Bogert, A.J., Hupperets, M., Schlarb, H., Krabbe, B.: Predictive musculoskeletal simulation using optimal control: effects of added limb mass on energy cost and kinematics of walking and running. J. Sports Eng. Technol. 226, 123–133 (2012)

[27] Barrett, R.S., Besier, T.F., Lloyd, D.G.: Individual muscle contributions to the swing phase of gait: An EMG-based forward dynamics modelling approach. Simulation Modelling Practice and Theory 15 (9), 1146–1155 (2007)

[28] Redl, C., Gfoehler, M., Pandy, M.G.: Sensitivity of muscle force estimates to variations in muscle-tendon properties. Human Movement Science 26(2), 306–319 (2007)

# A Comparative Study of Classification-Based Machine Learning Methods for Novel Disease Gene Prediction

Duc-Hau Le, Nguyen Xuan Hoai, and Yung-Keun Kwon

**Abstract.** Prediction of novel genes associated to a disease is an important issue in biomedical research. At early days, annotation-based methods were proposed for this problem. In next stage, with high-throughput technologies, data of interaction between genes/proteins has grown quickly and covered almost genome and proteome, and therefore network-based methods for the issue is becoming prominent. Besides those two methods, the prediction problem can be also approached using machine learning techniques because it can be formulated as a classification task of machine learning. To date, a number of supervised learning techniques and various types of gene/protein annotation data have been used to solve the disease gene classification/prediction problem. However, to the best of our knowledge, there has been no study on the comparison of these methods that work on comprehensive biomedical annotation data. In addition, it is generally true that no classifier is better than others for all classification problems. Therefore, in this study, we compare the performance of disease gene prediction of several supervised learning techniques that have been used in the literature such as Decision Tree Learning, k-Nearest Neighbor, Naive Bayesian, Artificial Neural Networks and Support Vector Machines. We additionally assess Random Forest, a relatively new decision-tree-based ensemble learning method. The simulation results indicate that Random Forest obtained the best performance of all. Also, all methods are stable with the change of known disease genes used as positive training samples.

**Keywords:** classification/supervised learning techniques, disease gene prediction, data integration.

Duc-Hau Le
Center for IT Services, Water Resources University,
175 Tay Son, Dong Da, Hanoi, Vietnam

Nguyen Xuan Hoai
Information Technology Research and Development Center,
Hanoi Univerisity, Vietnam

Yung-Keun Kwon
School of Electrical Engineering, University of Ulsan, 93 Daehak-ro,
Nam-gu, Ulsan 680-749, Republic of Korea

© Springer International Publishing Switzerland 2015
V.-H. Nguyen et al. (eds.), *Knowledge and Systems Engineering*,
Advances in Intelligent Systems and Computing 326, DOI: 10.1007/978-3-319-11680-8_46

# 1 Introduction

Disease gene prediction, the task of identifying the most plausible candidate disease genes, is an important issue in biomedical research, and a variety of approaches have been proposed [1, 2]. Most of the early approaches including POCUS [3], SUS-PECTS [4], Endeavour [5] and ToppGene [6], have prioritized candidate genes by annotating them with respect to biological structures or functions, and comparing their annotations with those of already known disease genes. These annotation-based approaches are limited in that they fail to capture the indirect relationships between genes whose common features or functions are not yet annotated. To overcome this challenge, disease gene prediction methods guided by biological networks have recently been proposed [7] and they have shown better performance than the direct similarity-based methods. The emergence of such network-based methods is thank to large coverage of interactome data as recent high-throughput technologies has yielded large volumes of interaction data between cellular molecules covering most of genome and proteome.

Machine learning techniques has been successfully applied to solve various important biomedical problems [8, 9] such as genome annotation [10], pattern recognition [11], classification of microarray data [12], inference of gene regulatory networks [13], prediction of drug-target [14] and discovery of gene-gene interaction in disease data [15, 16]. In particular, they have been applied to identifying disease-associated genes [17 − 22]. First, the problem is formulated as a classification (supervised learning) problem, where the task is to learn classifiers from training data; then, the learned classifier is used to predict whether or not a test/candidate gene is a disease gene. Figure 1 shows a common scheme of classification-based approaches for disease gene prediction. Training data are usually known disease genes/proteins; however, some studies have also used unknown genes in the training task. These genes and unknown genes are annotated by -omics data.

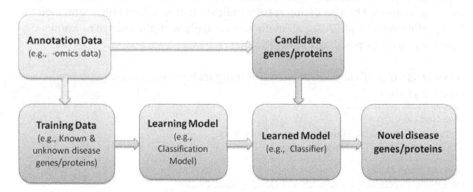

**Fig. 1** A common scheme of classification-based methods for disease gene prediction

To date, the binary-class classification techniques have largely been applied to the problems. The early applications of such techniques were of decision trees (DT) [17, 18] using distinctive sequence features of known disease proteins compared to all human proteins. With growth of interaction data between proteins, k-nearest neighbor algorithm (k-NN), an instance-based classifier, was introduced and it was based on topological properties of proteins on a human protein interaction network [19]. Naive Bayesian classifier (NB) was also used to identify human disease genes by integrating multiple types of genomic, phenotypic and interactomic data. In particular, a NB classifier was built based on eight different genomic datasets to identify human mitochondrial disease genes in [20]. The authors in [20] reported that their method outperformed a decision tree-based method (CART) [23] and a boosting algorithm (AdaBoost) [24]. Based on both interaction and sequence data of protein, support vector machine (SVM) was also used for the problem [21] and showed that SVM's performance was better than the k-NN. SVMs were subsequently used in a number of studies for disease gene prediction [25, 26]. Specifically, in addition to protein interaction and sequence data, protein functional information at the molecular level was employed in [25]. Moreover, unlike the methods mentioned above where the classifier was trained on all disease genes associated to all diseases, learning model was trained for each disease ontology term [25]. Likewise, the SVM classifier was used to identify genes associated to a specific disease (*i.e.*, primary immunodeficiency disease) in [26]. Furthermore, an artificial neural network (ANN) was proposed to identify novel disease genes for four complex diseases (*i.e.*, Cancer, Type 1 Diabetes, Type 2 Diabetes and Ageing) using eight topological features calculated from protein interaction network [22].

In summary, these aforementioned studies are not only different in the learning algorithms they used but also in the way they construct non-disease gene class. All of these approaches have been based on an assumption that disease genes and non-disease genes are clearly separate for the supervised training. This, however, is not entirely true. In reality, only disease genes are well known since their association with diseases can be collected from literature or public data sources such as OMIM [27] and GeneCards [28]. In contrast, we cannot be sure which genes are non-disease genes as there is no public data for them. For instance, studies in [17] and [18] obtained disease genes from OMIM as positive training samples, but negative training sets with the same size were randomly selected from the human genome and simply presumed not to be involved in disease. Also, the Bayesian predictor was built using positive training samples aggregated from known disease genes from GeneCards and OMIM, whereas the negative training samples were genes in the genome excluded known disease genes [29]. Based on the results in [30] that human genome may contain thousands of essential genes of which the features differ significantly from both disease genes and the other genes, the negative training set was constructed more precisely by randomly selecting proteins from human genome which excluded those belong to disease and essential genes[19, 21, 22].

Although binary-class classification techniques have extensively been used for disease gene prediction problem using many kinds of annotated genes/proteins data, to the best of our knowledge, there has not been a comparative study of those techniques that is based on the same training platform of comprehensive gene/protein annotation data. In this study, we present a first attempt to compare the performance of different classification techniques for disease gene prediction. The techniques tested in our study include those that have been used in the literature such as Decision Tree learning, Naive Bayesian, k-Nearest Neighbor, Support Vector Machines and Artificial Neural Networks. In addition, we also test Random Forest, a relatively new ensemble learning method, which has not previously used to solve the problem.

The paper is organized as follows. In the next section, we introduced about data preparation and training methods. Next, we described comparison results of the learning methods. Finally, in section Conclusions and Future Works, we present our conclusions and highlight some future directions as extensions of the work in this paper.

Interactomic, genomic and proteomic data was collected from various data sources and used to extract features to annotate genes/proteins. Only features, which have been indicated about the statistically significant difference between disease genes and non-disease genes from previous studies, were used. To construct the training set, we follow the same procedure as in [19] (see more detail in section Materials and Methods). We used 10-fold cross-validation method to measure the performance of each technique. The results showed that Random Forest outperformed other methods on both whole and subset of known disease genes used as positive training samples. Moreover, all methods are stable with the changing in number of known disease genes used as positive training samples.

## 2   Data Preparation and Learning Methods

### 2.1   Building the Feature Set

To build the feature set, we collected interactomic, genomic and proteomic data for genes/proteins. In particular, a protein interaction network consisting of 13,488 proteins and 217,893 interactions was collected from Online Predicted Human Interaction Database (OPHID) [31] and used to calculate topological features of proteins such as degree, 1-N index, 2-N index, distance to disease genes and positive topology coefficient. These topological properties were observed about their statistically significant difference between two groups of disease and non-disease genes [19]. In addition, it was reported that proteins with longer sequence may provide more opportunity for mutation [17, 18] and specific gene ontology terms may annotate disease information of proteins [3, 32]; therefore, we collected sequence data and gene ontology (GO) terms of proteins from UniProtKB [33] to calculate protein length and number of annotated gene ontology terms, respectively. Moreover, study in [34] showed that more domains and binding sites may allow mutation to more easily corrupt protein functions. Based on this observation, we used a database of protein, InterPro [35, 36] via BioMart tool [37] to calculate number of domains and number of binding sites for each protein. Final, based on an observation that

**Table 1** Features of a protein

| Feature | Category | Definition |
|---|---|---|
| Degree | Topology | Number of links to a protein |
| 1-N index | Topology | Proportion of links to disease proteins overall number of links to a protein |
| 2-N index | Topology | Proportion of links to disease proteins overall number of links to neighbors of a protein |
| Average distance to disease genes | Topology | The extent to which a protein communicate to all disease proteins in the network |
| Positive topology coefficient | Topology | A variant of classical topological coefficient, measure the degree of sharing partners between a protein and disease proteins |
| Length of protein | Sequence | Number of amino acids a protein contains |
| Number of domains | Structure | Number of distinct functional and/or structural units of a protein |
| Number of binding sites | Structure | Number of regions on a protein to which other molecules and ions form a chemical bond |
| Number of GO terms | Annotation | Number of GO terms annotating a protein |
| Evolution rate | Evolution | Inversion of conservation score, where conservation score is number of species in a homologene group |

evolution rate of gene may contribute to the likelihood of hereditary disease [18, 30], we used a database of homolog genes, NCBI's Homologene [38] to calculate the evolution rate of each protein. In summary, a total of ten features including topological, sequence, structural, annotation, and evolutionary properties of each protein were defined (see Table 1). They are all numerical data and normalized in the range of [0, 1] for all experiments in this paper.

The training set for binary-class classification problem typically includes positive and negative samples. The positive training samples were built on known disease genes collected from OMIM [27]. After collecting only corresponding proteins available in the protein interaction network, we obtained 2,365 known disease proteins. Currently, constructing a set of genes which are known not to be associated with disease is impossible task. Therefore, to build the "control set" (*i.e.*, non-disease proteins), we followed the method used in [19]. This set contains all proteins in the interaction network except the known disease proteins and essential proteins. In which, 2,056 essential proteins were collected from three data sources - DEG [39], BioMart [37] and DAVID [40]. As a result, our "control set" included

11,123 proteins. Finally, the negative training set was randomly selected from this "control set" and equal in size to the positive training set.

## 2.2 Classification Techniques, Parameter Settings and Validation

We tested the supervised learning techniques that have been used for disease gene prediction: i) Decision Tree Classifier is a simple and widely used classification technique which is capable of breaking down a complex decision-making process into a collection of simpler ones [41,42]; ii) K-Nearest Neighbors algorithm is a popular non-parametric method used for classification and regression [43]; iii) Naive Bayesian classifier is a simple probabilistic classifier based on applying Bayes' theorem with independence assumptions between predictors [44]; iv) Artificial Neural Networks are computational models inspired by animals' central nervous systems (in particular the brain), and v) Support Vector Machines are methods that attempt to map the input feature space into a new high dimensional feature space, and then finds an optimal separating hyper-plane, which can be used to discriminate between positive and negative samples [45]. In addition to these techniques, we also tested an ensemble learning method - Random Forest [46], which operates by constructing a multitude of decision trees at training time and outputting the class that is the mode of the classes predicted by individual trees. For the implementation of algorithms including DT, k-NN, NB, ANN and RF, we used WEKA [47], a Java-based machine learning tool. For SVM, we used the tools from LibSVM [48].

To determine the parameters for the model selection, we faithfully used the settings that have been reported in the literature. For instance, we selected a non-linear kernel function (*i.e.*, Radial Basic Function) and other parameters such as error penalty for miss-classed samples (C) and complexity of non-linear optimal separating hyper-plane ($\gamma$) as in [21]. For k-NN, the optimal number of neighbors (k) of 3 was selected as in [19]. For DT, two default parameters including confidence factor for pruning (C) and minimum number of instances per leaf (M) was set to 0.25 and 2, respectively. For ANN, number of input neurons was set to number of features (*i.e.*, 10), and number of hidden neurons was set to an empirical value (*i.e.*, average of number of features and number of classes (*i.e.*, 2)). In addition, sigmoid threshold function was used for every neuron. Finally, trees were built with unlimited depth for RF.

In order to evaluate and compare these supervised learning techniques for disease gene prediction, we used 10-fold cross-validation on the training set and measured the performance by F-measure [49]. This measure is harmonic mean of precision and recall, so it reflects an average effect of both precision and recall.

$$Precision = \frac{TP}{TP+FP}$$

$$, Recall = \frac{TP}{TP+FN}, Fmeasure = 2 \times \frac{Precision \times Recall}{Precision+Recall}$$

where TP, TN, FP and FN denote the quantity of true positive, true negative, false positive and false negative, respectively.

To further avoid sampling bias of negative training data and to reduce the performance variance, we repeated the sampling procedure 100 times. A classifier was trained and evaluated for each time, and finally the overall performance of each supervised learning technique was averaged over Precision, Recall and F-measure.

## 3 Results

### 3.1 Performance Comparison Overall Known Disease Genes

In this experiment, all known disease genes were used as positive training samples to train and evaluate the classifiers (see the previous section). Figure 2 shows the performance comparison of six supervised learning techniques for disease gene prediction. With respect to the precision, ANN achieved the best accuracy (*i.e.*, 75.7%) and k-NN performed worst (*i.e.*, 63.1%). However, when considering the recall, RF achieved the highest value (*i.e.*, 75.9%) and NB obtained the lowest (*i.e.*, 42.4%). In addition, Random Forest (RF) was also the best classifier in terms of F-measure (*i.e.*, 71.2%). Note that F-measure is the harmonic mean of precision and recall and it reflects the average effect of them. Therefore, when either precision or recall is small, F-measure will be small. By looking at DT and RF techniques, we can observe that their precision scores are very similar (*i.e.*, 69.6% for DT and 68.1% for RF). This indicates that they are similar in their prediction ability of true disease genes. However, the recall of RF is much higher than DT (*i.e.*, 75.9% and 69.4% for RF and DT, respectively) suggesting that RF is more sensitive in retrieving true disease genes. In addition, based on the F-measure, the performances of k-NN, SVM and ANN are similar (*i.e.*, 62.7%, 63.5% and 63.6%, respectively), whereas NB performed worst (*i.e.*, 52.3%).

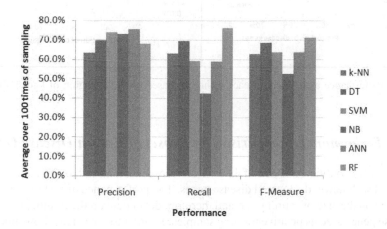

**Fig. 2** Performance comparison of six classification techniques for disease gene prediction

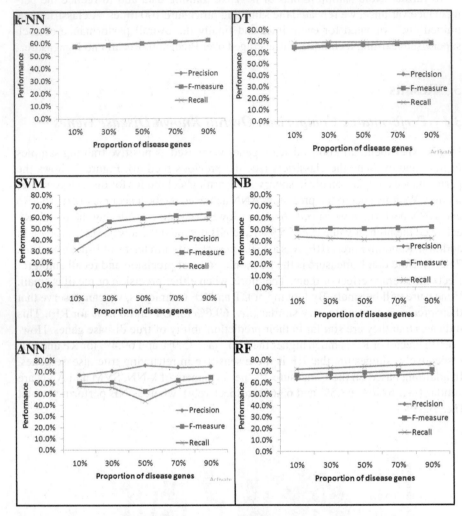

**Fig. 3** Performance assessment of each technique based on the changing of known disease genes

## 3.2  Performance Comparison on Subset of Known Disease Genes

OMIM is an online database of known disease genes and it has frequently been updated with newly discovered disease genes. The performance of a classifier is dependent on the size of training set, and therefore dependent on the number of known disease genes used as positive training samples. In order to investigate how much the number of known disease genes affects the overall performance of different classifiers, in the second experiment, we varied the positive training set from 10% to 90% of known disease genes. The remaining known disease genes were considered as "unknown novel disease genes" and merged with the "control set". After that, the same number of negative training samples was randomly selected from this "control set" to construct the training set. This training set was then used to train and evaluate the six classifiers based on the same procedure as for whole known disease genes. After that, precision, recall and F-measure of each classifier for each subset of known disease genes were recorded. The results are depicted in Figure 3. It could be seen from Figure 3 that most methods were stable with the change of known disease genes. The performance was generally better when the number of known disease genes used as positive training samples increased. This indicates that the performance of these supervised learning techniques was getting better when novel disease genes were discovered and included in the positive training set. As mentioned above, F-measure could be the most suitable measure since either too small precision or recall score is not good for disease gene prediction and this is reflected in small F-measure value. Therefore, based on F-measure, we additionally compared the performance of each method when the size of positive training is changed. Figure 4 also shows that RF still outperformed other methods, whereas NB performed worst.

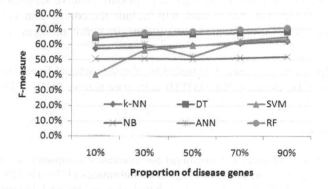

**Fig. 4** Performance comparison of six classification techniques based on the changing in number of known disease genes used as positive training samples

## 4 Conclusions and Future Works

Machine learning-based approaches usually treat disease gene prediction as a binary-class classification problem. In this study, we present some first experimental comparisons on the performance of five well known supervised learning techniques which have been used for the problem. In addition, we also tested Random Forest to measure how well an ensemble learning method could predict novel disease gene. We assessed the stability of these techniques when the number of known disease genes used as positive training samples was increased. Experiment results showed that Random Forest performed best based on F-measure and all methods are stable with the change in the size of the positive training set.

In our experiments, similar to most work in the literature, each binary-class classifier was trained and evaluated on a training set that includes both positive and negative training samples. The positive training samples are constructed from known disease genes, whereas negative training samples are usually selected from the remaining genes which are neither known to be associated to diseases (*i.e.*, unknown set) nor essential genes. This perhaps is the limitation of the machine learning approaches that are based on training binary-class classifiers for disease gene prediction problem, since the negative training set should ideally be actual non-disease genes. However, the construction of this set is nearly impossible in biomedical researches as it is often the case in biomedicine that not observing an association does not always imply the nonexistence of association (i.e. there is no proven negative sample). Therefore, to reduce this uncertainty, the binary-class supervised learning methods have to choose and were different in the way to construct the set of negative training samples. In reality, the unknown set may contain unknown disease genes; therefore, semi-supervised learning methods were recently proposed to solve the problem, where the classifier is learned from both labeled (*i.e.*, known disease genes) and unlabeled (*i.e.*, the unknown genes) data [50, 51]. Another recent approach to disease gene prediction is to use unary/one-class classification techniques, in which the classifier is learned from only positive samples (*i.e.*, known disease genes) [52]. Our future work will include the comparison and analysis of such techniques for the problem of disease gene prediction problem.

**Acknowledgements.** This research is funded by Vietnam National Foundation for Science and Technology Development (NAFOSTED) under grant number 102.01-2014.21.

# References

[1] Kann, M.G.: Advances in translational bioinformatics: computational approaches for the hunting of disease genes. Briefings in Bioinformatics 11, 96–110 (2009)
[2] Tranchevent, L.-C., et al.: A guide to web tools to prioritize candidate genes. Briefings in Bioinformatics 12, 22–32 (2010)
[3] Turner, F., et al.: POCUS: mining genomic sequence annotation to predict disease genes. Genome Biology 4, R75 (2003)
[4] Adie, E.A., et al.: SUSPECTS: enabling fast and effective prioritization of positional candidates. Bioinformatics 22, 773–774 (2006)
[5] Aerts, S., et al.: Gene prioritization through genomic data fusion. Nature Biotechnology 24, 537–544 (2006)
[6] Chen, J., et al.: Improved human disease candidate gene prioritization using mouse phenotype. BMC Bioinformatics 8, 392 (2007)
[7] Wang, X., et al.: Network-based methods for human disease gene prediction. Briefings in Functional Genomics 10, 280–293 (2011)
[8] Tarca, A.L., et al.: Machine learning and its applications to biology. PLoS Computational Biology 3, e116 (2007)
[9] Larrañaga, P., et al.: Machine learning in bioinformatics. Briefings in Bioinformatics 7, 86–112 (2006)

[10] Yip, K.Y., et al.: Machine learning and genome annotation: a match meant to be? Genome Biology 14, 205 (2013)

[11] de Ridder, D., et al.: Pattern recognition in bioinformatics. Briefings in Bioinformatics 14, 633–647 (2013)

[12] Basford, K.E., et al.: On the classification of microarray gene-expression data. Briefings in Bioinformatics 14, 402–410 (2013)

[13] Maetschke, S.R., et al.: Supervised, semi-supervised and unsupervised inference of gene regulatory networks. Briefings in Bioinformatics (2013)

[14] Ding, H., et al.: Similarity-based machine learning methods for predicting drug-target interactions: a brief review. Briefings in Bioinformatics (2013)

[15] Upstill-Goddard, R., et al.: Machine learning approaches for the discovery of gene-gene interactions in disease data. Briefings in Bioinformatics 14, 251–260 (2012)

[16] Okser, S., et al.: Genetic variants and their interactions in disease risk prediction - machine learning and network perspectives. BioData Mining (2013)

[17] Lospez-Bigas, N., Ouzounis, C.A.: Genome-wide identification of genes likely to be involved in human genetic disease. Nucleic Acids Research 32, 3108–3114 (2004)

[18] Adie, E., et al.: Speeding disease gene discovery by sequence based candidate prioritization. BMC Bioinformatics 6, 55 (2005)

[19] Xu, J., Li, Y.: Discovering disease-genes by topological features in human protein-protein interaction network. Bioinformatics 22, 2800–2805 (2006)

[20] Calvo, S., et al.: Systematic identification of human mitochondrial disease genes through integrative genomics. Nat. Genet. 38, 576–582 (2006)

[21] Smalter, A., et al.: Human disease-gene classification with integrative sequence-based and topological features of protein-protein interaction networks. In: IEEE International Conference on Bioinformatics and Biomedicine, BIBM 2007, pp. 209–216 (2007)

[22] Sun, J., et al.: Functional link artificial neural network-based disease gene prediction. In: Neural Networks, IJCNN 2009, pp. 3003–3010 (2009)

[23] Breiman, L., et al.: Classification and regression trees. Wadsworth & Brooks, Monterey (1984)

[24] Schapire, R.E.: A brief introduction to boosting. Ijcai 99, 1401–1406 (1999)

[25] Radivojac, P., et al.: An integrated approach to inferring gene-disease associations in humans. Proteins: Structure, Function, and Bioinformatics 72, 1030–1037 (2008)

[26] Keerthikumar, S., et al.: Prediction of candidate primary immunodeficiency disease genes using a support vector machine learning approach. DNA Research 16, 345–351 (2009)

[27] Amberger, J., et al.: McKusick's Online Mendelian Inheritance in Man (OMIM®). Nucleic Acids Research 37, D793–D796 (2009)

[28] Safran, M., et al.: GeneCards TM 2002: towards a complete, object-oriented, human gene compendium. Bioinformatics, 1542–1543 (2002)

[29] Lage, K., et al.: A human phenome-interactome network of protein complexes implicated in genetic disorders. Nat. Biotech. 25, 309–316 (2007)

[30] Tu, Z., et al.: Further understanding human disease genes by comparing with housekeeping genes and other genes. BMC Genomics 7, 31 (2006)

[31] Brown, K.R., Jurisica, I.: Online Predicted Human Interaction Database. Bioinformatics 21, 2076–2082 (2005)

[32] Freudenberg, J., Propping, P.: A similarity-based method for genome-wide prediction of disease-relevant human genes. Bioinformatics 18, S110–S115 (2002)

[33] The UniProt, C.: The Universal Protein Resource (UniProt) in 2010. Nucl. Acids Res. 38, D142–D148 (2010)

[34] Jonsson, P.F., Bates, P.A.: Global topological features of cancer proteins in the human interactome. Bioinformatics 22, 2291–2297 (2006)

[35] Apweiler, R., et al.: The InterPro database, an integrated documentation resource for protein families, domains and functional sites. Nucleic Acids Research 29, 37–40 (2001)

[36] Hunter, S., et al.: InterPro in 2011: new developments in the family and domain prediction database. Nucleic Acids Research 40, D306–D312 (2011)

[37] Smedley, D., et al.: BioMart - biological queries made easy. BMC Genomics 10, 22 (2009)

[38] Sayers, E.W., et al.: Database resources of the National Center for Biotechnology Information. Nucleic Acids Research 39, D38–D51 (2011)

[39] Luo, H., et al.: DEG 10, an update of the database of essential genes that includes both protein-coding genes and noncoding genomic elements. Nucleic Acids Research 42, D574–D580 (2014)

[40] Dennis, G., et al.: DAVID: Database for Annotation, Visualization, and Integrated Discovery. Genome Biology 4, R60 (2003)

[41] Quinlan, J.R.: Induction of decision trees. Machine Learning 1, 81–106 (1986)

[42] Olshen, L.B.J.H.F.R.A., Stone, C.J.: Classification and regression trees. Wadsworth International Group (1984)

[43] Altman, N.S.: An introduction to kernel and nearest-neighbor nonparametric regression. The American Statistician 46, 175–185 (1992)

[44] Rish, I.: An empirical study of the naive Bayes classifier. In: IJCAI 2001 Workshop on Empirical Methods in Artificial Intelligence, vol. 3, pp. 41–46 (2001)

[45] Cortes, C., Vapnik, V.: Support-vector networks. Machine Learning 20, 273–297 (1995)

[46] Breiman, L.: Random Forests. Machine Learning 45, 5–32 (2001)

[47] Hall, M., et al.: The WEKA data mining software: an update. ACM SIGKDD Explorations Newsletter 11, 10–18 (2009)

[48] Chang, C.-C., Lin, C.-J.: LIBSVM: a library for support vector machines. ACM Transactions on Intelligent Systems and Technology (TIST) 2, 27 (2011)

[49] Bollmann, P., Cherniavsky, V.S.: Restricted evaluation in information retrieval. ACM SIGIR Forum 16, 15–21 (1981)

[50] Mordelet, F., Vert, J.-P.: ProDiGe: Prioritization Of Disease Genes with multitask machine learning from positive and unlabeled examples. BMC Bioinformatics 12, 389 (2011)

[51] Yang, P., et al.: Positive-unlabeled learning for disease gene identification. Bioinformatics 28, 2640–2647 (2012)

[52] Yu, S., et al.: Gene prioritization and clustering by multi-view text mining. BMC Bioinformatics 11, 28 (2010)

# An Efficient Ant Colony Algorithm for DNA Motif Finding

Hoang X. Huan, Duong T.A. Tuyet, Doan T.T. Ha, and Nguyen T. Hung

**Abstract.** Finding motifs in gene sequences is one of the most important problems of bioinformatics and belongs to NP-hard type. This paper proposes a new ant colony optimization algorithm based on consensus approach, in which a relax technique is applied to find the location of the motif. The efficiency of the algorithm is evaluated by comparing it with the state-of-the-art algorithms.

**Keywords:** Ant Colony Optimization, MEMETIC, motif finding problem.

## 1 Introduction

Gene regulatory elements are called the DNA motifs (later we call it "motifs" in short), which contain a number of important biological information [1, 5, 12, 14, 18]. The identification of DNA motif is currently one of the most important problems in bioinformatics and is NP-hard (see [2, 10, 12, 16, 17, 19]). There are two main approaches to search for a motif: biological experiments and computing methods, i.e. bioinformatics. Due to the high cost and time consuming, biological experiments are not really effective, whereas computing methods are widely used to predict motifs. Researchers have made various definitions of motif, many statements for motif finding problem and also developed a number of algorithms for finding motif [1, 3, 5, 15]. One of the widely used approaches is to use an approximate algorithm to optimize consensus score or information content [2, 7, 10, 11, 15, 16, 19]. Recently, the methods that use ant colony optimization (ACO) have been applied effectively by several authors for this problem. For example, Bouamama et al. (2010) proposed MFACO algorithm [2] that uses consensus score to find motifs and information content to locate their appearances (binding sites) in each DNA sequence. Yang et al. (2011) proposed an algorithm [19], referred to from now on as EMACO,

Hoang X. Huan · Duong T.A. Tuyet · Doan T.T. Ha · Nguyen T. Hung
University of Engineering and Technology,
VNU, Hanoi, Vietnam
e-mail: {huanhx,tuyetdta_55,hadtt.mi13,hungnt_55}@vnu.edu.vn

© Springer International Publishing Switzerland 2015     589
V.-H. Nguyen et al. (eds.), *Knowledge and Systems Engineering,*
Advances in Intelligent Systems and Computing 326, DOI: 10.1007/978-3-319-11680-8_47

that combines ACO algorithm with Expectation Maximization (EM) to find the starting positions of motif in sequences. Liu et al. (2013) proposed ACRI algorithm [11] that uses information content as the objective function for the same purpose as EMACO.In this paper, we propose a new ACO algorithm called ACOMotif using the total Hamming distance score function of motif to DNA sequences for this problem. ACOMotif uses the structural graph as in MFACO but with different heuristics information, pheromone update rule, and local search technique. For each motif found, to locate the starting positions in the DNA sequences, the algorithm subsequently applies a relax technique and gives R-ACOMotif version for this goal. Runtime of ACOMotif is also compared with MotifSuite (2012) on a very large dataset obtained from [21] called SCPD. The efficiency of ACOMotif is indicated by the experiments on the same datasets published in three articles above and on SCPD. The rest of this paper is organized as follows. Section 2 states the DNA motif finding problem, followed by a brief introduction of ACO method and how it was applied in MFACO, ACRI, and EMACO algorithms.Our new algorithm will be introduced in Section 3. Section 4 describes the experiments comparing ACOMotif/R-ACOMotif with MFACO, EMACO, ACRI, and MotifSuite. Some conclusions are presented in the last section.

## 2 DNA Motif Finding Problem and Related Works

### 2.1 DNA Motif Finding Problem

DNA motif finding problem, from optimization perspective, can be described as follows [2, 19]: given a set of identical length DNA sequences $S = \{S_1, S_2, ..., S_N\}$, in which $S_i = s_1^i s_2^i ... s_n^i$, $s_j^i$ is one of the four characters in the alphabet $\Sigma = \{A, C, G, T\}$ for all $i, j$. For a given natural number $l$, there are two problems corresponding to two approaches:

Consensus Approach

Find a string $S_c$ of length $l$ and a set of subsequences $M = m_1, m_2, ..., m_N$, in which $m_i$ is a substring of length $l$ in $S_i$, such that they minimize the objective function:

$$H(S_c, M) = \sum_{i=1}^{N} d_H(S_c, m_i) \qquad (1)$$

where $d_H(S_c, m_i)$ is the hamming distance between string $S_c$ and substring $m_i$ defined by the number of positions at which the nucleotides in two string are different. Then $S_c$ is called a motif of $S$, each $m_i$ is called a motif instance (or instance in short) of $S_c$ in $S_i$. If we consider $\mathbf{M}$ as a matrix (called a consensus matrix) with the $i^{th}$ row being the string $m_i$ and denote $C(u, j)$ as the number of nucleotides $u$ in column $j$, the objective function $CSc(\mathbf{M})$ is formulated as:

$$CSc(\mathbf{M}) = \sum_{j=1}^{l} max(C(u, j)) : u \in \Sigma \qquad (2)$$

Then motif $S_c$ is a string of length $l$ and the letter at position $j$ is the nucleotide that occurs most often in the $j^{th}$ column of $\mathbf{M}$. Note that each $\mathbf{M}$ can have many motifs but we consider only one of them.

Positional Approach

Find a set of substrings $M = \{m_1, m_2, ..., m_N\}$ and a set of starting positions $A = \{a_1, a_2, ..., a_N\}$, in which, each instance $m_i$ is of length $l$ in $S_i$ corresponding to starting position $a_i$. In this approach, the objective function is information content:

$$IC = \sum_{j=1}^{l} \sum_{u \in \Sigma} Q(u, j) log_2 \frac{Q(u, j)}{p_u} \tag{3}$$

where $Q(u, j)$ indicates the frequency of nucleotide $u$ in column $j$ of the matrix $\mathbf{M}$, $p_u$ is the background frequency of $u$ in the entire set $S$. In reality, the location of $m_i$ on $S_i$ is called the binding site of DNA.

### 2.1.1 Remark

Note that it is not sure the optimal solutions of this objective function are real motifs. So, the more solutions and the closer to real motif's binding sites from locations of the instances, the better an algorithm is.

## 2.2 Ant Colony Optimization Method

Ant colony Optimization (ACO) proposed by Dorigo [6, 9] is a random meta-heuristic method to solve hard combinatorial optimization problems. This algorithm has been diversely improved in the literature and widely applied in many applications. Memetic scheme using population-based search technique was first proposed by Moscato [13] and applied for genetic algorithm. Today, it is incorporated with other algorithms [3, 8].

### 2.2.1 Memetic-ACO Algorithms

Our proposed method applies ACO with reinforcement search technique following a simple memetic scheme as described in Fig. 1 below. The Motif Finding problems are converted into the problems of finding solutions on the structural graph $G = (V, E, \Omega, \eta, T)$ where $V$ is the vertex set, $E$ is the edge set, and $\eta$ and $T$ are the set of heuristics information and pheromone trail, respectively, in reinforcement learning ; $\eta$ and $T$ can be placed on vertices or edges. An acceptable solution is a path satisfying the condition $\Omega$, starting from a vertex in a subset $C_0$ of $V$, then expanded by random to the next vertex based on heuristics information and pheromone trail. The ACO algorithm uses $N_{ants}$ artificial ants, in each iteration each ant finds a solution by a randomized procedure on the structural graph. Then all the ant solutions are assessed and choosing the best one to apply enhancing strategy or local search technique. Consequently, the obtained solutions will be evaluated again and the pheromone trail is updated as reinforcement learning information in the next iteration. Although many algorithms use the same graph $G(V,E)$, they use different heuristics information, pheromone update rules and localsearch techniques. From now on we will call $G(V,E)$ as the structural graph.

```
Procedure Memetic-ACO algorithms;
Begin
        Initialize; // initialize pheromone trail matrix and u ants;
    Repeat
            Construct solutions; // each ant constructs its own solution;
            Choose a subset Ω_il to evolve by local search;
            For each individual in Ω_il do
                    Run local search;
            End for;
            Update trail;
    Until Ending condition;
    Choose the solutions;
End;
```

Fig. 1 Specification of a simple memetic-ACO algorithm

Recently, some ACO-based algorithms following this scheme have been applied effectively for DNA motif finding:

**MFACO(2010)** proposed by Bouamama et al. [2], uses consensus score to find motifs and information content to determine their starting positions. Experiments showed that this algorithm obtains better results than the other best techniques: GS, BP, and MEME.

**EMACO (2011)** proposed by Do Yang et al. [19], combines ACO and Expectation Maximization (EM), in which EM is used to determine binding sites. Experiments revealed that this method is better than GAME and GALF.

**ACRI (2013)** proposed by Liu et al. [11], uses information content as the objective function to determine positions. Different from two methods above, this algorithm uses local search at two adjacent positions instead of using random search method. It is also experimentally proven to have better results than that of these algorithms: MEME, AlignACE, and Gibbs Sampler.

In ACO algorithms, there are four important factors affecting their performance: 1) structural graph, 2) heuristics information, 3) pheromone update rule, and 4) local search technique.

# 3 The Proposed Algorithm

The algorithm, named ACOMotif, takes into account total Hamming distance of motif to DNA sequences as the objective function. ACOMotif uses structural graph of MFACO but with different heuristics information, pheromone update rule, and local search technique. For each motif found, to locate binding sites in DNA sequences, it subsequently applies relax technique, which is why in this case ACOMotif is called R-ACOMotif.

## 3.1 ACOMotif

ACOMotif follows the scheme described in Fig. 1.The output is the set $Q$ including the motifs of length $l$ and the corresponding instances on the DNA sequences which have smallest hamming distance based on the found motifs. The detailed description of ACOMotif is as follows.

### 3.1.1 Structural Graph

To find a motif of length $l$, the algorithm's structural graph has $4 \times l$ vertices arranged in four rows and $l$ columns. Each vertex at position $(u, j)$ is labeled by the corresponding nucleotide $u$ as shown in Fig. 2. The labels of the vertices in each row are also used to refer to the rows. From left to right, edges connect vertices of two consecutive columns. We denote $e_j(u, v)$ as the edge connecting vertex $(u, j)$ to $(v, j + 1)$. Heuristics Information and pheromone trail are placed at the vertices of the first column and on the edges.

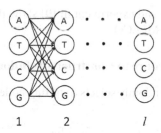

**Fig. 2** Structural graph for finding motif of length $l$

### 3.1.2 Heuristic Information

The heuristics information is placed at the first column vertices and on the edges. At $t$ vertices of the first column, heuristics information is the frequency of nucleotide in the entire dataset $S$. Heuristics information on edge $e_j(u, v)$ is the frequency of the couple $uv$ nucleotide in $S$. There are only 16 such quantities $\eta_{u,v}$ with $(u, v) \in \Omega \times \Omega$.

Remark

Note that, in MFACO, heuristics information at edges is computed by high-order
background model based onthe frequency of the motif pattern from the first column
to the current column in $S$. Since the appearance of these patterns in each DNA
sequence $S_i$ is rare, this kind of statistical information is limited.

### 3.1.3 Pheromone Update Rule

Our algorithm uses the SMMAS pheromone update rule (Smooth Max-Min Ant
System) [9]. Pheromone trails $\tau_u^1$ on each vertex $u$ of the first column) and $\tau_{u,v}^j$ (on
edges $e_j(u,v)$ are first initialized to a pre-determined value $\tau_{max}$. After each loop,
pheromone trail $\tau_u^1$ at each vertex $u$ is updated:

$$\tau_u^1 \leftarrow (1-\rho)\tau_u^1 + \Delta_u^1 \tag{4}$$

where

$$\Delta_u^1 = \begin{cases} \rho\tau_{max} & u \in \text{best solution} \\ \rho\tau_{min} & \text{otherwise} \end{cases}$$

$\tau_{max}, \tau_{min}$ and $\rho$ are predetermined parameters. Pheromone trail on edge $e_j(u,v)$ is
updated:

$$\tau_{u,v}^j \leftarrow (1-\rho)\tau_{u,v}^j + \Delta_{u,v}^j \tag{5}$$

where:

$$\Delta_{u,v}^j = \begin{cases} \rho\tau_{max} & uv \in \text{best solution} \\ \rho\tau_{min} & \text{otherwise} \end{cases}$$

Remark

The computational analysis and experiment in [9] show that this rule is better than
MMAS update rule used in MFACO.

### 3.1.4 Randomized Procedure to Construct a Solution

In each iteration, each ant randomly selects a starting node $u$ at the first column with
probability $P_u^1$:

$$P_u^1 = \frac{\tau_u^1 \times \eta_u}{\sum_{v \in A,C,G,T} \tau_v^1 \times \eta_v} \tag{6}$$

Then the ant randomly walks through all the columns sequentially with the probability of choosing the edge from vertex $u$ of column $j$ to vertex $v$ at column $j+1$
being $P_{u,v}^j$:

$$P_{u,v}^j = \frac{\tau_{uv}^j \times \eta_{u,v}}{\sum_{r \in A,C,G,T} \tau_{ur}^j \times \eta_{u,r}} \tag{7}$$

The path of the ant from starting vertex to the last column vertex identifies an acceptable solution for motif.

### 3.1.5 The Objective Function and Identification of Binding Sites

For each acceptable solution $S_c$, ACOMotif uses the total Hamming distance $H_d(S_c)$ from $S_c$ to DNA sequences in $S$ as the objective function:

$$H_d(S_c) = \sum i = 1^N d_H(S_c, S_i) \tag{8}$$

where $d_H(S_c, m) : m$ with length $l$ is a subsequence of $S_i$

$$d_H(S_c, S_i) = min \tag{9}$$

String $m_i$ (9) is an instance of $S_c$ and its position in $S_i$ is the binding site. Note that each can have multiple instances on $S_i$ with the same distance.

### 3.1.6 Local Search

ACOMotif applies hill-climbing technique for local search as described below. After all ants finish their paths through the graph, the solutions are formed with their corresponding total hamming scores; then the local search is applied on the solutions having smallest scores. For each potential motif $S_m$, use the set $Q(S_m)$ to contain search results, and the iteration procedure is carried out as follows:

    *Step 1:* Initialize $Q(S_m) = \{S_m\}$

    *Step 2:*

        **Repeat:**

            **For** each $i = 1, ..., l$ **do**:

                2.1. Replace letter at position $i$ of $S_m$ by one of three remaining letters consecutively in set $\Omega$ to get $S_p$;

                2.2. Compute $H_d(S_p)$;

                2.3. If $H_d(S_p) \leqslant H_d(S_m)$ then $S_m \leftarrow S_p$ and $Q(S_m) \leftarrow Q(S_m) \cup S_p$;

        **End for**;

        **Until** we can not improve the objective function anymore

After applying local search for potential motifs in each iteration, the sets $Q(S_m)$, consisting of candidates with the smallest or nearly smallest score, are combined into the set $Q$ containing all the best solutions up to that point, which have the same binding site (retaining only a motif). Based on set $Q$, the pheromone trail on the graph is updated according to Eq.(4) and Eq.(5). The algorithm stops when it finishes running a predefined number of loops. The binding sites associated with motifs in Qallow us to identify instances of motif. Remark: The local search algorithm used in this article generalizes that of ACRI in order to climb to the highest local position as much as possible while remaining the previous simple idea instead of using complicated Gibb sampling technique as in MFACO.

## 3.2 R-ACOMotif

ACOMotif additionlly uses relax technique to locate binding site of each motif. When ACOMotif deploys this technique, it is called R-ACOMotif. With each motif $S_c$ found in the set of solutions $Q$ of ACOMotif and given a number $\varepsilon \in (0, 1)$, relax technique finds set of instances $M = \{m_1, ..., m_N\}$ and set of starting positions $A = \{a_1, ..., a_N\}$ as follows:

   Step 1. //Expand the instances set and binding sites
        1.1. On each sequence $S_i$, finds locations $a_i^j$ so that for each
             substring $m_i^j$ of length $l$, Hamming distance from substring
             to $S_c$ is $d_H(S_c, S_i)$ or $d_H(S_c, S_i) + 1$. We get sets $M_i$ and $A_i$
             including $m_i^j$ and $a_i^j$ respectively;
        1.2. Compute the number of elements $n_i$ of sets $M_i$ respectively
             and $\sum_{i=1}^{N} n_i$;
   Step 2. // Filter to reduce the size of sets $M_i$ and $A_i$
        **Repeat**
        2.1. Rearrange the order of the set $M$ incrementally with respect
             to $n_i$ // later follows this new order;
        2.2. Determine the smallest number $k$ so that with every $i \geq k$
             then $n_i > 1$;
        2.3. **For** each $i = k$ to $N$ **do**
             2.3.1. For each $m_i^j \in M_i$, compute

$$H_t(m_i^j) = d_H(m_i^j, S_c) + \sum_{t=1}^{i} \sum_{k=1}^{|M_t|} d_H(m_i^j, m_t^k)$$

             2.3.2. Compute $g_i = min\{H_t(m_i^j)\}$;
             2.3.3. If $H_t(m_i^j) \geq (1 + \varepsilon)g_i$ then remove $m_i^j$ out of set $M_i$;
        2.4. Update $\varepsilon \leftarrow \varepsilon \times 0.95$ // reduce ;
        **Until** $\sum_{i=1}^{N} n_i$ is smaller than half of that value before the loop;
   Step 3. // find the best solutions
        3.1. Build all consensus matrices from the reduced set $M_i$
             and compute consensus score as in Eq. (2);
        3.2. Sort the matrices in step 3.1 and the corresponding locations
             of the instances in decreasing orderwith respect to their
             consensus;
   Step 4. The solution is the first tuple in the list in step 3.2 // can take
        more depending on priority computed.

## 4  Exprimental Results

The program was written in Perl, run on a desktop computer equipped with CPU Intel Core i5 2.5 GHz and 4 GB RAM, using Ubuntu 12.04 Operating System. Our experiments compare the new algorithm's efficiency with those of MFACO [2], EMACO [19], and ACRI [11] on the same datasets, using the same numbers of loops and ants as in the corresponding evaluations.Because of not having the programs of these algorithms, we cannot compare runtime on the same configuration machine, the results of the compared algorithms will be taken directly from the published articles. The runtime of ACOMotif/R-ACOMotif is in average. The parameters had been set as follows:

- $\tau_{max} = 1.0$; $\tau_{min} = \frac{1}{4^l}$
- $\rho$ is chosen depending on the algorithm's number of loops

| Number of loops | 10-100 | 100-300 | 300-600 | >600 |
|---|---|---|---|---|
| Coefficient $\rho$ | 0.03-0.05 | 0.02-0.03 | 0.01-0.02 | 0.005 |

To evaluate computation time, ACOMotif was compared with MotifSuite (2012) [20] on SCPD dataset [21]. The efficiency of ACOMotif was assessed by experiments on the same published dataset of three algorithms above and on SCPD.

## 4.1 Comparison with MFACO Using Consensus Approach

Experiments on H.sapiens used in [2] contain three small sets with the number of strings are 6, 9, and 12, respectively. Each of them has 3001 nucleotide in length. H.sapiens dataset did not have known actual motif biologically. Therefore, our experiments just compare the values of objective functions as computed in Eq. (1) and Eq. (2). We use notations HSc for total Hamming distance score and CSc for consensus score. Note that:

$$HSc = N \times l - CSc \qquad (10)$$

then smaller $HSc$ is equivalent to greater $CSc$; therefore, we only need to care about $HSc$. The experiments have been performed in the same way in [2]:each dataset is run three times with 50 loops, running time is in average. The result of MFACO is taken from [2], $N_{ants} = 8$. The experimental results for dataset 1 are presented in Table 1 and Table 2, which shows that ACOMotifis is considerably better than MFACO. Note that, although in Table 2, MFACO found three motifs whose $HSc$ score equals 3, while ACOMotif found only one, ACOMotif discovered much more motif candidates.

**Table 1** Comparison between ACOMotif and MFACO on H.sapiens 1: $\rho = 0.03$, $l = 7$; $N=6$, runtime 33s

| ACOMotif | CSc | HSc | MFACO | CSc | HSc |
|---|---|---|---|---|---|
| CCTCCCC | 42 | 0 | AAAAAAA | 42 | 0 |
| AAAAAAA | 42 | 0 | AGGAGGA | 42 | 0 |
| GCAGCGG | 42 | 0 | AAAAAAG | 42 | 0 |
| GCCGGGG | 42 | 0 | TAAAAAT | 42 | 0 |
| GCCGCCG | 42 | 0 | | | |
| AAAAAAG | 42 | 0 | | | |
| GCCTGTG | 42 | 0 | | | |
| TAAAAAT | 42 | 0 | | | |
| CGGCGCC | 42 | 0 | | | |
| GGGCCAG | 42 | 0 | | | |
| GGCCAGG | 42 | 0 | | | |
| GCGGGCG | 42 | 0 | | | |
| CCCGGGC | 42 | 0 | | | |

**Table 2** Comparison between ACOMotif and MFACO on H.sapiens 1: $\rho = 0.03$, $l = 13$, $N = 6$, runtime 57s

| ACOMotif | CSc | HSc | MFACO | CSc | HSc |
|---|---|---|---|---|---|
| AAAAAAAAAAAGA | 76 | 2 | AAAAAAAAAAAGA | 76 | 2 |
| AAAAAAAAAAAAG | 75 | 3 | AAAAAAAAAAAAG | 75 | 2 |
| GCTGAGGCAGGAG | 72 | 6 | AAAAAAAAAAAGT | 75 | 2 |
| GCCGCCGCCGCCG | 72 | 6 | AAAAAAAAAAAAG | 75 | 2 |
| CGCCGCCGCCGCC | 72 | 6 | | | |
| GAGGCTGAGGCAG | 71 | 7 | | | |

The experimental results for dataset 2 are represented in Table 3 and Table 4 and for dataset 3 are presented in Table 5 and Table 6.

**Table 3** Comparison between ACOMotif and MFACO on H.sapiens 1: $\rho = 0.03$, $l = 7$; $N=9$, runtime 29s

| ACOMotif | CSc | HSc | MFACO | CSc | HSc |
|---|---|---|---|---|---|
| CCCTCCT | 63 | 0 | CCCTCCT | 63 | 0 |
| CTCCCTT | 62 | 1 | CCCTCAG | 62 | 1 |
| GAGCAGG | 62 | 1 | GGGTTGG | 62 | 1 |
| GGGTTGG | 62 | 1 | GAGCAGG | 62 | 1 |
| GGGGCTG | 62 | 1 | | | |
| TGGGAGG | 62 | 1 | | | |
| GGCGGCC | 62 | 1 | | | |
| GGGGCTG | 62 | 1 | | | |
| CCCCTCC | 62 | 1 | | | |
| TTCCTGG | 62 | 1 | | | |
| CCCCTCC | 62 | 1 | | | |
| GGGCTGG | 62 | 1 | | | |
| CCTCCCT | 62 | 1 | | | |

**Table 4** Comparison between ACOMotif and MFACO on H.sapiens 1: $\rho = 0.03$, $l = 13$, $N = 9$, runtime 79s

| ACOMotif | CSc | HSc | MFACO | CSc | HSc |
|---|---|---|---|---|---|
| GCCGGCGGGCGCC | 102 | 15 | GCCGGCGGGCGCC | 102 | 15 |
| GGCCCCCGGGCGG | 101 | 16 | GCGGGCGGGCGCC | 101 | 16 |
| GGGGGAGCAGGAG | 101 | 16 | GCCGGCGGGCGGG | 100 | 17 |
| GGCCGGCGGGCGG | 100 | 17 | GCCGGAGGGCGCC | 100 | 17 |
| GCAGGGGCTGGGG | 100 | 17 | | | |
| GGCCAGGCTCGGC | 100 | 17 | | | |
| CCCCGCCCCCGGC | 100 | 17 | | | |

**Table 5** Comparison between ACOMotif and MFACO on H.sapiens 1: $\rho = 0.03$, $l = 7$; $N=12$, runtime 49s

| ACOMotif | CSc | HSc | MFACO | CSc | HSc |
|---|---|---|---|---|---|
| GGCGGGG | 123 | 3 | GGGGCGG | 123 | 3 |
| CTGAGGC | 123 | 3 | CCCAGCT | 123 | 3 |
| CCAGCTG | 123 | 3 | CCAGCTG | 123 | 3 |
| GAGGCAG | 123 | 3 | CTGAGGC | 123 | 3 |
| GGGGCGG | 123 | 3 | | | |

**Table 6** Comparison between ACOMotif and MFACO on H.sapiens 1: $\rho = 0.03$, $l = 13$, $N = 12$, runtime 103s

| ACOMotif | CSc | HSc | MFACO | CSc | HSc |
|---|---|---|---|---|---|
| GGGAGGCTGAGGC | 205 | 29 | GGGAGGCTGAGGC | 205 | 29 |
| CGGGAGGCGGAGG | 204 | 30 | CGGGAGGCGGAGG | 204 | 30 |
| GCTGAGGCAGGAG | 202 | 32 | GGAGGCTGAGGCA | 202 | 32 |
| GGAGGCTGAGGCA | 202 | 32 | CGGGAGGCGGGGG | 201 | 33 |
| GGCTGAGGCAGGA | 119 | 35 | | | |
| GGGCGGGGCGGGG | 119 | 35 | | | |

**Comment:** Table 3 shows that ACOMotif is considerably better in terms of number of found motifs with minimum score. Table 5 proves that ACOMotif discovered one more motif with minimum score 3. With table 6, ACOMotif still gave better results in terms of both number of found motifs and score.

## 4.2 Comparison with MFACO and ACRI Using Positional Approach

The experiment was carried on E.coli dataset: CRP binding sites used by both MFACO and ACRI in [2], [11] to compare discovered binding site. The dataset contains eighteen 105-nucleotite strings. The length of examined motif is 22 as the same in MFACO and ACRI. R-ACOMotif ran 20 times, each with 300 loops, $N_{ants}=10$, and $\rho = 0.02$. Experimental result is expressed in Table 7, which shows that R-ACOMotif and MFACO both discovered all the correct starting positions however ACRI had comparably high error.

**Table 7** Comparison result betweenR-ACOMotif and MFACO, ACRI algorithms

| Order | Position | ACRI | Error | MFACO | Error | R-ACOMotif | Error |
|---|---|---|---|---|---|---|---|
| 1 | 17,61 | 63 | 2 | 61 | 0 | 61 | 0 |
| 2 | 17,55 | 57 | 2 | 55 | 0 | 55 | 0 |
| 3 | 76 | 78 | 2 | 76 | 0 | 76 | 0 |
| 4 | 63 | 65 | 2 | 63 | 0 | 63 | 0 |
| 5 | 50 | 52 | 2 | 50 | 0 | 50 | 0 |
| 6 | 7,60 | 9 | 2 | 7 | 0 | 7 | 0 |
| 7 | 42 | 44 | 2 | 44 | 0 | 44 | 0 |
| 8 | 39 | 41 | 2 | 39 | 0 | 39 | 0 |
| 9 | 9,80 | 11 | 2 | 9 | 0 | 9 | 0 |
| 10 | 14 | 16 | 2 | 14 | 0 | 14 | 0 |
| 11 | 61 | 63 | 2 | 61 | 0 | 61 | 0 |
| 12 | 41 | 43 | 2 | 41 | 0 | 41 | 0 |
| 13 | 48 | 50 | 2 | 48 | 0 | 48 | 0 |
| 14 | 71 | 73 | 2 | 71 | 0 | 71 | 0 |
| 15 | 17 | 19 | 2 | 17 | 0 | 17 | 0 |
| 16 | 53 | 55 | 2 | 53 | 0 | 53 | 0 |
| 17 | 1,84 | 95 | 4 | 84 | 0 | 84 | 0 |
| 18 | 78 | 78 | 0 | 78 | 0 | 78 | 0 |

## 4.3 Comparison with EMACO Using Positional Approach

Experiments on two datasets, ERE and E2F, were carried on with EMACO [19]. Each of them includes 25 strings and 200 nucleotides per string; real motifs and its starting positionson each string are known in advance. The algorithms were run 20 times and compared their average values, using $N_{ants}=20$, 100 loops, and $\rho = 0.03$.According to [19], the discovered position is correct if it is at most 3 unit(s) away fromreal location.To assess the result, the study [4] proposed three measurements including precision, recall, and F-score:

$$Precision = \frac{n_c}{n_p}, Recall = \frac{n_c}{n_t}, F-score = \frac{2 \times Precision \times Recall}{Precision + Recall} \quad (11)$$

where $n_c$ denotes the number of correct predicted positions, $n_p$ indicates the total number of predicted positions and $n_t$ is the total number of real known positions. Experimental result comparing with EMACO is presented in Table 8.

**Table 8** Comparison between R-ACOMotif and EMACO on ERE and E2F datasets

| Data | R-ACOMotif | | | ACO combined with EM | | |
|---|---|---|---|---|---|---|
| | Precision | Recall | F-score | Precision | Recall | F-score |
| ERE | 0.89 ±0.005 | 0.83 ± 0.04 | 0.86±0.005 | 0.76±0.01 | 0.85±0.01 | 0.81±0.01 |
| E2F | 1±0 | 1±0 | 1±0 | 0.91±0.02 | 0.92±0.02 | 0.91±0.02 |

**Comment:** From the above table, we can see easily that on both two datasets, R-ACOMotif has significantly better measurements as against EMACO.In particular, the former's F-score ishigher than the latter's one. Thus, we can conclude that R-ACOMotif runs more efficient than ACO combined with EM algorithm in [19].

## 4.4    Comparison with MotifSuite

With Motif finding problem, the quality of solution is the most important factor. However, to get an evaluation of running time, ACOMotif is also compared with MotifSuite[20] in two data GCR1 and GCN4 taken from SCPD dataset [21]. To compare computation time and precision scores, ACOMotif and MotifSuite [20] were run on two datasets GCR1 and GCN4 taken from SCPD data [21]. GCR1 contains 6 strings with length 9050 each (DNA sequences), and real motif is CTTCC ($l = 5$). GCN4 contains 9 strings, and real motif is TGACTC ($l = 6$).

**Table 9** Comparison of runtime between ACOMotif and MotifSuite

|      | ACOMotif(s) | MotifSuite(s) |
|------|-------------|---------------|
| GCR1 | 148.8       | 501.8         |
| GCN4 | 314.9       | 891.9         |

## 5    Conclusion

Motif Finding is one of the challenges of bio-molecular. The application of ant colony optimization algorithm to address the problem hasshown its power, but each algorithm has its own advantages and disadvantages. Experiments prove that ACOMotif algorithm is superior in comparison with existing algorithms, and R-ACOMotif version allows us to find bindingsitesof real motifprecisely. This algorithm can be developed to apply to other types of motif problems,and can improve search techniques to enhance quality. When parallel processing is employed, the runtime will be lower.

**Acknowledgements.** This work was done during the stay of the first author in Vienamese Institute for Advanced Study in Mathematics (VIASM).

## References

[1] Bandyopadhyay, S., Sahni, S., Rajasekaran, S.: Pms6: A faster algorithm for motif discovery. In: Proceedings of the second IEEE Int. Conf. on Computational Advances in Bio and Medical Sciences, pp. 1–6 (2012)

[2] Bouamama, S., Boukerram, A., Al-Badarneh, A.F.: Motif finding using ant colony optimization. In: Dorigo, M., et al. (eds.) ANTS 2010. LNCS, vol. 6234, pp. 464–471. Springer, Heidelberg (2010)

[3] Chen, X.S., Ong, Y.S., Lim, M.H.: Research frontier: memetic computation - past, present & future. IEEE Computational Intelligence Magazine 5(2), 24–36 (2011)

[4] Claeys, M., Storms, V., Sun, H., Michoel, T., Marchal, K.: MotifSuite: workflow for probabilistic motif detection and assessment. Bioinformatics 28(14), 1931–1932 (2012), doi:10.1093/bioinformatics/bts293

[5] Dinh, H., Rajasekaran, S., Davila, J.: qPMS7: A fast algorithm for finding the (l; d)-motif in DNA and protein sequences. PLoS One 7(7), e41425 (2012)

[6] Dorigo, M., Stutzle, T.: Ant Colony Optimization. MIT Press, Cambridge (2004)

[7] Eskin, E., Pevzner, P.: Finding composite regulatory patterns in DNA sequences. Bioinformatics S1, 354–363 (2002)

[8] Hoang Xuan, H., Do Duc, D., Manh Ha, N.: An efficient two-phase ant colony optimization algorithm for the closest string problem, pp. 188–197 (2012)

[9] Xuan, H., NguyenLinh, T., Do Duc, D., Huu Tue, H.: Solving the traveling salesman problem with ant colony optimization: arevisit and new efficient algorithms. REV Journal on Electronics and Communications 2(3-4), 121–129 (2012)

[10] Keith, M.K.: A simulated annealing algorithm for finding consensus sequences. J. Bioinformatics 18, 1494–1499 (2002)

[11] Liu, W., Chen, H., Chen, L.: An ant colony optimization based algorithm for identifying gene regulatory elements. Comp. in Bio. and Med. 43(7), 922–932 (2013)

[12] Lo, N.W., Changchien, S.W., Chang, Y.F., Lu, T.C.: Human promoter prediction based on sorted consensus sequence patterns by genetic algorithms. Proc. of the Int. Congress on Biological and Medical Engineering D3I-1540, 111–112 (2002)

[13] Moscato, P.: Onevolution, search, optimization, genetic algorithms and martial arts: towards memeticalgorithms. Tech. Rep.Caltech Concurrent Computation Program, Report. 826, California Institute of Technology, Pasadena, California, USA (1989)

[14] Neuwald, A., Liu, J., Lawrence, C.: Gibbs motif sampling: detection of bacterial outer membrane protein repeats. Protein Science, 1618–1632 (1995)

[15] Pisanti, N., Carvalho, A.M., Marsan, L., Sagot, M.-F.: RISOTTO: Fast extraction of motifs with mismatches. In: Correa, J.R., Hevia, A., Kiwi, M. (eds.) LATIN 2006. LNCS, vol. 3887, pp. 757–768. Springer, Heidelberg (2006)

[16] Stormo, G.D., Hartzell, G.W.: Identifying protein-binding sites from unaligned DNA fragments. Proc. Natl. Acad. Sci. USA 86(4), 1183–1187 (1989)

[17] Stormo, G.D., Hartzell, G.W.: Identifying protein-binding sites from unaligned DNA fragments. Proc. Natl. Acad. Sci. USA 86(4), 1183–1187 (1989)

[18] Thompson, W., Eric, C.R., Lawrence, E.L.: Gibbsrecursive sampler: finding transcription factor binding sites. J. Nucleic Acids Research 31, 3580–3585 (2003)

[19] Yang, C.H., Liu, Y.T., Chuang, L.Y.: DNA motif discovery based on ant colony optimization and expectation maximization. In: Proc. of IMECS, pp. 169–174 (2011)

[20] http://bioinformatics.psb.ugent.be/webtools/MotifSuite/Index.htm#pub

[21] http://rulai.cshl.edu/SCPD/

# Entity Linking for Vietnamese Tweets

Duy K. Van, Huy M. Huynh, Hien T. Nguyen, and Vinh T. Vo

**Abstract.** We study the task of entity linking for Vietnamese tweets, which aims at detecting entity mentions and linking them to corresponding entries in a given knowledge base. Unlike authored news or textual web content, tweets are noisy, irregular, and short, which causes entity linking in tweets much more challenging. We propose an approach to build an end-to-end entity linking system for Vietnamese tweets. The system consists of two stages. The first stage is to detect mentions and the second one performs entity disambiguation. We create a dataset including 524 Vietnamese tweets with 1,061 mentions and evaluate the system on this dataset. Our system achieves 69.2% F1-score. In order to show that our system is language-independent, we evaluate the system on a public dataset including 562 English tweets. The experiment results show that our system achieves 54.5% F1-score and outperforms the state-of-the-art end-to-end entity linking methods for tweets. To the best of our knowledge, this is the first attempt to build an end-to-end entity linking system for Vietnamese tweets and the system achieves very encouraging performance.

**Keywords:** Entity Linking, Wikification, Entity Disambiguation.

## 1 Introduction

Nowadays, many popular online social networks (OSNs) such as Twitter and Facebook, or social media in general, has become a channel for users to share information with each other. The popularity of online social media together with their diversity has drastically changed the landscape of communications and information sharing over the internet. Due to the huge magnitude of online social network users, an enormous amount of data is created everyday. According to the Statistics in 2011,

Duy K. Van · Huy M. Huynh · Hien T. Nguyen · Vinh T. Vo
Faculty of Information Technology, Ton Duc Thang University, Vietnam
e-mail: {vankhanhduy,huynhminhhuy,hien,vothanhvinh}@tdt.edu.vn

© Springer International Publishing Switzerland 2015                              603
V.-H. Nguyen et al. (eds.), *Knowledge and Systems Engineering*,
Advances in Intelligent Systems and Computing 326, DOI: 10.1007/978-3-319-11680-8_48

the number of Tweets people sent on Twitter per day has been up to 140 million tweets[1]. Currently, social media analytics to explore hidden rules and semantics of the posting content on social media is a very hot topic.

In this paper, we aim at analyzing the semantics of postings on social media, focusing mainly on entity linking. The users on social media are different from users on traditional web content in that they are not only passive consumers but also create the content on their own. As a result, unlike authored news or textual web content, the contents on social media pose new challenges for analyzing the semantics of those contents due to their large-scale, noisy, irregular, short (a message posted on Twitter has maximum 140 characters), and temporal dynamics. Therefore, solving *entity linking* in order to analyze the semantics of postings on social media is a challenging task, but helping behavior modeling of users, community detection, misinformation detection, information diffusion, etc.

Entity linking (EL) is a task of identifying entity mentions in a document, disambiguating them and linking them to their corresponding entries in a given knowledge base. It is an essential component in information extraction, information retrieval, or natural language processing applications and a challenging task. Entity linking (EL) is challenging due to the ambiguity of surface forms. That is because one surface form may refer to different entities in different occurrences and one entity may be referred to by different surface forms in different contexts. For example, the surface form *Michael Jordan* in different occurrences may refer to the basketball player (who had ever played for Chicago Bulls), the professor working at UC Berkeley, etc; or surface forms *Michael Jordan* and *Jordan* in different contexts can referred to the same person.

This problem has recently emerged as a hot topic and attracted many research efforts. For the past eight years, many approaches have been proposed for entity linking [21, 22, 23, 24, 27, 29]. And, since 2009, Entity Linking shared task held at Text Analysis Conference (TAC) 2009 [15], 2010 [14], 2011 [13], 2012[2], and 2013[3] has attracted more and more attentions in linking entity mentions to knowledge base entries. However, those work above-mentioned mainly focuses on long texts and most of them assumes that string mentions were detected. Due to short, noisy, context-dependent, and dynamic nature, semantic interpretation of messages propagated over social media is an extremely challenging task [20]. Until now, there have been several publications on entity linking in microblogs [2, 3, 4, 16, 17, 18, 19, 26, 28], but as reported in [19], after investigating and employing proposed state-of-the-art entity linking methods for microblog posts, the authors demonstrated by experiments that those state-of-the-art entity linking methods are not sufficiently robust on noise and short informal texts in microblogs and entity linking in microblog is particularly challenging.

Even though those approaches to EL exploited diverse features and employed many learning models, EL is still a hot topic. Therefore, TAC 2014[4] and the

---

[1] http://blog.twitter.com/2011/03/numbers.html

[2] http://www.nist.gov/tac/publications/2012/index.html

[3] http://www.nist.gov/tac/2013/KBP/index.html

[4] http://www.nist.gov/tac/2014/KBP/index.html

workshop 2014 Entity Recognition and Disambiguation Challenge[5] (ERD 2014) held with SIGIR 2014[6] conference are calling for papers proposing new approaches to entity linking. Especially, ERD 2014 calls for end-to-end systems that analyze short documents, as well as informal texts, and find entity mentions of more diverse entity types, such as Books and TV Shows. For those systems, no string mention will be provided.

We propose an approach to build an end-to-end entity linking system for Vietnamese tweets. The system consists of two stages. The first stage is to detect mentions and the second one performs entity disambiguation. We create a dataset including 524 Vietnamese tweets with 1,061 mentions and evaluate the system on this dataset. The contributions of this paper are three-fold as follows: (i) we propose an end-to-end entity linking system for Vietnamese tweets, (ii) we build a Vietnamese dataset to evaluate the entity linking system for Vietnamese tweets, and (iii) we evaluate our end-to-end entity linking system on a public dataset of English tweets and show that it outperforms the state-of-the-art end-to-end entity linking methods for tweets. To the best of our knowledge, this is the first attempt to build an end-to-end entity linking system for Vietnamese tweets and the Vietnamese dataset to evaluate the performance of the entity linking system.

The rest of this paper is organized as follows. Section 2 presents related work. Section 3 presents our proposed approach. Section 4 presents datasets, experiments and results. Finally, we draw conclusion in Section 5.

## 2 Related Work

Even though there are many entity linking approaches proposed in literature; however, they have some drawbacks follows:

- Almost all entity linking systems proposed in literature are not end-to-end systems where no string mention will be provided.
- They focus mainly on long texts with rich context for analyzing. However, techniques that work well for long texts may not generalize to short documents.
- They focus mainly on three kinds of types – person, location, and organization, but the real world consists of more diverse entity types, such as Books and TV Shows.
- There are not any entity linking approaches proposed to short Vietnamese texts.

According to García *et al.* (2014), one can classify entity linking approaches in literature into two families: (i) bag-of-links approaches and (ii) graph approaches. The former select the most likely candidate by computing similarity between the context document and each candidate, and choose the most similar one [1, 2, 4, 5, 17, 21, 22, 23, 24]; or computing the popularity of each candidate and choose the most popular [12]; or using statistical methods based on a given context information [11]. The latter rely on building referent graph and then computing random walk or

---

[5] http://web-ngram.research.microsoft.com/erd2014/Default.aspx
[6] http://sigir.org/sigir2014/

applying PageRank to select the most likely candidates for mentions [3, 10, 16, 18, 28].

The most similar entity linking systems to ours were proposed in [2, 17, 28, 30]. In [2] and [30], the authors employed some supervised learning algorithms for entity linking, but without taking global evidence in consideration. In [17], the authors proposed an end-to-end entity linking method that jointly optimizes mention detection and entity disambiguation. This method not only used local features but also explored second order entity-entity relations as global evidence. In [28], the authors constructed a relational graph in which each node is a mention and its referent concept candidate and each edge is established based on local compatibility, coreference, or semantic relatedness between two corresponding nodes. They then propose a graph-based semi-supervised learning algorithm to jointly identification and linking mentions.

In this work, we propose a method that consists of two stages for mention detection and entity disambiguation respectively. In entity disambiguation stage, we employ the method proposed in [1] to rank candidates of a mention and choose the most likely one. In mention detection stage, a novel method is proposed for identification of mentions. In this work, we build an end-to-end entity linking system that can perform entity linking for both Vietnamese and English tweets.

## 3 Proposed Method

In this section, we firstly introduce the framework of our system that is adapted for Vietnamese tweets, and then the detail of each component in our system will be presented along with our contribution points.

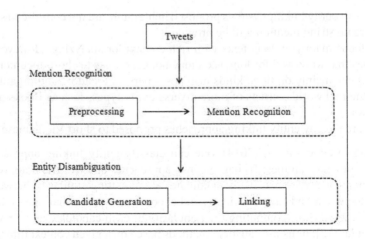

**Fig. 1** Our End-to-End Entity Linking System Architecture

As in Fig. 1, our framework consists of two main components, which are mention recognition and entity disambiguation. The first component is used for extracting possible entity mentions in tweets. Then those mentions are passed into the entity disambiguation component. For each mention, entity disambiguation is to generate and rank candidates, and then links the mention to the best one.

## 3.1 Mention Recognition

### 3.1.1 Preprocessing

The original text of a tweet can contain various noisy contents such as emotion symbols (e.g: ∧∧,...), hashtag symbol, etc. Those noisy words can affect the performance of the mention recognition algorithm. Therefore, we try to clean up those noisy symbols before proceeding to the mention recognition step.

One of Vietnamese distinguished characteristics is the diacritic (aka tone mark). There are six diacritics in total, which are *unmarked* (e.g: "*a*"), *grave accent* (e.g: "*à*"), *hook above* (e.g: "*ả*"), *tilde* (e.g: "*ã*"), *acute accent* (e.g: "*á*") and *dot below* (e.g: "*ạ*"). It is more difficult when there are two ways of marking those diacritics, which are *old style* and *new style*. The old style tries to put the diacritics as close to the center of words as possible (e.g: "*của*"), while the *new style* tried to place the marks at the main vowel (e.g: "*cuả*"). This also causes the difficulty for string matching in our mention recognition algorithm. Therefore, we need to normalize the tone marks. That is, all the diacritics of words are processed and changed to *old style* if necessary.

The last issue to consider is the character encodings in use while composing Vietnamese text. There have been various character encodings such as *VNQR, VNI, Precomposed Unicode, Decomposed Unicode, etc.* In this work, we choose to convert all of the encoding into *Precomposed Unicode* to take it easy for the string matching step in our mention recognition algorithm.

### 3.1.2 Mention Recognition Algorithm

We adapted a method called link-proba-bility-based [1, 9] to decide whether a term, which can contain multiple words, should be a mention or not. The link-probability-based method relies on the *link probability* (aka *keyphraseness*) of a term $t$, which is computed in the following formula:

$$Link\ Probability(t) = \frac{|Article(t_{link})|}{|Article(t)|} \tag{1}$$

where $Article(t_{link})$ is the set of distinct articles that contain $t$ as label of links and $Article(t)$ is the set of distinct articles that contain $t$. For example, there are 500 distinct articles in Wikipedia that contain the term "*Jordan*"; only 110 out of 500 articles, "*Jordan*" appears as the label of links. Then the link probability ("*Jordan*") = 110/500 = 0.22 (or 22%). In our link-probability-based method, a term needs to have its link probability value surpassing a threshold λ in order to be an entity mention[7] and are neither nested nor overlapped with each other. Moreover, due to the informal nature of tweets, the entity mentions can be written in various ways regardless the naming rules. The recurrence is irregular uncapitalized characters for names. For example, the following tweet *"Chị Cẩm ly hôm nay hát hay quá xá ∧∧ ...!"* (Mrs. Cẩm ly sing awesomely today ∧∧ ...!) the term *"Cẩm ly"* is a mention that refers to a Vietnamese pop singer named *"Cẩm Ly"*. The correct name *"Cẩm Ly"* is being written as *"Cẩm ly"* (the character *"l"* is not capitalized). Unfortunately, the term *"Cẩm ly"* does not appear as a label of any link in Wikipedia, which causes the link-probability-based method failing to recognize it as a mention. Therefore, the link probability (Formula 1) is adjusted to case-insensitive string matching instead of case-sensitive matching. In particular, a term that mined from a given document and labels of links in Wikipedia are all lowercased before attempting to do the string matching to calculate the link probability value.

To the best of our knowledge, the link-probability-based method is widely adapted in English entity linking system; however, for Vietnamese, it is first used in [8] in which they claimed their work is the first complete entity linking system for Vietnamese. The authors focus on long and formal text specifically written in Vietnamese (i.e.: news, online articles), whereas, ours deals with informal, short text (i.e., tweet) and is expected to work universally to either English or Vietnamese.

## 3.2 Entity Disambiguation

This section describes the method of mapping (aka linking) a mention to the corresponding entity in Wikipedia. In this component, there are two essential steps. The first step is candidate generation, in which a set of potential entities that the mention can refer to are generated by using the method introduced in [1][8]. The next step is to determine which candidate entity in the set is the most appropriate for the given mention. The most used approach is to rank the candidates, and then the top rank candidate is evaluated to see whether or not it is possibly the correct entity to map to.

The ranking method can be heuristics, statistics, machine learning, or hybrid. Long [8] used the hybrid approach that combine heuristics and statistics based on the work in [27]. To date, the learning-based approach is getting more popular and successful for coping with this problem in English as in [1, 5, 9]; however, it has not yet been tested and reported for Vietnamese entity linking system in general. Therefore, we employ the supervised binary classifier method as in [1, 6, 9] to rank the candidate set. Then the top rank candidate is evaluated by the score that is assigned by the supervised classifier. If the candidate has its score surpasses the given threshold $\alpha$[9], then it is chosen as the correct entity and mapped by the given

---

[7] We use $\lambda = 0.01$ as it gives the optimal performance under our observation

[8] Candidates entities must have their commonness surpass a certain threshold, which we choose 0.15 in experiments as it is an optimal value under our observation

[9] We use $\alpha = 0.3$ in experiments as it is an optimal value under our observation

mention. Otherwise, the mention map is considered un-answering and not mapped to any entity.

There are a vast amount of features existing that can be suitable for ranking; however, three features proposed by [1, 9] are provided to be sufficient to gain a good ranking performance. The three features are commonness, semantic relatedness and context quality, each of which is explained further below.

### 3.2.1 Commonness

The commonness denotes how often people refer a mention to an entity (i.e.: article) in Wikipedia regardless of the surrounding context. Give a mention $m$ and a candidate entity $e$, the commonness is computed as the following formula:

$$Commonness(m, e) = \frac{Count(m \rightarrow e)}{Count(m)} \tag{2}$$

where $count(m \rightarrow e)$ is the number of times $m$ is linked to $e$ in Wikipedia and $count(m)$ is the number of times $m$ appears as label of links in Wikipedia. For example, the mention "*mouse*" appears 1200 times in Wikipedia as label of links. 420 out of 1,200 times, the mention is linked to entity "*mouse (computing)*" (a pointing device used together with computers) and the remaining times (780 out of 1,200), it is linked to entity "*mouse*" (a small mammal). So, commonness ("*mouse*", "*mouse (computing)*") = 420/1,200 = 0.35 and commonness ("*mouse*", "*mouse*") = 780/1,200 = 0.65. That is, when a term "*mouse*" is mentioned, the entity "*mouse*" is likely a better choice than "*mouse (computing)*".

### 3.2.2 Semantic Relatedness

The semantic relatedness denotes how suitable a candidate entity to surrounding context. For example, when a document appears to have context relating to computer field, then a mention "*mouse*" appearing in the same document is likely referring to the entity "*mouse (computing)*" than the bio-"*mouse*".

The context, in this case, consists of "unambiguous" entities that mined from document. An entity $e$ is called unambiguous when the commonness of $e$ regard to a mention $m$ is 100% (in other words, every time $m$ appears as a label of a link in Wikipedia, $e$ is always the target entity).

To model how related a candidate entity $e$ to context $C$ of a given document; we use the weighted average of its semantic relatedness to each contextual entity in $C$. The semantic relatedness between two entities is computed based on the links in Wikipedia as the following formula:

$$Semantic\,Relatedness(e_1, e_2) = 1 - \frac{log(Max(|E_1|, |E_2|)) - log(|E_1 \cap E_2|)}{log(|W|) - log(Min(|E_1|, |E_2|))} \tag{3}$$

where $e_1$ and $e_2$ are two entities. $E_1$ and $E_2$ are two sets of articles, in which each article contains at least a link that its target is $e_1$ and $e_2$ respectively. $W$ is the set that consists of all articles in Wikipedia.

### 3.2.3 Context Quality

The context quality is used to balance commonness and relatedness features of each candidate entity. The higher the context quality is, the more consistent is the contextual entities (in other words, most of the contextual entities are relating to the same subject or field); thus, the relatedness is rated higher. In contrast, the lower the context quality, the more important the commonness is. The context quality is computed by sum all the weight value assigned to each contextual entity. The weight of a contextual entity $e$ is computed as an average of the link probability value of the mention $m$ referring to $e$ and the average semantic relatedness value of $e$ to each contextual entity.

## 4 Dataset and Evaluation

This section, we describe about datasets and experiments evaluating the performance of our system. Section 4.1 is about the knowledge base we use for our system and its detail information. Section 4.2 is about how we obtain and annotate the dataset that is used for evaluation. Section 4.3 is the performance we report for each main component as well as the end-to-end system.

### 4.1 Wikipedia

We use the processed dump of Vietnamese Wikipedia on March $10^{th}$, 2014 and English Wikipedia on March $4^{th}$, 2014 as the knowledge bases for our tweet entity linking system. The processed Vietnamese Wikipedia consists of 582,878 article entities and English Wikipedia consists of 3,843,699 article entities.

### 4.2 Dataset

To evaluate the system in English, we use the shared dataset provided by Meij [2]; For Vietnamese tweets, there have not yet any dataset that existed. Hence, we have to build one on our own. We collect 1,120 Vietnamese tweets by using an auto-crawler which automatically crawls tweets from random verified users on Twitter. The tweets' content varies in subject and field. We then asked for the help of 6 annotators to annotate the collected tweets. They are given the instructions on how to recognize mentions in tweets and find the correct article entities in Wikipedia. Out of 1,120 tweets, there are only 524 tweets reported by the annotators that are meaningful and contain mentions that the corresponding entities existing in Wikipedia. Thus, the 524 meaningful tweets are kept as golden dataset for Vietnamese tweet and the rest are discarded. There are total of 1,061 mentions in 524 tweets, make it 2.02 mentions per tweet on average.

## 4.3 Performance Evaluation

### 4.3.1 Mention Recognition Performance

The performance is evaluated based on standard precision, recall and $F_1$ measures. As shown in Table 1, the mention recognition performs better with case-insensitive string matching than case-sensitive in both datasets. Therefore, it proves that the irregular uncapitalization occurring quite often in tweets. However, the recall is still moderate mainly because it failed to recognize the informal mentions (i.e.: "Đà lạttt") or abbreviation (i.e.: "SG"), etc. In fact, these mentions have never appeared in Wikipedia.

**Table 1** Our mention recognition performance

| Language | Maching Type | Precision | Recall | F1 |
|---|---|---|---|---|
| English | Case-Sensitive | 0.384 | 0.396 | 0.390 |
| | Case-Insensitive | 0.343 | 0.536 | 0.418 |
| Vietnamese | Case-Sensitive | 0.230 | 0.232 | 0.231 |
| | Case-Insensitive | 0.415 | 0.630 | 0.500 |

### 4.3.2 Disambiguation Evaluation

We use the same method in [3] and [28], which also utilize precision, recall and $F_1$ measures for evaluation this stage. In order to evaluate this stage separately, the input mentions are supposed to be error-free. In other words, we isolate entity disambiguation stage to evaluate and the mention strings are provided as input for this stage. We choose random forest as supervised learning algorithm for the disambiguation ranker. Following [2], we perform a 5-fold cross validation on the datasets.

As shown in Table 2, the commonness alone yields a good disambiguation performance, while adding both SR and CQ gives a small boost in $F_1$. Therefore, the 3 features are used for the end-to-end system. Under our observation, mentions that are named entities (i.e.: mentions refer to person, location or organization) take a larger portion of the Vietnamese dataset than the English one and they are often less ambiguous than the abstract concepts (or common concepts). Consequently, the disambiguation works better on Vietnamese dataset is understandable.

**Table 2** Disambiguation performance with combination of features. CM, SR and CQ denotes commonness, semantic relatedness and context quality respectively.

| Language | Features | Precision | Recall | F1 |
|---|---|---|---|---|
| English | CM | 0.786 | 0.672 | 0.724 |
| | CM + SR | 0.807 | 0.647 | 0.718 |
| | CM + SR + CQ | 0.819 | 0.660 | 0.731 |
| Vietnamese | CM | 0.916 | 0.795 | 0.851 |
| | CM + SR | 0.939 | 0.799 | 0.863 |
| | CM + SR + CQ | 0.933 | 0.799 | 0.861 |

### 4.3.3 End-to-End System Evaluation

We also utilize standard measures precision, recall and $F_1$ to evaluate our end-to-end entity linking system. As shown in Table 3, the system archives high precision in exchange for the low recall.

**Table 3** Our end-to-end entity linking system performance

| Language | Precision | Recall | F1 |
|---|---|---|---|
| English | 0.845 | 0.402 | 0.545 |
| Vietnamese | 0.929 | 0.551 | 0.692 |

The reason that the system archives rather low recall is due to the insufficient recall of mention recognition. However, most of the times, our disambiguation does not map mentions that are incorrectly recognized by recognition component; hence, in spite of the low precision of mention recognition, we are able to archive such high precision and fairly reasonable $F_1$ on end-to-end system.

**Table 4** Disambiguation performance comparison

| Methods | Precision | Recall | F1 |
|---|---|---|---|
| Meij [2] | 0.734 | 0.632 | 0.679 |
| Liu [3] | 0.752 | 0.675 | 0.711 |
| Our entity disambiguation | 0.819 | 0.660 | **0.731** |

Table 4 shows the comparison between the performance of entity disambiguation in our system and that of state-of-the-art systems. The precision, recall and F1-score

**Table 5** End-to-end system performance comparison

| Methods | Precision | Recall | F1 |
|---------|-----------|--------|-----|
| TagMe [30] | 0.329 | 0.423 | 0.370 |
| Meij [2] | 0.393 | 0.598 | 0.475 |
| SSRegu [28] | 0.650 | 0.441 | 0.525 |
| Our system | 0.845 | 0.402 | **0.545** |

shown in Table 4 were reported in [3]. The results in Table 4 show that our entity disambiguation achieves the performance better than those of the proposed methods in [2] and [3] in the term of F1-score. Table 5 shows the comparison between the performance of our end-to-end entity linking system and state-of-the-art end-to-end entity linking systems. The precision, recall and F1-score shown in Table 5 were reported in [28]. The results in Table 5 show that our system gives the best performance among the state-of-the-art end-to-end entity linking systems in the term of F1-score.

## 5 Conclusion

We study the task of entity linking for tweets, which aims at detecting entity mentions and linking them to corresponding entries in a given knowledge base. Unlike authored news or textual web content, tweets are noisy, irregular, and short, which causes entity linking in tweets much more challenging. In this paper, we have developed the first end-to-end entity linking system for Vietnamese tweets. The system is language-independent and archives a high precision and a reasonable F1-score on the dataset including 524 Vietnamese tweets. It also gives the best performance among the state-of-the-art end-to-end entity linking systems in the term of F1-score when evaluating on a public dataset including English tweets. However, due to insufficient performance in mention recognition, the system suffers an overall decrease in end-to-end system performance. Our next interests are improving the mention recognition component to deal with information mentions that do not exist in Wikipedia.

## References

[1] Milne, D., Witten, H.I.: Learning to Link with Wikipedia. In: Proc. of the ACM Conference on Information and Knowledge Management, pp. 509–518 (2008)

[2] Meij, E., Weerkamp, W., Rijke, D.M.: Adding Semantics to Microblog Posts. In: Proc. of the Fifth ACM International Conference on Web Search and Data Mining (WSDM) (2012)

[3] Liu, X., Li, Y., Wu, H., Zhou, M., Wei, F., Lu, Y.: Entity Linking for Tweets. In: Proc. of the 51st Annual Meeting of the Association for Computational Linguistics (ACL 2013), pp. 1304–1311 (2013)

[4] Cassidy, T., Ji, H., Ratinov, L., Zubiaga, A., Huang, H.: Analysis and Enhancement of Wikification for Microblogs with Context Expansion. In: Proc. of the 23th International Conference on Computational Linguistics (COLING 2012), pp. 441–456 (2012)

[5] Ratinov, L., Roth, D., Downey, D., Anderson, M.: Local and Global Algorithms for Disambiguation to Wikipedia. In: Proc. of the 49th Annual Meeting of the Association for Computational Linguistics: Human Language Technologies, pp. 1375–1384 (2011)

[6] Huynh, H.M., Nguyen, T.T., Cao, T.H.: Using Coreference and Surrounding Context for Entity Linking. In: Proc. of the 10th IEEE RIVF International Conference on Computing and Communication Technologies (RIVF 2013) (2013)

[7] Sofean, M., Stewart, A., Denecke, K., Smith, M.: Medical Case-Driven Classification of Microblogs: Characteristics and Annotation. In: Proc. of IHI 2012 (2012)

[8] Truong, L.M., Cao, T.H., Dinh, D.: Towards vietnamese entity disambiguation. In: Van Huynh, N., Denoeux, T., Tran, D.H., Le, A.C., Pham, B.S. (eds.) KSE 2013, Part II. Advances in Intelligent Systems and Computing, vol. 245, pp. 299–310. Springer, Heidelberg (2014)

[9] Milne, D., Witten, H.I.: An open-source toolkit for mining Wikipedia. Artificial Intelligence 194, 222–239 (2012)

[10] Han, X., Sun, L., Zhao, J.: Collective Entity Linking in Web Text: A Graph-Based Method. In: Proc. of the 34th International ACM SIGIR Conference on Research and Development in Information Retrieval, pp. 765–774 (2011)

[11] Han, X., Sun, L.: A generative entity-mention model for linking entities with knowledge base. In: Proc. of the 49th Annual Meeting of the Association for Computational Linguistics: Human Language Technologies, vol. 1, pp. 945–954. Association for Computational Linguistics

[12] Hachey, B., Radford, W., Curran, J.R.: Graph-based named entity linking with wikipedia. In: Bouguettaya, A., Hauswirth, M., Liu, L. (eds.) WISE 2011. LNCS, vol. 6997, pp. 213–226. Springer, Heidelberg (2011)

[13] Ji, H., Grishman, R., Dang, H.T.: Overview of the TAC 2011 Knowledge Base Population Track. In: Proc. of Text Analysis Conference (2011)

[14] Ji, H., Grishman, R., Dang, H.T., Griffitt, K., Ellis, J.: Overview of the TAC 2010 Knowledge Base Population Track. In: Proc. Text Analysis Conference (2010)

[15] McNamee, P., Dang, H.T.: Overview of the tac 2009 knowledge base population track. In: Proc. Text Analysis Conference (2009)

[16] Shen, W., Wang, J., Luo, P., Wang, M.: Linking named entities in tweets with knowledge base via user interest modeling. In: Proc. of the 19th ACM SIGKDD International Conference on Knowledge Discovery and Data Mining, pp. 68–76 (2013)

[17] Guo, S., Chang, M.W., Kiciman, E.: To Link or Not to Link? A Study on End-to-End Tweet Entity Linking. In: Proc. of NAACL 2013 (2013)

[18] Murnane, E.L., Haslhofer, B., Lagoze, C.: RESLVE: leveraging user interest to improve entity disambiguation on short text. In: Proc. of the 22nd International Conference on World Wide Web, pp. 1275–1284 (2013)

[19] Derczynski, L., Maynard, D., Aswani, N., Bontcheva, K.: Microblog-Genre Noise and Impact on Semantic Annotation Accuracy. In: Proc. of 24th ACM Conference on Hypertext and Social Media (2013)

[20] Bontcheva, K., Rout, D.: Making sense of social media streams through semantics: a survey. Semantic Web Journal (2012)

[21] Jin, Y., Kiciman, E., Wang, K., Loynd, R.: Entity Linking at the Tail: Sparse Signals, Unknown Entities and Phrase Models. In: Proc. of The Seventh ACM International Conference on Web Search and Data Mining (WSDM 2014) (2014)

[22] Li, Y., Wang, C., Han, F., Han, J., Roth, D., Yan, X.: Mining evidences for named entity disambiguation. In: Proc. of the 19th ACM SIGKDD International Conference on Knowledge Discovery and Data Mining (KDD 2013) (2013)

[23] He, Z., Liu, S., Song, Y., Li, M., Zhou, M., Wang, H.: Efficient collective entity linking with stacking. In: Proc. of the 2013 Conference on Empirical Methods in Natural Language Processing (EMNLP 2013) (2013)

[24] He, Z., Liu, S., Li, M., Zhou, M., Zhang, L., Wang, H.: Learning entity representation for entity disambiguation. In: Proc. of the 51st Annual Meeting of the Association for Computational Linguistics (ACL 2013), pp. 30–34 (2013)

[25] Berners-Lee, T., Hendler, J., Lassila, O.: The Semantic Web. Scientific American, 34–43 (2001)

[26] Spina, D., Gonzalo, J., Amigó, E.: Discovering filter keywords for company name disambiguation in twitter. Expert Systems with Applications 40(12), 4986–5003 (2013)

[27] Nguyen, H.T., Cao, T.H.: Named Entity Disambiguation: A Hybrid Approach. International Journal of Computational Intelligence Systems 5(6), 1052–1067 (2012)

[28] Huang, H., Cao, Y., Huang, X., Ji, H., Lin, C.-Y.: Collective Tweet Wikification based on Semi-supervised Graph Regularization. In: Proc. of the 52nd Annual Meeting of the Association for Computational Linguistics (ACL 2014) (2014)

[29] Garcia, N.F., Fisteus, J.A., Fernández, L.S.: Comparative Evaluation of Link-Based Approaches for Candidate Ranking in Link-to-Wikipedia Systems. Journal of Artificial Intelligence Research 49, 733–773 (2014)

[30] Ferragina, P., Scaiella, U.: Tagme: on-the-fly annotation of short text fragments (by Wikipedia entities). In: Proc. of the 19th ACM International Conference on Information and Knowledge Management (CIKM 2010) (2010)

# Using Large N-gram for Vietnamese Spell Checking

Nguyen Thi Xuan Huong, Tran-Thai Dang, The-Tung Nguyen, and Anh-Cuong Le

**Abstract.** Spell checking is a process including detecting, correcting or providing spelling suggestions for misspelled words. In this paper, we present our spell checking system relied on the context and our experimental results when doing for Vietnamese. This system uses N-gram model with large corpus. N-grams is compressed to save the memory. Furthermore, we take the contexts in both sides of syllables to improve the system's performance. Our system got high accuracy approximate 94% F-score on the Vietnamese text.

## 1 Introduction

In the regulatory documents, spelling is very important because the misspelled words can make some misunderstandings. Hence, a spelling checking system is necessary to make these documents are correct. This is one of pre-processing steps for Natural Language Processing (NLP) problems.

A common spelling system usually includes two main components that are detection and correction. In several cases, the system provides some correction suggestions for the users to choose. The spelling errors are considered in previous researches includes: non-word spelling error and real-word spelling error.

- The non-word errors are strings which do not exist in dictionary. For example, the following sentence in Vietnamese: "Hôm nay tôi **ddi** học". The string "ddi" is a

Nguyen Thi Xuan Huong · Tran-Thai Dang · The-Tung Nguyen · Anh-Cuong Le
University of Engineering and Technology
Vietnam National University, Hanoi
144 Xuanthuy, Caugiay, Hanoi, Vietnam
e-mail: {cuongla,tungnt_55}@vnu.edu.vn,
        thaidangtran12@gmail.com

Nguyen Thi Xuan Huong
Haiphong Private University
36 Danlap, Duhangkenh, Lechan, Haiphong, Vietnam
e-mail: huong_ntxh@hpu.edu.vn

© Springer International Publishing Switzerland 2015                     617
V.-H. Nguyen et al. (eds.), *Knowledge and Systems Engineering,*
Advances in Intelligent Systems and Computing 326, DOI: 10.1007/978-3-319-11680-8_49

misspelled word, and it does not exist in Vietnamese dictionary (its corresponding correct word is "đi").

- The real word errors occurs in the document when the words are mistakenly used. For example, the following sentence in Vietnamese: "Quyển **xách** này rất hay", the word "xách" (carry) is incorrectly used while the corresponding correct one is "sách" (book).

In general, the real-word error is more difficult to detect and correct than the non-word error.

In Vietnamese, we investigate several spell errors in the documents that are instances of two type which are mentioned above:

**The non-word error:**

- Typing error.
  Example: "bof" -> "bò"

**The real-word error:**

- Tones making error.
  Example: "hõi" -> "hỏi"
- Initial consonant error.
  Example: "bức chanh" -> "bức tranh"
- End consonant error.
  Example: "bắt buột" -> "bắt buộc".
- Region error (Vietnam has many regions with various dialect, we need to change them to the common language).
  Example: "kím" -> "kiếm"

Although several spelling correction tools are used widely for Vietnamese, there is no official publication about this problem. In this paper, we present the context-based method to check spelling with large scale of N-gram model on Vietnamese text. In our work, the major clue helping us correct spelling errors is the surrounding syllables. We take the context at the left side, right side and both sides to make more clues for choosing the correct word.

In order to choose the best syllable in confusion set (the set of candidates for correcting), we need to measure the relation between a syllable and its neighbors. In this case, N-gram model is useful for modeling these relations. This model is created based on statistical approach. Hence, the distribution and variety of vocabulary on the training data have significantly affected to the system's performance. In order to deal with the spare data, a common problem of statistical methods, we used a large corpus for training. The large corpus are collected from many text resources with various topics to reduce number of unknown words. It also help us increase number of combinations of syllables, meaning that we are able to exploit more relations among syllables and their distribution. Instead of using linguistics information such as word segmentation, POS tags, syntactic parser, the large corpus provide us enough information to get the clues for correcting.

In theoretical term, the large corpus helps N-gram model is better, in fact, we investigate whether the more size of corpus rises, the more the performance rises.

That means we consider the correlation and the influence between the size of corpus and the system's performance. We also determine if there is a limitation of corpus size. We also implement a compression method to save the memory and reduce the running time to deal with the large corpus.

## 2 Related Work

Many researchers have been attempting to solve the spell checking problem and applied on many languages(English, French, Japanese, Chinese) such as winnow-based method was proposed by Golding in 1999 [8]; Zhang 2000 [15] introduced approximate word matching method. Iterative transformation approach was proposed by Cucerzan and Brill, 2004 [5].

Winnow-based method was one of the first approach using contextual information (context-sensitive). Using the neighboring words to detect and fix spelling errors is the main idea of context-sensitive method.

SCInsunSpell model, a spell checking model, was proposed by Li and Wang Xiaolong Juanhua in 2002. This is one of the first attempt to apply N-gram model to Chinese spell checking. The model is a combination of N-gram (trigram) model, Bayesian estimation methods and probability weighted distribution automation. In particular, the N-gram (trigram) was used for detecting phrases. It considers the link between the current word and four words surrounding it. The current word is called misspelled if its links with its four neighbors is less than a certain threshold.

IGHAN-7 is the first Chinese spell checking evaluation project. It includes two subtasks: error detection and error correction. The project was organized based on some research works (Wu et all., 2010 [14]); Chen et al., 2011 [4]; Liu et al., 2011[11]). It was received a lot of attentions from scientists. There are thirteen systems participated in the bake-off and four of them used N-gram model. This proved that despite of being an old approach, N-gram is still a powerful model to solve the spell checking problems.

In Vietnamese, Nguyen Duc Hai and Nguyen Pham Hanh Nhi, 1999 [18] proposed a spell checking model based on sentence parsing. Five years later, Nguyen Thai Ngoc Duy [19] developed a spell checking system based on word-network. These system's performance did not meet the user's requirement. In 2012, Nguyen Huu Tien Quang [20] proposed a spell checking system using N-gram model and Viterbi algorithm. This is one of the earliest study on applying N-gram model to Vietnamese spell checking problem and its results were much better than previous work. Although, there are some researches about this problem for Vietnamese, they just have been presented in bachelor thesis, and no official publication.

Basing on previous research for this problem, we also use context based approach with n-gram model to build the spell checking system. In our empirical work, there are some improvements which extend the context in both sides of syllable and use n-gram model with the large corpus.

## 3 Our Approach

Similarly to the previous works on English, Chinese, Japanese we use context based approach and N-gram model for our spell checking system. The measure of relation between a syllable and its neighbors is computed and evaluated to select the most likely correct syllable. In order to improve the performance of this system, we extended the context in both sides of syllable and used the large corpus. The N-gram compression is also implemented to optimize the size of memory for storage.

As previous mention, our System includes two main components: making confusion sets, calculating probabilities then comparing to choose a most likely candidate in the confusion set. Before checking, the data need to be processed such as normalization, sentence splitting, punctuation marks removal. Our system's architecture is illustrated in the figure 1.

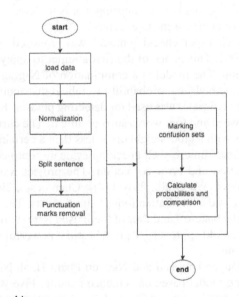

**Fig. 1** Our system's architecture

### 3.1 Pre-processing

Pre-processing stage contains three steps:

- **Step 1:** Recognizing the special syllables such as web address, email, number... and replacing them by special notation. For example:
  The number "5" is replaced by "NUMBER".
  The name of a website "http://facebook.com" is replaced by "WEBSITE".
  Replacing the special patterns by the predefined notations help to reduce the number of unknown words.
- **Step 2:** Splitting the document into sentences because two syllables in different sentences have no relationship with the other.

- **Step 3:** Removing all punctuation marks in the sentences because they do not have relation of meaning with the words.

## 3.2   Extending the Context in Both Sides

The main component of our spell checking system includes two steps:

- **Step 1:** Building the confusion set for each syllable based on edit distance and chosen Vietnamese characteristics.
- **Step 2:** Calculating the measure of relationship between a syllable and its neighbors based on N-gram model to decide whether the current syllable is incorrect or not then choose the most likely candidate to fix it.

The previous context-sensitive spelling corrections almost take the context in left side. We take the context at both sides of the syllable to improve the system's performance. Using the context in both sides help us get more clues to choose the best candidate in confusion set.

The context used in our system is the 2-radius window of syllables surrounding, which means if we denoted the current syllable as $w_0$ and its contexts are $w_{-1}$, $w_{-2}$, $w_1$, $w_2$. We can model the $w_0$'s dependency on its neighbors by the conditional probability below.

$$P(w_0 \mid w_{-2}, w_{-1}, w_1, w_2)$$

This probability above may be estimated by the following function:

$$P(w_0 \mid w_{-2}, w_{-1}, w_1, w_2) = f(P(w_0 \mid w_{-2}, w_{-1}), P(w_0 \mid w_{-1}, w_1), P(w_0 \mid w_1, w_2))$$

where f is a geometric mean function.

In order to calculate this probability, we need 5-gram and 4-gram. This is not feasible to carry out because there is a large number of combinations and the data is too spare. Therefore, instead of calculating $P(w_0 \mid w_{-2}, w_{-1}, w_1, w_2)$ we estimate it by three trigram probabilities: $P(w_0 \mid w_{-2}, w_{-1})$, $P(w_0 \mid w_{-1}, w_1)$, $P(w_0 \mid w_1, w_2)$. The N-gram scores $p$ is the geometric mean logarithms of these three probabilities. We choose the geometric mean because the name entity such as human name or organization can weaken the link of syllable with its context.

Positive errors occur when a syllable is determined as error, but it is actually spelling-correct. To reduce number of these errors, we used heuristic coefficients "error threshold" and "difference threshold". For short we denoted them as $e\_thresh$ and $d\_thresh$. Assume that our current syllable $w_0$ has N-gram score $p$ and one syllable from the error set $w_0'$ has N-gram score $p'$, $w_0'$ is considered to be "better" than $w_0$ if and only if it satisfies two inequalities below:

- $p' > e\_thresh$
- $p' > p + d\_thresh$

$e\_thresh$ is a constant determined by using development data, it ensures that if one syllable is going to be used to fix the current syllable, its probability must be higher than a certain threshold; this help us to reduce the false positive caused by named-entity.

## 3.3   Large Scale N-gram and Compression

Our system essentially uses N-gram model, the training data plays an important role which affects to the performance of the system. As the mention in section 3.2, we need to compute the probabilities of trigrams, thus the frequencies of bigrams and trigrams must be determined. The large corpus including many topics help us deal with the spare data problem without extra information such as word segmentation, part-of-speech, syntactic parser. The collected data from variety of resources and topics will reduce the number of unknown words. It also includes many possible combinations of syllables which help us exploit more information of relations among the syllables and their frequencies. We also use smoothing methods to compute probabilities of trigrams effectively in case of unknown words.

Using large corpus arises a problem that is how to compress the n-gram to save the memory when we load data. Moreover, the compression does not change the accuracy of N-gram model. In our system, we are going to implement a method to encode the n-grams by numbers without the loss of data. In encoding process, we collected the Vietnamese syllable dictionary including approximate 6800 syllables. Each syllable is represented by a number that start at 0. Firstly, we count the frequency of n-gram from the raw corpus and remove the n-grams which appear less than 5 times. Secondly, we start encoding n-gram. 1-gram (unigram) is encoded similarly to syllables because it contains one syllable. There are approximate 6800 syllables, so we need from 0 to 6800 to represent, each number need two bytes to store. For 2-gram (bigram), each syllable need 2 bytes for storing, so each bigram can be stored by 4 bytes (an integer). We also used *bit shifts* operator and *or* operator to encode bigram by an integer as the following:

int x = syllables[0]

x = x << 16

x = x | syllables[1]

Similarly, we need 6 bytes to encode trigram (3-gram).

In this paper, we also investigate the *"saturation"* of the training data to answer the question whether the more data rises, the more performance is improved. We will consider the influence of size of corpus to the performance and identify if there exist a limitation, which the data size reaches to the accuracy of the system does not increase.

## 4   Experiments

### 4.1   Experimental Data

#### 4.1.1   Training Data

In order to build N-gram model, we collected the data from various sources such as *Wikipedia.org, dantri.com.vn, vnExpress.net*. The data includes many topics such as mathematics, physics, science, literature, philosophy, history, economy, sport, law, news, entertainment... The size of our corpus is about 2GB. We counted frequencies of unigram, bigram, trigram then remove the n-gram which has frequency less than 5.

### 4.1.2 Testing Data

We created two test sets for evaluating our system. Firstly, we collected text from the Internet. In the first set, we manually check to ensure that there is no spelling error in it. After that we generated artificial spelling errors in the test set and marked these errors to evaluate the efficiency of our system. In the second set, we also found and marked spell errors. The first test set was used in experiment 1 and 2. Experiment 3 used the second set. The first test set contains 2500 sentences and second one contains 632 sentences.

## 4.2 Experimental Results

Before evaluating the performance of our spell checking system, we applied the n-gram compression method presented in section 3.3 to compress the training data after counting the frequency. The compression results are illustrated in table 1.

**Table 1** N-gram compression results

|  | # of elements | data size before encoding | data size after encoding |
|---|---|---|---|
| unigram | 6776 | 77.9 KB | 13.55 KB |
| bigram | 1208943 | 15.6 MB | 4.6 MB |
| trigram | 4886364 | 84 MB | 28 MB |

From table 1, we see that by using the compression, the size of memory reduces significantly. After using encoding, the sizes of bigram, trigram data are one-third of the original one.

In order to evaluate the spelling checking system, we used five metrics:
- Detection precision (DP).
- Detection recall (DR).
- Correction precision (CP).
- Detection F-score (DF).
- False positive rate (FPR).

The formula of each each metrics as the following:

- $DP = \dfrac{\#\ of\ correct\ detections}{\#\ of\ errors\ detected}$

- $DR = \dfrac{\#\ of\ correct\ detections}{\#\ of\ actual\ errors}$

- $CP = \dfrac{\#\ of\ correct\ fix}{\#\ of\ errors\ detected}$

- $DF = \dfrac{2 * DP * DR}{DP + DR}$

- $FPR = \dfrac{\#\ of\ incorrect\ detections}{\#\ of\ correct\ syllables}$

### 4.2.1 Experiment 1

We investigated the influence of the corpus size to the system's performance in order to answer the question whether there exits a limitation of corpus, when the size of data reach to this limitation, the performance does not rise or not. We split the corpus into fragments, each one is about 100 MB. A mini-corpus was created by joining these fragments together. For example: a 200 MB mini-corpus is the combination of two 100 MB fragments. Then each mini-corpus was created their corresponding n-gram dictionary (counting the frequencies of n-grams) and used in our system. The figure 2 illustrates the detection F-score of our system with each mini-corpus.

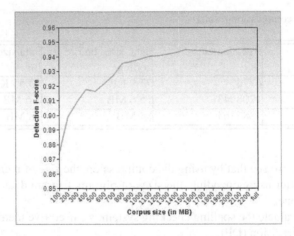

**Fig. 2** Influence of corpus size to the system's performance

From figure 2, we see that: when the size of corpus is greater than 1.5 MB the system's accuracy does not increase. In our work, we can use approximate 1.5 GB of corpus to train N-gram model. This experimental result also illustrates that there exist a threshold of corpus size and when the size of corpus reach to this threshold, the training data does not affect to the system's performance.

### 4.2.2    Experiment 2

We evaluated the influence of the context to the system's accuracy. The table 2 show
the evaluation results of each context.

**Table 2** Influence of context to the system's performance

| Context | DP | DR | CP | DF | FPR |
|---|---|---|---|---|---|
| $w_{-2}, w_{-1}$ | 89.42% | 52.22% | 97.31% | 65.93% | 0.12% |
| $w_{-1}, w_1$ | 94.04% | 91.53% | 98.26% | 92.76% | 0.11% |
| $w_1, w_2$ | 93.83% | 73.63% | 96.79% | 82.51% | 0.09% |
| $w_{-2}, w_{-1}, w_1, w_2$ | 94.68% | 94.26% | 99.32% | 94.46% | 0.1% |

From the table 2, we can easily observe that two-sided contexts give us much
better results than the one-sided contexts, and the context $w_{-2}, w_{-1}, w_1, w_2$ gives
the best result. Furthermore, the last column in table 2 shows that the false positive
rate is very low in our system, meaning that our method has reduced these errors
effectively.

### 4.2.3    Experiment 3

We compared our system with an other spell checking system for Vietnamese: copcon 5.0.3 beta [1]. We will compare the accuracy of their detection and their correction
on the second test set. The results are shown in table 3.

**Table 3** Comparison between our system and Copcon

| | DP | DR | CP | DF | FPR |
|---|---|---|---|---|---|
| Our system | 92.62% | 91.12% | 95.45% | 91.86% | 0.2% |
| Copcon 5.0.3 beta | 80.8% | 77.6% | 87.5% | 79.2% | 0% |

From the table 3, we see that the accuracy of detecting and correcting of our
system is greater than the copcon's accuracy.

## 5    Conclusion

This paper presents the context-sensitive spell checking system for Vietnamese
based on N-gram model with large scale. By extending the context at both sides

---

[1] link: http://chinhta.vn

of syllable, we improved the accuracy of spell checking system, and got higher performance than copcon, which is a spelling tool for Vietnamese. Our system just use N-gram model instead of extra linguistic information. Our empirical work also shows the influence of corpus size to the system's performance. In order to deal with the large corpus, a method to compress n-grams is implemented. Moreover, statistical approach is able to be applied to novel domains such as scientific papers, and spoken language in the forums...

**Acknowledgment.** This paper is supported by the projects QGTĐ.12.21 and QG.14.04 funded by Vietnam National University, Hanoi.

# References

[1] Blair, C.: A program for correcting errors. Information and Control, 60–70 (1960)

[2] Carlson, A., Rosen, J., Roth, D.: Scaling up context-sensitive text correction. In: Proceedings of the 13th Innovative Applications of Artificial Intelligence Conference, pp. 45–50 (2001)

[3] Carlson, A., Fette, I.: Memory-based Context-Sensitive Spelling Correction at Web Scale. In: Proceedings of the 6th International Conference on Machine Learning and Applications, pp. 166–171 (2007)

[4] Chen, Y.Z., Wu, S.H., Yang, P.C., Ku, T., Chen, G.D.: Improve the detection of improperly used Chinese characters in student's essays with error model. In: Int. J. Cont. Engineering Education and Life-Long Learning, pp. 103–116 (2001)

[5] Cucerzan, S., Brill, E.: Spelling correction as an iterative process that exploits the collective knowledge of web users. In: Proceedings of EMNLP, pp. 293–300 (2004)

[6] Damerau, F.: A technique for computer detection and correction of spelling errors. Communications of the ACM 7, 171–176 (1964)

[7] Deorowicz, S., Ciura, M.G.: Correcting Spelling Errors by Modelling Their Causes. International Journal of Applied Mathematics and Computer Science 15, 275–285 (2005)

[8] Golding, A., Roth, D.: A winnow-based approach to context-sensitive spelling correction. Machine Learning 34(1-3), 107–130 (1999)

[9] Islam, A., Inkpen, D.: Real-word spelling correction using googleweb 1t 3-grams. In: Proceedings of Empirical Methods in Natural Language Processing (EMNLP 2009), pp. 1241–1249 (2009)

[10] Liu, W., Allison, B., Guthrie, L.: Professor or screaming beast? Detecting words misuse in Chinese. In: The 6th edition of the Language Resources and Evaluation Conference (2008)

[11] Liu, C.L., Lai, M.H., Tien, K.W., Chuang, Y.H., Wu, S.H., Lee, C.Y.: Visually and phonologically similar characters in incorrect Chinese words: Analyses, identification, and applications. ACM Transactions on Asian Language Information Processing, 1–39 (2011)

[12] Verberne, S.: Context-sensitive spell checking based on word trigram probabilities. Master thesis, University of Nijmegen (2002)

[13] Whitelaw, C., Hutchinson, B., Chung, G.Y., Ellis, G.: Using the Web for Language Independent Spellchecking and Autocorrection. In: Proceedings of Conference on Empirical Methods In Natural Language Processing (EMNLP 2009), pp. 890–899 (2009)

[14] Wu, S.H., Chen, Y.Z., Yang, P.C., Ku, T., Liu, C.L.: Reducing the False Alarm Rate of Chinese Character Error Detection and Correction. In: Proceedings of CIPS-SIGHAN Joint Conference on Chinese Language Processing (CLP 2010), pp. 54–61 (2010)

[15] Zhang, L., Zhou, M., Huang, C.N., Pan, H.H.: Automatic detecting/correcting errors in Chinese text by an approximate word-matching algorithm. In: Proceedings of the 38th Annual Meeting on Association for Computational Linguistics, pp. 248–254 (2000)

[16] Li, J., Wang, X.: Combine trigram and Automatic Weight Distribution in Chinese Spelling ErrorCorrection. Journal of Computer Science and Technology Archive 17(6), 915–923 (2002)

[17] Mitton, R.: Ordering the Suggestions of a Spellchecker Without Using Context. Natural Language Engineering 15, 173–192 (2008)

[18] Hai, N.D., Nhi, N.P.H.: Syntactic parser in Vietnamese sentences and its application in Spell Checking. In: Vietnamese, bachelor thesis, in University of Science Ho Chi Minh city (1999)

[19] Duy, N.T.N., Dien, D.: An approach in Vietnamese spell checking. In: Vietnamese, bachelor thesis in University of Science Ho Chi Minh city (2004)

[20] Quang, N.H.T.: Language model and word segmentation in Vietnamese Spell Checking. In: Vietnamese, bachelor thesis in University of Engineering and Technology, Hanoi National University (2012)

# Automatically Learning Patterns in Subjectivity Classification for Vietnamese

Tran-Thai Dang, Nguyen Thi Xuan Huong,
Anh-Cuong Le, and Van-Nam Huynh

**Abstract.** Opinions are subjective expressions that describe people's viewpoints, perspectives or feeling about entities, events. They are essential information for sentiment analysis. Therefore, opinions detection, which is also called subjectivity classification, is an important task. In this paper, we propose a statistical method to automatically create the patterns for determining opinions from various resources on the web. The learned patterns are more flexible and adaptive to domain in comparison with manual creation. In this work, we obtained approximate 84% of accuracy when doing on Vietnamese comment data.

## 1 Introduction

Sentiment analysis process includes crawling, extracting, and analyzing people's opinions shared on forums, news portals, social networks... It helps manufacturers can gain real feedback to improve their products, or customers can get useful information to make decision when buying products.

After crawling data from the Internet, we have to determine which comment belongs to subjective comments (comments contain opinions) or objective comments

Tran-Thai Dang · Nguyen Thi Xuan Huong · Anh-Cuong Le
University of Engineering and Technology
Vietnam National University, Hanoi
144 Xuanthuy, Caugiay, Hanoi, Vietnam
e-mail: thaidangtran12@gmail.com, cuongla@vnu.edu.vn

Nguyen Thi Xuan Huong
Haiphong Private University
36 Danlap, Duhangkenh, Lechan, Haiphong, Vietnam
e-mail: huong_ntxh@hpu.edu.vn

Van-Nam Huynh
Japan Advanced Institute of Science and Technology
1-1 Asahidai, Nomi, Ishikawa, Japan
e-mail: huynh@jaist.ac.jp

© Springer International Publishing Switzerland 2015                                   629
V.-H. Nguyen et al. (eds.), *Knowledge and Systems Engineering,*
Advances in Intelligent Systems and Computing 326, DOI: 10.1007/978-3-319-11680-8_50

(comments just express the fact). Subjectivity classification is considered as the first step in sentiment analysis process. The subjective comments will be normally used in next step to determine which are positive, negative or neutral comments.

There are some introduced methods to find words, phrases which express opinions. Most previous works are carried out on English data. However, those methods are not effective totally when applying on Vietnamese data. In this paper, we focus on determining subjective comments in Vietnamese data.

Through investigating the comments on several Vietnamese forums and blogs, people usually use adjective and verb to express their opinions. For example, some common adjectives and verbs are used in people's comments such as: "đẹp" (nice), "xấu" (ugly), "tốt" (good), "mượt mà" (smooth), "thích" (like), "ghét" (hate), "cảm thấy" (feel), etc. Therefore adjectives and verbs are able to be strong clues which help to distinguish between subjective and objective comments.

The words and phrases express opinion can be extracted based on sentiment dictionary, n-gram or syntactic patterns. Among those ways, syntactic patterns are useful to enrich the set of features. The patterns can be created manually based on knowledge of specific language such as grammar, POS. For example we can build a pattern of POS as the following:

*"Con/Nu Nokia/Np này/P nhìn/V rất/R đẹp/A."* (This Nokia looks very nice)
*(Nu: Unit noun; Np: Proper noun; P: Pronoun; V: Verb; R: Adverb; A: Adjective)*

From above example, the phrase "nhìn rất đẹp" (look very nice) is a component that expresses opinion. This phrase can be extracted from the pattern: V-R-A.

The manual creation not only requires much time but also is difficult to cover all the rules to find out subjective features. In the Vietnamese forums, blogs people often use spoken language and slang which is short and informal. Hence, we need to propose suitable patterns, then investigate and evaluate their influence.

To deal with this problem, we introduce a statistical method to help the system learn syntactic patterns and evaluate these patterns from labeled training data. The learning processing includes two main steps such as patterns identification and evaluation. The system will determine whether the patterns are used to express opinions frequently or not. In our work, the training data are tagged by two labels "<sub>" (subjectivity), and "<obj>" (objectivity). After that, the system extract and evaluate the subjective patterns to build the features set. The patterns may be created by using syntactic parser tree or POS information. In our work, we chose POS for some reasons: firstly, people often use spoken language which includes incomplete sentences (the sentences lack subject or predicate), so it is difficult to obtain correct parsed tree; secondly, using the POS information is easier for adapting to domain than parser tree because we can use statistical approach to learn POS tags. This method is able to apply for many languages without deep knowledge of their syntactic information. Moreover, the learned POS patterns from the training data are more flexible and adaptive to domain than manual creation.

## 2  Related Work

The subjectivity classification focus on how to build the good features set to improve the system's performance. Janyce Wibe[6] identified strong clues of subjectivity based on distributional similarity, he use small seed manual annotation data to develop promising adjective features. In [1], Pang et al., used n-gram as features for polarity classification. An other way, to enrich feature set, Perter D.Turney [10] used patterns to extract phrases which contain adjectives or verbs. Similarly to English, people usually use adjectives and verbs to express their opinions in Vietnamese data, those are important information to extract features. E.Riloff et al.,[3] used two boot-strapping algorithms that extract patterns to learn set of subjective nouns. In [4], [5], [7] they used syntactic information to create the patterns. To create the patterns to extract features, we need linguistic knowledge, or small set of sentiment seed words [3]. In contrast, we propose a statistical method which learns the patterns from labeled training data. POS information is also used in various previous researches to determine features. Gamon [14] performed sentiment analysis on feedback data and analyzed the role of linguistic features like POS tags. Pak and Paroubek [13] reported that both POS and bigram help to perform subjectivity classification of tweets. Barbosa and Feng [15] proposed the use of syntax features of tweets like retweet, hashtags, link, punctuation and exclamation marks in conjunction with features like prior polarity of words and POS tags. Agarwal et al [16] extended Barbosa and Feng (2010) by using a combination of real-valued polarity with POS and reported POS features are important to the classification accuracy. In [9], M.Sokolova and G.Lapalme proposed a hierarchical text representation and built domain-independent rules that do not rely on domain content words and emotional words. Other observed characteristics may be used to recognize subjective expressions such as word lengthening in [12], emoticon in [11].

After extracting the features, they usually use many classification techniques to assign labels for comments. Pang and Lee used Support vector machines (SVM), Naive Bayes (NB), Maximum Entropy (ME) in [1]. Riloff used SVM in [4]. We also investigate some classification algorithms such as SVM, NB in our empirical work.

## 3  Our Approach

Motivating from using syntactical patterns and POS information to extract features on English data from previous works, we build a set of Vietnamese patterns that help to enrich the features set. Different from previous researches, in our approach POS information is chosen to build the patterns because it is easy to adapt to domain. In learning process, we determine the forms of patterns which are called as templates. This process is illustrated in figure 1.

In first stage of this process, we extract all patterns on the labeled training data from the predefined templates. The second stage, the patterns are evaluated to select the best set of patterns. The evaluation process contains two steps:

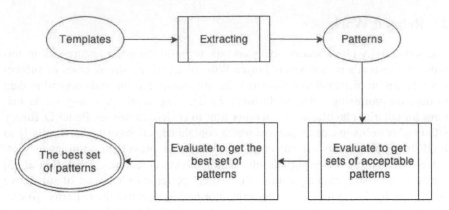

**Fig. 1** POS patterns learning process

. Evaluating to get sets of acceptable patterns. A pattern is acceptable if and only if it appears more frequently in subjective comments than in objective comments.

. Evaluating to get the best set of patterns. The best set of patterns is selected from the sets of acceptable patterns by classifying on the training data with set of features (adjectives and verbs) extracted from the patterns.

## 3.1 Training Data

In the training data, each comment are tagged by two labels which are "<sub>" (subjectivity) or "<obj>" (objectivity) and the number of subjective comments are equal to number of objective comments. For example:

<sub> Giá quá tốt (good price)
<sub> Xấu tệ hại (very ugly)
<sub> Tôi không thích thiết kế của con này lắm (I don't like the design of this machine)
<obj> Tôi đang tính mua em này thay cho em Qmobile M45 (I am considering buying this machine instead of Qmobile M45)
<obj> Lại đổi giao diện (change the interface)
<obj> Có rồi nhưng màu bạc thì không có hàng (got this product but there is no sliver product)

The training data is checked spell and segmented into words before tagging by the POS labels because we work on Vietnamese text.

## 3.2   Templates Definition

We mainly use adjectives and verbs as the features, so the templates are created by them and their surrounding POS labels. The surrounding POS may be noun (N), proper noun (Np), other adjective (A), or verb (V), adverb (R), coordinating conjunction (Cc), auxiliary (T).

We propose two types of templates to learn that are:

- **Type 1:** The templates are built to extract the patterns which contain only POS labels. We consider the verb and adjective with their surrounding POS labels in left side, right side, or both sides.
- **Type 2:** The type 2 is similar to type 1, but it is more specific than type 1 because we build the templates which extract the patterns which contain words (adjective or verb) and their surrounding POS labels.

We can use templates of type 1 or type 2. In this paper, we will show experimental results when applying two types of templates for extracting patterns. Table 1 and table 2 illustrate examples of the templates of two types.

**Table 1** Templates of type 1

| Template | Description |
|---|---|
| tag-tag[+1] | if the current tag is adjective or verb, considering the template which contains the current tag and one next tag. |
| tag-tag[-1] | if the current tag is adjective or verb, considering the template which contains the current tag and one previous tag. |
| tag-tag[-1] & tag[+1] | if the current tag is adjective or verb, considering the template which contains the current tag and one previous tag and one next tag (tags in both sides of the current tag) |
| tag-tag[+2] | if the current tag is adjective or verbs, considering the template which contains the current tag and two next tags. |
| tag-tag[-2] | if the current tag is adjective or verb, considering the template which contains the current tag and two previous tags. |

For example, the sentence in section 1 has one adjective, if we use the template in line 2 of table 1, we will extract the pattern R-A. We are able to expand the templates shown in two tables by increasing the number of surrounding POS labels.

Similarly to previous works N-gram (unigram, bigram) are used as the features to classify. N-gram of words (after segmenting the training data into words) are learned as same as the patterns. We also consider that N-gram is equivalent to the phrases extracted from the patterns.

**Table 2** Templates of type 2

| Template | Description |
|----------|-------------|
| word-tag[+1] | if the current tag is adjective or verb, considering the template which contains the current word and one next tag. |
| word-tag[-1] | if the current tag is adjective or verb, considering the template which contains the current word and one previous tag. |
| word-tag[-1] & tag[+1] | if the current tag is adjective or verb, considering the template which contains the current word and one previous tag and one next tag (tags in both sides of current word) |
| word-tag[+2] | if the current tag is adjective or verb, considering the template which contains the current word and two next tags. |
| word-tag[-2] | if the current tag is adjective or verb, considering the template which contains the current word and two previous tags. |

## 3.3 Extraction and Patterns Evaluation

The predefined templates (section 3.2) are applied on the training data (it is tagged by POS) to extract all possible patterns. After that we evaluate them to get the best set of patterns which is result of learning process. Evaluation process contains two steps which are mentioned in section 3.

### 3.3.1 Acceptable Patterns Evaluation

We aim to find the patterns in the training data which characterize subjective expressions. That means we only consider the patterns which satisfy the constrain:

- A pattern is believable to express subjectivity if and only if:
  $P (<\text{sub}> |pattern_i) > P (<\text{obj}> |pattern_i)$

The constrain means that: A pattern is acceptable if and only if it appears in subjective comments more frequently than in objective comments in the training data.

The formula below is proposed to get the sets of acceptable patterns:

$$\frac{P(<sub>|pattern_i)}{P(<sub>|pattern_i)+P(<obj>|pattern_i)} > \text{threshold}$$

In order to satisfy the constrain above, the threshold must be greater than 0.5. The threshold can be increased to get the different sets of acceptable patterns. The range of threshold is in [0.5, 1.0) ($0.5 \leq$ threshold $< 1.0$). In other word, we can generate a new set by changing the threshold. By increasing the threshold, the new set is generated whose number of patterns may be smaller than the old set, so the set of acceptable patterns will be narrowed.

This evaluating step is considered as the first filter in learning process. It help us remove a large amount of patterns that are unbelievable to express opinions. The best set of patterns is selected from the remainder of patterns.

### 3.3.2    The Best Patterns Set Evaluation

This evaluation step aims to get the best set of patterns from sets of acceptable patterns. We use the training data to evaluated. In this case, the training data plays development data role. A set of acceptable patterns is selected as the best set if it gets the highest accuracy by classifying on training data.

We assume that the best set of patterns on the training data will be the best set on the other data. That means other data is similar to the training data about their grammar of sentences, so the best training data must cover most syntactic structures of subjective sentences. However, in fact, set of patterns getting the best performance on the training data can be worse on other data because of the differences of the distribution and the grammar of sentences in data.

From each set of acceptable patterns, we extract phrases then take the adjectives and verbs in these phrases as the features. These features is used to classify on training data and evaluated by 10-fold cross-validation. After that, we select a set which has highest value.

We use the training data as the development data to evaluate acceptable patterns for adapting to domain and satisfying our assumption. This work helps us build the set of patterns which is more flexible and diverse. The quality of the patterns depends on the training data.

## 4    Experiment and Discussion

### 4.1    Experimental Data

Our experiment is conducted on technical product review data (review of mobile, laptop, tablet, camera, TV). We collected data from some Vietnamese technical forums such as *tinhte.vn, voz.vn, thegioididong.com* by scrapy framework[1]. After that, we remove the non-diacritic comments, then correct spell errors in the comments.

We labeled manually 9000 collected Vietnamese comments with two kinds of labels "<sub>" (subjective) and "<obj>" (objective) (in section 3.1). After that we divided this annotated data into two parts. The first part contains 3000 subjective comments and 3000 objective comments as the training data to learn the patterns. The remainder of comments (3000 comments) is used to test the quality of learned patterns. The training data and testing data are segmented into words and tagged by POS.

We used some classification tools in weka[2] for evaluating in learning process and evaluating the quality of the learned patterns on test data.

---

[1] http://scrapy.org
[2] http://www.cs.waikato.ac.nz/ml/weka/

## 4.2   Experimental Results

### 4.2.1   Learning Process

Firstly, we learned N-gram (unigram, bigram) of words from the training data with
threshold in range [0.5; 0.6; 0.7; 0.8; 0.9]. In order to reduce number of bigrams we
just use the bigrams which appear at least two times in training data to build features
set. The results of this process are illustrated in table 3. Unigram and bigram will
be used as the features to classify on the training data. We used liblinear library in
weka which implements SVM for classification. We implemented in 10-fold cross-
validation, then evaluated classification's performance.

**Table 3**  Classification results of unigram and bigram

| threshold | unigram | bigram |
|-----------|---------|--------|
| 0.5 | 82.59% | 72.93% |
| 0.6 | 83.14% | 73.52% |
| 0.7 | 83.47% | 75.52% |
| 0.8 | 83.29% | 77.47% |
| 0.9 | 81.82% | 79.27% |

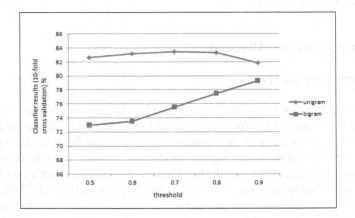

**Fig. 2**  Classification results of unigram and bigram (%)

The results of this experiment are shown in figure 2. From figure 2, we can see
that using unigram are better than bigram to build the set of features. Moreover, we
would like to investigate whether the combination of unigram and bigram can make
a better set of features or not. We combined the best set of unigram (threshold=0.7)
and the best set of bigram (threshold=0.9) into a features set and got 85.03% of

accuracy. Although using bigram without unigram make the system's performance decline, it can enrich the set of unigram features.

Secondly, we learned the POS patterns of two types which are mentioned in section 3.2. The predefined templates are applied on training data to extract all patterns. We also use threshold in range [0.5; 0.6; 0.7; 0.8; 0.9] to generate the sets of acceptable patterns. The patterns in each set are applied on the training data to extract phrases, adjectives, verbs as features for subjectivity classification. We also used liblinear in weka to evaluate set of patterns. The results are shown in table 4 (patterns from templates of type 1) and table 5 (patterns from templates of type 2).

In table 4, we got the template tag-tag[+2] with threshold is 0.5 as the best template. We extracted 72 patterns, some of them and their extracted phrases are illustrated in table 6.

**Table 4** Classification results of learning Patterns of type 1

| threshold | tag-tag[+1] | tag-tag[-1] | tag-tag[+2] | tag-tag[-1] & tag[+1] | tag-tag[-2] |
|-----------|-------------|-------------|-------------|-----------------------|-------------|
| 0.5 | 82.81% | 82.82% | **83.29%** | 82.66% | 82.62% |
| 0.6 | 81.16% | 79.54% | 81.41% | 81.46% | 81.67% |
| 0.7 | 75.47% | 79.41% | 80.56% | 78.27% | 80.99% |
| 0.8 | 67.92% | 50.03% | 76.19% | 76.17% | 76.81% |
| 0.9 | 50.03% | 50.03% | 71.32% | 72.74% | 68.95% |

**Table 5** Classification results of learning Patterns of type 2

| threshold | word-tag[+1] | word-tag[-1] | word-tag[+2] | word-tag[-1] & tag[+1] | word-tag[-2] |
|-----------|--------------|--------------|--------------|------------------------|--------------|
| 0.5 | 82.95% | 82.87% | 82.69% | 82.89% | 82.76% |
| 0.6 | 82.77% | 83.06% | 82.91% | 82.89% | 82.77% |
| 0.7 | 82.66% | 83.02% | 82.91% | 82.96% | 82.57% |
| 0.8 | 82.82% | 83.21% | 82.97% | 82.86% | 82.69% |
| 0.9 | 82.42% | **83.37%** | 83.06% | 82.82% | 82.71% |

**Table 6** Learned patterns of type 1

| patterns | phrases |
|----------|---------|
| V R A | nhìn cũng bóng_bẩy; chụp cũng tốt; nhìn khá lạ... |
| V A A | ốp chắc cực; nhìn đẹp thiệt; xài tốt hơn... |
| A R A | thực_sự rất ấn tượng; đen rất menly; tiện_-lợi lại sang_trọng... |

**Table 7** Learned patterns of type 2

| patterns | phrases |
|----------|---------|
| R yếu | quá yếu; không yếu; rất yếu... |
| R hay | khá hay; rất hay; không hay... |
| V xấu | trông xấu; thiết_kế xấu; nhìn xấu... |
| V yếu_ớt | nhìn yếu_ớt; thấy yếu_ớt... |
| N khỏe | cấu_hình khỏe; sóng khỏe; máy khỏe... |

In table 5, we got the template word-tag[-1] with threshold is 0.9 as the best case. We got 2175 patterns and some of them with their extracted phrases are illustrated in table 7. Note: A (adjective); V(verb); R (adverb); N(noun)

### 4.2.2 Testing Learned Patterns

We investigated the quality of features set which is n-grams, and words or words and phrases which are extracted from learned patterns. The results is shown in table 8.

**Table 8** Classification results on testing data

|  | # of features | SVM | Naive Bayes |
|--|---------------|-----|-------------|
| unigram | 3894 | 82.29% | 79.28% |
| bigram | 3210 | 64.25% | 59.24% |
| unigram + bigram | 7104 | 82.54% | 79.67% |
| words of patterns (type 1) | 2972 | 82.56% | 77.46% |
| words of patterns (type 2) | 1655 | 82.68% | 68.27% |
| words and phrases of patterns (type 1) | 4472 | 82.56% | 77.46% |
| words and phrases of patterns (type 2) | 3062 | 82.68% | 68.27% |
| words (type 1) + unigram + bigram | 9847 | 83.47% | 78.22% |
| words (type 2) + unigram + bigram | 8421 | 84.03% | 76.72% |

From table 8, we see that:

- Base on the results in line 1, 2, 3, 4, 5 we can compare the quality of the features from POS patterns and N-gram. When comparing with unigram or bigram or the combination of unigram and bigram, the features of POS patterns in two types are better, but it is not significant.
- The classification results in line 4 and 5 showed that the quality of patterns of type 1 and type 2 are similar. In fact, we can use one of them to extract the patterns.
- The results in line 6 and 7 show that the phrases extracted from these patterns seem to be not good features because it can not improve the performance of classification.

- Line 8, 9 contain the results of combination of unigram, bigram and features from POS patterns. We can see that the learned patterns help us enrich the features set. The patterns of type 1 provided more 2700 features and the patterns of type 2 provided more 1317 features. The new features from these patterns and N-gram help to increase accuracy of subjectivity classification (more 1.5% in comparison with just using patterns or N-gram).

However, the results are not really high for some reasons: Firstly, the features set has to suffer from the errors of spelling, words segmentation, POS tagging. The spelling errors can lead to the noise in features set; Secondly, the quality and variety of the training data also affect to the performance of system.

## 5 Conclusion and Future Work

This paper focuses on subjectivity classification problem in which we have proposed a new statistical method for enriching features set based on POS patterns. In our work, the patterns are built automatically from labeled training data. These patterns are more flexible and adaptive to domain. We used learned patterns to extract word collocations using two type of word such as adjective and verb. The Support Vector Machine (SVM) and Naive Bayes (NB) are applied to determine whether a given comment belongs to subjective or objective class. In our experiment, by combining unigram, bigram and words extracted from POS patterns, we obtained 84.04% by SVM as the best case on the Vietnamese technical product review data.

In future, we will extend the features set by using other POS tags. We also can exploit more templates to generate POS patterns.

**Acknowledgment.** This paper is supported by the project QGTĐ.12.21 funded by Vietnam National University, Hanoi.

## References

[1] Pang, B., Lee, L., Vaithyanathan, S.: Thumbs up? Sentiment classification using machine learning techniques. In: Proceedings of EMNLP 2002, pp. 79–86 (2002)
[2] Long, C., Zhang, J., Shut, X.: A review selection approach for accurate feature rating estimation. In: Proceedings of the 23rd International Conference on Computation Linguistics 2010, pp. 766–774 (2010)
[3] Riloff, E., Wiebe, J., Wilson, T.: Learning Subjective Nouns using Extraction Pattern Bootstrapping*. In: Proceedings of the Seventh coNLL Conference held at HLT-NAACL 2003, pp. 25–32 (2003)
[4] Riloff, E., Patwardhan, S., Wiebe, J.: Feature Subsumption for Opinion Analysis. In: Proceedings of the Conference on Empirical Methods in Natural Language Processing 2006, pp. 440–448 (2006)
[5] Xuan, H.N.T., Le, A.C., Nguyen, L.M.: Linguistic Features for Subjectivity Classification in Asian Language Processing (IALP). In: International Conference, pp. 17–20 (2012)

[6]  Wiebe, J.: Learning Subjective Adjectives from Corpora. In: Proceedings of the Seventeenth National Conference on Artificial Intelligence and Twelfth Conference on Innovative Application of Artificial Intelligence 2000, pp. 735–740 (2000)

[7]  Wiebe, J., Wilson, T., Bruce, R., Bell, M., Martin, M.: Learning Subjective Language. Journal Computational Linguistics 30(3), 277–308 (2004)

[8]  Taboada, M., Brooke, J., Tofiloski, M., Voll, K., Stede, M.: Lexicon-based methods for sentiment analysis. Journal Computational Linguistics 37(2), 267–307 (2011)

[9]  Sokolova, M., Lapalme, G.: Opinion Classification with Non-affective Adjective and Adverbs. In: Proceedings of the International Conference on Recent Advances in Natural Language Processing (RANLP 2009) (2009)

[10] Turney, P.D.: Thumbs up of thumbs down?: semantic orientation applied to unsupervised classification of reviews. In: Proceedings of the 40th Annual Meeting on Association for Computational Linguistics 2002, pp. 417–424 (2002)

[11] Rosenthal, S., McKeown, K.: Columbia NLP: Sentiment Detection of Subjective Phrases in Social Media. In: Conference on Lexical and Computation Semantics (2013)

[12] Brody, S., Diakopoulos, N.: Cooooooooooooooooollllllllllllll!!!!!!!!!!!!!!: using word lengthening to detect sentiment in microblogs. In: Proceedings of the Conference on Empirical Methods in Natural Language Processing 2011, pp. 562–570 (2011)

[13] Pak, A., Paroubek, P.: Twitter as a Corpus for Sentiment Analysis and Opinion Mining. In: Proceedings of the Seventh Conference on International Language Resources and Evaluation, LREC 2010, Valletta, Malta, European Language Resources Association ELRA (2010)

[14] Gamon, M.: Sentiment classification on customer feedback data: noisy data, large feature vectors, and the role of linguistic analysis. In: Proceeding of COLING 2004, the 20th International Conference on Computational Linguistics 2004, pp. 841–847 (2004)

[15] Barbosa, L., Feng, J.: Robust sentiment detection on Twitter from biased and noisy data. In: Proceedings COLING 2010 Proceedings of the 23rd International Conference on Computational Linguistics 2010, pp. 36–44 (2010)

[16] Agarwal, A., Xie, B., Vovsha, I., Rambow, O., Passonneau, R.: Sentiment analysis of Twitter data. In: Proceedings of LSM 2011 Proceedings of the Workshop on Languages in Social Media 2011, pp. 30–38 (2011)

# Question Analysis for a Community-Based Vietnamese Question Answering System

Quan Hung Tran, Minh Le Nguyen, and Son Bao Pham

**Abstract.** This paper describes the approach for analyzing questions in our community-based Vietnamese question answering system (VnCQAs), in which we focus on two subtasks: question classification and keyword identification. The question classification employs the machine learning approaches with a feature which represents a measure of similarity between two questions, while the keyword identification uses the dependency-tree-based features. Experimental results are promising, in which the question classification obtains the accuracy of 95.7% and the keyword identification gains the accuracy of 85.8%. Furthermore, these two sub-tasks help to improve the accuracy for finding the similar questions in our VnCQAs by 6.75%.

## 1 Introduction

Question answering systems usually have a module for analyzing questions in order to extract the important information such as keywords, question types or semantic constraints. In this research, we focus on two subtasks of question analysis: question classification and keyword identification. Identifying important words from a set of documents is an important task on information retrieval and question answering with two main approaches: using the corpus-based statistics for term weighting [7, 9] and employing the supervised methods [3, 18, 10]. The question classification aims to

Quan Hung Tran · Son Bao Pham
Faculty of Information Technology
University of Engineering and Technology
Vietnam National University, Hanoi
e-mail: {quanth_55,sonpb}@vnu.edu.vn

Minh Le Nguyen
School of Information Science
Japan Advanced Institute of Science and Technology
e-mail: nguyenml@jaist.ac.jp

© Springer International Publishing Switzerland 2015     641
V.-H. Nguyen et al. (eds.), *Knowledge and Systems Engineering,*
Advances in Intelligent Systems and Computing 326, DOI: 10.1007/978-3-319-11680-8_51

classify questions into several pre-defined classes for seeking the suitable answers. Li et al. [8] proposed a two-layer taxonomy with 6 coarse classes and 50 fine-grained classes, while Bu et al. [4] introduced a six-types taxonomy. Futhermore, regarding to the methods for classifying questions, some researches employed the rule-based approaches [6] and the machine learning algorithms [4, 1], while other researches considered on combining rule-based and machine learning-based techniques [5].

Recently, some question analysis techniques have been examined for Vietnamese [11, 13, 12, 16]. However, these researches experimented on a standard corpus, where the words' spellings are generally good. The question analysis for our VnC-QAs system, on the other hand, has to deal with noisy data from the community-based resources. In this paper, we propose the dependency-tree-based features in finding keywords. We also introduce a new feature called "similarity feature" for classifying questions. To the best of our knowledge, it is the first time the question analysis are adapted to the Vietnamese community data.

The paper was presented as follows: in section 2, we briefly describe the architecture of our VnCQAs system. Section 3 presents the overview of our approach for analyzing questions, while the question classification and the keyword identification are introduced in section 4 and 5 respectively. Section 6 gives the experimental results and the conclusion are shown in section 7.

## 2   The VnCQAs System Architecture

The architecture of our question answering system [17] is shown in Figure 1. It includes three modules: Database Construction, Question Analysis and Answer selection, in which the database construction module aims to build the database of question-answer pairs, while the question analysis module extract the useful information such keywords, question types and synonyms. The answer selection module finds the most similar questions for the input question from the database, in which each similar question corresponds to a candidate answer. The candidate answers then are processed to output the best answer.

In this paper, we focus on analyzing questions in the question analysis module with two main tasks: Question Classification and Keyword Identification. Furthermore, in this module, we also use a dictionary of 6626 entities to find the synonyms in the question. Figure 2 shows an example for the question analysis module with the question: *"Làm thế nào để tạo vùng nhớ ảo thay thế RAM"* (How to create virtual memory to replace RAM).

## 3   Question Classification

### 3.1   Question Types

We classify questions into three types: *Fact, Explanation* and *Solution* according to the main purpose of the questioner.

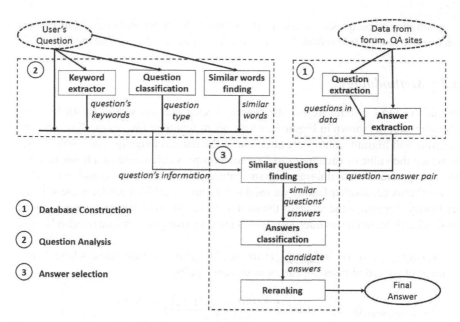

**Fig. 1** The system architecture

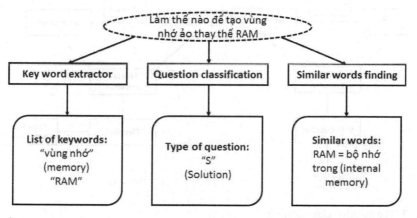

**Fig. 2** An example for the question analysis module

• *Fact*: The questions is only about objects and resources, the expected answer is about the general facts and/or attributes. E.g., with the question: *"Tấm dán màn hình từ tính là gì?" (What is magnetic screen stickers?)*, the object is *"Tấm dán màn hình từ tính" (the magnetic screen stickers)*, and the expected attribute in the answer is definition.

• *Explanation*: The questions require explanations or opinions e.g., *"Vì sao điện thoại của mình hay bị mất sóng?" (Why does my phone frequently lose signal?)*.

- *Solution*: The questions ask for the solution for a problem e.g.,*"Chỉ cho em cách vào facebook trên Iphone?" (How to access Facebook from Iphone?).*

## 3.2  Methodology

In the VnCQAs system, we use the support vector machines (SVMs) for learning classification (as shown in Figure 3) with a set of features: *Unigrams, Bigrams.* The unigram and bigram features are common in the natural language processing tasks, in which the value of each unigram/bigram feature is calculated as a boolean value indicating whether that unigram/bigram feature is included in the question or not.

Furthermore, another feature we used for training the SVM model is the *similarity* feature, for which the value of the similarity feature which represents a measure of similarity between two questions is estimated by using the phrasal overlap [2, 15]:

$overlap_{phrase}(s_1, s_2) = \sum_{i=1}^{n} \sum_{m} i^2$ for $m$ phrasal $n$-word overlaps, where $m$ is a number of $i$-word phrases appearing in sentence pairs.

$$sim_{overlap,phrase}(s_1, s_2) = tanh(\frac{overlap_{phrase}(s_1, s_2)}{|s_1| + |s_2|})$$

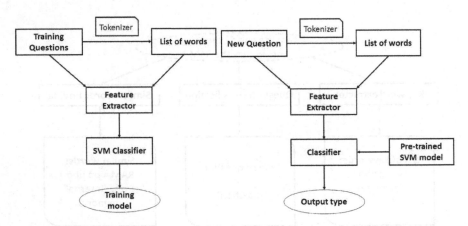

**Fig. 3** The question classification

## 4   Keyword Identification

This section describes the keyword identification by using the machine learning technique with the dependency-tree-based features.

## 4.1 Keyword Definition

We define the keywords as:

• The most informative words in the question is the set of keywords which contains most of the information (e.g., topics, main objects and actions).

• The words can be used to distinguish different questions, two questions that have the same set of keywords are likely to be similar.

E.g.,: The question: *"Hỏi cách xóa tin nhắn trên iphone?"* (*How to delete messages on iphone?*) have the keywords of: *"cách" (how), "xóa" (delete), "tin nhắn" (messages), "iphone"*. The main verb *"hỏi" (ask)* is not considered a keyword because this word does not represent any important information and it also cannot be used to distinguish among questions.

## 4.2 Methodology

We use the dependency tree to find keywords on the premise that is to identify a word as informative or not, we take into account the relationships of that word with other words in the question by using the Vietnamese dependency parser [14]. For each question, the dependency parser creates a tree that contains the tree structure, the relation of the words and several other information such as part-of-speech tag (as shown in Figure 4, the question is "How do I delete messages in Iphone?").

| 1 | hỏi | hỏi | V | V | - | 0 | root | - | - |
| 2 | cách | cách | N | N | - | 1 | dob | - | - |
| 3 | xóa | xóa | V | V | - | 2 | vmod | - | - |
| 4 | tin_nhắn | tin_nhắn | N | N | - | 3 | dob | - | - |
| 5 | trên | trên | E | E | - | 4 | loc | - | - |
| 6 | iphone | iphone | N | N | - | 5 | pob | - | - |

**Fig. 4** An example of the dependency tree

The features are then extracted from the tree and used for training the SVM model which is used to classify a word as a keyword or not (as presented in Figure 5). These features are grouped as follows:

• *The part of speech (POS) tag of each word: POSW*

• *The part of speech (POS) tag of the parent word of each word in the dependency tree: POSP*

• *Unigrams*

• *The common dependent words: CDW*

The POSW feature is used because words with certain POS tags (e.g., Noun, Verb, and Adjective) are more likely to be a keyword of a sentence. The POSP feature

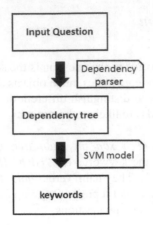

**Fig. 5** Dependency tree method work flow

helps to identify that words that are children of verbs are more likely to be the object of the question. The CDW feature is employed because the children of several words (e.g. *"cách" (solution)*) have a high chance of being keywords. During implementation for the CDW feature, we use a map that stores the words and an Integer that indicates the number of times that word is the parent of a keyword. A manual threshold is then used to identify which words have a high frequency of being the parent of keywords.

# 5  Experimental Results

## 5.1  Question Classification Evaluation

We use a set of 1013 manually tagged questions focusing on the technology domain with three mentioned types: Fact, Explanation, and Solution. The tagged questions come from the database construction module of our VnCQAs system. These questions are kept in its original form, no modifications are made. However, some of the questions in online forums are not understandable, they lack information or context to be understood. These questions are removed from the set of data. The question distribution is shown in Figure 6.

Regarding to learn the SVM model, we use the LIBLinear and SVM-SMO algorithms with 10-fold cross-validation scheme. The experiments were conducted on a Window PC with Core i7 CPU and 8GB of RAM. The highest accuracy (95.7%) is achieved with the combination of phrasal overlap, unigram and bigram features (as shown in Table 1).

Although, the accuracy of the question classification is around 95.7%, our corpus is different from the corpus of other published works, it is hard to directly compare our method to other available methods in the question classification task. To make a meaningful comparison of our method with other methods, we investigate on the

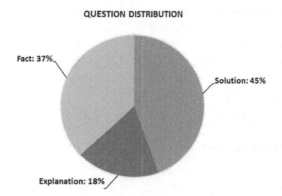

**Fig. 6** The distribution of questions in the tagged data set

**Table 1** The question classification accuracy on the community data

| Features | Accuracy (SVM - LIBLinear) | Accuracy (SVM-SMO) |
|---|---|---|
| Bigram | 89.8 (%) | 89.5 (%) |
| Unigram | 94.8 (%) | 95.2 (%) |
| phrasal overlap | 93.4 (%) | 92.9 (%) |
| phrasal overlap + Unigram + bigram | 95.6 (%) | **95.7 (%)** |

TREC corpus in Vietnamese [16]. Our obtained accuracy is comparable to the accuracy of the Tran et al. [16]'s approach with the same 10-fold cross-validation scheme (as shown in Table 2).

**Table 2** The question classification accuracy on the TREC data

| Classes | Our method | Tran's method |
|---|---|---|
| coarse classes classification | 85.0 (%) | 86.0 (%) |
| fine grain classes classification | 84.9 (%) | 84.7 (%) |

## 5.2 Keyword Identification Evaluation

To test the performance of the keyword identification, we use a set of 753 words tagged from the sentences in our database (as shown in Figure 7, the question is "How do I delete messages in Iphone?").

To make a comparison, we implement a baseline for identifying the keywords by using the using term frequency - inverse document frequency (TF-IDF) method.

| 1 | hỏi | hỏi | V | V | - | 0 | root | - | - | |
| 2 | cách | cách | N | N | - | 1 | dob | - | - | key |
| 3 | xóa | xóa | V | V | - | 2 | vmod | - | - | key |
| 4 | tin_nhắn | tin_nhắn | N | N | - | 3 | dob- | | - | key |
| 5 | trên | trên | E | E | - | 4 | loc | - | - | |
| 6 | iphone | iphone | N | N | - | 5 | pob- | | - | key |

**Fig. 7** An example of keywords

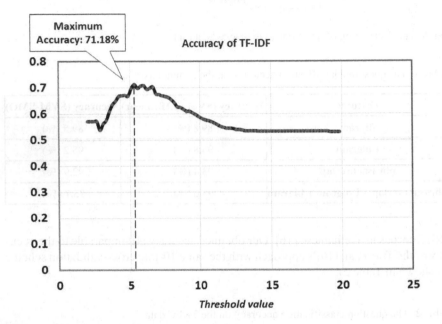

**Fig. 8** The TF-IDF method's accuracy

The TF-IDF score of each word in a sentence will be calculated, and a threshold is chosen to identify whether a word is a keyword or not. The accuracy results of the TF-IDF method are presented in figure 8.

Our method outperforms the TF-IDF method as we can see from the obtained accuracies in Table 3).

## 5.3 Question Analysis Evaluation

In this section, we evaluate the contribution of the question analysis to our VnC-QAs system by measuring the improvement in finding similar questions. To evaluate the ability of the VnCQAs system to find similar questions, we use a set of

**Table 3** The accuracy of the keyword identification

| Features | Accuracy (SVM - LIBLinear) | Accuracy (SVM-SMO) |
|---|---|---|
| POSW | 79.1 (%) | 79.1 (%) |
| POSW + POSP | 81.4 (%) | 81.1 (%) |
| POSW + POSP + BOW | 85.4 (%) | 85.4 (%) |
| POSW + POSP+ BOW+CDW | **85.8 (%)** | 85.4 (%) |

1704 questions which are checked by hand to ensure that each question is not similar to other questions. Then we paraphrase each question into 3 different versions. The paraphrased questions must have close meaning to the original questions.

We use 1704 original questions as the input questions for testing our system performance, if the returned question is one of 3 paraphrased questions, we evaluate this as a good result, and otherwise we count it as a bad result. Table 4 shows the accuracy improvement in finding the similar questions.

**Table 4** The question analysis evaluation

| Method | Accuracy) |
|---|---|
| **Cosine similarity** | 80.86 (%) |
| **Cosine Similarity + Question Analysis** | **87.61 (%)** |

## 6 Conclusion

In this paper, we described the question analysis module of our VnCQAs system on two subtasks: question classification and keyword identifications. We classify questions into three types: *Fact, Explanation* and *Solution* by using the support vector machines (SVMs) for learning classification with a set of features: *unigrams, bigrams, and similarity.* Our classification accuracy is high even though we have to deal with noisy community data. Furthermore, on the Vietnamese TREC corpus, we gain the competitive accuracy results. For the keyword identification subtask, we used the machine learning method with the dependency tree-based features and achieved the accuracy of 85.8% which outperforms the TF-IDF method.

In the future, we will improve the size and quality of data set used in both subtasks above. We also will examine other methods for further improving the performance accuracy in analyzing questions.

**Acknowledgment.** This work is partially supported by the Research Grant from Vietnam National University, Hanoi No. QG.14.04.

# References

[1] Paliwal, M., Kumar, U.A.: Neural networks and statistical techniques: A review of applications. Expert Systems with Applications 36(1), 2–17 (2009)

[2] Banerjee, S., Pedersen, T.: Extended gloss overlaps as a measure of semantic relatedness. In: Proceedings of the 18th International Joint Conference on Artificial Intelligence, IJCAI 2003, pp. 805–810 (2003)

[3] Bendersky, M., Croft, W.B.: Discovering key concepts in verbose queries. In: Proceedings of the 31st Annual International ACM SIGIR Conference on Research and Development in Information Retrieval, SIGIR 2008, pp. 491–498 (2008)

[4] Bu, F., Zhu, X., Hao, Y., Zhu, X.: Function-based question classification for general qa. In: Proceedings of the 2010 Conference on Empirical Methods in Natural Language Processing, EMNLP 2010, pp. 1119–1128 (2010)

[5] Huang, Z., Thint, M., Qin, Z.: Question classification using head words and their hypernyms. In: Proceedings of the Conference on Empirical Methods in Natural Language Processing, EMNLP 2008, pp. 927–936 (2008)

[6] Hui, Z., Liu, J., Ouyang, L.: Question classification based on an extended class sequential rule model. In: Proceedings of 5th International Joint Conference on Natural Language Processing, Chiang Mai, Thailand, pp. 938–946. Asian Federation of Natural Language Processing (November 2011)

[7] Lan, M., Tan, C.L., Su, J., Lu, Y.: Supervised and traditional term weighting methods for automatic text categorization. IEEE Transactions on Pattern Analysis and Machine Intelligence 31(4), 721–735 (2009)

[8] Li, X., Roth, D.: Learning question classifiers. In: Proceedings of the 19th International Conference on Computational Linguistics, COLING 2002, vol. 1, pp. 1–7 (2002)

[9] Luhn, H.P.: A business intelligence system. IBM J. Res. Dev. 2(4), 314–319 (1958)

[10] Luo, X., Raghavan, H., Castelli, V., Maskey, S., Florian, R.: Finding what matters in questions. In: Proceedings of the 2013 Conference of the North American Chapter of the Association for Computational Linguistics: Human Language Technologies, Atlanta, Georgia, pp. 878–887. Association for Computational Linguistics (June 2013)

[11] Nguyen, D.Q., Nguyen, D.Q., Pham, S.B.: A Vietnamese Question Answering System. In: Proceedings of the 2009 International Conference on Knowledge and Systems Engineering, KSE 2009, pp. 26–32 (2009)

[12] Nguyen, D.Q., Nguyen, D.Q., Pham, S.B.: A Semantic Approach for Question Analysis. In: Jiang, H., Ding, W., Ali, M., Wu, X. (eds.) IEA/AIE 2012. LNCS, vol. 7345, pp. 156–165. Springer, Heidelberg (2012)

[13] Nguyen, D.Q., Nguyen, D.Q., Pham, S.B.: Systematic Knowledge Acquisition for Question Analysis. In: Proceedings of the International Conference Recent Advances in Natural Language Processing 2011, pp. 406–412 (2011)

[14] Nguyen, D.Q., Nguyen, D.Q., Pham, S.B., Nguyen, P.-T., Le Nguyen, M.: From treebank conversion to automatic dependency parsing for vietnamese. In: Métais, E., Roche, M., Teisseire, M. (eds.) NLDB 2014. LNCS, vol. 8455, pp. 196–207. Springer, Heidelberg (2014)

[15] Ponzetto, S.P., Strube, M.: Knowledge derived from wikipedia for computing semantic relatedness. Journal of Artificial Intelligence Research 30(1), 181–212 (2007)

[16] Tran, D., Chu, C., Pham, S., Nguyen, M.: Learning based approaches for vietnamese question classification using keywords extraction from the web. In: Proceedings of the Sixth International Joint Conference on Natural Language Processing, pp. 740–746. Asian Federation of Natural Language Processing (October 2013)

[17] Tran, Q.H., Nguyen, N.D., Do, K.D., Nguyen, T.K., Tran, D.H., Le Nguyen, M., Pham, S.B.: A Community-based Vietnamese Question Answering System. In: Proceedings of the 2014 International Conference on Knowledge and Systems Engineering, KSE 2014 (2014)

[18] Zhao, L., Callan, J.: Term necessity prediction. In: Proceedings of the 19th ACM International Conference on Information and Knowledge Management, CIKM 2010, pp. 259–268 (2010)

# Using Dependency Analysis to Improve Question Classification

Phuong Le-Hong, Xuan-Hieu Phan, and Tien-Dung Nguyen

**Abstract.** Question classification is a first necessary task of automatic question answering systems. Linguistic features play an important role in developing an accurate question classifier. This paper proposes to use typed dependencies which are extracted automatically from dependency parses of questions to improve accuracy of classification. Experiment results show that with only surface typed dependencies, one can improve the accuracy of a discriminative question classifier by over 8.0% on two benchmark datasets.

## 1 Introduction

Question answering (QA) has been an important line of research in natural language processing in general and in human-machine interface in particular. The ultimate goal of a QA system is to provide a concise and exact answer to a question asked in a natural language. For example, the answer to the question *"What French city has the largest population?"* should be *"Paris"*.

Open-domain QA is a challenging task because the research and validation of a precise answer to a question require a good understanding of the question itself and of the text containing the potential answer to the question. Typically, we need to

Phuong Le-Hong
University of Science
Vietnam National University, Hanoi
e-mail: phuonglh@vnu.edu.vn

Xuan-Hieu Phan
University of Engineering and Technology
Vietnam National University, Hanoi
e-mail: hieupx@vnu.edu.vn

Tien-Dung Nguyen
FPT Research, FPT Corp., Hanoi, Vietnam
e-mail: dungnt64@fpt.com.vn

© Springer International Publishing Switzerland 2015      653
V.-H. Nguyen et al. (eds.), *Knowledge and Systems Engineering*,
Advances in Intelligent Systems and Computing 326, DOI: 10.1007/978-3-319-11680-8_52

carry out both syntactic and semantic analysis in order to fully understand a question and pinpoint an answer. This is much more difficult than common information retrieval task, where one only needs to present a ranked list of documents in response to a question, which can be efficiently performed by available search engines.

The first step of understanding a question is to perform question analysis. Question classification is an important task of question analysis which detects the answer type of the question. Question classification helps not only filter out a wide range of candidate answers but also determine answer selection strategies. In the example question above, if one knows that the answer type is *city*, one can restrict candidate answers as cities instead of consider every noun phrase of a document providing the answer.

At first glance, one may think that question classification can be framed as a text classification task. However, there exists characteristics of question classification that distinguish it from the common task. First, a question is relatively short and contains less word-based information than an entire text. Second, a short question needs a deeper-level analysis to reveal its hidden semantics. Therefore, application of text classification algorithms *per se* to question classification could not result in a good result. Furthermore, natural language is inherently ambiguous; the question classification is not trivial, especially for *what* and *which* type questions. For example, "*What is the capital of France?*" is of location (city) type, while "*What is the Internet of things?*" is of definition type. Consider also examples: *(1) What tourist attractions are there in Reims? (2) What do most tourists visit in Reims? (3) What are the names of the tourist attractions in Reims? (4) What attracts tourists to Reims? (5) What is worth seeing in Reims?* [1]; all these questions are of the same answer type location. Different wording and syntactic structures make it difficult for classification [2].

This paper focuses on the question classification task. The main contribution of the paper is to show that accuracy of question classification can be improved by using dependency analysis of questions: even with two simple typed dependencies extracted from dependency parses, we can improve the accuracy of question classifiers on two different question classification datasets by over 8.0%, a significant improvement, given that no hand-crafted rules are required as in many existing works.

The rest of this paper is structured as follows. Section 2 gives an overview of existing works on question classification. Section 3 presents briefly some backgrounds for the work, including two classification models in use, namely naive Bayes models and maximum entropy models, which are typical instances of generative and discriminative classifiers, respectively; and dependency parsing of natural languages. Section 4 describes datasets, experimental results and discussions. Finally, Section 5 concludes the paper.

# 2    Related Works

Early works in question classification used rule-based approaches to map a question to a type, which require expert labor for manually constructing rules. This makes rule-based approaches not only inefficient in maintain but also difficult to upgrade or port to different domains.

With the increasing popularity of statistical approaches to natural language processing in general and to question classification in particular, recent years have seen many machine learning approaches which have been applied to question classification problem. The main advantage of machine learning approaches is that one can learn a statistical model using useful features extracted from a sufficiently large set of labeled questions and then use it to automatically classify new questions. In this section, we summarize existing machine learning approaches to question classification and their results.

Li and Roth [3] developed the first machine learning approach to question classification which uses the SNoW learning architecture. They have created the UIUC question classification dataset[1] containing 5,952 manually labeled questions of 6 coarse-grained classes and 50 fine-grained classes (see Table 2). Using the feature set of lexical words, part-of-speech tags, chunks and named entities, they achieved 78.8% of accuracy for 50 fine-grained classes. When augmented with a hand-built dictionary of semantically related words, they were able to reach 84.2% of accuracy.[2]

**Table 1** Accuracy of question classifiers on the UIUC dataset

| Model | 6 classes | 50 classes |
|---|---|---|
| Li and Roth, SNoW | – | 78.8% |
| Zhang and Lee, Linear SVM | 87.4% | 79.2% |
| Zhang and Lee, Tree SVM | 90.0% | – |
| Hacioglu and Ward, SVM+ECOC | – | 82.0% |
| Krishnan et al., SVM+CRF | 93.4% | 86.2% |
| Nguyen et al., ST-Boost+ME | 91.2% | 83.6% |
| Huang et al., Linear SVM | 93.4% | 89.2% |
| Huang et al., MaxEnt | 93.6% | 89.0% |

The UIUC dataset has inspired many follow-up works on question classification. Zhang and Lee [4] used linear support vector machines (SVM) with all question $n$-grams and obtained 79.2% of accuracy. Hacioglu and Ward [5] used linear SVM with question bigrams and error-correcting codes and achieved 82.0% of accuracy. Krishnan et al. [6] also used linear SVM with contiguous subsequence of question words detected by a Conditional Random Field (CRF) and achieved 86.2% of

---

[1] Available at http://cogcomp.cs.illinois.edu/Data/QA/QC/
[2] However, follow-up works did not use this dictionary.

**Table 2** Distribution of question types in the UIUC dataset

| Category | # Train | # Test | Category | # Train | # Test |
|----------|---------|--------|----------|---------|--------|
| ABBREVIATION | 86 | 9 | term | 93 | 7 |
| abb | 16 | 1 | vehicle | 27 | 4 |
| exp | 70 | 8 | word | 26 | 0 |
| DESCRIPTION | 1162 | 138 | HUMAN | 1223 | 65 |
| definition | 421 | 123 | group | 47 | 6 |
| description | 274 | 7 | individual | 189 | 55 |
| manner | 276 | 2 | title | 962 | 1 |
| reason | 191 | 6 | description | 25 | 3 |
| ENTITY | 1250 | 94 | LOCATION | 835 | 81 |
| animal | 112 | 16 | city | 129 | 18 |
| body | 16 | 2 | country | 155 | 3 |
| color | 40 | 10 | mountain | 21 | 3 |
| creative | 207 | 0 | other | 464 | 50 |
| currency | 4 | 6 | state | 66 | 7 |
| dis.med. | 103 | 2 | NUMERIC | 896 | 113 |
| event | 56 | 2 | code | 9 | 0 |
| food | 103 | 4 | count | 363 | 9 |
| instrument | 10 | 1 | date | 218 | 47 |
| lang | 16 | 2 | distance | 34 | 16 |
| letter | 9 | 0 | money | 71 | 3 |
| other | 217 | 12 | order | 6 | 0 |
| plant | 13 | 5 | other | 52 | 12 |
| product | 42 | 4 | period | 27 | 8 |
| religion | 4 | 0 | percent | 75 | 3 |
| sport | 62 | 1 | speed | 9 | 6 |
| substance | 41 | 15 | size | 13 | 0 |
| symbol | 11 | 0 | temp | 8 | 5 |
| technique | 38 | 1 | weight | 11 | 4 |

accuracy on the UIUC dataset over fine-grained question types. Li and Roth [1] used more syntactic and semantic features including chunks, named entities, Word-Net senses, class-specific related words and distributional similarity and obtained 89.3% of accuracy.[3] Nguyen et al. [7] used maximum entropy and boosting models and achieved 83.6% of accuracy. Most recently, Huang et al. [2] used SVM and maximum entropy models with question head words and their hypernyms and obtained 89.2% of accuracy, which is the highest reported accuracy on this dataset. Table 1 shows the summary of classification accuracy of all models which were tested on the UIUC dataset.

---

[3] Nevertheless, their model was applied on a larger dataset comprising of 21,500 training questions and 1,000 test questions.

## 3  Background

In this study, we employ both generative and discriminative learning approaches for question classification, typified by two common machine learning models: naive Bayes model and maximum entropy model [8]. For ease of exposition, in this section, we first briefly present these two classification models. We then give a short introduction of dependency analysis of natural languages for those who are not familiar with this particular topic.

### 3.1  Naive Bayes Classifier

Generative models in general and Naive Bayes (NB) models in particular learn a model of the joint probability $P(\mathbf{x}, y)$ of the observation $\mathbf{x} \in X$ and the label $y \in \mathcal{Y}$, and make their predictions by using Bayes rules to calculate the posterior $P(y|\mathbf{x})$, then choosing the most probable label $y$. Let $=\{0,1\}^D$ be the $D$-dimensional input space, let the output labels $\mathcal{Y} = \{1, 2, \ldots, K\}$ and let $\mathcal{D} = \{(\mathbf{x}_1, y_1), \ldots, (\mathbf{x}_N, y_N)\}$ be a training set of $N$ independent and identically distributed examples, where $\mathbf{x}_i = (x_{i1}, x_{i2}, \ldots, x_{iD}) \in X$. The generative Bayes classifier is parameterized as follows:[4]

$$\theta_k = P(y = k), \forall k = 1, 2, \ldots, K$$
$$\theta_{j|k} = P(x_j = 1 | y = k), \forall j = 1, 2, \ldots, D; \forall k = 1, 2, \ldots, K.$$

It uses $\mathcal{D}$ to calculate the maximum likelihood estimates $\hat{\theta}_{j|k}$ and $\hat{\theta}_k$ as follows:[5]

$$\hat{\theta}_k = \frac{\sum_{i=1}^N \delta(y_i = k)}{N},$$

$$\hat{\theta}_{j|k} = \frac{\sum_{i=1}^N \delta(x_{ij} = 1 \text{ and } y_i = k) + \alpha}{\sum_{i=1}^N \delta(y_i = k) + \alpha K},$$

where $\delta(\cdot)$ is the identity function

$$\delta(b) = \begin{cases} 1, & \text{if } b = \text{true,} \\ 0, & \text{if } b = \text{false.} \end{cases}$$

Using the naive Bayes assumption of independent features, the posterior probability is computed as

$$P(y = k | \mathbf{x}) = \frac{P(\mathbf{x} | y = k) P(y = k)}{P(\mathbf{x})}$$

$$= \frac{\Pi_{j=1}^D P(x_j = 1 | y = k) P(y = k)}{P(\mathbf{x})} = \frac{\Pi_{j=1}^D \theta_{j|k} \theta_k}{P(\mathbf{x})}.$$

Since $y$ does not depend on $P(\mathbf{x})$, the classification rule for an observation $\mathbf{x}$ is simply

---

[4] This is indeed the Bernoulli NB model, not a multinomial NB model.
[5] We use Laplace smoothing technique with constants fixed at $\alpha = 1$.

$$y = \operatorname*{arg\,max}_{k=1,2,\ldots,K} P(y = k \mid \mathbf{x}) = \operatorname*{arg\,max}_{k=1,2,\ldots,K} \prod_{j=1}^{D} \theta_{j \mid k} \theta_k.$$

Or, by using the logarithmic transformation:

$$y = \operatorname*{arg\,max}_{k=1,\ldots,K} \left( \sum_{j=1}^{D} \log \theta_{j \mid k} + \log \theta_k \right).$$

In the question classification problem, each question is represented by an observation $\mathbf{x}$, which is a binary feature vector representing the presence or absence of particular features, for instance the presence or absence of words (unigram features).

## 3.2 Maximum Entropy Classifier

Maximum Entropy (ME) models (a.k.a multinomial logistic regression model) is a general purpose discriminative learning method for classification and prediction which has been succesfully applied to many problems of natural language processing. In contrast to generative classifiers, discriminative classifiers model the posterior $P(y \mid \mathbf{x})$ directly. One of the main advantages of discriminative models is that one can integrate many heterogeneous features for prediction, which are not necessarily independent. Each feature corresponds to a constraint on the model. In ME models, the conditional probability of a label $y$ given an observation $\mathbf{x}$ is defined as

$$P(y \mid \mathbf{x}) = \frac{\exp(\theta \cdot f(\mathbf{x}, y))}{\sum_{y \in \mathcal{Y}} \exp(\theta \cdot f(\mathbf{x}, y))},$$

where $f(\mathbf{x}, y) \in \mathbb{R}^D$ is a real-valued feature vector[6], $\mathcal{Y}$ is the set of labels and $\theta \in \mathbb{R}^D$ is the parameter vector to be estimated from training data. This form of distribution corresponds to the maximum entropy probability distribution satisfying the constraint that the empirical expectation of each feature is equal to its true expectation in the model:

$$\hat{\mathbb{E}}(f_j(h, t)) = \mathbb{E}(f_j(h, t)), \qquad \forall j = 1, 2, \ldots, D.$$

The parameter $\theta \in \mathbb{R}^D$ can be estimated using iterative scaling algorithms or some more efficient gradient-based optimization algorithms like conjugate gradient or quasi-Newton methods [9]. In this paper, we use the L-BFGS optimization algorithm and $L_2$-regularization technique to estimate the parameters of the ME models, with smooth term is fixed at 1.

## 3.3 Dependency Analysis

Constituency structure and dependency structure are two types of syntactic representation of a natural language sentence. While a constituency structure represents a nesting of multi-word constituents, a dependency structure represents dependencies between individual words of a sentence. The syntactic dependency represents the

---

[6] However, feature vectors of large scale ME models are typically sparse binary ones, indicating the presence or absence of corresponding features.

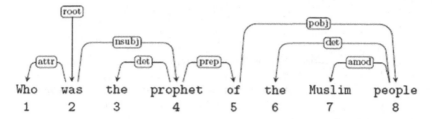

**Fig. 1** Dependency analysis of an English sentence

fact that the presence of a word is licensed by another word which is its governor. In a typed dependency analysis, grammatical labels are added to the dependencies to mark their grammatical relations, for example *subject* or *indirect object*.

Recently, there have been many published works on syntactic dependency analysis both for well-studied languages, such as English [10] or French [11], and for less-studied ones like Vietnamese [12, 17]. It has been shown that syntactic dependencies are useful for semantic dependency analysis, where semantic dependencies are understood in terms of predicates and their arguments, which aimed at natural language understanding applications.

Our motivation of using dependency analysis in question classification stems from the idea that a question can be classified more exactly if the meaning of the question can be determined at some level, even at a surface one. Recently, it has been shown that dependency structure of a sentence can be used to automatically learn its semantics [12]. Figure 1 shows the dependency analysis of a question in the UIUC dataset *"Who was the prophet of the Muslim people?"*, represented by using the Stanford Dependency scheme [13]. Intuitively, if we know that the subject of the question is *prophet* and its semantically associated word *people* – the head of the following prepositional phrase, then it is easier to classify the question. Similarly, consider the question *"What U.S. state lived under six flags?"* whose dependency structure is shown in Figure 2; knowledge of the subject *U.S. state* and the prepositional object of the question *flags* is useful for prediction of its category.

## 4 Experimental Results

### 4.1 Datasets

The experiments reported in this work were conducted on two datasets: the UIUC question classification dataset presented previously and a Vietnamese question classification dataset created by FPT Research[7] in an ongoing larger project whose aim is to develop an open domain question answering system. For the UIUC dataset, we use the standard split of training and test set as was used in previous studies:

---

[7] http://tech.fpt.com.vn/

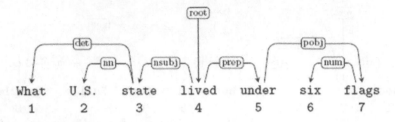

**Fig. 2** Dependency analysis of another English sentence

the training set is composed of 5,500 questions and the test set is composed of 500 questions from TREC 10 [14].

The FPT question classification dataset is currently composed of 1,000 questions containing 13 coarse categories which focus on questions about the FPT Corporation. The complete list of categories is: ACTION (action), CONC (concept), DESC (description), DTIME (datetime), EVT (event), HUM (human), LOC (location), NET (internet), NUM (number), ORG (organization), OTHER (other), THG (thing), YESNO (yes/no). Since each coarse-grained category can contain an overlapping set of fined-grained categories, in this study we report only the results of this dataset on the coarse-grained categories. Since this dataset has a relatively small size, we do not split it into fix training and test sets but use a 5-fold cross-validation technique for evaluation. The average accuracy is then computed.

## 4.2  Feature Sets

In this subsection, we present the binary feature sets that are used in the experiments, namely *question wh-word*, *unigrams* and *typed dependencies*.

Question wh-words

The wh-word feature is the question wh-word in a given question. For example, the wh-word of question *Which city has the oldest relationship as a sister city with Los Angeles?* is *what*. We have used 7 question wh-words for English, including *what*, *who*, *when*, *which*, *where*, *why* and *how*, and 12 question wh-words for Vietnamese, including *gì*, *ai*,..., *bao nhiêu*.

Unigrams

Unigrams are bag-of-word features which take into account all single words of a question. N-gram features in general and unigram features in particular play an important role in question classification since they provide word sense for question.[8]

---

[8] Note that stop words are also important for question classification, in contrast to other problems like text categorization.

**Table 3** Accuracy using individual feature sets on the UIUC dataset

| Feature Set | 6 class | | 50 class | |
|---|---|---|---|---|
| | NB | ME | NB | ME |
| wh-words | 46.20 | 46.20 | 46.80 | 46.80 |
| unigrams | 71.20 | 85.00 | 16.00 | 77.00 |

We have adopted the same unigram selection method as presented in [15] as they can give better performance while reducing the size of the feature space. In particular,

- All numbers are converted to the same feature, say, for example "*1960*", "*1972*", "*401*" are changed to "*1000*";
- Auxiliary verbs are changed to their infinitive forms. More precisely, all the tokens "*am*", "*is*", "*are*", "*was*", "*were*", and "*been*" are converted to "*be*"; all the tokens "*does*", "*did*", "*don*", "*done*" are converted to "*do*"; and "*has*", "*had*" are converted to "*have*".

Typed Dependencies

As discussed in the previous section, we propose to use typed dependencies extracted from the dependency analysis of a question as features. We are interested in subjects and prepositional objects which can be automatically extracted using *nsubj* and *pobj* dependencies. For instance, the typed dependency features of the question in Figure 1 are *nsubj=prophet* and *pobj=people* and those of the question in Figure 2 are *nsubj=state* and *pobj=flags*. To generate all typed dependencies of a sentence, we adopt the Stanford Parser [16] to parse all the questions of the UIUC dataset and use a modified version of vnLTAGParser [17] to parse all the questions of the FPT dataset. In addition, Vietnamese questions are tokenized using vnTokenizer, a highly accurate word segmentation tool of Vietnamese [18].

## 4.3 Results

We first compute the accuracy of NB and ME models using individual feature sets for 6 coarse and 50 fine classes on the UIUC dataset. All the models are trained on the training set of 5,500 questions and tested on the test set of 500 questions. The results are shown in Table 3.

Some interesting remarks can be drawn from these results. First, we can achieve a reasonable accuracy by using only wh-words. There is no difference in accuracy between NB models and ME models. They give 46.20% and 46.80% of accuracy on the coarse-grained and fine-grained categories respectively. Also, the wh-word feature set produces very compact models which have only 8 features. Second, unigrams give significantly higher accuracy over wh-words, nevertheless the improvement is far better in coarse-grained classification than in fine-grained classification.

**Table 4** Accuracy using incremental feature sets on the UIUC dataset

| Feature Set | 6 class | | 50 class | |
|---|---|---|---|---|
| | NB | ME | NB | ME |
| wh-words + deps | 49.00 | 59.20 | 17.80 | 61.60 |
| unigrams + deps | 66.80 | 87.60 | 11.40 | 78.40 |

**Table 5** Accuracy using different feature sets on the FPT dataset

| Feature Set | NB | ME |
|---|---|---|
| wh-words | 47.50 | 51.50 |
| unigrams | 57.60 | 78.40 |
| wh-words + deps | 59.70 | 69.60 |
| unigrams + deps | 58.80 | 80.50 |

Third, discriminative models are much more superior than generative models on the unigram feature set, which includes a large number of features.

Table 4 shows the accuracy of the models when typed dependency features are included. It is shown that typed dependencies are very informative features for question classification. Integration of these features helps improve largely the accuracy of the ME classifier on both of the feature sets, in coarse-grained or fine-grained classification, with a net average improvement of 7.5% for coarse-grained categories and 8.1% for fine-grained respectively. However, in general, typed dependencies make NB models perform worse than without using them. This fact can be explained by the assumption of independence of features in NB models, which would treat additional features as noisy information, especially when the domain dimension of the classification problem at hand is large, as in the case of using unigram features.

Table 5 reports the accuracy of the models on the FPT dataset using individual feature sets (first half) and incremental feature sets (second half).

It seems that the improvement when integrating typed dependencies is more significant for Vietnamese than for English. The ME models give an average improvement of 10.1% and that of naive Bayes models is about 5.5%. These results also demonstrate a net benefit of typed dependencies in the question classification for Vietnamese, a language of different family from English.

## 4.4   Discussion

The idea of using syntactic parsing to improve the performance of question classification is not new. For example, Nguyen et al. [7] proposed to use subtrees extracted from the constituency parses of questions in a boosting model with ME classifier and achieved the accuracy of 91.2% and 83.6% for coarse-grained and fine-grained categories respectively. However, this approach not only makes use of very rich feature space (the generated subtrees can be extremely numerous) but also employs a sophisticated technique for subtree selection for boosting. In contrast, we propose a compact yet effective feature set of typed dependencies extracted from dependency parses of questions to achieve a good improvement of classification accuracy.

Huang et al. [2] proposed to use *head words* and their hypernyms as important features to achieve the accuracy of 89.00% over the UIUC dataset. Head word is one single word specifying the object that the question seeks. For example, in the example question: *What is a group of turkeys called?*, the head word is *turkeys*. To obtain the head word, they have to go through a quite complicated procedure. First the Berkeley parser is used to get the constituency parse of a question; then a modified version of Collins head rules is used to extract the semantic head word; finally, a set of manually compiled tree patterns and regular expression patterns is used to re-assign the head word resulting from the previous step so as to fix it if necessary. For example, the initial head word extracted from the question *What is the proper name for a female walrus?* is *name* should be fixed to *walrus* since it matches a pre-compiled tree pattern. It is interesting to note that our approach of using typed dependencies is able to automatically identify the correct head word without having to go through the complex procedure as discussed above. In particular, the dependencies extracted from the questions above are *pobj=turkeys*; and *nsubj=name, pobj=walrus* respectively. Our approach is thus more general than that of Huang et al. and provides a principled, fully automatic way to identify semantic features that are informative for question classification.

In a recent paper, Tran [19] *et al.* has built a Vietnamese question classification system which employed some machine learning techniques. In addition to the bag-of-word features, they use a keyword corpus extracted from the Web. Their experiments are carried out on a translated version the UIUC dataset to Vietnamese, which are essentially different from our Vietnamese corpus. Therefore, the reported results are not comparable.

It is not our goal to argue against the use of manually compiled features in high-performance question classification. It has been demonstrated that head words and their hypernyms are useful in resolving classes of questions, and a question classifier should make use of such information where possible. We focus here on not only using automatically extracted features, but also trying to improve the performance of question classification for less-resource languages like Vietnamese where semantic information source like WordNet is not currently available. We see this investigation as only one part of the foundation for state-of-the-art question classification. We believe that the integration of typed dependencies, manually compiled regular expressions and word hypernym features will result in high-performance question classifiers for natural languages. This is a direction of our future work. Finally, semi-supervised learning for question classification has been shown to provide promising results [15]. We plan to investigate this approach with additional dependency features and a boosting technique to improve further the results. This line of research is another direction for our future work.

## 5 Conclusion

In contrast to many existing approaches for question classification which make use of very rich feature space or hand-crafted rules, we propose a compact yet effective feature set. In particular, we propose to use typed dependencies as semantic features. We have shown that by integrating only two simple dependencies of type nominal subject and prepositional object, one can improve the accuracy of question classification by over 8.0% using common statistical classifiers over two benchmark datasets, the UIUC dataset for English and a recently introduced FPT question dataset for Vietnamese. With unigram feature and typed dependency feature, one can obtain accuracy of 87.6% and 80.5% using maximum entropy for the UIUC and FPT question dataset respectively.

**Acknowledgment.** This work is partly supported by the FPT Technology Research Institute. We are grateful to anonymous reviewers for helpful comments on the draft.

## References

[1] Li, X., Roth, D.: Learning question classifiers: the role of semantic information. Natural Language Engineering 12(3), 229–249 (2006)

[2] Huang, Z., Thint, M., Qin, Z.: Question classification using head words and their hypernyms. In: Proceedings of the 2008 Conference on EMNLP, pp. 927–936 (2008)

[3] Li, X., Roth, D.: Learning question classifiers. In: Proceedings of COLING, pp. 556–569 (2002)

[4] Zhang, D., Lee, W.S.: Question classification using support vector machines. In: Proceedings of the 26th ACM SIGIR, pp. 26–32 (2003)

[5] Hacioglu, K., Ward, W.: Question classification with support vector machines and error correcting codes. In: Proceedings of NAACL/HLT, pp. 28–30 (2003)

[6] Krishnan, V., Das, S., Chakrabarti, S.: Enhanced answer type inference from questions using sequential models. In: Proceedings of EMNLP, pp. 315–322 (2005)

[7] Minh, N.L., Thanh, N.T., Akira, S.: Subtree mining for question classification problem. In: Proceedings of IJCAI, pp. 1695–1700 (2007)

[8] Ng, A.Y., Jordan, M.I.: On discriminative vs. generative classifiers: A comparison of logistic regression and naive Bayes. In: Advances in NIPS (2001)

[9] Andrew, G., Gao, J.: Scalable training of $l_1$-regularized log-linear models. In: ICML, pp. 33–40 (2007)

[10] Kübler, S., McDonald, R., Nivre, J.: Dependency Parsing. Morgan & Claypool Publishers (2009)

[11] Candito, M., Crabbé, B.: Improve generative statistical parsing with semi-supervised word clustering. In: Proceedings of the Eleventh International Conference on Parsing Technologies (IWPT) (2009)

[12] Liang, P., Jordan, M.I., Klein, D.: Learning dependency-based compositional semantics. Computational Linguistics 39(2), 389–446 (2012)

[13] de Marneffe, M.-C., MacCartney, B., Manning, C.D.: Generating typed dependency parses from phrase structure parses. In: Proceedings of LREC 2006, Italy, (2006)

[14] Voorhees, E.M.: Overview of TREC 2001 question answering track. In: Proceedings of The Tenth Text REtrieval Conference. NIST (2001)

[15] Thanh, N.T., Minh, N.L., Akira, S.: Using semi-supervised learning for question classification. Journal of Natural Language Processing 3(1), 112–130 (2008)

[16] Klein, D., Manning, C.D.: Accurate unlexicalized parsing. In: Proceedings of ACL, Sapporo, Japan, pp. 423–430 (July 2003)

[17] Le-Hong, P., Nguyen, T.M.H., Azim, R.: Vietnamese parsing with an automatically extracted tree-adjoining grammar. In: Proceedings of the IEEE RIVF, HCMC, Vietnam (2012)

[18] Hông Phuong, L.ê., Thi Minh Huyên, N., Roussanaly, A., Vinh, H.T.: A hybrid approach to word segmentation of Vietnamese texts. In: Martín-Vide, C., Otto, F., Fernau, H. (eds.) LATA 2008. LNCS, vol. 5196, pp. 240–249. Springer, Heidelberg (2008)

[19] Tran, D.H., Chu, C.X., Pham, S.B., Nguyen, M.L.: Learning based approaches for Vietnamese question classification using keywords extraction from the web. In: Proceedings of the IJCNLP, Nagoya, Japan, pp. 740–746 (2013)

# Unequal Clustering Formation Based on Bat Algorithm for Wireless Sensor Networks

Trong-The Nguyen, Chin-Shiuh Shieh, Mong-Fong Horng,
Truong-Giang Ngo, and Thi-Kien Dao

**Abstract.** Prolonging the lifetime of the network is an important issue in the design and deployment of wireless sensor networks (WSN). One of the crucial factors to prolong lifetime of WSNs is to reduce energy consumption. In this study, Bat algorithm (BA) is used to find out an optimal cluster formation trying to minimize the total communication distance. Taking into account the hot spot problem in multi-hop WSNs, the communication distance is modeled by bat's loudness parameter in our scheme. The communication distance and energy consumption of each node in the cluster can then be optimized with the bat algorithm. The experimental results show that this approach achieves 3% improvement of convergence and accuracy in comparison with Particle swarm optimization.

**Keywords:** Bat Algorithm, Wireless Sensor Network, Unequal Clustering Formation.

## 1 Introduction

Grouping sensor nodes into clusters has been widely paid attention by the research community in order to achieve the network scalability objective. Every cluster would have a leader, often referred to as the cluster-head (CH). Although many clustering algorithms have been proposed in the literature for ad-hoc networks [1],[18],[19], the objective was mainly to generate stable clusters in environments with mobile nodes. Many of such techniques care mostly about node reachability and route stability, without much concern about critical design goals of WSNs such as

Chin-Shiuh Shieh · Mong-Fong Horng · Thi-Kien Dao
Department of Electronics Engineering,
National Kaohsiung University of Applied Sciences, Taiwan

Trong-The Nguyen · Truong-Giang Ngo
Faculty of Information Technology, Haiphong Private University, Vietnam
e-mail: {vnthe,giangnt}@hpu.edu.vn

© Springer International Publishing Switzerland 2015                          667
V.-H. Nguyen et al. (eds.), *Knowledge and Systems Engineering,*
Advances in Intelligent Systems and Computing 326, DOI: 10.1007/978-3-319-11680-8_53

network longevity and coverage. Recently, a number of clustering algorithms have been specifically designed for WSNs [18],[5],[22]. These proposed clustering techniques widely vary depending on the node deployment and bootstrapping schemes, the pursued network architecture, the characteristics of the CH nodes and the network operation model. A cluster-head may be elected by the sensors in a cluster or preassigned by the network designer. A cluster-head may also be just one of the sensors or a node that is richer in resources. The cluster membership may be fixed or variable. CHs may form a second tier network or may just ship the data to interested parties, e.g. a base-station or a command center. Besides supporting network scalability, there are numerous advantages of clustering. It can localize the route set up within the cluster and thus reduce the size of the routing table stored at the individual node [12]. Clustering can also conserve communication bandwidth since it limits the scope of inter-cluster interactions to CHs and avoids redundant exchange of messages among sensor nodes [21]. Moreover, clustering can stabilize the network topology at the level of sensors and thus cuts on topology maintenance overhead [4].

Sensors would care only for connecting with their CHs and would not be affected by changes at the level of inter-CH tier [10]. The CH can also implement optimized management strategies to further enhance the network operation and prolong the battery life of the individual sensors and the network lifetime [2]. A CH can schedule activities in the cluster so that nodes can switch to the low-power sleep mode most of the time and reduce the rate of energy consumption. Sensors can be engaged in a round-robin order and the time for their transmission and reception can be determined so that the sensors reties are avoided, redundancy in coverage can be limited and medium access collision is prevented [3].

Clustering can greatly reduce communication cost of the nodes because they only need to send data to the nearest cluster-head. However, CH expends more energy than ordinary nodes communicating with the sink. A rotation of CHs in each round of communication is identified by an election process influenced by randomness [9]. A centralized solution assuming that the position of all nodes are known in advance and powerful computer perform the computation and inform all the nodes about their respective cluster-heads [8]. Particle Swarm Optimization (PSO) applied to identify in ad hoc sensor networks [17]. However, the main aim was to reduce an intra-cluster distance by completely ignoring the distance to the sink. The optimization for distances in clustering formation using an evolutionary computing as PSO [6]. However, these methods just applied in uniformly distributed of sensors, not considering about unequal clustering, this leads to the hot spot in wireless sensor networks.

In this paper, the minimized total communication distances in unequal clustering WSNs by optimization based on Bat algorithm is proposed. A distance in clustering criterion for optimizing is adopted based on loudness parameter of Bat algorithm for considering to hot spot problem in multi-hop WSNs.

The rest of the paper is organized as follows: Section 2 reviews certain works related to this study. Section 3, our approach is explained in great details. The simulation results are given in Section 4. Finally, Section 5 concludes the paper.

## 2 Related Works

In 2010, Xin-SheYang proposed a new optimization algorithm, namely, Bat Algorithm (BA), based on swarm intelligence and the inspiration form observing the bats [20]. BA simulates parts of the echolocation characteristics of the micro-bat in the simplicity way. It is potentially more powerful than particle swarm optimization and genetic algorithms as well as Harmony Search. The primary reason is that BA uses a good combination of major advantages of these algorithms in some way. Moreover, PSO and Harmony Search are the special cases of the Bat Algorithm under appropriate simplifications. Three major characteristics of the micro-bat are employed to construct the basic structure of BA. The used approximate and the idealized rules in Xin-SheYang's method are listed as follows:

. All bats utilize the echolocation to detect their prey, but not all species of the bat do the same thing. However, the micro-bat, one of species of the bat is a famous example of extensively using the echolocation. Hence, the first characteristic is the echolocation behavior. The second characteristic is the frequency. The sending frequency of micro-bat is fixed with a variable wavelength $\lambda$ and the loudness $A_0$ to search for prey.
. Bats fly randomly with velocity $v_i$ at position $x_i$. They can adjust the wavelength (or frequency) of their emitted pulses and adjust the rate of pulse emission $r \in [0, 1]$, depending on the proximity of their target;
. There are many ways to adjust the loudness. For simplicity, the loudness is assumed to be varied from a positive large $A_0$ to a minimum constant value, which is denoted by $A_{min}$.

In Yang's method, the movement of the virtual bat is simulated by Eq.(1) - Eq.(3):

$$f_i = f_{min} + (f_{max} - f_{min}) * \beta \tag{1}$$

$$v_i^t = v_i^{t-1} + (x_i^{t-1} - x_{best}) * f_i \tag{2}$$

$$x_i^t = x_i^{t-1} + v_i^t \tag{3}$$

where $f$ is the frequency used by the bat seeking for its prey, $f_{min}$ and $f_{max}$, represent the minimum and maximum value, respectively. $x_i$ denotes the location of the $i$-th bat in the solution space, $v_i$ represents the velocity of the bat, $t$ indicates the current iteration, $\beta$ is a random vector, which is drawn from a uniform distribution, and $\beta \in [0, 1]$, and $x_{best}$ indicates the global near best solution found so far over the whole population. In addition, the rate of the pulse emission from the bat is also taken to be one of the roles in the process. The micro-bat emits the echo and adjusts the wavelength depending on the proximity of their target. The pulse emission rate is denoted by the symbol $r_i$, and $r_i \in [0, 1]$, where the suffix $i$ indicates the $i$-th bat. In every iteration, a random number is generated and is compared with $r_i$. If the random number is greater than $r_i$, a local search strategy, namely, random walk, is detonated. A new solution for the bat is generated by Eq.(4):

$$x_{new} = x_{old} + \varepsilon A^t \tag{4}$$

where $\varepsilon$ is a random number and $\varepsilon \in [-1,1]$, and $A^t$ represents the average loudness of all bats at the current time step. After updating the positions of the bats, the loudness $A_i$ and the pulse emission rate $r_i$ are also updated only when the global near best solution is updated and the random generated number is smaller than $A_i$. The update of $A_i$ and $r_i$ are operated by Eq.(5) and Eq.(6):

$$A_i^{t+1} = \alpha A_i^t \tag{5}$$

$$r_i^{t+1} = r_i^0 [1 - e^{\gamma t}] \tag{6}$$

where $\alpha$ and $\gamma$ are constants. In Yang's experiments, $\alpha = \gamma = 0.9$ is used for the simplicity. The process of BA is depicted as follows:

- *Step* 1. Initialize the bat population, the pulse rates, the loudness, and define the pulse frequency.
- *Step* 2. Update the velocities to update the location of the bats, and decide whether detonate the random walk process.
- *Step* 3. Rank the bats according to their fitness value, find the current near best solution found so far, and then update the loudness and the emission rate.
- *Step* 4. Check the termination condition to decide whether go back to step 2 or end the process and output the result.

Moreover, an others related works with this paper is following: Particle Swarm Optimization (PSO) is an evolutionary computing technique based on principle such as bird flocking [14]. In PSO a set of potential solutions are called particles that are initialized randomly. Each particle will have a fitness value, which will be evaluated by the fitness function to be optimized in each generation. Each particle knows its best position *pbest* and the best position so far among the entire group of particles *gbest*. The particle will have velocities, which direct the flying of the particle. In each generation the velocity and the position of the particle will be updated. The velocity and the position update equations are given below as (8) and (8) respectively.

$$x_i^{j+1} = x_i^j + v_i^{j+1} \tag{7}$$

where $x_i^j$ is position of the particle i at iteration $j$, $x_i^{j+1}$ is position of the particle $i$ at iteration $j+1$, and $v_i^{j+1}$ is given as:

$$v_i^{j+1} = v_i^j + c_1 \times r_1 \times (pbest_i - x_i^j) + c_2 \times_2 (gbest - x_i^j) \tag{8}$$

where, $v_i^j$ is velocity of particle $i$ at iteration $j$, $v_i^{j+1}$ is velocity of particle $i$ at iteration $j+1$, $w$ inertia weight, $c_{1,2}$ are acceleration coefficients, $pbest_i$ is *pbest* of particle $i$, *gbest* is best of the group. PSO-Time Varying Inertia Weight (PSO-TVIW) [15] is the basic PSO algorithm with inertia weight varying with time from 0.9 to 0.4 and the acceleration coefficient is set to 2.

$$v_i^{j+1} = w \times v_i^j + c_1 \times r_1 \times (pbest_i - x_i^j) + c_2 \times r_2 \times (gbest - x_i^j) \tag{9}$$

PSO-Time varying acceleration coefficients (PSO-TVAC) [13], applied the time varying acceleration coefficient (TVAC); such as the $c_1$ varies from 2.5 to 0.5 and the $c_2$ varies from 0.5 to 2.5.

## 3 The Approach of Clustering Formation Based on BA

The energy dissipation to transmit and receive the data from a node is measured by using the simple radio model. The energy dissipation for transmitting $l$ bits to $d$ distance is given as equation (10).

$$E_{Tx(l,d)} = E_{Tx-elec}(l) + E_{Tx-amp}(l,d) = \begin{cases} l \times E_{elec} + l \times \varepsilon_{fs} \times d^2 & \text{if } d \leqslant d_0 \\ l \times E_{elec} + l \times \varepsilon_{mp} \times d^4 & \text{if } d > d_0 \end{cases} \quad (10)$$

where $E_{Tx(l,d)}$ is the transmitting energy dissipation. There is the free space and multipath fading model have been considered. If the distance $d$ is less than or equal to a threshold $d_0$, the free space model ($d^2$ power loss) is used; otherwise, the multi path model ($d^4$ power loss) is used. It means that for short distance transmission, such as intra-cluster communication, the energy consumption by an amplified transmission is proportional to $d^2$ and for long distance transmission, such as inter-cluster communication, the energy consumption is proportional to $d^4$. The distance $d$ is separated by the threshold transmission distance $d_0$ given as being calculated equation (11).

$$d_0 = \sqrt{\frac{\varepsilon_{fs}}{\varepsilon_{mp}}} \quad (11)$$

The energy dissipation for receiving this message, the radio expects:

$$E_{Rx}(l) = E_{Rx-elec}(l) = l \times E_{elec} \quad (12)$$

where $E_{elec}$ is the energy consumed per bit to run the circuitry of the transmitter and receiver $\varepsilon_{fs}$ and $\varepsilon_{mp}$ are the power loss of free space and multipath models, respectively, which depend on a chosen acceptable bit-error rate. The electronics energy $E_{elec}$ depends on factors such as the digital coding, modulation, filtering, and spreading of the signal whereas the distance to the receiver and the acceptable bit-error rate. The cluster formation algorithm was created to ensure that the expected number of clusters per round is $k$, a system parameter. The optimal value of the expected number of clusters per round $k$ in cluster formation can be analytically determined by using the computation and communication energy models. Assume that there are $N$ nodes distributed in an $M \times M$ region. If there are $k$ clusters, each cluster has $n_j$ sensors, where $l < j$. Considering the intra-cluster, $d_{ij}$ is distance between normal node to the cluster head ($CH_j$), and $D_j$ is distance between the cluster head to the sink or distance of hops to hop or to base station BS (sink) in heterogeneous sensor network. There are on average $N/k$ nodes per cluster (one cluster head and $(N/k) - 1$ non-cluster head nodes). Each cluster head dissipates energy receiving signals from the nodes, aggregating the signals, and transmitting the aggregate signal to the BS. Since the BS is far from the nodes, presumably the energy dissipation follows the multi path model. Therefore, the energy dissipated in the cluster head node during a single frame is [9],[16].

$$E_{CH} = l \times E_{elec} \times \left(\frac{N}{k} - 1\right) + l \times E_{DA} \times \frac{N}{k} + l \times E_{elec} + l\varepsilon_{mp} \times D_j^2 \qquad (13)$$

where $E_{CH}$ energy dissipated in the cluster head node during a single frame, $E_{DA}$ energy dissipated of aggregating the signals, $l$ is the number of bits in each data message, $D_j^2$ is the distance from the cluster head to hop node or to the BS, data aggregation is assumed perfectly. Each non-cluster-head node only needs to transmit its data to the cluster head once during a frame. Presumably the distance to the cluster head is small, so the energy dissipation follows the free-space model $d^2$ power loss). Thus, the energy used in each non-cluster head node is:

$$E_{non-CH} = l \times E_{elec} + l \times \varepsilon_{fs} \times d_{toCH}^2 \qquad (14)$$

where $d_{toCH}$ is the distance from the node to the cluster-head. The area occupied by each cluster is approximately $M^2/k$. In general, this is an arbitrary-shaped region with a node distribution $\rho(x,y)$. The expected squared distance from the nodes to the cluster head (assumed to be at the center of mass of the cluster) is given by

$$E_{d_{toCH}^2} = \int\int (x^2 + y^2)\rho(x,y)dxdy = \int\int r^2\rho(r,\theta)drd\theta \qquad (15)$$

The energy dissipated in a cluster during the frame is

$$E_{cluster} = E_{CH} + \left(\frac{N}{k} - 1\right) \times E_{non+CH} = \varepsilon_{mp} \times \sum_{j=1}^{k}\sum_{i=1}^{n_j}\left(d_{ij}^2 + \frac{D_j^2}{n_j}\right) \qquad (16)$$

And the total energy for the frame is

$$E_{frame} = k \times E_{cluster} \qquad (17)$$

Energy consumption of a wireless sensor node transmitting and receiving data from another node at a distance d can be divided into two main components: Energy used to transmit, receive and amplify data $E_{frame}$ and energy used for processing data $E_p$, mainly by the microcontroller. The total energy loss $E$ of a sensor is equal $E_{frame}$ plus $E_p$.

$$E_{total} = E_p + E_{frame} \qquad (18)$$

It is clear that the only component that can be optimized independent from parameters related to the microcontroller and the supply voltage used. Consequently, $E_{frame}$ was used as the energy loss based on the distance for formation of clusters, named as $E_{distance}$ for proof of referencing [7].

$$E_{distance} = \sum_{j=1}^{k}\sum_{i=1}^{n_j}\left(d_{ij}^2 + \frac{D_j^2}{n_j}\right) \qquad (19)$$

The hot spots problem in multi-hop WSNs makes the cluster-heads near the base station to die out earlier than the cluster-heads farther away from the base station. Because the traffic relay load of the cluster-heads near the base station is heavier than those cluster-heads farther away from the base station. In order to avoid this problem, the network should be partitioned into clusters with different sizes known as "unequal clustering method" [5],[11]. The clusters close to the base station are smaller than those clusters far from the base station.

The loudness $A_i^0$ of bats as seen in Eq.(4) and Eq.(5) is assigned to correspond to the radius of the cluster size changing. The time varying loudness is mathematically represented as follows

$$A_i^0 = (A_{max} - A_{min}) \times \frac{(MaxIteration - iter)}{MaxIteration} + A_{min} \qquad (20)$$

where $Max - Iteration$ is number of iteration, the maximum iteration allowed, $iter$ is the current iteration number, and $A_{max}, A_{min}$ are constant set to 0.5 and 0.25 respectively.

## 4 Optimization for Cluster Formation in WSNs

This section presents simulation results and compares the proposed of Bat algorithm for wireless sensor network (BA-WSN) with that applied PSO in cases of PSO with time varying inertia weight (PSO-TVIW) and PSO with time varying acceleration co-efficient (PSO-TVAC), in terms of solution quality and in the number of function evaluations taken. Eq.(19) can be optimized for mining the energy dissipation in a sensor network. Here is not only considered the intra-cluster distance, but also the distance between the cluster-head and base station $BS$. Simulation of network with 100-node are distributed in a 2-Dimensional problem space [0:100,0:100]. The impact of base station location on the fitness value of the algorithms is observed. In one set of simulations is considered the base station to be located remotely at random.

In another set of simulations is considered the base station to be located at (50, 50), i.e., at the center of the network. The simulation the nodes are clustered for taking into consideration 2 cases: the first is without consideration the hot spot problem and the second, the loudness $A_i$ parameter of Bat algorithm is considered for adjusting in hot spot problem in WSNs. All experiments are averaged over different random seeds with 25 runs. The optimization goal for fitness functions is to minimize the outcome of the distances between normal nodes to cluster-head, CH to Cluster-head, and CHs to BS.

### 4.1 Without Consideration the Hot Spot Problem

Sensors are clustered using entirely distance based on equation (21) with $k$ a given number of clusters. The fitness function for this case is as following:

$$F_{itness} = min(\sum_{j=1}^{k} \sum_{i=1}^{n_j} (d_{ij}^2 + \frac{D_j^2}{n_j})) \qquad (21)$$

where $\sum_{j=1}^{k} n_j + k = N$ is the number of nodes in a networks. The simulation initial parameters setting for Bat algorithm for wireless sensor network (BA-WSN) are the initial loudness $A_i^0$=0.25, pulse rate $r_i^0$=0.5 the total population size $N = 20$ and the dimension of the solution space $M = 30$, frequency minimum $fmin = the lowest of initial range function$ and frequency minimum

$fmax = thehighestofinitialrangefunction$, in this case is 0 to 100 respectively (Reader may reference further the initial setting in [20]. The parameters setting for PSO are the same population size $N = 20$ and the dimension of the solution space $M = 30$. The initial inertia weight $W$ is varying with time from 0.9 to 0.4 and the acceleration coefficient is set to 2 (the initial setting in [15]). The coefficient of learning factors $c_1$ varies from 2.5 to 0.5 and the $c_2$ varies from 0.5 to 2.5 (the initial setting in [13]). Here the cognitive component is reduced and social component is increased by changing $c_1$ and $c_2$ in PSO. Fitness function contains equation (21) the full iterations of 1000 is repeated by different random seeds with 25 runs. The final result is obtained by taking the average of the outcomes from all runs. The results are compared with that running in PSO as PSO-TVIW and PSO-TVAC.

Figure 1 compares the proposed of clustering formation base on Bat algorithm for WSN with that an applied of PSO time varying inertia weight (PSO-TVIW). It is clearly seen that the proposed of BA-WSN line (the star solid red line) is better than that in PSO-TVIW (the doted blue line) in terms of convergence. The average of fitness values of 25 runs in BA-WSN is 190.51 and that in PSO-TVIW is 194.91 respectively. It is reaching nearly at 2% of BA-WSN for improvement of convergence and accuracy in compare with PSO-TVIW.

Figure 2 illustrates that the proposed of clustering formation base on Bat algorithm for WSN in comparison with that an applied of PSO with time varying acceleration co-efficients (PSO-TVAC). It is clearly seen that the proposed line of BA-WSN ( the star solid red line) is better than that in PSO-TVAC (the doted blue line) in terms of convergence. The average of fitness values of 25 runs is 190.031 for BA-WSN and 196.92 for PSO-TVAC respectively. It is reaching nearly at 3% of BA-WSN for improvement of convergence and accuracy in compare with PSO-TVAC.

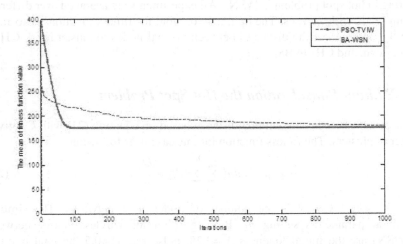

**Fig. 1** The new proposed of BA-WSN compares with PSO-TVIW for average fitness value of 25 runs

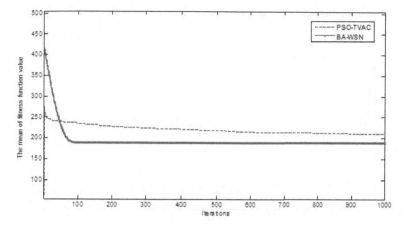

**Fig. 2** Comparison of two approaches: BA-WSN and PSO-TVAC for clustering formation optimization by average fitness values of 25 runs

## 4.2 Consideration the Hot Spot Problem in WSNs

To balance the energy consumption of sensor nodes, the clustering protocol is used. If the location of the base station is not taken into consideration, the hot spots problem in multi-hop WSNs will be happened. The cluster-heads near the base station die out earlier, because they will be subject to heavier relay traffic load than those cluster-heads farther away from the base station. In order to avoid this problem, the network is partitioned into clusters with different sizes. The clusters close to the base station are smaller than those clusters far from the base station. Applying the parameter loudness of Bat algorithm $A_i^0$ for varying with time from 0.5 to 0.25 is set. The time varying loudness is mathematically represented in equation (20) for Bat algorithm with consideration the hot spot problem in WSN (BA-WSNHS).

Figure 3 compares two lines of average fitness value in 25 trials of BA-WSN (doted green line) the normal distribution and BA-WSNHS (solid red line) with consideration the hot spot problem in WSN. It is easy to see that the BA-WSNHS is more con-vergence the BA-WSN.

Figure 4 compares four bars of four different methods for clustering formation namely BA-WSN, BA-WSNHS, PSO-TVIW and PSO-TVAC in average fitness values in 25 trials. It is easy to see that the BA-WSNHS is smallest of average fitness values.

## 5  Conclusion

In this paper, the results of the evaluated performance of optimization for the total communication distance based on Bat algorithm were presented. These results of experiment compared with methods of PSO-TVIW and PSO-TVAC for cluster optimization show that the proposed approach was more convergence and accurate than the methods of PSO-TVIW and PSO-TVAC up to 3%. The distances between

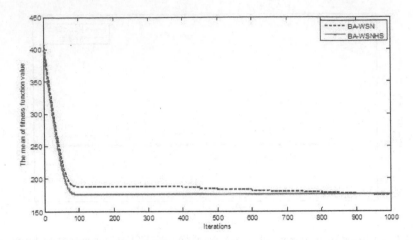

**Fig. 3** Comparison of two cases of optimization: The normal distribution and with consideration hot spot problem in WSNs

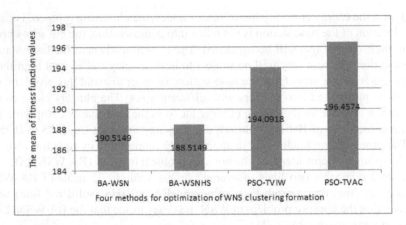

**Fig. 4** The mean values of fitness functions of four approaches in being executed with 25 runs. Each run is repeated 1000 iterations.

hop to hop and sink in consideration hot pot problem was modified for suitable with the loudness of Bats in Bat algorithm (BA-WSNHS). Thus, the BA-WSNHS is the smallest bar of average fitness values. It means BA-WSHS is better optimization than other methods because of it is the most convergence in this work.

**Acknowledgement.** This work is partly supported by the Ministry of Science and Technology, Taiwan, ROC, under the grant numbers NSC 102-2218-E-151-005- and NSC 102-2221-E-151-004-.

# References

[1] Akyildiz, I., Su, W., Sankarasubramaniam, Y., Cayirci, E.: A survey on sensor networks. IEEE Communications Magazine 40(8), 102–114 (2002)

[2] El-Aaasser, M., Ashour, M.: Energy aware classification for wireless sensor networks routing. In: 2013 15th International Conference on Advanced Communication Technology (ICACT), pp. 66–71 (January 2013)

[3] Gao, C., Kivela, I., Tan, X., Hakala, I.: A transmission scheduling for data-gathering wireless sensor networks. In: 2012 9th International Conference on Ubiquitous Intelligence Computing and 9th International Conference on Autonomic Trusted Computing (UIC/ATC), pp. 292–297 (September 2012)

[4] Guan, Q., Feng, S., Ma, Y.: A network topology clustering algorithm for service identification. In: 2012 International Conference on Computer Science Service System (CSSS), pp. 1583–1586 (August 2012)

[5] Gupta, I., Riordan, D., Sampalli, S.: Cluster-head election using fuzzy logic for wireless sensor networks. In: Proceedings of the 3rd Annual Communication Networks and Services Research Conference 2005, pp. 255–260 (May 2005)

[6] Guru, S., Halgamuge, S., Fernando, S.: Particle swarm optimisers for cluster formation in wireless sensor networks. In: Proceedings of the 2005 International Conference on Intelligent Sensors, Sensor Networks and Information Processing Conference, pp. 319–324 (December 2005)

[7] Guru, S.M., Hsu, A., Halgamuge, S., Fernando, S.: An extended growing self-organizing map for selection of clusters in sensor networks. International Journal of Distributed Sensor Networks 2(1), 227–243 (2005)

[8] Heinzelman, W., Chandrakasan, A., Balakrishnan, H.: An application-specific protocol architecture for wireless microsensor networks. IEEE Transactions on Wireless Communications 1(4), 660–670 (2002)

[9] Heinzelman, W., Chandrakasan, A., Balakrishnan, H.: Energy-efficient communication protocol for wireless microsensor networks. In: Proceedings of the 33rd Annual Hawaii International Conference on System Sciences 2000, vol. 2, p. 10 (January 2000)

[10] Hou, Y., Shi, Y., Sherali, H., Midkiff, S.: On energy provisioning and relay node placement for wireless sensor networks. IEEE Transactions on Wireless Communications 4(5), 2579–2590 (2005)

[11] Kim, J.M., Park, S.H., Han, Y.J., Chung, T.M.: Chef: Cluster head election mechanism using fuzzy logic in wireless sensor networks. In: 10th International Conference on Advanced Communication Technology, ICACT 2008, vol. 1, pp. 654–659 (February 2008)

[12] Lambrou, T., Panayiotou, C.: A survey on routing techniques supporting mobility in sensor networks. In: 5th International Conference on Mobile Ad-hoc and Sensor Networks, MSN 2009, pp. 78–85 (December 2009)

[13] Ratnaweera, A., Halgamuge, S., Watson, H.: Self-organizing hierarchical particle swarm optimizer with time-varying acceleration coefficients. IEEE Transactions on Evolutionary Computation 8(3), 240–255 (2004)

[14] Shi, Y., Eberhart, R.: A modified particle swarm optimizer. In: The 1998 IEEE International Conference on Evolutionary Computation Proceedings, IEEE World Congress on Computational Intelligence, pp. 69–73 (May 1998)

[15] Shi, Y., Eberhart, R.: Empirical study of particle swarm optimization. In: Proceedings of the 1999 Congress on Evolutionary Computation, CEC 1999, vol. 3, p. 1950 (1999)

[16] Su, C.C., Chang, K.M., Kuo, Y.H., Horng, M.F.: The new intrusion prevention and detection approaches for clustering-based sensor networks (wireless sensor networks). In: 2005 IEEE Wireless Communications and Networking Conference, vol. 4, pp. 1927–1932 (March 2005)

[17] Tillett, J., Rao, R., Sahin, F.: Cluster-head identification in ad hoc sensor networks using particle swarm optimization. In: 2002 IEEE International Conference on Personal Wireless Communications, pp. 201–205 (December 2002)

[18] Xu, R., Wunsch, D.I.: Survey of clustering algorithms. IEEE Transactions on Neural Networks 16(3), 645–678 (2005)

[19] Yan, J.F., Liu, Y.L.: Improved leach routing protocol for large scale wireless sensor networks routing. In: 2011 International Conference on Electronics, Communications and Control (ICECC), pp. 3754–3757 (September 2011)

[20] Yang, X.-S.: A new metaheuristic bat-inspired algorithm. In: González, J.R., Pelta, D.A., Cruz, C., Terrazas, G., Krasnogor, N. (eds.) NICSO 2010. SCI, vol. 284, pp. 65–74. Springer, Heidelberg (2010)

[21] Younis, M., Youssef, M., Arisha, K.: Energy-aware routing in cluster-based sensor networks. In: Proceedings of the 10th IEEE International Symposium on Modeling, Analysis and Simulation of Computer and Telecommunications Systems, MASCOTS 2002, pp. 129–136 (2002)

[22] Younis, O., Fahmy, S.: Heed: a hybrid, energy-efficient, distributed clustering approach for ad hoc sensor networks. IEEE Transactions on Mobile Computing 3(4), 366–379 (2004)

# A New Approach to Multi-variable Fuzzy Forecasting Using Picture Fuzzy Clustering and Picture Fuzzy Rule Interpolation Method

Pham Huy Thong and Le Hoang Son

**Abstract.** In this paper, a new approach to multi-variable fuzzy forecasting using picture fuzzy clustering and picture fuzzy rule interpolation techniques is proposed. Firstly, we partition dataset into clusters using picture fuzzy clustering algorithm. Secondly, we construct picture fuzzy rules based on given clusters. Finally, we determine the predicted outputs based on the picture fuzzy rule interpolation scheme. Our proposed approach is applied to forecast the Taiwan Stock Exchange Capitalization Weighted Stock Index (TAIEX) data. The experimental results indicate that our method predicts better forecasting results than some relevant ones.

**Keywords:** Fuzzy forecasting, Picture Fuzzy rule interpolation method, Multi-variable fuzzy forecasting, Picture fuzzy clustering, Stock prediction.

## 1 Introduction

Improving forecasting accuracy, especially in time series forecasting, plays an important role in our daily life, typically in weather forecasting, economic growth rate forecasting, stock market predicting, etc. Fuzzy logic which is based on fuzzy set and was first introduced by Zadeh [30] is an effective choice to deal with this problems [30, 31]. Since then, there have been many approaches to generate fuzzy rules from fuzzy logic proposed to solve the forecasting problems. Most of proposed methods were based on fuzzy clustering and neural network [2, 6, 7, 10, 12, 22, 23, 24, 27, 28, 29]. In [6], a method to predict the newspaper demand based on fuzzy clustering and fuzzy rules was proposed by Cardoso and Gomide. [7], another method to predict automobile fuel consumptions and stock prices

Pham Huy Thong · Le Hoang Son
VNU University of Science, Vietnam National University
334 Nguyen Trai, Thanh Xuan, Hanoi, Vietnam
e-mail: {thongph, sonlh}@vnu.edu.vn

© Springer International Publishing Switzerland 2015                    679
V.-H. Nguyen et al. (eds.), *Knowledge and Systems Engineering,*
Advances in Intelligent Systems and Computing 326, DOI: 10.1007/978-3-319-11680-8_54

based on the fuzzy system modeling using an improved fuzzy clustering algorithm was introduced by Celikyilmaz and Turksen. Chang and Chen [10, 12] presented some methods for temperature and stock predictions based on fuzzy clustering and fuzzy rule interpolation techniques. Some prediction methods using the neural network were introduced in [2, 10, 22, 23, 24, 27, 28, 29]. Armano et al [2] presented a hybrid genetic-neural architecture for stock indexes forecasting. Yu and Huang [23] proposed a multivariate heuristic model to forecast fuzzy time series. They also applied neural networks to fuzzy time series forecasting and proposed bivariate models to improve the quality of forecasting [22, 28, 29]. Another way to use neural network for time series forecasting was introduced by Khashei and Bijari [24]. They proposed a novel hybridization of artificial neural networks and the ARIMA model for time series forecasting. Lately, Wei et al [27] introduced a hybrid ANFIS based on n-period moving average model to forecast TAIEX stock. Besides that, there were some other methods to solve time series problem. Chen et al [14, 15, 16] presented a series methods using fuzzy time series, automatically generated weights of multiple factors, particle swarm optimization techniques and support vector machine. However, their methods were mostly based on fuzzy sets which might be underestimated in reality, because there may be some hesitation [20]. Although the existing forecasting methods are useful and have realistic applications, they need improving to increase predicting accuracy rates.

In this paper, we present a new approach to multi-variable fuzzy forecasting using the combination of picture fuzzy clustering and picture fuzzy rule interpolation method. Our approach is an expansion of Chen's method [12] and it is applied to forecast the Taiwan Stock Exchange Capitalization Weighted Stock Index (TAIEX) [26] for experiment. Firstly, we cluster the dataset using picture fuzzy clustering algorithm. Secondly, we construct picture fuzzy rules based on given clusters. Finally, based on the picture fuzzy rules, we predict the output results. The proposed method will be validated on the TAIEX dataset in terms of accuracy.

The rest of the paper is organized as follows. In section 2, we present a new picture fuzzy clustering algorithm and an innovated picture fuzzy rule interpolation method. In section 3, we present a solution of combining the two methods proposed in section 2 for solving multi-variable fuzzy forecasting. In section 4, we apply our solution to predict the TAIEX stock price and make a comparision with the existing methods. Finally, the conclusions are covered in Section 5.

## 2   Proposed Works

In this section, we introduce the picture fuzzy sets [18], the picture fuzzy clustering and the picture fuzzy rule interpolation scheme.

## 2.1  Picture Fuzzy Clustering

A *Picture Fuzzy Set* (PFS) [18] in a non-empty set $X$ is,

$$A = \{\langle x, u_A(x), \eta_A(x), \gamma_A(x)\rangle | \forall x \in X|\} \tag{1}$$

where $u_A(x)$ is the positive degree of each element $x \in X$, $\eta_A(x)$ is the neutral degree and $\gamma_A(x)$ is the negative degree satisfying the constraints,

$$u_A(x), \eta_A(x), \gamma_A(x) \in [0,1], \forall x \in X \tag{2}$$

$$0 \le u_A(x) + \eta_A(x) + \gamma_A(x) \le 1, \forall x \in X \tag{3}$$

The refusal degree of an element is calculated as,

$$\xi_A(x) = 1 - (u_A(x) + \eta_A(x) + \gamma_A(x)), \forall x \in X \tag{4}$$

Basically, picture fuzzy sets based models may be adequate in situations when we face human opinions involving more answers of type: yes, obtain, no, refusal [18]. In cases $\gamma_A(x) = 0$, PFS returns to the traditional Intuitionistic Fuzzy Sets (IFS), which has been introduced by Atanassov [3, 5] as an extension of original fuzzy sets. Obviously, we recognize that PFS is an extension of IFS where the neutral degree is appended to the definition.

Based on PFS, we propose a new Fuzzy Clustering method on Picture Fuzzy Sets (FC-PFS). Suppose we have a dataset $X$ consisting of $N$ data points in $d$ dimensions. Let us divide the dataset into $C$ groups satisfying the objective function below.

$$J = \sum_{k=1}^{N} \sum_{j=1}^{C} (u_{kj}(2 - \xi_{kj}))^m \|X_k - V_j\|^2 + \sum_{k=1}^{N} \sum_{j=1}^{C} \eta_{kj}(log\eta_{kj} + \xi_{kj}) \to min, \tag{5}$$

Where $X_k$ is the $k^{th}$ element on $X$, $V_j$ is the center of $j^{th}$ cluster, $m$ is the fuzzifier parameter. Some constraints are defined as follows,

$$u_{kj}, \eta_{kj}, \xi_{kj} \in [0,1], \tag{6}$$

$$u_{kj} + \eta_{kj} + \xi_{kj} \le 1, \tag{7}$$

$$\sum_{j=1}^{C} (u_{kj}(2 - \xi_{kj})) = 1, \tag{8}$$

$$\sum_{j=1}^{C} (\eta_{kj} + \frac{\xi_{kj}}{C}) = 1, (k = \overline{1,N}, j = \overline{1,C}). \tag{9}$$

FC-PFS is based on picture fuzzy sets, which is expansion of IFS. The advantages of FC-PFS are the clustering proccess would achive more details with four values: degree of positive, degree of negative, degree of neutral and degree of refusal. The more detail the clustering algorithm is, the more accuracy the partitioning have. However, because of the more details than others, FC-PFS is more complex in computing and evaluating than others. The proposed model in equations (5-9) relies on the principles of the PFS set. Now, we summarize the major points of this model as follows.

- The proposed model is the generalization of the intuitionistic fuzzy clustering model [8, 9] since when $\xi_{kj} = 0$ and the condition (9) does not exist, the proposed model returns to the intuitionistic fuzzy clustering model.
- When $\eta_{kj} = 0$ and the conditions above are met, the proposed model returns to the fuzzy clustering model [4].
- Equation (8) implies that the "true" membership of a data point $X_k$ to the center $V_j$, denoted by $u_{kj}(2 - \xi_{kj})$ still satisfies the sum-row constraint of memberships in the traditional fuzzy clustering model.
- Equation (9) indicates that the constrains of the neutral and refusal degrees, defined the limit of sum of $(\eta_{kj} + \frac{\xi_{kj}}{C})$ are 1, it means that at least $\eta_{kj}$ and $\xi_{kj}$ are greater than zero so that the neutral and refusal degrees always exist in the model since we are working on PFS sets.

We use the Lagranian method to determine the optimal solutions of the model. The optimal solutions of the systems (5-9) are:

$$V_j = \frac{\sum_{k=1}^{N}(u_{kj}(2 - \xi_{kj}))^m X_k}{\sum_{k=1}^{N}(u_{kj}(2 - \xi_{kj}))^m}, (j = \overline{1,C}) \tag{10}$$

$$u_{kj} = \frac{1}{\sum_{i=1}^{C}(2 - \xi_{kj})\left(\frac{\|X_k - V_j\|}{\|X_k - V_i\|}\right)^{\frac{2}{m-1}}}, (k = \overline{1,N}, j = \overline{1,C}) \tag{11}$$

$$\eta_{kj} = \frac{e^{-\xi_{kj}}}{\sum_{i=1}^{C} e^{-\xi_{ki}}}(1 - \frac{1}{C}sum_{i=1}^{C}\xi_{ki}), (k = \overline{1,N}, j = \overline{1,C}) \tag{12}$$

$$\xi_{kj} = 1 - (u_{kj} + \eta_{kj}) - (1 - (u_{kj} + \eta_{kj})^\alpha)^{\frac{1}{\alpha}}, (k = \overline{1,N}, j = \overline{1,C}) \tag{13}$$

In (13), we employ the Yager generating operator [5] to calculate the refusal degree of an element through the positive and neutral degrees. Notice that $\alpha \in (0,1]$ is an exponent coefficient used to control the refusal degree in PFS sets. Now, we present in details the FC-PFS algorithm.

### *Fuzzy Clustering Method on Picture Fuzzy Sets*

**I:** Data $X$ whose number of elements ($N$) in $d$ dimensions;
Number of clusters ($C$); the fuzzifier $m$, Threshold $\varepsilon$,
maximum iteration $maxSteps > 0$.
**O:**Matrices $u$, $\eta$, $\xi$ and centers $V$;

### *FC-PFS Algorithm:*

1: $t = 0$
2: $u_{kj}^{(t)} \leftarrow random$; $\eta_{kj}^{(t)} \leftarrow random$; $\xi_{kj}^{(t)} \leftarrow random$
   $k = \overline{1,N}$, $j = \overline{1,C}$, satisfy equations (2, 3)
3: Repeat
4:      $t = t + 1$
5:      Calculate $V_j^{(t)}$, $(j = \overline{1,C})$ by equation (10)
6:      Calculate $u_{kj}^{(t)}$, $((k = \overline{1,N}, (j = \overline{1,C})$ by equation (11)
7:      Calculate $\eta_{kj}^{(t)}$, $((k = \overline{1,N}, (j = \overline{1,C})$ by equation (12)
8:      Calculate $\xi_{kj}^{(t)}$, $((k = \overline{1,N}, (j = \overline{1,C})$ by equation (13)
9: Until $\|u^{(t)} - u^{(t-1)}\| + \|\eta^{(t)} - \eta^{(t-1)}\| + \|\xi^{(t)} - \xi^{(t-1)}\| \leq \varepsilon$ or $maxSteps$
   has reached

## 2.2 The Picture Fuzzy Rule Interpolation Scheme

Based on the clusters, in this section, we establish picture fuzzy rule interpolation scheme. Zadeh proposed a fuzzy rule which is an IF-THEN rule involving linguistic terms in [30, 32]. The most widely used membership function for fuzzy rules is the triangular membership function [25]. Based on the triangular intuitionistic membership function in [1], we innovate it to adapt with the PFS set and call the new function as the triangular picture fuzzy numbers (TPFN). The TPFN $A$ is described by five real numbers $(a', a, b, c, c')$ with $(a' \leq a \leq b \leq c \leq c')$ and two triangular functions shown in equations (14-15) and Fig. 1 as follows.

$$u = \begin{cases} \frac{x-a}{b-a}, a \leq x \leq b, \\ \frac{c-x}{c-b}, b \leq x \leq c, \\ 0, otherwise. \end{cases} \tag{14}$$

$$\eta + \xi = \begin{cases} \frac{b-x}{b-a'}, a' \leq x \leq b, \\ \frac{x-b}{c'-b}, b \leq x \leq c', \\ 0, otherwise. \end{cases} \tag{15}$$

Integrate (14-15) with (13) and denote $L = \eta + \xi$, the values of the neutral degree and the refusal degree are calculated in equations (16-17).

$$\eta = (1 - (1 - u - L)^\alpha)^{\frac{1}{\alpha}} - u, \tag{16}$$

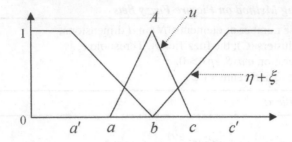

**Fig. 1** A triangular picture fuzzy set $A$

$$\xi = 1 - (u + \eta) - (1 - (u + \eta)^\alpha)^{\frac{1}{\alpha}}. \tag{17}$$

The defuzzified value $DEF(A)$ of the TPFN $A$ shown in Fig. 1 is calculated as follows.

$$DEF(A) = \frac{a' + 2a + 3b + 2c + c'}{9}. \tag{18}$$

The closest fuzzy rules with respect to the input observation are utilized to produce an interpolated conclusion for sparse fuzzy rule-based systems. The following picture fuzzy rule interpolation scheme illustrates that:

**Rule 1: If** $x_1 = A_{1,1}$ and $x_2 = A_{2,1}$ and ... and $x_k = A_{k,1}$ Then $y = B_1$
**Rule 2: If** $x_1 = A_{1,2}$ and $x_2 = A_{2,2}$ and ... and $x_k = A_{k,2}$ Then $y = B_2$
...
**Rule $q$: If** $x_1 = A_{1,q}$ and $x_2 = A_{2,q}$ and ... and $x_k = A_{k,q}$ Then $y = B_q$
**Observations:** $x_1 = A_1^*$ and $x_2 = A_2^*$ and ... and $x_d = A_d^*$

**Conclusion:** $y = B^*$

Where Rule $j$ is the $j^{th}$ fuzzy rule in the sparse fuzzy rule base, $x_k$ denotes the $k^{th}$ antecedent variable, $y$ denotes the consequence variable, $A_{k,j}$ denotes the $k^{th}$ antecedent fuzzy set of Rule $j$, $B_j$ denotes the consequence fuzzy set of Rule $j$, $A_k^*$ denotes the $k^{th}$ observation fuzzy set for the $k^{th}$ antecedent variable $x_k$, $B^*$ denotes the interpolated consequence fuzzy set, $d$ is the number of variables appearing in the antecedents of fuzzy rules, $q$ is the number of fuzzy rules, $k = \overline{1,d}$, and $j = \overline{1,q}$.

## 3   Multi-variable Fuzzy Forecasting Using Picture Fuzzy Clustering and Picture Fuzzy Rule Interpolation Method

In this section, we present a method to combine the FC-PFS algorithm with picture fuzzy rule interpolation method and call is as *Multi-variable picture fuzzy forecasting algorithm* (MVPFFA) in order to solve the multi-variable fuzzy forecasting problems. Suppose that we have a data set with $d$ input time series $T_1(t), T_2(t), ..., T_d(t)$, and one output time series $M(t), t = \overline{0,N}$. Our method is described as follows.
**Step 1:** We split the original dataset into the training and testing samples. The

training sample is pre-processed by equation (19) based on the variation rates $R_k(i), i = \overline{1,N}$ of the $k^{th}$ input time series $T_k(i)$ at time $i$, where $k = \overline{1,d}$.

$$R_k(i) = \frac{T_k(i) - T_k(i-1)}{T_k(i-1)} \times 100\% \qquad (19)$$

Based on (19), the variation rates $R_1(i), R_2(i), ..., R_d(i)$ of the input time series $\{T_1(t), T_2(t), ..., T_d(t)\}, t = \overline{1,N}$, at time $i$ are determined. The output variation rates $R_0(i)$ of the output time series at time $i$ is subsequently computed.

We construct $N$ training samples $\{X_1, X_2, ..., X_N\}$, where $X_i$ is represented by $\{R_1(i), R_2(i), ..., R_d(i), R_0(i)\}, i = \overline{1,N}$. Denote

$X_i = \{I_i^{(1)}, I_i^{(2)}, ..., I_i^{(d)}, O_i\} = \{R_1(i), R_2(i), ..., R_d(i), R_0(i)\}$, where $I_i^{(k)}(O_i)$ is the $k^{th}$ input (output) of $X_i$, $k = \overline{1,d}$.

**Step 2:** We use the FC-PFS algorithm presented in section 2.1 to partition the training sample into $C$ clusters $\{P_1, P_2, ..., P_C\}$. The center $V_j$ of cluster $P_j$, the positive degree $u_{ij}$, the neutral degree $\eta_{ij}$ and the refusal degree $\xi_{ij}$ of $X_i$ are calculated by equations (10-13), $j = \overline{1,C}, i = \overline{1,N}$.

**Step 3:** Based on the clusters $\{P_1, P_2, ..., P_C\}$, we construct picture fuzzy rules using TPFN described in section 2.2, where rule $j$ corresponds to $P_j$, shown as follows.

**Rule** $j$: If $X1 = A_{1,j}$ and $X2 = A_{2,j}$ and ... and $Xk = A_{k,j}$ Then $y = B_j$

where *Rule* $j$ is the fuzzy rule corresponding to the cluster $P_j$, $x_k$ is the $k^{th}$ antecedent variable, $A_{k,j}$ is the $k^{th}$ antecedent fuzzy set of *Rule* $j$, $y$ is the consequence variable, $B_j$ is the $k^{th}$ consequence fuzzy set of *Rule* $j$, $j = \overline{1,C}, k = \overline{1,d}$, and the real numbers $(a', a, b, c, c')$ of TPFN $A_{k,j}$ are calculated in (20-24) with $U_{ij} = \frac{u_{ij}}{(1+\eta_{ij})(1+\xi_{ij})}$

$$a'_{k,j} = \min_{i=1,2,...,n} I_i^{(k)}, \qquad (20)$$

$$c'_{k,j} = \max_{i=1,2,...,n} I_i^k, \qquad (21)$$

$$b_{k,j} = I_t^{(k)}, where\ U_{j,t} \times I_i^{(k)} = \max_{1 \leq i \leq n}(U_{i,t}), \qquad (22)$$

$$a_{k,j} = \frac{\Sigma_{i=1,2,..n\ and\ I_i^k \leq b_{k,j}} U_{i,j} \times I_i^{(k)}}{\Sigma_{i=1,2,..n\ and\ I_i^{(k)} \leq b_{k,j}} U_{i,j}}, \qquad (23)$$

$$c_{k,j} = \frac{\Sigma_{i=1,2,..n\ and\ I_i^k \geq b_{k,j}} U_{i,j} \times I_i^{(k)}}{\Sigma_{i=1,2,..n\ and\ I_i^{(k)} \geq b_{k,j}} U_{i,j}}, \qquad (24)$$

Where $I_i^{(k)}$ is the $k^{th}$ input of the training sample $X_i$, $j = \overline{1,C}, k = \overline{1,d}$. The real numbers $(a', a, b, c, c')$ of TPFN $B_j$ of *Rule* $j$ are described in (25-29).

$$a'_j = \min_{i=1,2,...,n} O_i, \qquad (25)$$

$$c'_{k,j} = \max_{i=1,2,...,n} O_i, \qquad (26)$$

$$b_j = O_k, \text{ where } U_{j,t} = \max_{1 \le i \le n} (U_{i,j}), \tag{27}$$

$$a_j = \frac{\sum_{i=1,2,..n \text{ and } I_i^k \le b_j} U_{i,j} \times O_i}{\sum_{i=1,2,..n \text{ and } I_i^{(k)} \le b_j} U_{i,j}}, \tag{28}$$

$$a_j = \frac{\sum_{i=1,2,..n \text{ and } I_i^k \ge b_j} U_{i,j} \times O_i}{\sum_{i=1,2,..n \text{ and } I_i^{(k)} \ge b_j} U_{i,j}}, \tag{29}$$

Where $O_i$ is the desired output of $X_i$ and $j = \overline{1,C}$. Based on equations (20-29), TPFN of the fuzzy rules are constructed.

**Step 4:** If some picture fuzzy rules are activated by the inputs of the $i^{th}$ sample $X_i$ that means $min_{1 \le k \le d} U_{A_{k,j}}(I_i^{(k)}) > 0$ then calculate the inferred output $O_i^*$ in equation (30) and move to Step 6. Otherwise go to Step 5.

$$O_i^* = \frac{\sum_{j=1}^q min_{1 \le k \le d} U_{A_{k,j}}(I_i^{(k)}) \times DEF(B_j)}{\sum_{j=1}^q min_{1 \le k \le d} U_{A_{k,j}}(I_i^{(k)})}, \tag{30}$$

$U_{A_{k,j}}(I_i^{(k)})$ plays as the membership value of the input $I_i^{(k)}$ belonging to the triangular picture fuzzy set $A_{k,j}$, $j = \overline{1,q}$ and $k = \overline{1,d}$. It is calculated based on the triangular picture fuzzy function in (10, 16, 17) with $q$ being denoted the number of activated picture fuzzy rules and $DEF(B_j)$ being the defuzzified value of the consequence picture fuzzy set $B_j$ of the activated picture fuzzy rule $j$, $j = \overline{1,q}$, $i = \overline{1,N}$.

**Step 5:** If there does not exist any activated picture fuzzy rule, calculate the weight $W_j$ of *Rule j* with respect to the input observations $x_1 = I_i^1$, $x_2 = I_i^2,...,x_d = I_i^d$ by equation (31) and compute the inferred output $O_i^*$ by equation (32). $r^*$ denotes the input vectors $\{I_i^{(1)}, I_i^{(2)}, ..., I_i^{(d)}\}$, $r_j$ denotes the vector of the defuzzified value of the antecedent fuzzy sets of *Rule j* - $\{DEF(A_{1,j}), DEF(A_{2,j}),...,DEF(A_{d,j})\}$. $\|r^* - r_j\|$ is the Euclidean distance between the vectors $r^*$ and $r_j$. The constraints of the weights are: $0 \le W_j \le 1$, $j = \overline{1,C}$ and $\sum_{j=1}^C W_j = 1$. $DEF(B_j)$ is the defuzzified value of consequence picture fuzzy sets $B_j$.

$$W_j = \frac{1}{\sum_{j=1}^C (\frac{\|r^* - r_j\|}{\|r^* - r_k\|})^2}, \tag{31}$$

$$O_i^* = \sum_{j=1}^C W_j \times DEF(B_j), \tag{32}$$

**Step 6:** Finally, we calculate the forecasted value $M_{Forecasted}(i)$ at time $i$ based on the predicted variation rate $O_i^*$, where $M(i-1)$ is the actual value at time $i-1$.

$$M_{Forecasted}(i) = M(i-1) \times (1 + O_i^*). \tag{33}$$

## 4 Experiments

In this section, we apply our proposed method to forecast the TAIEX dataset using two heuristic variables: NASDAQ and Dow Jones. The dataset includes the daily closing index of the TAIEX, the NASDAQ and the Dow Jones from 1999 to 2004 in [21] including the training sample with the first ten months of each year and the testing sample with the last two months of each year.

Table 1 shows the historical closing index of the TAIEX, the Dow Jones and the NASDAQ for instance in 2014 from Jan 2, 2004 to Dec 31, 2004. In order to evaluate the quality of the predicting results for the TAIEX, we use the root mean square error (RMSE) as follows.

$$RMSE = \sqrt{\frac{\sum_{i=1}^{n}(T_{Forecasted}(i) - T_{Actual}(i))^2}{n}}, \quad (34)$$

Where $T_{Forecasted}(i)$ and $T_{Actual}(i)$ denotes the forecasted TAIEX and the actual TAIEX of day $i$ and $n$ denotes the number of forecasted days.

Table 1 The historical closing index of the TAIEX, the Dow Jones and the NASDAQ in 2004

| Date | TAIEX | Dow Jones | NASDAQ |
|---|---|---|---|
| Jan 2, 2004 | 6041.56 | 10409.85 | 2006.68 |
| Jan 5, 2004 | 6125.42 | 10544.07 | 2047.36 |
| Jan 6, 2004 | 6144.01 | 10538.66 | 2057.37 |
| ... | ... | ... | ... |
| Oct 28, 2004 | 5695.56 | 10004.54 | 1975.74 |
| Oct 29, 2004 | 5705.93 | 10027.47 | 1974.99 |
| Nov 1, 2004 | 5656.17 | 10054.39 | 1979.87 |
| Nov 2, 2004 | 5759.61 | 10035.73 | 1984.79 |
| ... | ... | ... | ... |
| Dec 29, 2004 | 6088.49 | 10829.19 | 2177 |
| Dec 30, 2004 | 6100.86 | 10800.3 | 2178.34 |
| Dec 31, 2004 | 6139.69 | 10783.01 | 2175.44 |

Based on equation (34), we have the experimental results in Table 2. They were conducted as the average RMSE value of 50 execution times with 28 clusters, 28 generated fuzzy rules [12] and $\alpha = 0.5$. From Table 2, our proposed method produce the results with a smaller average RMSE compared to the existing methods. Therefore, it proposes better forecasting result than other methods in [11, 12, 13, 19, 22, 23, 27, 28, 29] for predicting the TAIEX.

## 5 Conclusion

In this paper, we have proposed a new method for multi-variable fuzzy forecasting using picture fuzzy clustering and picture fuzzy rule interpolation technique.

**Table 2** The historical closing index of the TAIEX, the Dow Jones and the NASDAQ in 2004

| Methods | Year | | | | | | Average |
|---|---|---|---|---|---|---|---|
| | 1999 | 2000 | 2001 | 2002 | 2003 | 2004 | RMSE |
| Huarng[23](NASDAQ) | N/A | 158.71 | 36.49 | 95.15 | 65.51 | 73.57 | 105.88 |
| Huarng[23](Dow Jones) | N/A | 165.81 | 38.25 | 93.73 | 72.95 | 73.49 | 108.84 |
| Huarng[23](NASDAQ&Dow Jones) | N/A | 157.6 | 131.9 | 93.48 | 65.51 | 73.49 | 104.42 |
| Chen[11, 28, 29] | 120 | 176 | 148 | 101 | 74 | 84 | 117.17 |
| U_R Model [28, 29] | 164 | 420 | 1070 | 116 | 329 | 146 | 374.17 |
| U_NN Model [28, 29] | 107 | 309 | 259 | 78 | 57 | 60 | 145 |
| U_NN_FTS Model [22, 28, 29] | 109 | 255 | 130 | 84 | 56 | 116 | 125 |
| U_NN_FTS_S Model [22, 28, 29] | 109 | 152 | 130 | 84 | 56 | 116 | 107.83 |
| B_R Model [28, 29] | 103 | 154 | 120 | 77 | 54 | 85 | 98.83 |
| B_NN Model [28, 29] | 112 | 274 | 131 | 69 | 52 | 61 | 116.5 |
| B_NN_FTS model [28, 29] | 108 | 259 | 133 | 85 | 58 | 67 | 118.33 |
| B_NN_FTS_S Model [28, 29] | 112 | 131 | 130 | 80 | 58 | 67 | 96.33 |
| Wei[27](ANFIS+Adaptive(1)) | 103 | 126 | 122 | 64 | 52 | NA | 93.4 |
| Wei[27](ANFIS+Adaptive(2)) | 103 | 125 | 120 | 63 | 52 | NA | 92.6 |
| Chen[12](NASDAQ) | 123.64 | 131.11 | 15.08 | 73.06 | 66.36 | 60.48 | 94.95 |
| Chen[12](Dow Jones) | 101.97 | 148.81 | 13.70 | 79.81 | 64.08 | 82.32 | 98.46 |
| Chen[12](NASDAQ&Dow Jones) | 106.34 | 130.11 | 13.33 | 72.33 | 60.29 | 68.07 | 91.75 |
| Chen[13](Dow Jones) | 115.47 | 127.51 | 21.98 | 74.65 | 66.02 | 58.89 | 94.09 |
| Chen[13](NASDAQ) | 119.32 | 129.81 | 23.12 | 71.01 | 65.14 | 61.94 | 95.07 |
| Chen[13](DowJones&NASDAQ) | 116.64 | 123.61 | 23.85 | 71.98 | 58.06 | 57.73 | 91.98 |
| Chen[16](Dow Jones) | 102.34 | 131.21 | 13.62 | 65.77 | 52.23 | 56.16 | 86.89 |
| Chen[16](NASDAQ) | 102.11 | 131.31 | 13.83 | 66.45 | 52.83 | 54.17 | 86.78 |
| **MVPFFA**(Dow Jones) | 109.9 | 138.6 | 92.47 | 64.17 | 60.42 | 55.64 | 86.87* |
| **MVPFFA**(NASDAQ) | 107.09 | 138.5 | 94.71 | 64.92 | 55.89 | 53 | 86.28* |
| **MVPFFA**(NASDAQ & Dow Jones) | 109.62 | 138.6 | 93.59 | 63.14 | 59.44 | 58.13 | 86.55* |

Our method uses FC-PFS algorithm to partition dataset in to clusters and then it builds the picture fuzzy rules for forecasting within determining the weight of each picture fuzzy rule. Our proposed method is applied to the TAIEX data with better predicting results than others. In future, we will improve the FC-PFS algorithm to determine the number of clusters automatically, combine our method with other optimal method to get more accuracy and make the picture fuzzy rules based on other membership functions such as trapezoidal function, Gaussian function, etc.

**Acknowledgements.** This work is sponsored by a project of VNU University of Science under contract No. TN-14-25.

# References

[1] Albeanu, G., Popentiu-Vladicescu, F.L.: Intuitionistic fuzzy methods in software reliability modeling. Journal of Sustainable Energy 1(1) (2010)

[2] Armano, G., Marchesi, M., Murru, A.: A hybrid genetic-neural architecture for stock indexes forecasting. Information Sciences 170(1), 3–33 (2005)

[3] Atanassov, K.T.: Intuitionistic fuzzy sets. Fuzzy Sets and Systems 20, 87–96 (1986)

[4] Bezdek, J.C.: Pattern recognition with fuzzy objective function algorithms. Kluwer Academic Publishers (1981)

[5] Burillo, P., Bustince, H.: Entropy on intuitionistic fuzzy set and on interval-valued fuzzy set. Fuzzy Sets and Systems 78, 305–316 (1996)

[6] Cardoso, G., Gomide, F.: Newspaper demand prediction and replacement model based on fuzzy clustering and rules. Information Sciences 177(21), 4799–4809 (2007)

[7] Celikyilmaz, A., Turksen, I.B.: Enhanced fuzzy system models with im-proved fuzzy clustering algorithm. IEEE Transactions on Fuzzy Systems 16(3), 779–794 (2008)

[8] Chaira, T.: A novel intuitionistic fuzzy C means clustering algorithm and its application to medical images. Applied Soft Computing 11(2), 1711–1717 (2011)

[9] Chaira, T., Panwar, A.: An Atanassov's intuitionistic Fuzzy Kernel Clustering for Medical Image segmentation. International Journal of Computational Intelligence Systems, 1–11 (2013)

[10] Chang, Y.C., Chen, S.M.: Temperature prediction based on fuzzy clustering and fuzzy rules interpolation techniques. In: IEEE International Conference on Systems, Man and Cybernetics, SMC 2009, pp. 3444–3449. IEEE (2009)

[11] Chen, S.M.: Forecasting enrollments based on fuzzy time series. Fuzzy Sets and Systems 81(3), 311–319 (1996)

[12] Chen, S.M., Chang, Y.C.: Multi-variable fuzzy forecasting based on fuzzy clustering and fuzzy rule interpolation techniques. Information Sciences 180(24), 4772–4783 (2010)

[13] Chen, S.M., Chen, C.D.: TAIEX forecasting based on fuzzy time series and fuzzy variation groups. IEEE Transactions on Fuzzy Systems 19(1), 1–12 (2011)

[14] Chen, S.M., Chu, H.P., Sheu, T.W.: TAIEX forecasting using fuzzy time series and automatically generated weights of multiple factors. IEEE Transactions on Systems, Man and Cybernetics, Part A, Systems and Humans 42(6), 1485–1495 (2012)

[15] Chen, S.-M., Kao, P.-Y.: Forecasting the TAIEX based on fuzzy time series, PSO techniques and support vector machines. In: Selamat, A., Nguyen, N.T., Haron, H. (eds.) ACIIDS 2013, Part I. LNCS (LNAI), vol. 7802, pp. 89–98. Springer, Heidelberg (2013)

[16] Chen, S.M., Manalu, G.M.T., Pan, J.S., Liu, H.C.: Fuzzy forecasting based on two-factors second-order fuzzy-trend logical relationship groups and particle swarm optimization techniques. IEEE Transactions on Cybernetics 43(3), 1102–1117 (2013)

[17] Cheng, C.H., Cheng, G.W., Wang, J.W.: Multi-attribute fuzzy time series method based on fuzzy clustering. Expert Systems with Applications 34(2), 1235–1242 (2008)

[18] Cuong, B.C., Kreinovich, V.: Picture Fuzzy Sets - a new concept for computational intelligence problems. In: Proceeding of 2013 Third World Congress on Information and Communication Technologies (WICT), pp. 1–6 (2013)

[19] Egrioglu, E., Aladag, C.H., Yolcu, U., Uslu, V.R., Erilli, N.A.: Fuzzy time series forecasting method based on Gustafson–Kessel fuzzy clustering. Expert Systems with Applications 38(8), 10355–10357 (2011)

[20] Ejegwa, P.A., Akubo, A.J., Joshua, O.M.: Intuitionistic fuzzy set and its application in career determination via normalized Euclidean distance method. European Scientific Journal 10(15) (2014)

[21] Google Finance, https://www.google.com/finance/

[22] Huarng, K., Yu, T.H.K.: The application of neural networks to forecast fuzzy time series. Physica A: Statistical Mechanics and its Applications 363(2), 481–491 (2006)

[23] Huarng, K.H., Yu, T.H.K., Hsu, Y.W.: A multivariate heuristic model for fuzzy time-series forecasting. IEEE Transactions on Systems, Man, and Cybernetics, Part B: Cybernetics 37(4), 836–846 (2007)

[24] Khashei, M., Bijari, M.: A novel hybridization of artificial neural net-works and ARIMA models for time series forecasting. Applied Soft Computing 11(2), 2664–2675 (2011)

[25] Pedrycz, W.: Why triangular membership functions? Fuzzy Sets and Systems 64(1), 21–30 (1994)

[26] TAIEX Website, http://www.twse.com.tw/en/products/indices/tsec/taiex.php

[27] Wei, L.Y., Cheng, C.H., Wu, H.H.: A hybrid ANFIS based on n-period moving average model to forecast TAIEX stock. Applied Soft Computing 19, 86–92 (2014)

[28] Yu, T.H.K., Huarng, K.H.: A bivariate fuzzy time series model to forecast the TAIEX. Expert Systems with Applications 34(4), 2945–2952 (2008)

[29] Yu, T.H.K., Huarng, K.H.: Corrigendum to "A bivariate fuzzy time series model to forecast the TAIEX". Expert Systems with Applications 34(4), 2945–2952 (2010); Expert Systems with Applications 37(7), 5529 (2010)

[30] Zadeh, L.A.: Fuzzy sets. Information and Control 8, 338–353 (1965)

[31] Zadeh, L.A.: Toward a generalized theory of uncertainty (GTU) – an outline. Information Sciences 172(1-2), 1–40 (2005)

[32] Zadeh, L.A.: Is there a need for fuzzy logic? Information Sciences 178(13), 2751–2779 (2008)

# Author Index

Akagi, Masato   129
Akama, Seiki   209
André, Étienne   339, 473

Bui, Quang Hung   247
Bui, Thang H.   513
Bui, The Duy   129, 487

Cabot, Jordi   219
Choppy, Christine   339, 473
Christine Ho Ba Tho, Marie   553, 567
Chujyou, Shinobu   541

Dang, Duc-Hanh   219
Dang, Tran-Thai   617, 629
Dao, Minh-Son   79
Dao, Thi-Kien   667
Dao, Tien Tuan   553, 567
Diab, Ahmad   527
Dinh Thanh, Pham   367
Dinh Viet, Nguyen   61
Dinh, Duy   195
Dinh, Quan Dang   261
Do, Kien Duc   117
Do, Quang Hung   381
Duc, Thanh Ngo   195
Duong, Duc Anh   195
Duy Phuong, Nguyen   273
Duy Tan, Nguyen   61
Duy Truong, Cao   155

Fujita, Hamido   3

Gamet, Didier   553

Ha, Doan T.T.   589
Ha, Quang-Thuy   233
Hai Anh, Ngo   183
Hassan, Mahmoud   527
Hoai Viet, Vo   447
Hoang Le, Thai   393
Hoang Son, Le   679
Hoang, Van Thang   247
Horng, Mong-Fong   667
Huan, Hoang X.   589
Hung, Nguyen T.   589
Huy Thong, Pham   679
Huynh, Huy M.   603
Huynh, Khai T.   513
Huynh, Van-Nam   629

Ichise, Ryutaro   413
Ikeda, Kokolo   325
Inuiguchi, Masahiro   15

Jiang, Frank   261

Karlsson, Brynjar   527
Kato, Hiroshi   541
Kieu, Binh Thanh   413
Kim Anh, Nguyen   93
Ko, Yu-Chien   3
Kwon, Yung-Keun   577

Laforêt, Jérémy   527
Lathuilière, Stéphane   313
Le, Anh-Cuong   617, 629
Le, Chi-Luan   499
Le, Duc Minh   433

Le, Duc-Hau   577
Le, Duy Hanh   143
Le, Hai-Son   299, 459
Le, Thi-Lan   313
Le-Hong, Phuong   653
Luong, Chi Mai   49
Luu, Viet Hung   247
Ly, Ngoc Quoc   105

Ma, Thi Chau   487
Mahdi Benmoussa, Mohamed   473
Man, Duc Chuc   247
Marque, Catherine   527
Meesad, Phayung   287
Minh Phuong, Tu   35, 353
Minoro Abe, Jair   209

Nakamatsu, Kazumi   209
Ngo, Thi Duyen   129, 487
Ngo, Truong-Giang   459, 667
Ngoc Diep, Nguyen   353
Nguyen, Dang Tuan   405
Nguyen, Duc-Dung   299, 459
Nguyen, Duc-Duy   79
Nguyen, Hien T.   603
Nguyen, Hoai-Xuan   425
Nguyen, Hua Phung   143
Nguyen, Huong   233
Nguyen, Huy   325
Nguyen, Huy-Quang   425
Nguyen, Minh Le   117, 641
Nguyen, Ngoc-Bao   195
Nguyen, Nien Dinh   117
Nguyen, Phuc Tri   405
Nguyen, Quang-Trung   499
Nguyen, Quoc Bao   49
Nguyen, The-Tung   617
Nguyen, Thi Nhat Thanh   247
Nguyen, Thi Xuan Huong   617
Nguyen, Thinh Khanh   117
Nguyen, Tien-Dung   653
Nguyen, Trong-The   667
Nguyen, Truc-Vien T.   79
Nguyen, Vinh-Van   425
Nguyen, Vu-Hoang   195
Nguyen, Xuan Hoai   577
Nhat Quang, Tran   273
Noulamo, Thierry   339

Pham, Anh-Tuan   425

Pham, Cuong   353
Pham, Ngoc-Quan   299, 459
Pham, Ngoc Hung   499
Pham, Son Bao   117, 413, 641
Phan, Xuan-Hieu   653
Pouletaut, Philippe   553

Quan, Tho T.   513
Quoc Ngoc, Ly   447

Shieh, Chin-Shiuh   667

Tan Hung, Dinh   313
Thai Son, Tran   447
Than, Khoat   93
Thanh Binh, Nguyen   169
Thanh Giang, Pham   183
Thanh Tung, Khuat   169
Thi Lien, Do   273
Thi My Hanh, Le   169
Thi Thanh Binh, Huynh   367
Thi Xuan Huong, Nguyen   629
Thien Van, Hoang   393
Thu Lam, Bui   367
Tien Lan, Nguyen   183
Tojo, Satoshi   31
Tran, Dang Hai   117
Tran, Dinh Huy   143
Tran, Hoang-Viet   499
Tran, Quan Hung   117, 641
Tran, Quang-Anh   261
Tran, Thanh-Hai   313
Tran, Viet-Hong   425
Tuan Anh, Duong   155
Tuyet, Duong T.A.   589

Van Hieu, Duong   287
Van Linh, Ngo   93
Van, Duy K.   603
Viennot, Simon   325
Viet Hung, Pham   35
Vo, Anh   105
Vo, Vinh T.   603
Vu, Hai   313
Vu, Tat Thang   49
Vu, Thanh   233
Vu, Xuan-Son   233

Yoshitaka, Atsuo   541